NATURAL SCIENCES IN AMERICA

NATURAL SCIENCES IN AMERICA

Advisory Editor
KEIR B. STERLING

Editorial Board
EDWIN H. COLBERT
EDWARD GRUSON
ERNST MAYR
RICHARD G. VAN GELDER

AMERICAN NATURAL HISTORY
PART I: MASTOLOGY

and

RAMBLES OF A NATURALIST

John D. Godman

ARNO PRESS
A New York Times Company
New York, N. Y. • 1974

Reprint Edition 1974 by Arno Press Inc.

Reprinted from copies in the Princeton
 University and The American Museum of
 Natural History Libraries.

NATURAL SCIENCES IN AMERICA
ISBN for complete set: 0-405-05700-8
See last pages of this volume for titles.

Manufactured in the United States of America

Library of Congress Cataloging in Publication Data

Godman, John Davidson, 1794-1830.
    American natural history.

    (Natural sciences in America)
    Reprint of two works, the 1st published in 1826-28 by
H. C. Carey & I. Lea, Philadelphia; the 2d published
in 1833 by T. T. Ash, Philadelphia.
    1. Mammals--North America. 2. Natural history.
I. Godman, John Davidson, 1794-1830. Rambles of a
naturalist. 1974. II. Title. III. Title: Rambles
of a naturalist. IV. Series.
QL715.G6 1974   599'.09'7   73-17821
ISBN 0-405-05737-7

AMERICAN

NATURAL HISTORY.

Engr.d by Peter Maverick from a sketch in oil by In.º Neagle.

PETALESHAROO,

Son of Latelesha, Knife Chief of the Pani-Loups: in full dress.

AMERICAN
Natural History

JOHN D. GODMAN, M.D.

Drawn by J. Peale Jr. *Engraved by F. Kearney*

Philadelphia

H. C. CAREY & I. LEA, CHESTNUT STREET.

1826.

AMERICAN
NATURAL HISTORY.

VOLUME I.

PART I.—MASTOLOGY.

BY JOHN D. GODMAN, M.D.

PROFESSOR OF NATURAL HISTORY IN THE FRANKLIN INSTITUTE OF PENNSYLVANIA; ONE OF THE PROFESSORS OF THE PHILADELPHIA MUSEUM; MEMBER OF THE AMERICAN PHILOSOPHICAL SOCIETY; OF THE PHILADELPHIA ACADEMY OF NATURAL SCIENCES, &C.

PHILADELPHIA:
H. C. CAREY & I. LEA—CHESTNUT-STREET

R. WRIGHT, PRINTER.

1826.

Eastern District of Pennsylvania, to wit:

    BE IT REMEMBERED, That on the twenty-sixth day of June, in the fiftieth year of the Independence of the United States of America, A. D. 1826, Robert Wright, of the said district, hath deposited in this office the title of a book, the right whereof he claims as proprietor, in the words following, to wit:

"American Natural History. Volume I. Part I. Mastology. By John D. Godman, M. D. professor of Natural History in the Franklin Institute of Pennsylvania; one of the Professors of the Philadelphia Museum; Member of the American Philosophical Society, of the Philadelphia Academy of Natural Sciences, &c.

In conformity to the Act of the Congress of the United States, intituled, "An Act for the encouragement of learning, by securing the copies of Maps, Charts, and Books, to the authors and proprietors of such copies, during the times therein mentioned"—And also to the Act entitled, An Act supplementary to an Act, entitled, An Act for the encouragement of learning, by securing the copies of Maps, Charts, and Books to the authors and proprietors of such copies, during the times therein mentioned, and extending the benefits thereof to the arts of designing, engraving, and etching historical and other prints."

               D. CALDWELL,
         Clerk of the Eastern District of Pennsylvania.

TO

ROBERT OLIVER,

OF

BALTIMORE, MARYLAND,

WHOSE NAME

IS ONLY ANOTHER EXPRESSION

FOR

LIBERALITY AND MUNIFICENCE,

THIS WORK IS RESPECTFULLY

AND GRATEFULLY INSCRIBED,

BY HIS FRIEND,

THE AUTHOR.

PREFACE.

This work was begun in the spring of 1823, under a belief that the whole of the first part would be published within a year, or at farthest eighteen months from its commencement. Experience soon proved that the difficulties of this enterprize had not been correctly estimated, and that a vast labour remained to be performed after all the facts and observations were systematised, which remained to be gleaned from works that professedly or incidentally treated of American Natural History. How meagre and unsatisfactory the best of these books are, can only be imagined by those persons who have been obliged to examine them carefully; by them alone can an adequate idea be formed of the toil and disgust to be endured by whoever makes such search in hopes of collecting original observations, statements of facts worthy of repetition, or remarks properly illustrative of the manners and habits of our animals.

To account for the delay which has inevitably occurred in the preparation of this work, it may be sufficient to state that it has been frequently necessary to suspend it for weeks and months, in order to procure certain animals, to observe their habits in captivity, or to make daily visits to the woods and fields for the sake of witnessing their actions in a state of nature. On other occasions we have undertaken considerable journies, in order to ascertain the correctness of statements, or to obtain sight of an individual subject of description. It would be far more agreeable thus to obtain materials for the whole work from nature, than to depend in the slightest degree upon books; but a long lifetime spent in this way on such a work, would not be too much to give it the requisite degree of perfection.

Another cause of delay has been the necessity we have frequently been under of collecting materials for the second and third volumes, when very solicitous to expedite the first; but as similar opportunities might not again occur, it was imperatively necessary to profit by them. This retardation of the first will operate, however, in equal proportion to increase the value and hasten the publication of the second and third volumes, which will be completed with as much speed as is consistent with propriety.

Our great aim in preparing this work has been to render it as useful and agreeable as possible, and to this end we have freely drawn upon all accessible authentic sources, with due acknowledgment for the benefits received. In addition to the references made in the body of the work, we shall give in the third volume an alphabetical catalogue of all the books whence we have derived assistance, or which we have consulted with advantage. This we do not only because we believe that there is no crime more despicable than an attempt to deprive another of the fruits of his intellectual exertions, but in order to facilitate the labours of those who may be desirous of extending their researches through the most authentic works.

We should act with injustice to our own feelings if we omitted to avow the liberal and valuable assistance we have received from one whose name is sufficient to justify any encomiums on expanded views and zeal in the cause of scientific truth. The prince of Musignano, CHARLES L. BONAPARTE, has at all times thrown open to our use his rich library and cabinets, and still farther enhanced this kindness by contributing in numerous instances the result of his own scientific observations. Similar liberality in the cause of science has long since secured to him the esteem and respect of those who are devoted to its cultivation, and the warm admiration of all who have the advantage of his acquaintance.

To our distinguished countryman SAY, we are indebted for numerous excellent suggestions and much interesting information. Mr. GEORGE ORD, the respected vice-president of the Academy of Natural Sciences, has on several occasions yielded the most acceptable assistance, and allowed us the use of his note-book, by which we have profited considerably. From Mr. TITIAN PEALE, who in practical acquaintance with the natural history of his country has few equals, we have received frequent aid. To our friend ROBERT BEST, M. D. of Lexington, who was formerly conservator of the Western Museum, we feel especially grateful for the communication of his most interesting notes and observations, made on the animals of the western country during a residence of twenty-five or thirty years in the state of Ohio.

The kindness of the venerable librarian of the Philosophical Society, JOHN VAUGHAN, Esq. has been amply exercised towards us in furnishing the freest opportunities of examining the admirable library of that respectable institution. Although he considers himself sufficiently repaid by the consciousness of having discharged what he is pleased to call his duty, we feel none the less grateful for his urbanity and the extent of his good will.

We have also received manifold acts of kindness from all the respectable naturalists of our acquaintance, both in this city and New York. All of these have exhibited the most gratifying willingness to aid in the advancement of our undertaking, and we beg them to accept of our sincerest thanks for the spirit in which they have contributed to its success.

The following circumstance, to be enumerated among the services rendered to this work, is one which we feel sure the reader will unite with us in considering as worthy of the highest commendation, and evincing a spirit which every generous mind must regard with unmingled pleasure. An American artist

PREFACE. ix

had painted two original portraits of distinguished Indian chieftains, which were regarded by all who saw them as admirable for their force and truth of character. He was repeatedly solicited to sell them at his own price, but uniformly refused; neither would he sell *copies* of them, although on one occasion they were sought by a foreign gentleman in order to be sent to Russia. On learning that an American work was about to be published, for which illustrations of aboriginal physiognomy and costume were desirable, the artist presented both these pictures, rejecting every offer of compensation with a feeling truly American. These interesting specimens of the talents and liberality of Mr. JOHN NEAGLE are given in this volume, the full length being a likeness of PETALESHAROO,* son of the Knife-chief of the Pani-Loups,

---

* " Almost from the beginning of this interesting fete our attention had been attracted to a young man who seemed to be the leader or partizan of the warriors. He was about twenty-three years of age, of the finest form, tall, muscular, exceedingly graceful, and of a most prepossessing countenance. His head-dress of war-eagles' feathers descended in a double series upon his back, like wings, down to his saddle croup; his shield was highly decorated, and his long lance was ornamented by a plaited casing of red and blue cloth. On inquiring of the interpreter, our admiration was augmented by learning that he was no other than PETALESHAROO, with whose name and character we were already familiar. He is the most intrepid warrior of the nation, eldest son of LATELESHA, [the Knife-chief] destined as well by mental

VOL. I.—B

and the bust a portrait of ONGPATONGA, or Big Elk, distinguished as the great chief and orator of the Omawhaws. The first was engraved by MAVERICK of New York, the second by BRIDPORT of Philadelphia.

---

and physical qualifications as by his distinguished birth, to be the future leader of his people.—The name of PETALESHAROO is connected with the abolition of a custom formerly prevalent in this nation, at which humanity shudders.

"An Ietan woman, brought captive into the village, was doomed to the Great Star by a warrior whose property she had become by the fate of war; she underwent the usual preparations, and on the appointed day was led to the cross amidst a great concourse of people, as eager perhaps as their civilized fellow men to witness the horrors of an execution. The victim was bound to the cross with thongs of skin, and the usual ceremonies being performed, her dread of a more terrible death was about to be terminated by the tomahawk and arrow. At this critical juncture PETALESHAROO stepped forward into the area, and in a hurried but firm manner declared that it was his father's wish to abolish this sacrifice; that he presented himself for the purpose of laying down his life upon the spot or of releasing the victim. He then cut the cords which bound her, carried her swiftly through the crowd to a horse which he presented to her, and having mounted another himself, conveyed her beyond the reach of immediate pursuit."—*Long's Exped. to the Rocky Mountains, vol. i. p. 357.*

This distinguished man, (together with ONGPATONGA and various other chiefs,) was in Philadelphia a few years since, on his way to the city of Washington, accompanied by

PREFACE. xi

The vignette which ornaments our engraved title page is from an original design by JAMES PEALE, jr. an amateur who unites to the correct execution of a professed artist the happiest talent for perceiving and delineating the picturesque and beautiful.

In relation to the animals described in this work, it has been our constant aim to give none but such as certainly belong to this country, being much more desirous of presenting a faithful account of those known to inhabit it, than to produce an imposing catalogue of " NEW SPECIES," which at best might be little better than a string of barbarous new names applied to old and well known things.* In our third volume we shall give a complete synopsis of the mammalia inhabiting this country, with distinctive specific phrases, drawn up by the distinguished author of the recent splendid work on American

---

Major O'FALLON. Through the friendship of that gentleman Mr. NEAGLE enjoyed the most excellent opportunities of obtaining the portraits which adorn this volume.

* In consequence of having mislaid the manuscript containing the description of the wild cat or bay lynx, (*Felis Rufa,*) that species was accidentally omitted, and the inadvertence not perceived until it was too late to rectify it at the proper place. As this species will be fully described in the appendix, and we give a good figure of it on the same plate with the Canada lynx, (*Felis Canadensis,*) which it closely resembles in habits and manners, the omission will not be productive of inconvenience or injury to the reader.

Ornithology. In this synopsis all the doubtful and ridiculous species hitherto proposed as inhabitants of this country will be noted and rejected. Although we are conscious of having used every exertion to render this book correct, we are far from believing that it is free from error; it has been prepared under too many disadvantages to allow us for a moment to hope that it will not be obnoxious to some censure. We regard it rather as a groundwork upon which we hope hereafter to erect a more perfect superstructure, having determined henceforth zealously to continue our efforts to obtain the materials necessary to give it the degree of permanence and beauty of which the subject is in every respect worthy; we shall therefore thankfully receive every suggestion made by candid critics for its improvement.

INTRODUCTION.

IN this work the systematic arrangements of LINNE and CUVIER are principally followed; the LINNEAN names of the orders, genera, &c. have uniformly been retained whenever they are entitled to such preference by *right of priority*. As long as every succeeding writer is at liberty to overturn systems and change names according to his own views or caprice, the most serious injury will result to natural science, and her votaries be effectually deterred from engaging in studies to which every avenue is barricaded by aggregations of learned lumber. We are happy already to have an excellent precedent in this country for adhering to original nomenclature; SAY in all his scientific writings has evinced the same determination to give credit to whom credit is due, regardless of all arbitrary changes.

We have introduced the dental systems from F. CUVIER, with the view of enabling the reader more fully to appreciate the true importance of this part of animal organization, which indicates the *relations* in *point of regimen* existing between animals, *not their places* in a SYSTEM of classification. The peculiarities of the teeth have immediate reference to the food upon which animals subsist, and by consequence to the general character of their digestive organs; by comparing different genera we trace their degree of relationship in this respect, and form a better idea of their natural affinities. But if we arrange animals according to their proximity of dentition and regimen, we adopt *a method* (and there may be as many methods as organs) which will separate to great and unnatural distances animals whose striking similarity in all other respects would render such separation absurd.*

Beginners of the study of natural history are generally liable to form erroneous conclusions, among which none is more common and prejudicial than that of mistaking the system of classification for the subjects classed, or in other words, the arrangement of the names for the things themselves; nomenclature for natural history. The best system of classifica-

---

* According to the dentition, F. Cuvier properly places the cat first and the bear last of the carnivorous animals; the walrus after the ruminant animals, and the beaver among the last of the gnawers instead of the first. Various genera, well separated by other characters, are by their dentition reduced to mere species or varieties of the same genus. Thus, while the *system* of Cuvier arranges animals according to the closeness of their approach to the structure of the human body, which determines their comparative perfection, the *method* of his brother, having merely reference to their dentition, throws them into situations both curious and unnatural, because dependant upon the comparison of a single set of organs.

tion in the nature of things must be in a great degree arbitrary and imperfect, and so far from being natural history, is but a summary of distinctive epithets and characters to aid in the arrangement of knowledge, which can only be derived from a proper observation of natural objects. Had these obvious truths been attended to, we should never have had so great a number and such a farrago of new systems as have been offered to the world, neither would the study of natural history have been so long regarded by the mass of mankind as the study of any thing rather than the study of nature.

A good system is an invaluable assistant; an imperfect one is infinitely preferable to the entire absence of such an aid. The system, however, should always be secondary to the science, whose object is to teach the general and individual characters of living beings from an actual observation of their peculiarities of form and modes of living. The mistake above pointed out is continually urging many who would be esteemed naturalists to the formation of new genera and species, founded on trivial, accidental, or imperfectly noted differences between creatures which, to all rational observers, appear the same.— This retards science, and misleads individuals as to the character and objects of natural history, which, judged by the conduct of some who are regarded as authorities, would appear to be the science of magnifying trifles and bewildering the understanding.

In natural history, as in other departments of human knowledge, none but sciolists are pedants; such persons struggle to impart to their implements the dignity and importance that should belong to the work alone, and " in self adoring pride securely mailed" seek but to glorify themselves, considering the interests of science as nothing when weighed against the gratification of their own vanity.*

---

* "New nominal species perplex the student, increase the labours of the critical naturalist, and render the study of natural history tedious and difficult. If it was generally understood that it is more meritorious to extinguish a single nominal species than to establish a dozen new ones, it would effectually check the present mania for making new species often on slight foundations. This also leads to an overweening anxiety to secure priority; and hence descriptions are liable to be drawn up in a crude and hasty manner, without reference to the co-ordinate characters." DE KAY's address to the New York Lycæum, p. 76.

ERRATA

Page 33, first line, for " are desirous,"—read *or are desirous*.
" 35, next to last line of second paragraph, for " occasions"—read *occasion*.
" 46, in the foot-note, for " order"—read *family*.
" 49, third line from top, for " quantity,"—read *number*.
  [Page 80, the figure of the third species of shrew to have been marked thus, * could not be procured.]
" 117, second line, second paragraph, for " Missouri"—read *Mississipi*.
" 129, fourth line, second paragraph, for " strgggle"—read *struggle*.
" 255, note, fourth line from bottom, for " neminiunquam"—read *nemini unquam*. Same line, for " speiem"—read *speciem*.
" 291, fifth line from top, for " which are so circumstanced"—read *they being so circumstanced*.

AMERICAN NATURAL HISTORY.

CHAPTER I.

CLASS I.—MAMMALIA.

ORDER I. *Primates.*—FAMILY I. *Bimana.*

GENUS I. MAN; *Homo;* L.—SPECIES I. *H. Sapiens;* L.

VARIETIES, a. *Caucasian;* fair or white, originally from Europe.
b. *Mongolian;* Esquimaux; dark olive or swarthy, from the north of Asia.
c. *American;* red, indigenous?
d. *Ethiopian;* black, from Africa.

SECTION I.

Origin of the American Variety.

AMERICA, although undiscovered until near the conclusion of the fifteenth century of the christian era, must have been for ages previous the residence of an extensive and increasing population, since great numbers of native inhabitants were found on the southern portions of this continent by the adventurous voyagers who, under the guidance of CRISTOFORO COLOMBO,* first visited these shores. What must

---

* Latinized, *Columbus.*

VOL. I.—C

have been the mutual surprise of the inhabitants of the old and new worlds at this first meeting? The Europeans, astonished to encounter a numerous and eager crowd where they had anticipated one vast solitude—and the aborigines, lost in amazement, at the unimagined spectacle of a huge body which had slowly risen before their eyes from the remotest verge of ocean, and borne to their soil a strange assemblage of men differing from them in colour, language and apparel.

The origin of the North American Indians has justly attracted the attention of philosophers, and produced many interesting researches, as well as fruitless speculations. So long as those engaged in this investigation were content with mere theory without established data, or speculation without fact, no result was obtained except the useless multiplication of words; but, hen the geography of the country, the nature of the climate, and the history, manners and polity of the various tribes were studied, the mystery involving the subject gradually lessened; so that at present, without much difficulty or error, we may come to a satisfactory conclusion relative to the manner in which this continent was peopled.

Preliminary to our investigation we must refer to the fact, that the laws of nature, governing the continuance of different races of animals profusely multiplied over the earth, are fixed and immutable, and what we observe of Nature's regular modes of operating at one period, is unquestionably true of all preceding times. Animals which are of different kinds, or generically distinct, are incapable of producing offspring together, but animals of the *same kind*, though

AMERICAN, OR RED VARIETY. 19

of different *species*, may and do produce offspring resembling both parents, by their union; yet this confusion ceases with the first product, inasmuch as these hybrids, or *mules*, are universally sterile, or incapable of propagating their similitude. This circumstance furnishes the most satisfactory and unequivocal means of deciding whether any beings we examine are specifically distinct or not, since, if they are merely *varieties* of the *same species*, they are capable of producing offspring in illimitable progression; but, if they are of *different species*, the first offspring terminates the race.

By the application of this test, we are able to pronounce with certainty, that the human race, wherever found, or however different in colour, are merely varieties of the same species, and evidently descended from the same parents. In all countries the marriage of Europeans with the natives, whether Asiatics, Africans, or Indians, is followed by children more or less resembling their parents, and this offspring is perfectly capable of continuing the race.

If there be any mode of accounting for the arrival of even a single male and female on this continent, we shall find no difficulty in understanding how so many nations became distributed over this vast region, nor can we, on an unprejudiced view of the whole subject, find any difficulty in believing that the myriads of human beings, that have lived from the beginning of time to the present hour, have all descended from two individuals. The history of the world, as presented to us by the most authentic records, or by the voice of universal tradition, leads us inevitably to conclude that from some point on the Eastern continent the human race originated, and gradually extended in various directions, subject to the influence of all accidents, of place, climate, disease, and facility or difficulty of procuring food: hence, notwithstanding that the connexion of many nations with the parent stock is entirely lost, there is not the slightest evidence that such nations are derived from any but the source we have stated; neither, when philosophically considered, is there any necessity that they should have originated in a different manner, since the cause is perfectly adequate to the effect; and where one sufficient cause is given no other should be sought.

Under the operation of different motives we find the scattered members of the human family removing by degrees from the centre towards the extremes of the old continent, and subsisting in such remote situations until the disposition or ability to return was entirely lost, and they became inured to the climate, however dreadfully inclement.

Though the human race always remains specifically unchanged in every condition, yet the action of external causes is capable of producing considerable variations in the appearance of individuals or tribes exposed to their influence. Thus we find those who reside in uniformly warm and spontaneously productive countries, of a slender frame, a relaxed and delicate habit, and of a sallow or tawny complexion. The natives of Africa, who are exposed to the most intense heat of the sun, are full framed, robust and vigorous, being endowed with short, crisped and coarse hair, and a skin whose colour shields them from the destructive fierceness of the solar rays. In the middle latitudes, where the means of subsistence are readily procured, and the vicissitudes of season are never remarkably severe, we find the human frame in every variety of development, and distinguished by fairness and delicacy of complexion. But on leaving these favoured regions behind us, and visiting the far northern portions of the earth, we see man, like most of the other productions of nature, stunted and dwarfish, displaying little or no mental energy, barely capable of securing the scanty subsistence allowed him by the rigours of his situation, and maintaining an existence scarcely superior to that of the whale or seal, the hunting of which constitutes his highest ambition, as their flesh and oil are his greatest luxuries.

Since it is not only possible, but unquestionable, that the whole human race are varieties of the same species, most probably descended from one male and female, it remains for us to show in what manner the descendants of this stock may have reached America, and whether our observations can be supported by arguments drawn from the condition of the new world. A reference to a map of the globe will show us that immediately within the arctic circle, the eastern extremity of the old continent is separated from the new by a strait which is but thirty-nine miles across, and this is solidly frozen over during the severities of winter. Kamtschatka, the extremity of Asia, situated between the fortieth and fiftieth degree of north latitude, is peopled by natives who are thoroughly accustomed to endure all the rigours of this climate, and is provided with many animals equally capable of existing through all its inclemencies. Under such circumstances we can see no difficulty in concluding, that from the eastern extremity of Asia, both men and animals have passed to America, and subsequently been multiplied over the whole continent. In regard to man, it is not necessary to insist that he passed to the American shore during winter, since the distance is not too great for us to believe that even the rudest navigators, when driven by stress of weather from their own coast, (as often happens to the Eskimaux,) could, with little difficulty, reach this continent, where they would be compelled to remain by necessity, or induced by a disposition either to extend their acquaintance with a strange land, or to seek for a more agreeable place of abode.

The Aleutian islands, which are very numerous, and form an almost perfectly continuous chain, beginning with Behring s island, and extending from opposite to Kamtschatka, in about the fifty-fifth degree of north latitude, to Alyaska, the same parallel in America, may have afforded a much easier and more certain approach, and that without appearing at all extraordinary to the voyagers themselves, who might pass from one isle to another without having any idea of the land to which they were going. These islands are in the same parallels of latitude as the greater part of Labrador, Hudson's bay, &c. where even Europeans are able to endure the climate during the severest seasons. There is, in fact, the most irrefragable testimony to prove that the rein-deer cross over in vast herds on the ice, subsisting on the moss found in these islands during their passage.

In strictness of reasoning we have nothing to make it absolutely necessary that we should refer the peopling of America to so recent a period as the separation of the old and new continents by Behring's strait. There is neither extravagance nor impropriety in the opinion that the two continents were originally one, and being continuous, the only difficulty is removed that could be urged against the approach of population from the extremity of Asia. But, in addition to all the reasons that can be urged in support of the doctrine, we maintain it should not be forgotten that there are strong evidences derived from astronomical and geological observations, proving the axis and poles of our globe to be not now precisely where they originally stood. It is therefore very unfair to decide against the probability of peopling America from the extremity of Asia, if we reason from the existing climate of the countries adjacent to East cape, or cape Prince of Wales, the two nearest points of Asia and America.

The greatest difficulty thrown in the way of this opinion, was thought to be the striking difference between the Eskimaux and the common Indian, seeming to prove that they were derived from different races or kinds. We are informed in Crantz's History of Greenland, that the Moravian missionaries who visited the countries inhabited by the Eskimaux, were much surprised to find that they were in all respects similar to the Greenlanders, and made use of the same language; showing that the Eskimaux had sprung from the same race, and had gradually reached their present residence from the extreme northern parts of Europe. This fact, now rendered undeniable by more recent researches, entirely invalidates the conclusion that the Eskimaux were derived from another species The resemblance existing between these people and the Siberians, Kamtschadales, Tunguse, &c. is manifest, and notwithstanding they differ in many respects from other inhabitants of the new world, they are undeniably descended from the same parent stock, coming from different parts of the globe. The copper-coloured natives of America, who are the most numerous of the aborigines, approach more closely to the Asiatic Tartars in colour and stature, and this because they are descendants of that race arriving in America from the extremity of Asia.

The land animals *common* to the old and new world doubtless reached America by the same route with the human race; but, the species which are *peculiar* to America were originally placed on this soil, as we find no traces of their existence in Asia or elsewhere. The first inference is drawn because the community of species renders it necessary: the second is a fair and natural induction from the exclusive existence of certain species in this country, as we see no reason why *animals* may not have been from the beginning peculiar to America, as that creatures of a singular construction should be found exclusively pertaining to New Holland. This last named country, differing from all others in its animal and vegetable productions, is peopled by human beings, degraded and abject it is true, but still a variety of the common stock whence all mankind have sprung.

Those who endeavour to dispute the correctness of the doctrine we support, state that if America were peopled from the shores of Asia, many thousand years must have elapsed subsequent to the creation, before the population of the old world could have become sufficiently numerous to extend to its remote borders, and thence attain the American continent. It is also repugnant to their ideas that so large a part of the globe should have remained during " so long a time" unpeopled, or only tenanted by inferior animals. This is truly a convenient mode of objecting, but unfortunately for the theorist, duration of time is a very immaterial circumstance in the great operations of nature. If we may credit the testimony of our senses, and rely on our reason when guided by the clearest lights of geological science, many ages elapsed after the creation of our globe, and numerous races of inferior animals, previous to the existence of man. In very ancient strata, forming the crust of the earth, organic remains of various animals are preserved, but not a single relic of the human kind has yet been obtained from similar situations. He certainly forms a poor idea of DEITY who attempts to measure HIS power or works by notions drawn from human art, or supposes, because one part of the globe must have remained even ten thousand years " in one vast uninhabited solitude," that it is therefore repugnant to all the operations of the wonderful system of nature. With as much correctness might he conclude that the time required by the planet Herschell to describe his orbit around the sun should be reduced to the same duration as that necessary for the Earth, or Mars, or Venus—because to his comprehension the orbit of Herschell is almost unimaginable.

Another objection, founded on a similar mode of

viewing the subject, has led Mr. JEFFERSON and others to believe that the number of different languages spoken in North and South America is incompatible with the idea of so recent an arrival on this continent as even three or four thousand years. "How many ages," says Mr. J. "have elapsed since the English, the Dutch, the Germans, the Swiss, the Norwegians, Danes and Swedes, have separated from their common stock, and yet how many more must elapse before the proofs of their common origin will disappear? A separation into dialects may be the work of a few ages, but for two dialects to recede from each other until they have lost all vestiges of their common origin, must require an immense course of time, perhaps not less than many people give to the age of the world."*

Granting, as we are perfectly willing to do, the great lapse of time which would be requisite for the production of such radical changes, we do not think the objection derived from the languages more solid than those heretofore mentioned. As far as the researches of philologers have extended, we do not find that there is so much difference in the dialects of our aborigines as the arguments of these objectors would imply. Throughout a large mass of this native population a very perceptible connexion of language is apparent, and the relation to a parent stock is fairly evident. Even allowing that the amount of difference is as great as could be desired by our opponents, the comparison of the aboriginal dialects with those of European nations is by no means a correct mode of de

---

* Notes on Virginia, p. 148.

ciding the point. If, according to our idea, people reached this country at different times from the extreme north of Europe or the north-east of Asia, the immense extent of country they were gradually to be scattered over, the new situations they were placed in, the new objects by which they were surrounded, and the new modes of life they assumed, would all conspire to produce a change in their language in a much shorter time than could take place on the old continent, where their wanderings must have been, not only comparatively circumscribed, but their modes of living subject to very few variations. A reference to well authenticated and recently observed facts, will show how great an influence is exerted over language by these causes. Indian nations, which have commenced their migrations in the northern and eastern parts of this continent, and journeyed to the western regions, have on their route detached various colonies from their main body, and these, in many instances, now differ so much in language from their parent stock, as to exhibit none but faint traces of relationship. If changes of this kind can be produced under such circumstances, what difficulty is there in believing, that still greater could occur, when the whole extent of this vast continent was before the original adventurers, and the last comers might not only be separated from the first by thousands of miles, but live under other skies, and be surrounded by natural objects of a totally dissimilar character.

In the present condition of our knowledge, we have no right to state that the traces of affinity between the American dialects are entirely obliterated; it would be far more correct to say that we do not possess the means of making the necessary inquiries and decisions; our knowledge of their languages is confined to a few meagre vocabularies, frequently derived from persons whose statements cannot be implicitly relied on, however correct their intentions may have been, to say nothing of the almost insuperable difficulty of writing such languages from the hearer's idea of their pronunciation. We may with sufficient correctness trace the descent of words in our own language from the Hebrew, Sanscrit, &c. because we have established signs to indicate the ideas, and we have no doubt but that the same could be done to nearly an equal extent with our aboriginal tongues, provided we enjoyed a similar advantage of written characters, a proper knowledge of their languages, and a better acquaintance with the natural and other objects most frequently the subjects of their conversation.

Considering all the essential circumstances which are entirely wanting in this inquiry, we can place very little reliance on inferences from the aboriginal languages, more especially such as were drawn by a late writer, respecting the affinities of dialect between some of our Indians, and that of the Yolofs, the blackest of the African tribes. We must believe that these affinities were either totally accidental, or founded in misconception, arising from the nature of the subject, or rather from want of necessary intimacy with the languages examined. It may be taken as a very safe rule of judgment, that a man, whose knowledge of any language is derived exclusively from books, however perfectly he may be able to judge of its philosophy or grammar, can have but

AMERICAN, OR RED VARIETY.

few and faint ideas of the nice shades of distinction in the value and application of a very large proportion of words in such tongue, and by consequence, is very little qualified to do more than *conjecture* their affinities with words used by a people living under totally different circumstances.

The learned *Pennant,* in treating of this subject, expresses his belief that the inhabitants of the American continent were originally derived from Eastern Asia, and supports this conclusion by an examination of the customs common to the inhabitants of both continents.

"The custom of scalping, says he, was a barbarism in use with the Scythians, who carried about them at all times this savage mark of triumph: they cut a circle round the neck, and stripped off the skin, as they would that of an ox.* A little image, found among the Kalmucs of a Tartarian deity, mounted on a horse, and sitting on a human skin, with scalps pendent from the breast, fully illustrates the custom of the Scythian progenitors, as described by the Greek historian. This usage, as the Europeans know by horrid experience, is continued to this day in America. The ferocity of the Scythians to their prisoners, extended to the remotest part of Asia. The Kamtschadales, even at the time of their discovery by the Russians,† put their prisoners to death by the most lingering and excruciating inventions; a

---

* *Herodotus,* lib. iv.—Compare the account given by the historian with the Tartarian *icunculus* in Dr. PALLAS' Travels, i. tab. x. *a.*
† Hist. Kamtschat. 57.

practice in full force to this very day among the aboriginal Americans. A race of the Scythians were styled Anthropophagi* from their feeding on human flesh. The people of Nootka Sound still make a repast of their fellow creatures,† but what is more wonderful, the savage allies of the British army have been known to throw the mangled limbs of the French prisoners into the horrible cauldron, and devour them with the same relish as those of a quadruped.‡

"The Scythians were said, for a certain time, annually to transform themselves into wolves, and again to resume the human shape.§ The new discovered Americans about Nootka Sound disguise themselves in dresses made of the skins of wolves and other wild beasts, and wear even the heads fitted to their own.|| These habits they use in the chase to circumvent the animals of the field. But would not ignorance or superstition ascribe to a supernatural metamorphosis these temporary expedients to deceive the brute creation? In their march the Kamtschadales never went abreast, but followed one another in the same track. ¶ The same custom is exactly observed by the Americans.

"The Tungusi, the most numerous nation resident in Siberia, prick their faces with small punctures,

---

* Mela, lib. ii. c. i.
† Voyage ii.
‡ Colden's Five Nations, i, 155.
§ Herodotus, lib. iv.
|| Voyage ii, 311, 329.—A very curious head of a wolf, fitted for this use, is preserved in the *Leverian* museum.
¶ Hist. Kamtsch. 61.

AMERICAN, OR RED VARIETY. 31

with a needle, in various shapes; then rub charcoal into them, so that the marks become indelible.* This custom is still observed in several parts of America. The Indians on the back of the Hudson's Bay to this day perform the operation exactly in the same manner, and puncture the skin into various figures, as the natives of New Zealand do at present, and as the ancient Britons did with the herb *glastum*, or woad,† and the Virginians, on the first discovery of that country by the English.‡ Herodian delivers down to us this custom of the Britons:—He says that they painted their bodies with figures of all sorts of animals, and wore no clothes lest they should hide what was probably intended to render themselves more terrible to their enemies.

" The Tungusi use canoes made of birch bark, distended over ribs of wood and nicely sewed together.§ The Canadian and many other American nations use no other sort of boats. The paddles of the Tungusi are broad at each end; those of the people near Cook's river and of Oonalaska are of the same form.

" In burying of the dead many of the American nations place the corpse at full length, after preparing it according to their customs; others place it in a siting posture, and lay by it the most valuable clothing, wampum, and other matters. The Tartars did the same, and both people agree in covering the whole with earth, so as to form a *tumulus*, barrow,

---

* Bell's Travels, i. 240, 8vo.
† Herodian in Vita Severi.
‡ De Bry, Virginia, tab. iii. 111.
§ Isbrandt-Ides in Harris' Coll. ii. 919.

or carnedd.* In respect to the features and form of the human body, almost every tribe found along the western coast has some similitude to the *Tartar* nations, and still retain the little eyes, small noses, high cheeks, and broad faces. They vary in size from the lusty Calmucs to the little Nogaians. The internal Americans, such as the *Five Indian* Nations, who are tall of body, robust in make, and of oblong faces, are derived from a variety among the Tartars themselves. The fine race [tribe] of *Tschutski* seem to be the stock from which those Americans are derived. The Tschutski again from that fine race of Tartars, the *Kabardinski*, or inhabitants of Kabarda."†

Independent of all other arguments in favour of the Asiatic origin of the aboriginals of America, the circumstance of but *one species* of the human race existing throughout the world is sufficient to reduce us to the necessity of acknowledging that mankind have descended from one parent stock, however their external appearance may have been modified by accident, disease, or situation. We are aware that some persons talk of the possibility of there having been various centres of creation to the human race, as among inferior animals; but we consider it very unphilosophical to suppose the existence of various centres of *creation* for the *same* species. Occasionally we hear still more ridiculous opinions advanced by persons who have not been at the trouble of examining the facts which have been collected on the subject,

---

* Compare Colden, i. 17, Lafitau, i. 416, and Archæologia, ii. 252, tab. xiv.

† Pennant's Introduction to the Arctic Zoology, p. 260.

are desirous of rendering themselves notorious by supporting any opinion, however absurd.

Thus far we have paid a deference to those who are unwilling to suppose that this continent was peopled from the old, and we have bestowed on their arguments a sufficient degree of attention.* But, as we have already hinted, all this discussion, relative to the human inhabitants of this continent, may be dispensed with; first, because the human race, from the equator to the poles, are one and the same, without presenting a single *specific* difference; and, in the second place, because a very adequate and perfectly natural means of approach is given, by which all the results desired could be readily produced. We have shown that limitation of time relates merely to our own narrow conceptions of its duration, and has reference neither to the DEITY nor to the order of nature; nor is it rendered necessary by any knowledge we possess relative to the creation of the world. Even allowing a most immense lapse of ages to have intervened from the creation to the peo-

---

* The theory of Clavigero, which supposes that a country fifteen hundred miles in length, and of an unknown breadth, was sunk between America and the old continent, and that by this land the human race anciently passed to this country, is too extravagant and unfounded to require more than a passing notice. Instead of having islands or regions sunk in the neighbourhood of the American shores, they are continually forming and increasing with great rapidity, being almost uniformly founded by the labours of the coral molluscæ, and after being built up by them from great depths to the surface of the ocean, collect weeds, sand, and other matters, for the commencement of a soil.

pling of this continent, we should rather discover in it a proof of the correctness of our position, than a circumstance repugnant to the plan of nature. Had a race of men been created on this continent simultaneous with that established in the old world, the vast increase of population would have long since required more than the ordinary devastation of human life, by pestilence, famine, and murderous war.

An idea of creation, more consonant with enlightened intelligence than the one recently glanced at, is that which considers the ETERNAL as having given existence to a few laws, or rules of action, which, through his omniscience, comprise all subsidiary operations, and by their influence the whole admirable system becoming in due succession developed and perfected—each joined to each in proper corelation, and all approaching his immediate presence by a point too ineffably distant to be appreciated by finite comprehension. Such an evaluation of the plan of divine providence, or creative power, would shame us out of theories in which we attempt to reduce infinitude to our own standard, and mete out the operations of the mighty system of nature by our own miserable span of three score and ten years!

SECTION II.

General Character of the American Indian.

In various situations the North American Indians exhibit very considerable differences in stature, colour, and physiognomy; their medium height may be stated to correspond with that of the Europeans.

ONGPATONGA.

Big Elk
Chief of the Omauhaws.

Engraved by Hugh Bridport from an original Painting by J. Neagle.

though many individuals may be found in various tribes far exceeding the ordinary size. Their colour varies from a cinnamon brown to a deep copper colour, and some that I have seen were rather of an olive yellow hue. They almost universally have black, straight and rigid hair, though it frequently appears more harsh from their modes of dressing it, than it otherwise would be.

The features of the face are all large and strongly marked, if we except the eyes, which are generally deep seated, or sunk in large sockets, and placed rather horizontally; in this respect, and in general beauty of person, they more nearly resemble the European than any other variety of the human race. The forehead is most commonly low, somewhat compressed at the sides, and slightly retreating from the perpendicular. The facial angle is about eighty degrees. The nose is generally elevated from the face, and sometimes prominent, or even arched; the cheek bones are high and widely separate; the angle of the jaw is broad and the chin square. These latter circumstances give a peculiar fullness to the lower part of the face, and occasions much of the remarkable expression of the Indian countenance.

The Indians have been often supposed to be naturally destitute of hair on their bodies, but this deficiency is produced by the sedulous attention with which they eradicate the hair from the chin and other parts of the body. The hair of the head is also in great part removed, a small lock being usually left on the centre or crown, which is commonly decorated with feathers, porcupine quills, or other ornaments.

The habit of painting the body, either on occasions of ceremony or preparatory to battle, is very universal among the Indians, and hence vermilion has always been a substance of great value to them. Under ordinary circumstances, where this pigment is not to be obtained, they employ various coloured clays, charcoal, &c. which are smeared over different parts of the skin in fantastic figures.

To estimate the moral character of the Indians correctly, our inferences must be drawn from tribes undebased by their proximity to the whites, or from periods which preceded the introduction of European vices and corruptions amongst them. It is thus that the venerable and excellent Heckewelder gave his valuable recollections of the Lenapes or Delawares, not as they were at the time he wrote, but as they had been when he first knew them, many years before. Born and nurtured in the most uncontrolled liberty, the restraints of civilized life have as yet only served to bring the Indian still lower than the quadruped tenants of the forest that have been subdued by the white man. Instead of displaying the energies of nature, improved by cultivation, the civilized aboriginal has sunk into a state of hopeless apathy, incapable of any thing better than an imitation of the worst vices of the worst of men.

But, when free, in his native wilds, the American displayed a form worthy of admiration, and a conduct which secured him respect. Brave, hospitable, honest and confiding, to him danger had no terrors, and his house was ever open to the stranger. Taught to regard glory as the highest reward of his actions, he became a stoic under suffering, and so far subjugated his feelings as to stifle the emotions of his soul, allowing no outward sign of their workings to be perceived. His friendships were stedfast, and his promises sacredly kept; his anger was dreadful; his revenge, though often long cherished, was as horrible as it was sure; necessity and pride taught him patience—habitual exercise made him vigilant and skilful; his youth was principally spent in listening to the recital of his father's and ancestors' deeds of renown, and his manhood was passed in endeavouring to leave for his children an inducement to follow his example.

Grave, dignified and taciturn, under ordinary circumstances—in the assembly of his nation, the Indian frequently became fluent, impassioned, eloquent, sublime. With few words, and no artificial aid, drawing his images exclusively from surrounding objects, and yielding to the influence of his own ardent impulses, he roused his friends to enthusiasm, or inspired his enemies with dread, as he depicted, with few and rapid touches, the terrors of his vengeance, or the horrible carnage of his battles.

An Indian suffering with hunger complained not—nor when long absent from home expressed emotion on his return.—"I am come," would be his simple salutation—"it is well,"·the only reply. When refreshed by eating and smoking, he related the story of his enterprise to his assembled friends, who listened in respectful silence, or only testified their interest in his narrative by a single ejaculation.

The Indians almost universally revere the aged, and are exceedingly indulgent to their offspring.

whom they rarely chastise, unless by casting cold water on them. They are not so kind to their women, who, as a general rule, are treated rather as domestic animals than as companions, and are seldom exempted from severe toils, even when about to give birth to their children. Notwithstanding this, the women appear contented with their situation, and not unfrequently exhibit excellent traits of character. At times their jealousy, or other depressing passions, lead them to the commission of suicide, which is particularly frequent among some of the tribes. Indian habits of thinking, varying with their modes of education, differ very much in different nations. The want of chastity before marriage is not universally considered as a loss of character, neither is incontinence in the female, after marriage, regarded as a crime, provided the husband gives his consent. Yet the same people will treat as infamous, and even put to an ignominious death, a woman who receives the addresses of another man without permission of the husband. The number of wives taken by the men is most commonly limited only by their ability to maintain them, as almost all Indians are polygamous. Their wandering modes of living and precarious subsistence, render increase of population far inferior among them to what it is among the whites.

The government to which they submit, is that exercised by their chiefs, who are, with very few exceptions, chosen in consequence of their superior courage, physical strength, or great experience and wisdom. The deference paid to them is not at all to be compared with that manifested by Europeans to their rulers; it is a respect for qualification and standing, but confers no other privilege than that of leading them to battle, or directing the movements of their camp; it does not entitle the individual to interfere with the rights of others of his tribe, nor can his will be carried into effect, unless it be supported by the general opinion of his people.*

The most universal and enduring passion among the Indians is that for warlike glory. The earliest language he hears is the warrior's praise—the first actions he is taught to perform have for object the eventual attainment of this distinction, and every thought is bent towards the achievement of heroic deeds. Hence death is despised, suffering endured, and danger courted: the song of war is more musical to his ear than the voice of love, and the yells of the returning warrior thrill his bosom with pleasing anticipations of the time when *he* shall leave blood and ashes where the dwelling of his enemy stood, and hear the triumphant shouts of his kinsmen, responsive to his own returning war cry!

If we except their skill in hunting, and the great excellence of observation, by which they can detect the footsteps of game or of their enemies, we must admit that the Indians have but little knowledge, and their acquaintance with mechanic arts is still less perfect. They construct lodges with skins, bark, or earth, sustained by rude poles; make canoes of

---

* TACITUS, in his excellent account of the ancient Germans, informs us that their leaders were chosen in a similar way. The reader may dérive much pleasure from examining his 4th Chap. de Moribus Germaniæ, beginning " Reges ex nobilitate duces ex virtute sumunt." &c.

bark, shape bowls out of wood with vast labour, by the aid of sharp flints and other stones; make a rude and sun-dried pottery, fashion tobacco pipes of clay or stone, dress the skins of animals by rubbing them, when moistened with brains, until they are pliable, and from these skins form moccasins, pouches, &c. variously ornamented with porcupine quills, which they know how to dye of several brilliant colours.

They cannot be said to have any acquaintance with astronomy, if we except the ability some of them possess to guide their course by the polar star; their knowledge of medicine and surgery is exceedingly imperfect and rude, or more properly it consists of a very few actual remedies and a great deal of juggling mummery.*

Their modes of living vary throughout the countries they inhabit, according to the peculiar nature of circumstances. Those who reside where game is plenty, live entirely by hunting—others, in the neighbourhood of lakes and rivers, derive their support principally from fishing; many tribes raise small quantities of maize and tobacco. The Indians who live on the prairies or in level countries, are fond of horses, and are excellent horsemen, while such as frequent the forests are more remarkable for the celerity and sagacity with which they travel on foot.

Their ideas of Deity are very rude and imperfect.

---

* LUCRETIUS gives a very interesting description of savage man, which may in a great degree be applied to the aboriginal inhabitants of America.—*See his* 6*th Book, line* 920, *beginning* " et genus humanum multo fuit illud in arvis."

though they all seem to have an idea of a future state, as well as of a great Spirit and Director of the universe. Many tribes have some notion of rewards and punishments in a future life; their ideas on this subject are necessarily founded on their appreciation of what is at present agreeable or disagreeable to themselves. They believe in bad as well as good spirits, and are as much addicted to the worship of the former from fear, as they are to adore the latter from love and respect.

The Eskimaux, who inhabit the most northern parts of North America, differ considerably from the aboriginals generally diffused over this country, as they are far inferior in stature, and the features of their faces are extremely harsh and disagreeable to Europeans. Their cheek bones are very prominent, their cheeks tumid and somewhat globose, their noses small, flat, or sunk, and their whole physiognomy resembling considerably that of the most ill-favoured Tartar tribes.

The appearance of the Eskimaux varies from Prince Williams' sound, where they are of the largest size; as they extend to the more northern regions, to the coast of the Icy sea and the maritime parts of Hudson's bay, Greenland and Terra de Labrador, they become dwarfish, in comparison with the European, and have heretofore given rise to accounts of pygmies inhabiting these icy regions.

In the Eskimaux we have an admirable exemplification of the effects of severe climate on the human race, for the extreme cold seems in them to have repressed all superfluity of growth, as if to accommodate the body to a situation where food and raiment cannot be procured without great difficulty and danger.

SECTION III.

Dental System of the Human Race.

32 Teeth
- 16 *Upper*
  - 4 Cutting
  - 2 Canine
  - 10 Grinders.
- 16 *Lower*
  - 4 Cutting
  - 2 Canine
  - 10 Grinders.

In the upper jaw, the first cutting tooth is terminated inferiorly by a straight line; it is shaped like a wedge, and is larger than the second. This latter is straighter than the first cutting tooth, and terminates by two lines, which form between them an open angle, that is to say, its point is obtuse. Both these teeth are rounded externally, and their internal surface is slightly excavated. The canine has exteriorly the form of the second grinder, but differs from it, because its internal surface is salient instead of being depressed; this gives it a thickness not to be found in the other.

The first and second grinders (which are called *false molars*, on account of their slenderness when compared with the thickness of the true grinders,) resemble each other entirely in form and size. Externally examined, they are but slightly distinguished from the second cutting, and canine tooth; but they have on the crown two very thick and very obtuse tubercles, one on their internal, the other on their external border, separated by a deep groove.

The next tooth is a true grinder,* and is the largest of all the teeth in the upper jaw. On its external edge it is divided into two equal parts, forming two very obtuse tubercles, by a deep groove which only comes to the middle of the crown, and when there, separates into two very slight branches, which form, with the principal groove, nearly the same angle as they form with one another. On its internal edge this tooth is also divided, but unequally, by a groove situated much nearer its posterior than its anterior edge, so that the tubercle produced by this part of the groove is stronger than that of the other portion: the tubercles on this inner edge have their summits much nearer to the middle of the tooth than those on the opposite border.

The two succeeding grinders are of the same size and form, having on their external edge two equal tubercles, formed by a groove which divides the tooth to the middle of the surface, where it parts in two branches, like the groove of the great grinder; but these branches sometimes extend even to the anterior and posterior edges of the tooth. Their internal edge is composed of a single but very obtuse tubercle, and these anterior and posterior parts are separated by a deep depression.

IN THE LOWER JAW, the first cutting tooth is one-third straighter than that in the upper jaw, and is

---

* True grinders, or simple grinders, are all teeth covered with tubercles, evidently fit for crushing or triturating food.

equally terminated by a straight line; the second is nearly of the same size as the first, and terminates by a point analogous to that of the upper jaw, but this point is much nearer to the first cutting tooth than to the canine.

The canine resembles the one described in the upper jaw, except that it is not so thick.

The two false grinders have also much resemblance to those of the upper jaw, except that they are somewhat smaller, their external tubercle is much thicker than the internal, and a projection on their middle divides the groove which separates them into two parts, and forms their two principal tubercles.

The succeeding grinder, which is also the largest of this jaw, is divided into four principal parts, or four large tubercles, by two grooves which cross each other at right angles at the middle of the tooth; and these tubercles present irregular inequalities, produced by some isolated depressions, and also by some slight branches of the principal grooves. The two succeeding grinders are smaller than the preceding, and present the same principal divisions, that is, they have four tubercles and two grooves; but the groove which cuts the tooth transversely is deeper than that running from behind forwards, which makes it sometimes scarcely perceptible on the posterior half of these teeth. The three last teeth are true grinders.

IN THEIR RECIPROCAL POSITION, the lower teeth, as far as the middle cutting teeth, are more advanced anteriorly than the upper ones, that is to say, the posterior part of the lower teeth correspond to the anterior of their analogues in the opposite jaw, which appears to show the reason of the narrow dimensions in the lower middle cutting teeth, compared with the upper, and all are opposed to each other crown to crown, except the cutting teeth, which stand face to face; the lower by their anterior face to the posterior face of the upper.*

---

* The *Dental Systems* of the different genera described in this work, are translated from FREDERICK CUVIER's celebrated work, "Des Dents des Mammiféres Considerées comme Caractéres Zoologiques."

CHAPTER II.

ORDER III.* FERÆ.—*Beasts of Prey.*

SECTION I.

UNDER this order are arranged the animals which are unguiculated or provided with claws, are destitute of hands to their anterior extremities, and possess the three sorts of teeth called carnivorous, false molars, and tuberculous; those which have the teeth either entirely or partially tuberculous, feed to a greater or less degree on vegetable matter, while such as have their teeth studded with conical points, feed principally on insects. In proportion to the sharpness of the teeth we may decide whether or not they are exclusively carnivorous.

In all the species belonging to this order, the articulating or condyloid process of the lower jaw is semicylindric and transversely placed, corresponding so precisely with the glenoid cavity of the temporal bone, that it can only be moved in one direction, or is incapable of any motion except that of opening and shutting.

In some species the zygomatic arches are very large, and the skull, especially at the posterior part,

---

* The second order, (quadrumana, or four handed animals,) is not found in North America.

is much compressed, which gives the space requisite for the large and powerful muscles concerned in the act of mastication. Their brain is destitute of the third lobe, although sufficiently furrowed, and does not cover the cerebellum. The orbit of the eye is not separated by bone from the temporal fossa or hollow, in the skeleton.

Beasts of prey possess the sense of smelling in a high degree of perfection, and their olfactory nerves are generally spread out over very numerous plates of convoluted bony texture. They can turn the fore-arm, but by no means with the same facility as is done by the quadrumanous animals, and their fore-limbs are uniformly destitute of thumbs capable of antagonizing the other fingers.

The stomachs of beasts of prey are, generally speaking, simple, and their intestines are less voluminous than those of other quadrupeds, on account of the greater degree of facility with which the digestion of animal matter can be effected.

The greatest part of these creatures are forced, by the necessity of procuring animal food, to attack and destroy the lives of other animals, which they are well qualified to do by their great muscular strength, and the offensive armour of teeth and nails with which they are provided.

As the forms and particulars of construction are very various in these animals, we may expect corresponding variations in their habits and actions. Hence they have been arranged in different families, connected with each other by various relations.

SECTION II.

Family I.—Cheiroptera; *Wing-handed.*

The animals comprised in this family are generally known by the name of bats, and are strikingly characterised by the manner in which the skin of the body is extended between and connects the anterior and posterior extremities, and is also prolonged over the bones of the fingers, so as to form a large and efficient wing. Their clavicles, or collar-bones, are necessarily strong, and the scapulæ, or shoulder-blades, are large, in order that the shoulder-joint may have the necessary degree of solidity. The fore-arm is incapable of supination,* as the power of flying would thereby be impaired. Bats have two teats, which are situated on the chest.†

Dental System of the Bat.

This family has been divided into many genera on account of the differences of the cutting teeth, and the modifications observed in the organs of sense and motion. In fact these are almost the only particulars in which bats differ from each other; in all, without

---

* The hand is said to be *prone*, when the palm is turned downwards, or towards the earth; *supine*, when the palm is turned upwards.

† Penis illis, generis humani more, propendens: character profecto talibus animantibus mirum.

exception, the true grinders and canine teeth are of the same form and number; yet they differ in the quantity of their false molars, which do not always correspond with the number of their cutting teeth, and in other modifications on which the genera of this family are founded.

IN THE UPPER JAW, the canine is strong and angular, having the general form of this sort of tooth, and receives by an anterior, and sometimes very deep depression, as well as an internal very deep depression, a triangular form. In some species there is a salient point on the exterior of this tooth, which in all the cheiroptera seems to be a very important instrument in securing their prey, or as a weapon of attack and defence.

The false molars, which are most developed, and may be considered regular in these animals, have a point, and a base which extends from their internal and posterior side, sometimes producing a small point at its anterior part; these teeth always have two roots. The molars are three in number, the first and second having the same form, and differing but little in size. They present on their external surface two parallel prisms, a section of which is terminated by a point at each of its angles. These two prisms rest upon a base which is developed on the interior of the tooth, and is composed on its anterior part of a slightly salient and triangular tubercle, and at its posterior part, by a small simple point. The last molar, one half smaller than the others, appears to be one of these first teeth, obliquely truncated at its external and posterior part on account of the sudden termination of the maxillary bone, as if the half of the posterior prism was removed, as well as the like posterior point.

In the lower jaw, the canines, equally strong as those of the upper jaw, are rounded in front, but flattened on the posterior surface, and strongly grooved at their base on this part.

The regular false molars are slender, with a middle principal tubercle, that is to say, having all the general characters of these teeth. The molars, three in number, are composed of the two prisms, which we have seen form an essential part of the molars in the opposite jaw; but where they present one of their faces outward in the upper jaw, in the lower they present one of their angles, and the point of this angle is commonly stronger than those of both the others. The two first of these teeth are of equal size, and the third a little smaller, because the posterior angle is not entirely developed.

In their reciprocal position, the inferior canines are in advance of the superior, as in the carnivorous animals, and the projections of the grinders interlock with the hollows of the opposite teeth.*

---

* The author has been informed by Prince Musignano, that the celebrated Temminck of Amsterdam, has made some recent discoveries relative to the dental system of the Bat, which will most probably produce a considerable change in the classification of these animals. Should the observations of Temminck be published in time, we shall give them in the appendix to this work.

Genus II.—Bat; *Vespertilio;* L.

Gr. Νυκτερισ.
Germ. Fledermaus, Speckmaus.
Dutch. Vlarmuis, Vledermaus.
Swed. Flædermus, Læderlapp.
Russ. Letukscka, Neotopyr.

Fr. (*anc.*) Chaude souris, Ratepenade. (*Mod*,) chauve souris.
Ital. Pipistrello, nottola, sportiglione, Rattopennago, &c.
Span. Murciegalo, Murceguillo.
Portug. Morcego.

GENERIC CHARACTERS.

Dental Formula.

32 Teeth.	14 Upper	4 Incisive 2 Canine 8 Molar.	2 False Molars 6 Molars.	2 regular 0 anomalous
	18 Lower	6 Incisive 2 Canine 10 Molar.	4 False Molars 6 Molars.	2 regular 2 anomalous

The anterior limbs, or arms and fore-arms of Bats, are peculiarly elongated, and the bones of the fingers are still more lengthened, exceeding the total length of the arm. It is over the finger-bones, especially, that the skin of the body is extended, in the form of a soft and delicate membrane, which thus constitutes wings as extensive and effectual as those of many birds. The pectoral muscles, which are the principal agents in moving the wings, are thick and strong, to correspond with the nature of the service they are to perform, and the middle of the sternum, or breast-bone, has a projection to give these muscles a large surface of attachment, resembling considerably a similar part in birds. In consequence of these arrangements they are able to fly with rapidity, and at great heights in the air.

These curious wings are provided with a short thumb, armed with a crooked nail, which serves them to hang by when they wish to repose, or to climb with against the sides of caverns or other places. Their hind feet are weak and divided into five equal toes, all armed with nails. The eyes of these animals are quite small, and their ears are often very large; these, together with their wings, form a vast membranous surface, almost naked, and exquisitely sensible.

The perfect wisdom of the Author of Nature becomes more clearly evident, in proportion as we carefully study the relations existing between his creatures, the situations they occupy, and the offices they are intended to perform. As our inquiries are extended, we feel continually incited to express our admiration at the excellent adaptation of means to the end, and are induced to consider with pleasure, from their appropriateness, beings which, from a superficial view, would be thought frightful or disgusting. This may be well exemplified by the animals now to be investigated; a slight glance at their mere external form and appearance presents them as disagreeable, almost deformed, or most probably useless. A judicious observation of their structure and modes of living causes admiration at the excellent contrivance of their organization, and convinces us that they may be eminently serviceable, even to mankind.

The Bat certainly seems to occupy a very equivocal rank in creation, since, though bearing a marked resemblance to a quadruped, a great part of his life is spent in the air like a bird. Yet it is only in the latter circumstance that he can be compared with the feathered tribes, being not only destitute of beak, plumes and talons, but suspending himself in air by means of a velvet or leather-like membrane.

When the gray and dusky twilight succeeds the departing glories of the sun, myriads of insects, warmed into life and activity by his heat, take wing in search of their females, to increase the innumerable hosts of their own race. At the same moment the Bats, which have shrouded themselves during the glare of daylight, emerging from subterraneous recesses, or the gloomy vaults of time-worn ruins, speed with rapid flight along, glutting a voracious appetite on insects, which, but for their exertions, conjoined with those of other creatures, would soon swarm so profusely as to render the earth loathsome or uninhabitable.

The advantages of the Bat's peculiar structure are now seen—his soft and velvet wings, though plied with vigorous celerity, stir the air, but make no sound; their peculiarly delicate sensibility enables him to feel the proximity of every object, and unerringly directs his flight; his large ears catch every hum produced by the motions of his destined prey, and he noiselessly flits through the gloom, gathering a plenteous meal, and destroying great numbers of insects. His strong sharp teeth and powerful jaws are employed in seizing and crushing his prey with slight effort, nor does he relinquish the chase until the night is far advanced and the cravings of hunger are entirely satisfied.

The Bat flies with a tremulous flickering movement

of its membranous wings, and its progression is ir regular, now rising with swiftness, then suddenly darting downwards, or to one side, with apparent capriciousness, though it is engaged in seizing its prey, which it distinguishes with great quickness. This disposition to dart at any object seen in the air is often employed for the destruction of Bats, as they are shot or struck down with a long switch or whip at the moment they descend to examine objects thrown into the air. In general the Bat flies at no great distance from the earth, though it occasionally ascends above the tops of trees or houses, and even much higher. During the time of feeding it appears to be continually in motion, searching for food with much diligence, as if conscious that the opportunity of procuring it would soon pass by.

When on the ground, it is very evident that the Bat is a quadruped, and what are commonly called wings are entirely analogous to the members of other four footed animals, though varying greatly from the original type. "The elbow is found near the knee; the fore arm is very long, and obliquely extended from above downwards, and from behind forwards, as far as the nose of the animal. The wrist is placed against the ground, and there is but one finger on the anterior extremity, which is the thumb. The knee is raised as high as the lower part of the rump, and the five toes of the hind feet are of equal length, and turned outwards. The arm is extended horizontally from the front to the back part, and the thigh vertically from above downwards; the arm is concealed behind the fore arm, and the thigh behind the leg; they are moreover enveloped in the membranes which conceal the tail and all the hinder parts of the body. Besides the thumb seen on the anterior extremities, there are four other very long fingers, extending from the fore-arm, enveloped in the membrane, and folded near the elbow by their extremity."*

From these singularly lengthened extremities, the animal derives little or no assistance while on the ground, as it rests on the breast and belly, rather propped up than standing. In this situation the motions of the Bat are slow and heavy, more resembling the dragging along of the body and limbs than a fair and regular act of progression. It is by no means easy for the Bat to take wing, and some of the species cannot readily escape if placed on the ground; but it is an error to suppose that it is impossible for them to take to flight, while in this situation. Their mode of alighting or resting is, by fixing the hook of their hind feet to the projections of caves or old buildings, with their wings folded on their bodies, and their heads hanging down, until forced by hunger to resume their flight. This singular inverted position, is one which enables them to take wing with ease, inasmuch as they can launch themselves in air with great facility, merely by letting go their hold. In order to void its excrement, which is very frequently found in vast heaps in caves much frequented by them, they bend their bodies upwards, and extend the wing until they can lay hold of a projection with the hook on

Daubenton.

the thumb. When this is accomplished they relinquish their hold with the posterior claws, and thus perform their evacuations, resuming their original inverted position, by renewing their hold with the claws of the hind feet.

The Bat is entitled to the place it holds in our systematic arrangements, from the circumstance of having paps or teats placed on the chest, analogous to those of the human race. They suckle their young, who remain firmly attached to the teat during the flight of the parent, until they attain a considerable size.

These creatures are not deficient in those affections, which, in other animals, are supposed to denote much sensibility, and always excite the sympathies of mankind. The Bat has been known to exhibit the most devoted attachment to her young, and forego all efforts at self-preservation, in order to be near when she could not release her captive progeny. The following circumstance made known by that enterprising naturalist, my friend TITIAN PEALE, will show to what an extent they may evince this feeling:

" In June 1823, the son of Mr. Gillespie, keeper of the city square, caught a young red Bat, (*Vespertilio Nov-Eboracensis*, L.) which he took home with him. Three hours afterwards, in the evening, as he was conveying it to the Museum in his hand, while passing near the place where it was caught, the mother made her appearance, followed the boy for two squares, flying around him, and finally alighted on his breast, such was her anxiety to save her offspring. Both were brought to the Museum, the young one firmly adhering to its mother's teat. This faithful creature lived two days in the Museum, and then died of injuries received from her captor. The young one, being but half grown, was still too young to take care of itself, and died shortly after."

We have already glanced at the singular fact, that Bats have the power of directing their flight with perfect correctness, even when deprived of their sight. In 1793, Spallanzani put out the eyes of a Bat, and observed that it appeared to fly with as much ease as before, and without striking against objects in its way, following the curve of a ceiling, and avoiding, with accuracy, every thing against which it was expected to strike. Not only were blinded Bats capable of avoiding such objects, as parts of a building, but they shunned, with equal address, the most delicate obstacles, even silken threads, stretched in such a manner as to leave just space enough for them to pass with their wings expanded. When these threads were placed closer together, the Bats contracted their wings, in order to pass between them without touching. They also passed with the same security between branches of trees placed to intercept them, and suspended themselves by the wall, &c. with as much ease as if they could see distinctly. Similar experiments were made by JURINE of Geneva, who attempted to account for the fact by attributing it to the delicacy of the nerves expanded about the muzzle, ears, &c. Mr. Carlisle, who experimented in England with the large-eared Bat (*V. Auritus*) concluded that this faculty was owing to extreme acuteness of hearing, as the Bat, when its ears were covered, flew against objects. as if unconscious of their presence; it is probable, however, that there was some unobserved source of fallacy in this experiment.

A much more satisfactory and philosophical mode of explaining this curious circumstance was offered by the celebrated CUVIER, who sheds light wherever he directs his attention. In a paper read May, 1796, this naturalist referred it to the exquisite sense of touch resident in the membranous skin forming the wings, ears, &c. as had been previously hinted at by Odier. During the flight of the blinded Bat, whenever it approaches any object, the air set in motion by its wings reacts against their surface with a greater or less degree of force, and being in this manner warned of the proximity of the object, it avoids injury by changing its course.

Immediately preceding thunder-storms, Bats have been known to take shelter in dwelling-houses in great numbers; no less than thirty were recently captured in the house of a friend, where they had thus entered for refuge against an approaching gust.

The Bat brings forth in the month of June and July, generally from one to three young at a birth, which are carefully suckled by the parent until they grow to a considerable size.

As soon as the cold weather approaches, the Bat retires to places of security, and frequently gathers in large clusters, apparently for the sake of warmth. They gradually become perfectly torpid, and in this state the circulation of the blood and other functions of the animal economy seem to be entirely suspended. With the return of the warm season they slowly acquire the ability to move, and when their torpor has entirely passed away, they again sally forth to renew their wonted destruction of the insect tribe.

The singular structure and habits of the Bat have long since afforded the poets an emblem of darkness and terror, and induced them to consecrate this creature to Proserpine, their queen of Hades. Æsop, with his usual shrewdness of observation, has turned the Bat to good account in his fable of the war between the birds and beasts, in which he severely reproves those who in important affairs are disposed to belong to no party. He represents the Bat as unwilling to declare for either host, but to hover between both during the fight, in consequence of which it was no longer considered either as bird or beast, and was obliged to avoid appearing abroad until other animals had gone to repose.

The fact, as has already been shown, is, that the Bat is one of a large number of animals whose structure is adapted for activity and usefulness only when the light is feeble, and food is to be obtained.—However amusing it may be in poetry or apologue to consider such creatures as choosing night for their appearance from a desire of concealment, it is by no means allowable for students of natural history to forget that all beings must live in conformity to the laws of their organization, that the perfection of every species is relative to the situation in which it exists, and that our notions of beauty and deformity are neither true tests of the excellence nor importance of any inferior animal.

SECTION II.

Sometime after the preceding observations were written, I received a paper on Bats, by the distinguished Geoffroy St. Hilaire, published in the French Dictionary of Natural Sciences. As this learned man has devoted a great deal of time to the study of these creatures—as he has had the fullest opportunities of making observations—and as he may be relied on, I shall, for the benefit of my readers, introduce some of his remarks from the paper just mentioned.*

" The writings of naturalists attest the ignorance which formerly existed relative to Bats. Aristotle defined them as birds with skinny wings; he was not positive they were volatile, on account of their feet: but on the other hand, he could not resolve to view them as quadrupeds, since they were not provided with four distinct feet. His reflections on their want of tail and rump led him to theoretical notions, which were not based on any positive observation.

" Pliny speaks of Bats only to remark, that they are birds which bring forth their young alive, and suckle them.

" At the restoration of learning in Europe, naturalists confined themselves to copying the ancients,

---

* We anticipate, with great satisfaction, a paper on the subject of the bats of this country from that indefatigable and distinguished naturalist Capt. J Le Conte, who, during many years past, has been engaged in making the necessary observations. We understand that he has many new and very interesting species.

Aldrovandus first began to advance farther in relation to Bats: yielding uniformly to the prejudices of his time, he placed them in the same family with the ostrich, and the reason he gives is, that these two species of *birds* partake *equally* of the nature of *quadrupeds!*

" Scaliger makes out the Bat to be a perfectly marvellous being; he finds in it two and four feet; it walks without paws, and flies without wings—sees when there is no light, and becomes sightless when the dawn appears. It is, adds he, the most singular of all birds, because it has teeth, and is without a beak.

" The Bat has, like all viviparous quadrupeds, a double heart,* cellular lungs, suspended and surrounded by the pleura, a muscular diaphragm, interposed between the cavity of the chest and that of the belly, an ample and solid brain, and a skull composed of the usual number of pieces, joined to each other by sutures.

" They have the same sentient system, and the same organs of digestion and secretion. Their teeth are also of three sorts; their bodies equally covered with hair, and, as was long since known, without leading to the proper conclusion, their young are brought forth alive, and suckled at their teats, exactly as in all viviparous quadrupeds. Such is the degree of resemblance, that the smallest details of their organization suffice alone, and separately, to show that

---

* That is, with two auricles and two ventricles: the right auricle and ventricle to throw the blood through the lungs; the left, through the general circulation of the body.

they are true mammalia,* and must be classed with them.

" The Bat, though accustomed to move in the air like birds, nevertheless is sustained by very different instruments; hence all the anomalies observed in its structure are derived from the mammalious type.

" The parts which in birds answer to the fingers of the human body are almost effaced; they only exist in a rudimental form, attenuated and solidified with each other, whence it results that the hand of birds is nothing but a stump. The wing exists beyond this, supported and adjusted on the extremity of the member, and consisting of long terminal quills, that is to say, on a fair analysis, the most useful portion is actually composed of shoots or elements belonging to the epidermal or scarf-skin system.

" In the Bat, on the contrary, it is the fore-limbs themselves, and principally the hands, which are enlarged. If we can imagine the hand of an ape drawn out (as if through a wire-drawer's plate) and spreading from the wrist like rays through the segment of a circle, we shall have a clear notion of the construction of a Bat's hand.

" The thumb alone does not suffer the same modifications: it remains short, free from all connexions, and capable of very various movements; in this it resembles the thumb of monkeys, since it is not employed as an organ of flight; that it may preserve

---

* Or animals giving suck. It is to be regretted that we cannot express this term in English without circumlocution; the Germans have a word "saeugthiere" which conveys the true idea.

its ordinary function, and remain a finger as to use, it is maintained in its integrity—that is to say, it is provided with a last phalanx and nail.

"The four fingers, on the contrary, which by their excessive length change to instruments of flight, passing to so strange an employment, are no longer susceptible of the accustomed action; it is with much trouble and fatigue that the Bat occasionally is merely able to use them for the purpose of dragging itself along a horizontal plane, or to secure its young.

"Another anomaly renders these four fingers parcularly worthy of attention; they are not complete, being destitute of nails, and, what is remarkable, the terminal phalanx, which in all other instances is formed to suit the nail, is wanting also.

"These long finger-bones of the Bat answer the same end in their wings that the sticks of a parachute do to that instrument, that is, they are supports designed to enable a material to resist the air. This material in the Bat is a prolongation of the skin of the flanks: the back and the belly each furnish a layer, as we may ascertain by separating in two equal thicknesses the membrane of the wings. Notwithstanding that this membrane is formed thus, it always appears to us as a thin, transparent and slight network.

" As the bones of the hand diminish in thickness in proportion as they are increased in length, the integuments are, in like manner, thinned in proportion to their extension. It is, moreover, remarkable that this, which is here an effect of a general law of organization, wonderfully completes the means of flight in the Bat, since more compact bones, or a thicker and denser membrane, especially at so great a distance from the moving power, would have increased the weight of the animal so much as to render its flight impossible.

"We may judge how much the anterior extremities are enlarged, by comparing them with the posterior, which are of ordinary dimensions. These are only partially included in the lateral membrane, the feet being free. The membrane has its last attachments on the tarsus, or instep, the little bones of which, projecting in front, in form of a spine, give the interfemoral membranes a firm hold.

"The toes of the hind feet are small, compressed, equal in length, and always five in number. They are all terminated by claws, or small horny plates, making a quarter circle, very sharp at the points, and remarkable for their equality and parallelism.

"The whole of the functions ordinarily performed by fingers seem concentrated in the hinder toes of the Bat, which are invariable in their forms, or they are, in reality, the only fingers possessed by this genus. We have heretofore shown that only one is found on the anterior extremities, the rest being virtually nothing but solid supports to extend or fold the membranous wings.

"When the Bat is not on the wing, we could scarcely imagine how it would use its limbs, in order to move on the ground; yet, when it is necessary, they employ them advantageously for the purpose. The folded wings then become fore-legs; they are sustained on four feet; they advance, and draw themselves along with sufficient quickness to justify us in saying that they run fast.

"To effect this, how much trouble, how many efforts, and how many different actions are necessary! At first, they reach forward, and slightly to one side, the extremity or stump of the wing, fastening the claw of the thumb on the ground; retaining thus a strong hold, they draw up the hind legs under the belly, and to start from this crouching position they raise themselves on their hinder parts, and tumble the whole body forwards. But as they only take hold of the ground with the claw of one wing, they advance diagonally, throwing themselves towards the fixed point; to make the next step, they use the other wing, and tumbling in the contrary direction, in spite of these alternate deviations, they move forwards in a right line. As this exercise is very fatiguing, they do not attempt it unless when in perfect security, or else when they have accidentally fallen on a level surface.

"In the latter condition, the Bat makes an effort to escape from it as speedily as possible, because, while thus situated, it is almost impossible to raise itself in the air. The wings are too extensive, and the struggles made, in general, merely produce a new fall; if, on the contrary, it can gain a high place, a tree, or a stump, it speedily makes its escape."*

Professor JACOB GREEN has kindly furnished us with the following note of his observations, made on

---

* See Dict. des Sciences Naturelles, Art. Cheiroptéres.

the Bat, in a cavern explored by him on the 1st of November, 1816.

"I this day visited an extensive cavern, about twelve miles south of Albany, N. Y. I did not measure its extent into the mountain, but it was at least three or four hundred feet. There was nothing remarkable in this cave, except the vast multitudes of Bats which had selected this unfrequented place, to pass the winter. They did not appear to be much disturbed by the light of the torches carried by our party; but, upon being touched with sticks, they instantly recovered animation and activity, and flew into the dark passages of the cavern. As the cave was, for the most part, not more than six or seven feet in height, they could very easily be removed from the places to which they were suspended, and some of the party, who were behind me, disturbed some hundreds of them at once, when they swept by me in swarms to more remote, darker, and safer places of retreat. In flying through the caves they made little or no noise; sometimes upon being disturbed in one place they flew but a few yards, and then instantly settled in another, in a state of torpor apparently as profound as before. These Bats, in hibernating, suspended themselves by the hinder claws, from the roof or upper part of the cave; in no instance did I observe one along the sides. They were not promiscuously scattered, but were collected into groups or clusters, of some hundreds, all in close contact. On holding a candle within a few inches of one of these groups, they were not in the least troubled by it: their eyes continued closed, and I could perceive no signs of respiration. On opening the stomach of one of these Bats, it was found entirely empty; the species, I believe, was the *V. Noveboracensis.*

"While making some experiments in my laboratory at Princeton, in a cold dark afternoon in December, a good fire being in the stove, and the room being warm, a small reddish coloured Bat, (probably the *V Noveboracensis,*) which had secreted itself behind some of the cases which held the philosophical apparatus, made its appearance. It flew a short time about the room, and then retired; I saw it, however, once or twice afterwards, in the course of the winter."

Species I.—*Carolina Bat.*

Vespertilio Carolinensis.—Geoff.

The essential characters of this species are the following:—the ears are oblong and of the length of the head; and partly velvet-like. The auricle or projecting portion of the ear, is half-heart shaped. The pelage is chesnut brown coloured above, and yellowish below.

This Bat is most like the common or murine Bat of Europe, (*V. Murinus*, L.) The ears are of middling size, showing no folds on their internal larger edge, and having the first half of their external surface covered with fine hairs. The auricle is almost heart-shaped; the tail is altogether enveloped by the interfemoral membrane, excepting a small point of its extremity, which is free. The hair is, at base. of a blackish ash colour.

As the name implies, this Bat is a native of Carolina, where it is found in the vicinity of Charleston. The first specimens were sent to Europe by Mr. Bosc, and described by Geoffroy, in the eighth volume of the Annals of the Museum of Natural History.

Species II.—*New York Bat.*

Vespertilio Noveboracensis. L.

Red Bat of Pennsylvania, figured in Volume VI. of Wilson's Ornithology.

Essential characters. The ears are short, broad, and rounded, and the whole of the tail is comprised in the interfemoral membrane.

The nose of this species is short, and rather pointed; the interfemoral membrane, which completely includes the tail, is velvet-like above, and of the same reddish brown colour as the back and upper part of the neck. The belly is of a paler colour, and there is a white spot at the origin of the wings. The hair covering the body is soft and furry.

This species is found in the state of New York, Pennsylvania, New Jersey, on the Missouri, &c.

Species III.—*Hoary Bat.*

Vespertilio Pruinosus.—Say.

Long's Expedition to the Rocky Mountains, Vol. i. p. 168.

The ears of this species are large and short not equalling the length of the head, and are hairy on their outside, for more than half their length. The tragus* is very obtuse at tip, and arcuated or bow-shaped. The nostrils open at a distance from each other; the canine teeth are large and prominent; there is but one distinct cutting tooth on each side, placed very near the canine, and almost on a line with it; it is of a conical form, and furnished with a tubercle on its exterior base. The brachial membrane is densely hairy on the anterior margin beneath, and the interfemoral membrane is covered with fur.

The fur of the back is long, and of a black brown colour at base, then of a pale brownish yellow, then blackish, and then white; towards the rump dark ferruginous takes the place of the brownish yellow. The colours beneath, are similar to those of the back, but on the anterior portion of the breast the fur is not tipped with white, and on the throat it is of a dull yellowish white, and is dusky at base.

This Bat is nearly four and a-half inches long, and was common in the vicinity of Engineer Cantonment, where the expedition to the Rocky Mountains wintered. Mr. THOMAS NUTTALL, the justly distinguished botanist, observed it also at Council Bluffs. A specimen captured near Philadelphia, was presented to the Philadelphia museum, by the late professor BARTON.

---

* That part of the external ear, which corresponds to the projection on the human ear, situated immediately next the face, over the joint of the jaw-bone; the antitragus is immediately opposite, and a little lower down, on the outer portion of cartilage, leaving, between the two, the little gutter directly above the lobule or fleshy part of the ear.

This fine large species, remarkable for its variously coloured fur, was first scientifically described by SAY, in the work above quoted, and we are indebted to him for all we have said relative to it. He states it to have much affinity with the New York Bat, before described, but is of twice the size, and distinguished from that species by various minor characters.

Species IV.—*Arcuated Bat.*

Vespertilio Arcuatus.—SAY.

Long's Expedition to the Rocky Mountains, Vol. i. p. 168.

The head of this species is large, and the ears, rather shorter than the head, are wide, rounded at tip and hairy at base. The posterior edge has two slight and very obtuse emarginations: the anterior base is distant from the eye. The tragus is arcuated, and obtuse at tip; the interfemoral membrane is naked, including all the tail, except one half of the joint next the last. The total length of this species is five inches; the tail being one inch and a-half long. In expansion it is more than thirteen inches. The upper cutting teeth, like those of various species of our Bats, are not prominent, but much inclined forwards, not rising at their tips above the intervening callosity.

MR. SAY remarks, that "this Bat might readily be mistaken for the Carolina Bat (*V. Carolinensis,* GEOFF.) which it resembles in colour, but differs from it in being of a larger size, the ears broader and proportionally shorter, and an arcuated tragus curving in an almost uniform manner towards the anterior portion of the ear, like that of the *V. Serotinus*, DAUB. GEOFF. though not so broad."

This Bat was obtained from the same vicinity as the preceding.

SPECIES V.—*Subulate Bat.*

Vespertilio Subulatus.—SAY.

Long's Expedition to the Rocky Mountains, Vol. ii. p. 63.

The ears are longer than they are broad, and nearly as long as the head; the half nearest the head is hairy, a little ventricose on the anterior edge, and extending near to the eye. The tragus is elongated and subulate. The hair above is blackish at base, and at tip dull cinereous; the interfemoral membrane hairy at base, the hair of one colour; a few are also scattered over its surface, and along its edge, as well as that of the brachial membrane. Beneath, the hair is black, the tip yellowish white. The hind feet are rather long, a few bristles extending over the nails; only a minute portion of the tail protrudes beyond the membrane.

The total length of this species is two inches and nine-tenths; the tail being one inch and one-fifth long.

On this species Mr. SAY makes the following remarks, " It appears to be an immature specimen, as the molars are remarkably long and acute; the canines are very much incurved, and the right one is singularly bifid at tip, the division resembling short bristles. This species is, beyond a doubt, distinct from the Carolina Bat, (*V. Carolinensis*, Geoff.) with which the ears are in proportion equally elongated, and, as in that Bat, a little ventricose on the anterior edge, so as almost to extend over the eye; but the tragus is much longer, narrower, and more acute, resembling that of the *Emarginatus*, Geoff., as well in form as in proportion to the length of the ear."*

---

* The Prince of Musignano has several new species of American Bats, the descriptions of which he at present defers publishing, in order to gain more information relative to them. Should he publish his descriptions in time, we shall insert them in our appendix.

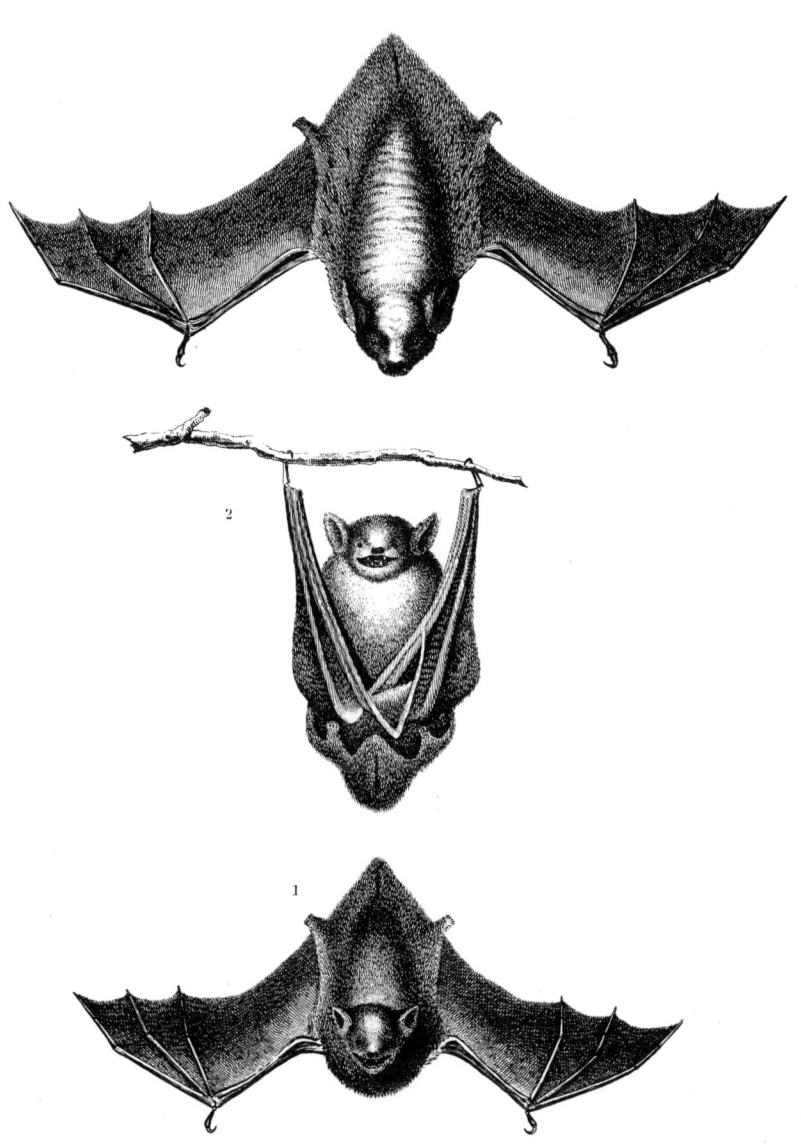

1,2. New York Bat. 3. Hoary Bat.

CHAPTER III.

FAMILY II.—INSECTIVORA; *Insect-Eaters.*

THE designation by which this family is known, is not intended to imply that the animals it comprehends feed more exclusively on insects, than those treated of in the preceding chapter; but the division, by aiding the memory, serves the purpose of facilitating our studies. The members of this family resemble the cheiroptera in the form of their molar teeth and in the nature of their food, but they are destitute of every thing like the lateral membrane, which in the former supplies the place of wings.

Their chief food is insects, and most of them in cold countries pass the winter in a state of torpidity. They all have clavicles; but their limbs are short, and their motions feeble. Their paps, or teats, are mostly placed on the belly; the intestines are destitute of cæcum.* They tread on the whole sole of the foot, pass a great part of their lives beneath the surface of the earth, and most generally leave their subterranean retreats only at night.

---

* Illis invaginatus est penis.

Genus V. Shrew; *Sorex*, L.

Gr. Μυγαλη.
Ang. Sax. Screawa, Scirfemus.
Germ. Die Spitzmaus, Reutmaus. Zismaus, Müger, Mützer, Bissammaus.
Dutch. Spitzmuis.
Dan. & Norse. Spidsmuus, Nebbemuus, Angelmus, Museskiær.
Swed. Näbbemus.
French. (*Ancient*) Muserain; Muzeraigne, Muset, Musette, Sery, Sri. (*Modern*) La Musaraigne.
Ital. Toparagno.
Span. Musgano, Musaraño Murgaño.
Port. Musaranho, Muferanho.
Polish. Keret.
Hungar. Patkaani. *Lap.* Zibak.

GENERIC CHARACTERS.

The head of the shrew is elongated, conical, and the eyes are visible, though extremely minute; the snout is long and moveable. The ears are short, rounded and broader than they are long; the body is covered with short fine hair. The fingers and toes are very small and feeble, stand separate from each other, and are provided with hooked nails, which are unfit for burrowing. The teats, from six to ten in number, are partly placed on the chest, and partly on the belly. The feet are short, having five toes with slight nails.

Dental System.

30 Teeth:
- 18 upper
  - 2 Incisor
  - 0 Canine
  - 16 Molar
    - 10 False } Molars.
    - 8 True
- 12 lower
  - 2 Incisor
  - 0 Canine
  - 10 Molar.
    - 4 False } Molars.
    - 6 True

IN THE UPPER JAW there is a very strong hooked cutting-tooth, terminating in a point, and strengthened at its base posteriorly by a strong edge, which is itself divided in two by a furrow on its internal face.

which forms another uneven edge at that part. Immediately next this, is a regular very strong false molar, which is followed by two others, both of the same size, equally regular in form, but one-half smaller than the first. The fourth false molar is rudimental, and concealed between the third and the fifth. The latter is very large, cutting, with a flat and salient part at its internal basis, which causes it to resemble the tearing grinders of the most carnivorous animals. The three following molars are similar to those thus far described, only that the base on which the prisms are placed are composed anteriorly of a pointed tubercle, and posteriorly of a smooth and flattened part: moreover, the anterior prism of the first is less developed than the other, and the posterior prism of the last is not at all perceptible.

IN THE LOWER JAW there is a strong, long and crooked cutting-tooth, terminating in a point, and reclining in front; next to this follow two regularly formed false molars, the first a little smaller than the second; the three true molars resemble those of the Shrew-mole and Star-nose, except that the last has only the anterior point of its posterior prism.

In their reciprocal position, the lower cutting-teeth correspond, by their points, with the inside of the upper ones, and fill up the furrow which separates the principal part of these teeth from that which exists at their base, posteriorly. The false molars leave a large space between them, except the two last, which touch each other, the anterior edge of the upper being applied to the posterior edge of the lower. The anterior face of the first lower molar, is opposed to the internal surface of the last upper false molar, which we have seen is small; it is between these two teeth that the cutting is performed by this system of dentition. The other teeth have the same relations as those heretofore described.

The shrews belonging to this country are remarkable for their diminutive size and apparent helplessness, and were we to judge of their utility or capacity for enjoyment by their size or strength, we should conclude that their usefulness is very equivocal, and their lives exceedingly comfortless. But their small bodies and slight strength only fit them more perfectly for the places they inhabit, and their senses being planned on a scale relative to their condition, leave us every reason to infer, that they may derive as much pleasure from their peculiar food, and the society of their mates, as the largest animals in existence, whose massive frames seem to make earth tremble beneath their footsteps.

Shrews are most generally found in the country, where their residence is either in burrows, or among heaps of stones, or in holes made by other animals: near dung heaps, hay ricks, or privies, they are more numerous than elsewhere. Insects are their principal subsistence, but they seem no less fond of grain, putrid flesh, and filth of various sorts, as they have been occasionally seen rooting in ordure in a manner similar to that of the hog.

These animals rarely come out in the day time, and are so small as to require very close attention to observe their modes of living. As the autumnal season advances, they are found resorting in considerable numbers to the barns and granaries, where they find a large supply of food and more comfortable quarters for their winter's sleep.

A very ridiculous notion formerly existed relative to the bite of these inoffensive creatures. It was thought they bit and poisoned cattle; hence we find, among other appellations, the name of "poisonous mouse" has been bestowed. This notion most probably originated in consequence of associating the existence of shrews in the vicinity of stables, with the appearance of some disease on the extremities of horses, &c. As the mouth of the shrew can scarcely be opened wide enough to grasp the doubling of skin necessary to allow of a bite, we must perceive that the accusation of poisoning is erroneous.

Cats and other animals will hunt and destroy shrews, but cannot sufficiently surmount the disgust caused by their offensive odour, to eat them. The cause of their peculiar smell, has been discovered by the celebrated GEOFFROY, who gives a description of the secretory apparatus in the Memoirs of the Museum of Natural History, vol. i, 1815.

These odoriferous glands, unlike those of various other animals, are in the shrew found on the side, nearer to the anterior than the posterior extremities, and are of an oval form. Externally, they are rendered visible by an oval-shaped disk, composed of two ranges of short, stiff hairs. The ranges in crossing each other are placed back to back, and are thus retained; they are constantly moistened by a viscous humour which is furnished by the internal organ, and gives them a greasy or oily appearance. The situation of these glands is rendered still more distinct by a circle round them, caused by the nakedness of the parts.

The results of numerous observations induced

Geoffroy to conclude, that this glandular apparatus is not in the same state of enlargement throughout every season of the year. It is more remarkable in the males than in females, and in the former it is much larger at the approach of the breeding season. From this circumstance, the French naturalist has concluded, that the only use of this gland is to furnish a guide to conduct the sexes to each other, during the season of their loves, through the long subterranean galleries they inhabit.

If we consider the long and pointed nose of the shrew, which is extended considerably beyond the jaws, we perceive a distant resemblance to the scalops or, shrew-mole. while its legs and tail give a slight resemblance to the mouse. This comparison does not, however, bear examination in reference to either animal, as will be fully seen when we come to speak of them, especially in reference to the eyes, ears and hands of the shrew-mole, and the teeth, &c. of the mouse. Nevertheless there is no other animal we can well compare it with, and the first view of one of the shrews does not fail to excite a recollection of the mouse, if not of the shrew-mole.

Species I.—*Small Shrew*.

Sorex Parvus.—Say.

Long's Expedition to the Rocky Mountains, vol. i. p. 163.

The small shrew is of a brownish ash colour on the body above, and cinereous beneath. Its head is elongated, having the eyes and ears concealed; the whiskers are long, the longest reaching nearly to the back of the head. The nose is naked and emarginated: the front teeth are black, and the lateral ones piceous. The feet are whitish and five toed; the nails prominent, acute, and white; the tail is short, subcylindric, moderately thick, slightly thicker in the middle, and whitish beneath.

The length from the tip of the nose to the root of the tail is two inches and three-eighths. The tail measures three-fourths of an inch. The length from the upper teeth to the tip of the nose is three-twentieths of an inch.

This species was obtained at Engineer Cantonment, on the Missouri, where it was caught in a pitfall set for a wolf by Mr. TITIAN PEALE. It may properly be considered as one of the smallest mammiferous animals belonging to this continent.

SPECIES II.—*Short-tail Shrew.*

Sorex Brevicaudus.—SAY.

Long's Expedition to the Rocky Mountains, vol. i. p. 164.

The body of the small shrew, when seen from before, is of a blackish lead colour above; when viewed from behind, it is of a silvery plumbeous hue; the fur, which is dense and rather long, is of a paler colour beneath. The head is large, the eyes very minute; the ears are white, entirely concealed beneath the fur. The passage to the internal ear is very large, with two distinct half divisions (tragus and antitragus?) which are at tip sparsely hairy. The mouth is short with a slightly impressed abbreviated line above. The nose is of a livid brown colour, and emarginated: the mouth is margined with whitish and with scattered short hairs; the teeth are of a pitchy black at tip. The feet are white, the second, third, and fourth toes being subequal; the first and fifth are shorter, the former rather shortest, the anterior having but few hairs and nearly naked.

The nails are nearly as long as the toes; the tail is covered with scattered hairs: it is of nearly an equal diameter, but slightly thickest in the middle, depressed, and nearly as long as the posterior feet.

The total length of this shrew is four inches and five-eighths; length of the tail one inch; from the upper teeth to the tip of the nose one-eighth.

This species bears a close resemblance to the one first described, but it is proportionally much larger: the head is much larger and more elongated, the tail is more robust, and the inferior anterior pair of incissors are similar to those of the Sorex Constrictus of GEOFFROY. Mr. Say inclines to the opinion that this is the same species as that mentioned by Barton as " the black shrew."

SPECIES III.—[*marked thus *, in the plate.*]

The description of this species is postponed for the present, as Mr. TITIAN PEALE, to whom it belongs, has not yet decided on its name. It has, however, been described by a recent writer, as the *Sorex Araneus* of Europe, with the description of which it by no means agrees. We hope to give the proper designation and description in our appendix.

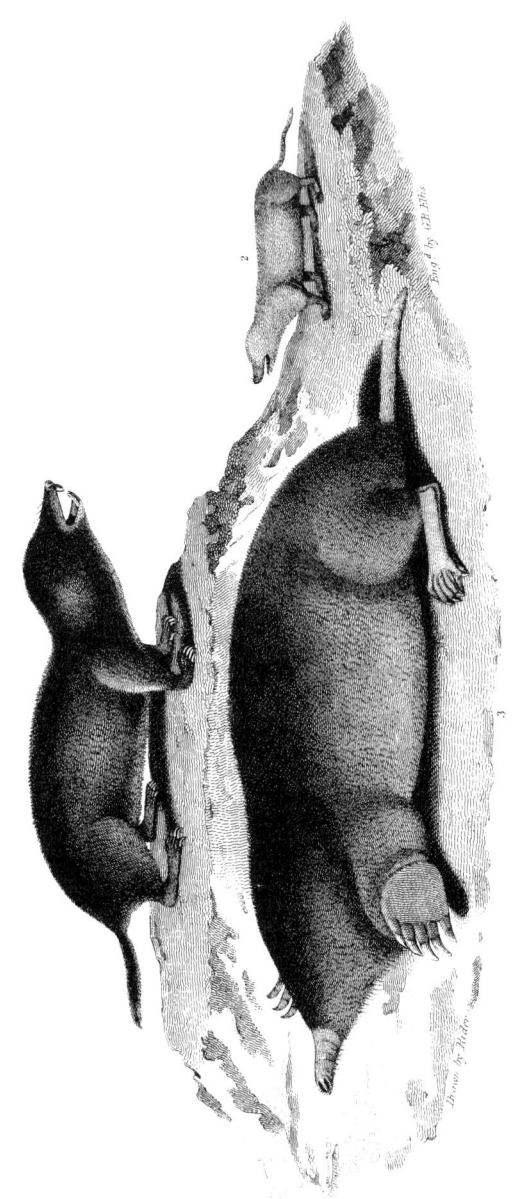

1. Short Tail Shrew. 2. Small Shrew. 3. Shrew Mole.

CHAPTER IV.

GENUS VI. SHREW-MOLE; *Scalops,* CUV.

Fr. Musaraigne Taupe. *Germ.* Wassermaus.
Ital. Scalopo.   *Eng.* Brown Mole.

GENERIC CHARACTERS.

THE head is elongated, and terminates in an extended and cartilaginous snout, which is very flexible. The eyes are exceedingly small and entirely concealed by the hair, requiring the closest attention for their detection. Not only is the auricle or external cartilaginous part of the ear entirely wanting, but the integument of the head nearly covers over the cartilaginous tube leading to the internal ear, and its situation is only known by a small naked spot, in which there is a minute opening.

The feet are very short and five toed; the anterior terminate in a remarkably large hand, having the fingers joined together by the integuments up to the last phalanx. These fingers are armed with long, flat and linear nails, increasing in length from the thumb to the third finger; the two others are less in size, and the external is smallest of all. The extraordinary breadth of the hand is produced, as we shall hereafter see, by a singular supplementary bone, &c. The hind feet are very delicate, and the toes are provided with small hooked nails.

Dental System.

$$36\text{ Teeth} \begin{cases} 20\text{ Upper} \begin{cases} 2\text{ Incisive} \\ 0\text{ Canine} \\ 18\text{ Molar.} \end{cases} \begin{cases} 12\text{ False} \\ 6\text{ True} \end{cases} \text{Molars.} \\ 16\text{ Lower} \begin{cases} 4\text{ Incisive} \\ 0\text{ Canine} \\ 12\text{ Molar.} \end{cases} \begin{cases} 6\text{ False} \\ 6\text{ True} \end{cases} \text{Molars.} \end{cases}$$

IN THE UPPER JAW we find an incisor with a rounded, cutting edge, its front surface being also rounded, and its posterior one very flat. There is much analogy between the incisors of this animal and those of the *gnawers*, more especially as they stand side by side on the same line. Behind these teeth come six false molars; next to these, two small ones resembling threads, such is their tenuity; afterwards there is another much larger, cylindrical and pointed, and after that a fourth, equally cylindrical and pointed, but much smaller. The fifth is obliquely truncated at its summit, from before backwards, the section resembling the head of a lance, the point being bent backwards; the sixth is entirely similar to the preceding, except that it is twice the size. The three molars are, in general, similar to those of the bat; the only difference is, that the anterior prism of the first molar is imperfect, its anterior half not being developed, and the same is the case with the posterior prism of the last one, by the obliteration of the posterior half of this prism. Finally, the interior projection of each of these three teeth is simple, and consists of nothing but a tubercle at the base of the anterior prism.

IN THE LOWER JAW there are two incisors; the first is very small and cutting, the second pointed, and slightly hooked, leaning forward, and destitute of roots, properly speaking, like the tusks of certain animals, in which the dental capsule remains always free: we call them incisors, only because they act in mastication against the upper incisors. The three succeeding false molars have a single point, with a small notching posteriorly, slightly inclined forwards, and resembling each other, except in size, the first being the smallest and the third the largest. The three molars are exactly similar to those of the bat, that is to say, composed of two parallel prisms, terminated each by three points, and presenting one of their angles on the outer side, and one of their faces on the internal surface. The two first are of the same size; the last somewhat smaller.

In their reciprocal position, the lower incisors correspond to the internal face of the upper; the false molars are alternate, and the molars are so related, that the anterior prisms of the lower, fill the hollow found between two teeth, and the posterior prism that which exists between the two prisms of a single upper tooth. The lower are the thickness of one prism in advance of the upper molars.

Species I.—*Shrew-Mole.*

Scalops Canadensis.—Cuv.

Sorex Aquaticus, *Lin. Syst. Nat. Vol. i. p.* 112, *Sp.* 3.
Musaraigne brune, *Enc. plate* 30, *fig.* 2.
Talpa Fusca, *Penn. Quad. p.* 314.
Scalops Canadensis, *Desm. Mam. p.* 155.

The Shrew-Mole when at rest bears more resemblance to a small stuffed sack than to a living animal, its head being entirely destitute of external ears, and elongated nearly to a point, and its eyes so extremely small and completely hidden by the fur, that it would not be surprising should a casual observer conclude this creature to be blind. But we must be continually guarded against hasty conclusions, or idle conjectures, drawn from slight observations; this apparently shapeless mass is endowed with great activity and a surprising degree of strength, and is excellently suited for deriving enjoyment from the peculiar life it is designed to lead.

The shrew-mole is found abundantly in North America, from Canada to Virginia; often living at no great distance from water-courses, or in dykes thrown up to protect meadows from inundation. But so far from exclusively inhabiting such places, as stated in various books, I have found them in far greater numbers at a very considerable distance from any water-course, and in high oftener than low grounds. In the country they frequent the gardens, where their subterranean galleries are sometimes productive of vexation to the farmer, especially as the animal occasionally courses along the rows of pea-vines, &c. apparently for the purpose of feeding on their roots. This, we shall hereafter learn, is most probably an error, and we may find good reasons for believing that the shrew-mole should be considered rather as a benefactor than a depredator.

The shrew-mole burrows with great quickness, and travels under ground with much celerity: nothing can be better constructed for this purpose than its broad and strong hands, or fore-paws, armed with long and powerful claws, which are very sharp at their extremities, and slightly curved on the inside. These are thrust forward so as to be even with the extremity of the flexible snout, and the earth to be removed is pressed outwards, and at the same time thrown backwards with remarkable quickness. The soft and polished fur with which this animal is covered, preventing a great degree of friction, tends to facilitate its subterranean march.

Numerous galleries, communicating with each other, enable the shrew-mole to travel in various directions, without coming to the surface, which they appear to do very rarely, unless their progress is impeded by a piece of ground so hard as to defy their strength and perseverance. The depth of their burrows depends very materially on the character of the soil, and the situation of the place: sometimes we find them running for a great distance, at a depth of from one to three inches, and again we trace them much deeper; after following such a gallery for several yards, it occasionally communicates with another going deep into the earth.

The most remarkable circumstance connected with these burrows, is the number of hills of loose dirt which are frequently formed over the surface of them. These hills of loose earth are usually found in considerable numbers, at a distance of two feet or a little more apart, being from four to six inches high, and about the same in diameter. I have often examined these eminences, in the neighbourhood of Philadelphia, and have never been able fully to understand how they are formed; a slight motion is observed at the surface, and presently this loose earth is seen to be worked up through a small orifice, whence, falling on all sides, by its accumulation the hills just mentioned are produced. It seems to be brought from some distance, for on breaking up the gallery, it was evident that more earth had been thrown out than could have been removed in excavating the immediately adjoining portions of the burrow. In one instance I have seen the shrew-mole show the extremity of its snout from the centre of one of these loose hills, where it had come at mid-day, as if for the purpose of enjoying the sunshine, without exposing its body to the full influence of the external air.

Under ordinary circumstances the burrows are simply oval arched galleries, running forward either straight or in gentle curvatures, at the depth heretofore mentioned, and they are most regular in soils abounding in earth-worms. In the dry and sandy soil of Jersey, I have found them very irregular in direction and depth, and in the woods, uniformly leading round the roots of trees, under which

OF THE SHREW-MOLE.

large excavations are frequently to be traced. We can readily understand the object of these excavations when we recollect that the ants very often have their nests in such situations, and their larvæ, or eggs, constitute a favourite food of the shrew-mole. The burrows made by this animal are sometimes found to terminate under large stones, where it resorts to gather the insects which are numerous in such situations. I have traced a burrow of this sort close to a barn-wall, and then following it nearly around the whole house, have found that it passed under every large stone in its vicinity, although not directly in the general course of the gallery, the cavity being much larger beneath the stones than elsewhere.

The favourite food of the shrew-mole is the earth-worm; grubs and insects of various kinds he destroys in great quantities, and it may fairly be questioned whether the good done in this way does not more than overbalance any evil attendant on his presence. It is true that this animal is accused of eating grass roots, and roots of succulent vegetables, and may thus be productive of some mischief in gardens, but scarcely to so great a degree as to constitute a serious evil. The presence of the shrew-mole in fields of Indian corn appears to be decidedly advantageous from the destruction of great numbers of slugs and worms; but in dry seasons these animals, if numerous, may injure small grain or grasses to a considerable extent, not only by the wounds they inflict on the root with their sharp claws, but by raising the sod while forming their burrows, so as to withdraw the roots from the influence of the moist soil below.

It is remarkable how unwilling they are to relinquish a long frequented burrow; I have frequently broken down or torn off the surface of the same burrow for several days in succession, but would always find it repaired at the next visit. This was especially the case with one individual whose nest I discovered, which was always repaired within a short time, as often as destroyed. It was an oval cavity, about six or seven inches in length by three in breadth, and was placed at about eight inches from the surface in a stiff clay. The entrance to it sloped obliquely downwards from the common gallery, about two inches from the surface; three times I entirely exposed this cell by cutting out the whole superincumbent clay with a knife, and three times a similar one was made a little beyond the situation of the former, the excavation having been continued from its back part. I paid a visit to the same spot two months after capturing its occupant, and breaking up the nest, all the injuries were found to be repaired, and another excavated within a few inches of the old one. Most probably numerous individuals, composing a whole family, reside together in these extensive galleries. In the winter they burrow closer to the streams, where the ground is not so deeply frozen.

The shrew-mole is not only able to make his way rapidly under ground, but can run quite fast when on the surface, notwithstanding the apparent disadvantages under which he labours. When placed on a smooth path-way, or a floor, and especially if alarmed, he runs with far more speed than could possibly be anticipated from the structure of his limbs. In attempting to escape by running his motions are very similar to those performed while burrowing; the broad fore paws are placed on edge, with the thumb to the ground, and both fore and hind feet are moved in rapid succession, the body being trailed along with a slight undulatory motion.

The strength of the shrew mole is really surprising, and altogether beyond what we should deem possible in so small a creature. One which we had in a basket on the mantlepiece of a parlour, made its escape, and fell to the hearth; apparently it sustained little injury by the fall, but hurried on until it reached the wall, where it began to travel round the room. Whenever its course was impeded by the feet of the chairs, which were of large size, it would not go round them, but wedging itself between them and the wall, pushed them with apparent ease far enough to obtain a free passage, and it thus continued to move several in succession. What was more astonishing, it passed in a similar manner behind the legs of a small mahogany breakfast-table, and pushed it aside in the same way it had done the chairs, finally hiding itself behind a pile of quarto volumes, more than two feet high, which it also moved out from the wall.

When endeavouring to escape a pursuer, while in his burrow, the shrew-mole displays his utmost strength. In this case, although you may have succeeded in catching him by his posterior extremities, it is exceedingly difficult to draw him from the hole without violence. His broad and strong fore paws are then struck outwards against the sides of the excavation, with all the energy of despair, and when the animal is finally dragged from his retreat, he frequently inflicts a severe bite on his disturber.

Shrew-moles are most active early in the morning, at mid-day, and in the evening; after rains they are particularly busy in repairing their damaged galleries; and in long continued wet weather we find that they seek the high grounds for security. The precision with which they daily come to the surface at twelve o'clock is very remarkable, and is well known in the country. In many instances when we have watched them, they appeared exactly at twelve, and at this time only have we succeeded in taking them alive, which is easily done by intercepting their progress with a spade, broad knife-blade, &c. and throwing them on the surface. These animals do not appear to be well suited for living in the open air, especially if it be somewhat cool: for, after being a few minutes exposed, we have always observed them shiver as if from the change of temperature.

That an animal of this kind should be domesticated with facility would seem hardly possible, yet our friend TITIAN PEALE tamed a very fine one, which he caught while we were together examining their modes of burrowing. This shrew-mole is kept in a box containing some loose earth and dried grass for his bed; he eats considerable quantities of fresh meat, either cooked or raw, drinks freely, and is remarkably lively and playful, following the hand of his feeder by the scent,—burrowing for a short distance in the loose earth, and after making a small circle, returning for more food. When engaged in eating he employs his flexible snout in a singular manner to thrust the food into his mouth, doubling it under so as to force it directly backwards. When he has obtained one piece of meat, he will not relinquish it even for the sake of earth-worms, or other favourite food; he is also fond of burying himself when he has received any thing, in order to eat it undisturbed.

The shrew-mole is covered with a soft, glossy fur, which is about half an inch in length, and of a uniform colour over the whole of the body. This general tint is bright plumbeous, having silvery reflexions when viewed from the front of the animal, and appearing of a darker hue when seen from the posterior part, with faint purple reflexions, varying according to the incidence of light. The fur is very closely set, and the direction of it is in all parts from before backwards. The only places where any differences of colour are distinguishable, is around the base of the snout, where it is faintly ferruginous, and this ferruginous tint may be traced for a short distance on the base of the fur towards the situation of the ear, which, though destitute of external appendages, may be discovered on the lateral part of the head, by carefully turning away the surrounding fur. Immediately posterior to the chin, or in the fold of the neck, there is a narrow but well marked ferruginous horizontal streak for about half an inch, and a faint trace of the same colour may be distinguished along the centre of the belly.

The whole fore-arm of the shrew-mole is concealed by the skin, so as to give no external mark of its figure, leaving the broadly expanded palm, with its long nails, resting on its internal or inferior edge, projecting from the fore part of the body in an awkward manner, if it be compared with the anterior extremity of other animals. All the joints of the fingers are moveable, and the carpus is articulated with the fore-arm, so as to be flexed with much force by a strong muscle, whose tendon is broadly expanded to be inserted into the extremities of the phalanges. The nails of the hands are strong, nearly straight, and edged at their extremities, being convex externally, and rather flattened on the posterior or inner side. In addition to the bones of the hand, which the shrew-mole has in common with various animals, it is provided with an additional bone on the radial edge of the hand, or exterior to the thumb, articulated with the wrist, and of a semi-circular, or rather scythe-like figure. There is also a rudiment, somewhat similar in character, exterior to the little finger. To the larger one first mentioned, a small tendon from the muscles is attached, and this bone, although entirely covered by the skin of the palm, appears to serve the purpose of very much increasing its breadth, and adding to the usefulness of the hand as an instrument for burrowing. An extension of the skin of the palm on the outside or under edge of this bone, so as to form a small fold, still farther increases the breadth of the hand. The palms of the hands are seven-eighths of an inch in breadth, and of a light flesh-colour when the animal is first caught.

The soles of the hind feet are placed on the earth, as in other insectivorous animals; the toes are delicate, and the joints moveable, having small curved claws: from the heel to the base of the claws, the distance is five-eighths of an inch.

The snout is composed of a cartilage articulated with the premaxillary bones, and moved in various directions by muscles situated on the side of the head; it is naked, and of a very light flesh-colour when the animal is first exposed to the open air; it is half an inch long, and feels somewhat horny at the extremity. The under surface is at the end slightly prolonged, or projects a little beyond the nostrils, which open on its upper surface, and are rather oblong. The mouth of the shrew-mole is comparatively large, and the tongue is considerable in size. The roof of the mouth is marked by nine transverse projections, or ridges and hollows, which are beautifully distinct. The eyes of this creature are exceedingly small, and very difficult to be discovered externally without a good glass. They are entirely concealed by the surrounding hair, which being cleared away, leaves a naked space equal in circumference to the head of a middling sized pin: in the centre of this space a small dark speck may be distinguished with the naked eye. On examining one adult individual with a microscope, (No. 9,) this proved to be a number of hairs arranged in a semi-elliptical manner, the eye-ball being scarcely discoverable. When the skin was carefully removed, the eye-balls were found corresponding to the situation of these hairs, and were less in size than a grain of mustard-seed. On holding the skin up to the light, and examining it attentively, the exceedingly small aperture, or separation of the eye-lids, was visible; it would allow the passage of an ordinary sized human hair, and possibly of a very fine horse hair. Hence we perceive that the vision of this animal must be extremely limited, as the focus of so microscopic an eye is almost inconceivably small. It seems to be barely sufficient to give the shrew-mole an intimation of light, without allowing it to distinguish the figures of bodies; this conclusion is further supported by the careful manner in which the eye is concealed by the fur, and the minute aperture through the lids, or the skin. The orifice leading to the internal ear is curiously situated, being placed about three-fourths of an inch behind the eye, and opening nearly over the shoulder-joint of the animal. In the specimen first examined, this opening was so small that it was not detected, and I thought the skin continuous over the cartilaginous tube, but a recent search enabled me to ascertain its position. The aperture is nearly circular, and would admit an ordinary pin; the cartilaginous tube, leading to the internal ear, from immediately within the skin, where it is expanded somewhat trumpet-form, to its entrance into the skull, is five-sixteenths of an inch long.

When the skin of the animal is removed, a much better idea of its peculiar adaptation to its mode of life is obtained. The muscles surrounding the shoulder-blade, arms, and fore-arm, are of great size, and occupy all the space anterior to the greater convexity of the chest, so as to destroy all appearance of neck, giving to the fore part of the body a robust and clumsy appearance. The head is, however, capable of some motion, both laterally and vertically. All the appearances of strength seem concentrated about the fore shoulders, where the muscles swell out with strong fibres, and of a deep red colour. The posterior extremities are delicate and slender when compared with the anterior, the thighs being flattened and thin, as if of less importance to the motion of the animal.

Every circumstance seems to be studied in the shrew-mole with a view to facilitate its progression under the surface of the earth. We observe this attention not only in its silky and polished fur, its pointed head, broad and powerful hands, and muscular limbs, but all its internal structure seems equally to co-operate. The pelvis is a single, slight and flattened bone; the pubic portions, instead of uniting in the centre, and projecting as in other animals, are very small, rise but slightly, and do not unite with each other; but from their extremities the crura penis, with the erector muscles, take their origin. The penis itself, after passing forwards for half an inch, as it approaches the surface is curved backwards, that it may present no resistance to the motion of the body through the ground. Neither do the testes present externally; a small portion, about the size of a large pin's head, is exterior to the abdominal ring, and this part corresponds very closely with the head of the epididymis in the human subject, but the body of the testis is within the cavity of the abdomen, being half an inch long, and one-fourth of an inch broad. Near the anus, and situated between the upper part of the thigh and tail, we find on each side an odoriferous gland, which is about half an inch long, of a dark greenish tint, and having a strong musky odour, which imparts to the animal, during life, a very peculiar smell.

The total length of the shrew-mole, from the point of the snout to the beginning of the tail, is five inches, and the tail one inch long. The longest fur on the body was half an inch in length. The specimen from which this description is made was a fine adult male, which, with the assistance of my friend W. W. Wood, was caught on the banks of the Delaware, and kept alive for some time. We have since had an opportunity of examining several specimens, and find no difference between them in colour, except that the ferruginous marking on the head and neck does not appear to be a constant character. A living specimen, kept for many weeks in a room, was nearly as tame as the one already mentioned; this individual spent the greater part of the day asleep, and was very active at night. He could not see in any light, as he uniformly ran his nose with some violence against every obstacle several times, before he learned to avoid those that were permanent.

We have, at the suggestion of a scientific friend, applied to this animal the name of *Shrew-mole*, which is a translation of the French designation, rather than run a risk of having it confounded with the European genus TALPA, by calling it simply "*the mole*," a name by which it is popularly known.

CHAPTER V.

Genus VII. Condylura; *Illiger*.

Fr.—La Taupe du Canada.   *Germ.*— Die Haarnase Spitzmause.

GENERIC CHARACTERS.

The head is long, conical, terminating in a snout, which is encircled by a cartilaginous fringe or stellated disk, having about twenty points. There are no external ears; (but a large meatus externus, or auricular orifice, with a division analogous to a tragus and antitragus, leads to the internal ear;) the eyes are very small, and concealed by fur. The anterior extremities are very short; the paws, or fore-feet, are broad, five-toed, (the toes being united by the integuments as far as the second phalanx,) covered by corneous scales, and terminated by long, straight, and robust nails. The posterior extremities are one-third longer than the anterior, having five slender toes, (the phalanges of which are not kept together by the integuments,) provided with small hooked nails. The body is thick, covered with fine, short, soft fur, arranged perpendicular to the length of the body: the tail has sixteen joints or vertebræ; it is compressed near the body, then swells slightly, and tapers thence to its extremity, showing no inequalities caused by the vertebræ.

Dental System.

40 Teeth:	20 Upper	6 Incisor 6 Conical, 8 Molar.	(which may be considered as canine, or false molar teeth.)
	20 Lower	4 Incisor 10 Conical, (as above) 6 Molar.	

IN THE UPPER JAW there are six anomalous cutting teeth, situated in the premaxillary bones; the two intermediate are very large, contiguous, and ranged along the whole border of the jaw, hollowed in the form of a spoon, having a slightly oblique cutting edge, and the angle by which they touch more salient externally than internally. The next cutting-tooth on each side touches the intermediary, and resembles a very long canine tooth, being conical and slightly triangular at its base, where it has two very small tubercles, the one before and the other behind. The external or lateral incisor, the smallest of all the teeth in this jaw, is simply conic, a little compressed, slightly curved forwards at its point, and placed at some distance from the incisor, in the manner of a canine tooth. There are seven molars on the right and left, the three first of which are smaller than the posterior, separated from each other, all three moderately large, and furnished with a small pointed lobe at the front of their base, and another behind. The four last molars are larger than the anterior ones, and each of them composed of two folds of enamel, forming two acute tubercles on the inner side, and obliquely hollowed to a gutter on the outside: there is a projection hollowed to a cupola at the internal bases of these teeth. The most anterior of these four last molars, and the smallest, are placed on a level with the commissure of the lips; the following one is higher, the third still more so, and the last smaller than the third.

IN THE LOWER JAW, which is very delicate, there are four flattened cutting teeth, reclined in the form of a spoon or ear-pick, the lateral ones in part horizontally inclined on the intermediary, and rising slightly at their external edge. Five teeth, with many lobes, then follow on each side, and may be considered as false molars, as much separated from each other as those of the upper jaw, the first being much larger than the others, and in this alone resembling a canine, having three lobes, the principal of which is intermediate, the second very much effaced, and the third slightly salient. The second is nearly similar, but shorter and more compressed, having the posterior lobe more apparent than in the preceding. The third has four lobes, the anterior of which is the least, the second the largest and most apparent of all, and two small posterior ones; the fourth is nearly similar to the third, with this difference, that the first posterior lobe is more internal, and this tooth consequently thicker: the fifth only differs from the fourth by its greater width, and is almost equal to the first true molar. There are only three molars in this jaw, presenting, like those of the upper jaw, two folds of enamel, forming a point; but these folds are inverted, the points being external instead of internal; the grooves, on the contrary, are internal, and the lower part of the tooth, instead of exhibiting the whole projection, presents a perpendicular wall, and has two depressions at its summit, each of these depressions corresponding to the groove that descends from one of the two points.*

Species I.—*Star-nose Mole.*

Condylura Cristata; Illiger.

Taupe du Canada; Delafaille, Ess. sur L'hist. Nat. de la Taupe, fig. 1769.
Sorex Cristatus; Linn. Erx.
Radiated Mole; Penn. Syn. Quad.
Condylure a Museau Etoilé; Desm. Mam. p. 157.

The Star-nose mole frequents the banks of rivulets, and the soft soil of adjacent meadows, where their burrows are most numerous, and apparently interminable; in many places it is scarcely possible to advance a step without breaking down their galleries, by which the surface is thrown into ridges, and the surface of the green sward in no slight degree disfigured. The excavations which are most continuous, and appear to be most frequented, are placed at a short distance below the grass roots, on the banks of small streams; these are to be traced along their margins, following every inflexion, and making frequent circuits in order to pass large stones or roots of trees, to regain their usual proximity to the surface nearest the water.

---

* This dental system is from Desmarest. See his excellent note on the genus Condylura, in the Journal de Physique for September, 1819.

The Star nose Mole
End of the Nose magnified.

The form of the burrow does not perceptibly differ from that made by the shrew-mole; but very few hills are to be found in the localities inhabited by the star-nose. The chamber-cell resembles that described in the last chapter, being a space of several inches dug out of some spot where the clay is tenacious, and the cell least exposed to injury from the weather or other accidents.

The system of dentition peculiar to this genus, would lead to the inference that the quality of its food must in some respects differ from that used by the shrew-mole; but on this point it is not easy to say more, than that as the star-nose prefers moist and low situations, and the shrew-mole is most frequently found in dry and rather elevated spots, they feed on the larvæ and insects proper to such places, which are doubtless of dissimilar kinds. In a state of captivity both animals feed readily on flesh, either raw or cooked, and neither seem to show any fondness for, or willingness to eat, vegetable matter.

The star-nose mole is about four inches in length, and of a blackish-gray colour; its pelage being short and very fine. Its head is much elongated, and the snout is distinguished by a remarkable disk, or naked cartilaginous fringe, which surrounds the nostrils. This disk has about twenty points or rays, the two superior and the four inferior intermediate of which are united at their bases, and are situated on a plane slightly in advance of the others: the surface of these fringes is granulated, and somewhat of a rose colour. The neck is not distinguishable in consequence of the position and great size of the muscles that are destined to move the anterior extremities, which are very short, broad, covered with scales, and provided with large straight nails, the shortest of which is on the finger corresponding to the thumb; the second, third, and fourth are successively and proportionally longer than each other: the nail on the superior or little finger is exactly of the same size as that of the second or index finger: all the fingers are united as far as the second phalanx. The hind feet are a third longer than the fore ones, being slender, delicate, and weak; the phalanges are separated from each other throughout, and have small, curved, and sharp nails.

The situation of the eyes is marked by three or four equal hairs, which may be readily discovered, and are not so stiff or large as those of the whiskers, the direction of which is not horizontal and lateral, like most other mammiferous animals, but raised nearly parallel, and turned towards the snout. Seven transverse wrinkles occupy the space in the palate between the cutting teeth and the first three molars.

There are several very interesting external characters peculiar to the star-nose, which have been much overlooked by those who have hitherto written on this subject; we will introduce them in this place, as they may be serviceable in enabling us to compare the present genus with some others.

The star-nose is destitute of an auricle projecting above the level of the skin, but, nevertheless, has a large auricular orifice. This meatus externus is half an inch long, having a distinctly marked tragus and anti-tragus, and is situated at a short distance from the shoulder, in the broad triangular fold of integument connecting the fore-arm and head. From the meatus, the course of the cartilaginous tube is obliquely downwards, forwards, and inwards, until it terminates in a delicate bony tube, previous to reaching the tympanum, which is large and composed of a very delicate membrane.

The scales on the anterior and posterior extremities have been mentioned in general terms by several writers, especially by DESMAREST, who gave the first correct description of this animal. But these scales are so peculiar and uniform in their position, that a naturalist should not pass over the particulars of their arrangement in silence.

On the anterior extremities, the superior or ulnar edge of the hand has on its anterior surface, (regarding the position of the animal) a row of corneous scales, about nine in number, which are broadest midway from the carpus to the first phalanx of the fifth finger. Another row of scales commences on the inferior part of the little finger, becoming broader and of a semilunar figure as they extend towards the metacarpus; between these two a much smaller row is placed. The fourth finger has a single row of small scales on its upper posterior side, and a large one extending along the back of the finger to the metacarpus; the middle finger has a small central row, which is just distinguishable; that on the fore finger is still more faint: the thumb has none but very small ones on its central posterior part, but on its inferior posterior part, or radial edge, it has one scale of considerable size on the phalanx, and four or five between this part and the carpus; the two nearest the scale on the phalanx are largest.

The surface of the palm of the hand is covered with small circular scales, extending most numerously, and of a darker colour, from opposite the root of the thumb, obliquely outward to the basis of the little finger.

On the inferior extremities, the whole of the superior surface of the foot is covered with minute, blackish, circular scales, which increase slightly in size as they approach the toes. On the anterior part of the fourth toe is a large central row of black scales, and on the fifth a rather smaller one; hence these toes have a very considerable resemblance to the toes of a bird. The other toes of the hind foot being applied with their anterior surfaces to the ground, have the scales very minute and almost colourless.

The colour of the scales varies on different parts of the hand. On so much of the back of the hand as is formed by the fourth and little fingers, the scales are a very dark blue, approaching to black, in the living animal; hence to the large scales of the thumb the colour changes to a faint purplish blue, which is little more than distinguishable.

Two other excellent characters belonging to the palm of the hand have been neglected; the first is the enlargement of the carpal edge of the palm by an elongation of the integuments; this, in addition to the row of bristles that margins all the rest of the palm, has two distinct bristly hairs at its superior and inferior edge, more than one-eighth of an inch long. The second character is still more striking; it is a process of the palmar cuticle on the superior edge of the thumb and three succeeding fingers,

These processes are separated and directed obliquely upwards and outwards; the serrations on the thumb being two, and on the three succeeding fingers three in number.

On the soles of the (posterior) feet another character is found, which consists of five circular, distinct spots, so arranged that the two nearest the body are parallel with each other, opposite the commencement of the first toe, counting, as in the human subject, from the one nearest the median line of the body; the superior spot is nearly in a line with the fourth toe, and larger and darker coloured than the inferior; the two succeeding spots (nearer the extremity of the toes) are also parallel with each other; the exterior one is the largest of all these plantar scales, and placed nearly over the extremity of the metatarsal of the fourth toe; the fifth, or single scale, is placed in advance of all the rest, and is situated immediately over the centre, and behind the separation of the third and fourth toes. A very analogous arrangement may be observed in the soles of the feet of the *Sigmodon Hispidum;* ORD.

By comparing the condylura with the scalops, we are led to several interesting observations. We have seen that the condylura has a remarkable and large external ear, though it is destitute of a projecting auricle. The scalops has no auricle, and but an extremely small meatus externus opening on the side of the head near the shoulder.

The hand of the scalops is peculiar for its great breadth and strength: the extraordinary breadth is produced by an additional metacarpal bone, inferior or external to the thumb, articulated with the carpus, and having a tendon for moving it from the common flexor of the fingers.* On the superior or ulnar edge of the hand, there is a cartilaginous additament, connected with the little finger by a tendon. The condylura has the additional metacarpal bone, but rather like a rudiment, and has not the cartilaginous additament at the superior edge of the hand; hence the very great difference in breadth in the hands of the two genera. The scalops has a slight process or elongation, not at the carpal extremity of the palm, but on the inferior or outer edge of the supplementary bone.

---

* This structure resembles that of the *Talpa Europea*, or common mole of Europe, which has recently been asserted to inhabit Pennsylvania, on the authority of the MS. notes of the justly celebrated WILLIAM BARTRAM. These notes having been made long before the genera Scalops and Condylura were established, can have no weight, unless along with the name *Talpa Americana*, Bartram had given such a description as to convince us that it was not the *scalops* he observed, of which we have little doubt. However this may be, we shall continue to discredit the existence of Talpa Europea in this country, until more positive testimony is adduced.

CHAPTER VI.

Family III.—Carnivora; *Flesh-Eaters.*

The animals belonging to this family are certainly not the only ones which feed on flesh, since all others provided with claws, and the three sorts of teeth, in different degrees feed on animal matter. The creatures now about to become the subject of our attention, are fairly and fully entitled to the appellation of carnivorous, as nature has endowed them with sanguinary appetites and ferocious dispositions, and supplied the strength and weapons necessary to their gratification.

They have four large and long canine teeth, which are separated in each jaw by six incisors, the second of which, in the lower jaw, is always more deeply set than the others. The jaw-teeth are uniformly either entirely trenchant, or partly supplied with blunt tubercles, and never with conical points. The anterior molars of these animals are the most trenchant; to these succeed a larger molar, which commonly has an additional tuberculous point, varying in size; and behind this tooth we find one or two small entirely flat teeth. These small teeth at the back of the mouth, enable dogs to chew the grass which they occasionally eat. To these three sorts of teeth the following names have been appropriated by Frederic Cuvier: the large molar in the upper and lower jaw he calls *carnivorous;* the anterior pointed jaw-teeth, *false molars,* and the posterior blunt ones, *tuberculous.* We may readily and correctly decide on the degree of exclusiveness with which the animals of this family feed on flesh, by observing the proportion between the trenchant and tuberculous surfaces of their teeth.

Many genera comprised in this family apply the whole sole of the foot to the ground while walking or standing erect, as may be perceived by the nakedness of the inferior surface of the hind feet. A much larger number walk on the tips of the toes; hence their speed is much greater, and their general habits are also very different. All of them are equally destitute of clavicles, having in its stead nothing but a bony rudiment.

Tribe I.—Plantigrada; *Plantigrade Animals.*

The individuals of this tribe have five toes on the fore and hind feet, and in walking they place the whole sole of the foot on the ground, which enables them to walk or stand erect better than any other beasts of prey. They are destitute of a cæcum; and partake of the sluggish gait and nocturnal habits of the insect-eaters. The greater part of those found in cold countries pass the winter in a state of torpidity.

Treading on the whole sole of the foot.

CHAPTER VII.

Genus VIII. *Ursus*, L; Bear.

Gr. Αρκτος
Ital. Orso.
Swed. Biörn.

Germ. Der Bär.
Sp. Orso.
Fr. L'ours.

GENERIC CHARACTERS.

The head is large, the muzzle varying in length, and terminating in a moveable cartilage. The eyes and ears are small, and the tongue smooth. The body and limbs are large, powerful, and covered with a thick woolly hair. The teats are six in number; four of them are placed on the chest, and two on the belly. The nails are incurved, very large and strong; the soles of the hind-feet are callous, and the tail is short.

The Bear is an animal of great strength and ferocity of disposition; slow in his movements, and of sluggish habits. His teeth being most fitted for subsisting on fruits and vegetable matters, he does not frequently attack other animals, unless impelled by necessity. During winter bears generally pass a great portion of time in a state of inaction and torpidity. They are most generally found in the remote and mountainous districts of North America, and are gradually becoming more scarce as population increases. As we shall give a full account of the species, we refer the reader to what is there said for a better understanding of the character and habits of this genus.

Dental System.

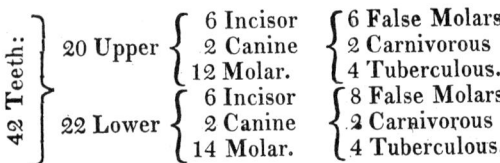

IN THE UPPER JAW the two first incisors resemble those of the dog, but the middle lobe almost entirely effaces the lateral, both being very small. Internally they are divided into two parts by a transverse depression, and the internal, much less salient than the opposite part, is itself divided into two lobes by a depression which is perpendicular to the transverse furrow. The third incisor is divided into two parts by an oblique furrow, and its hooked form gives it some relation to the canine.

The canine comes next, after a small unoccupied interval; it is conical, slightly hooked, and has longitudinally from before backwards a cutting edge. Immediately at the base of the canine is a rudimental false molar; then at a short distance we find a second, which sometimes falls out with age; and after another gap there is a third at the base of the carnivorous, very slightly developed, but sometimes provided with two roots. The carnivorous is reduced to its smallest dimensions; exteriorly we may recognize the middle tubercle, proper to this species of tooth, and the posterior tubercle, but the anterior lobe is almost effaced; at its internal side we find posteriorly a tubercle much smaller than the preceding, which increases its thickness. This particular position of the internal tubercle, which we always find at the anterior part of the upper carnivorous, while at the same time it is at the posterior part that the false molars become tuberculous, induces us to consider the tooth just described as only a false molar; but the upper carnivorous has entirely disappeared, and the only regular false molar existing, supplies its place.

The next tooth has at its extreme edge the two principal tubercles of the first tuberculous teeth; at its internal edge there are two tubercles parallel to the two first, but separated from each other by a much smaller tubercle. This tooth is nearly twice as long as it is broad.

The last molar is one-third larger than the preceding, but its proportions are the same in relation to length and breadth; it has on its external edge, at the anterior part, two tubercles, which seem analogous to those of the preceding tooth, but rather smaller. At the internal border of the same part, there is a crest divided in three, by two small grooves. The posterior part is a projection or spur, making nearly a third of the extent of the tooth, and bordered by a crest irregularly divided by three principal grooves; all the interior of the crown is covered by small furrows and asperities, which are peculiar to the bears.

In the lower jaw the incisors are bilobated like those of the dog, and the canine are shaped at the sides like those of the upper jaw. The false molars are two or three in number, and sometimes four; the first are at the base of the canine, the others are separated by an unoccupied interval, and have some relation to the true molars.

The first is larger than the second, and remains in the adult animal; the second extremely small, falls out with age, and in these different relations the third resembles it; the fourth alone has the regular form.

After this comes a tooth, narrow in proportion to its length, but not trenchant. We observe a tubercle on its anterior part, another on its external surface, and two smaller on its internal face, opposite the preceding. These four tubercles form nearly one-half of the tooth; to these succeed a deep groove, and the tooth terminates behind a pair of tubercles. The next molar, which is the largest tooth of this jaw, is very irregular in relation to the disposition of its grooves, tubercles, and the hollows or depressions which separate them. We may distinguish, however, two principal tubercles at its anterior half, one on its internal, and the other on its external face, which are united by a transverse crest; but these tubercles are subdivided, especially the internal one, by small depressions which separate it into two or three others. We may say the same of the posterior part, and, in fact, the engraving alone can give a clear idea of it, because it is much more irregular than the other. The last tooth is still less susceptible of a detailed description than the preceding; it is smaller, has an elliptic form, is bordered at its circumference by an irregularly notched crest, and is internally marked by still more irregular rugosities.

In their reciprocal position, all these teeth are opposed crown to crown, except the first lower molar.

whose external edge, at its anterior part, is in relation with the internal edge of the superior carnivorous: these are the only teeth belonging to animals of this genus suited to the comminution of flesh, which, in fact, they can but imperfectly perform.

[*The Brown Bear.*

Ursus Arctos: Lin. Erxleb. Bodd.
Ours: Buff. tom. 8, pl. 31, Briss. Reg. An. p. 258.
Alpine Bear of Europe.

This animal has so frequently been described as a native of this country, that persons unacquainted with the manner in which writers have copied each other, in relation to American natural history, may be surprised that we entirely reject the species as an inhabitant of the northern part of this continent.

It is true that various travellers have made an occasional mention of "brown bears," but there is abundant reason to believe that they have mistaken young or adult black bears, in a particular state of pelage, for them; this is rendered the more probable from the fact that no real "brown bear" has yet been seen by any of the expeditions which have traversed the vast forests, plains, and mountains of the western regions, where they would almost certainly have been encountered had they existed. Lewis and Clark, in several instances, speak of "Brown bears;" but these attentive observers expressly state, that they were uniformly found in the same districts, and were specifically the same as the Grizzly, white or variegated bear, which we shall hereafter describe. We have made many inquiries of persons who have resided in parts of the country where the brown bear would most probably be found, if it were a native, but have not yet met with an individual who has seen any other species than the common black, or American bear, and the great grizzly bear of the West.

Taking all circumstances into consideration, we feel authorised to believe that the Ursus Arctos is not found in America, and in this belief we shall remain, at least until there is unequivocal testimony adduced to establish the contrary.]

Species I.—*American, or Black Bear.*

Ursus Americanus.—Pallas.

Ours D'Amerique: Cuv. Men. du Mus.
Ours Gulaire: Geoff. Coll. du Mus.

This bear is found throughout North America, from the shores of the Arctic Sea, to its most southern extremity. That they must have existed in vast numbers throughout this great extent of country, previous to its settlement by Europeans, we may readily conceive, from the immense number of skins of this animal which are procured even at the present day. From the year 1798 to 1802, one hundred and ninety-two thousand four hundred and ninety-seven bear skins were exported from Quebec, and in the year 1822, the Hudson's Bay company alone exported three thousand skins of the black bear.

THE BLACK BEAR. 115

Captains Lewis and Clark observed black bears on the wooded portions of the rocky mountains, and subsequently found them on the great plains of Columbia, and in the tract of country lying between these plains and the Pacific Ocean. They are occasionally found throughout the territories of the United States, in the wooded and mountainous regions, and in unsettled districts, where their skins are of great value to the inhabitants as a substitute for blankets and other manufactured woollens.

The black bear, under ordinary circumstances, is not remarkably ferocious, nor is he in the habit of attacking man without provocation. But when wounded, he turns on the aggressor with great fury, and defends himself desperately. This disposition is more fully manifested during the coupling season, because the males are then highly excited, and are not so inert and clumsy as in the autumn, when they are exceedingly fat.

If taken when young, this bear is readily domesticated, and taught numerous tricks; we see him frequently exhibited by itinerant showmen, as a "learned" bear, though it requires a long continuance of severe and cruel discipline to bring him to this state of "improvement." In captivity they are always remarkable for the persevering manner in which they keep moving backward and forward at the extremity of their chain, thus expressing their impatience of confinement, or rather, as if solicitous to take exercise.

This feeling of the necessity for exercise is manifested in an especial manner when the animal is confined in a very small cage, where he has not room even to turn entirely round. Under such circumstances he perseveringly moves himself in every direction that his narrow limits will allow, stepping with his fore feet first to one side and then to the other, and finally, by raising and depressing his body quickly, as if jumping from the ground, he gives his whole frame a degree of exercise, which must tend to the preservation of his health and strength.

When the winters are severe at the north, and they find a difficulty of procuring food, they travel to the southern regions in considerable bodies. Dr. Libley states, in his report to the secretary of war, relative to the territory bordering on Red River, that from all the information he could gain, immense and almost incredible numbers of these animals descended the mountains, and passed southwardly into the timbered country.

The sight and hearing appear to be the most acute of the senses in this bear, as well as in those hereafter to be described. Although he kills many small animals, yet he does not follow them by the smell. When he walks, his gait is heavy and apparently awkward, and when running is not much less so, but his strength of body enables him to move with considerable celerity, and for a long time.

The females bring forth their young in the winter time, and exhibit for them a degree of attachment which nothing can surpass. They usually have two cubs, which are suckled until they are well grown. The fondness existing between the mother and cubs seems to be mutual, and no danger can separate her from them, nor any thing, short of death itself, induce her to forsake them.

"Near the old village of *Catharine*, in the state of New York, a young man of seventeen passing through the oods early in the morning, met with a young cub, which he pursued and caught, and seizing by the heels, swung it against a log repeatedly, to kill it. The noise it made alarmed the dam, and the lad, lifting his eyes, saw a large bear making towards him with great fury. Dropping the cub, he seized his gun in time to discharge the contents, which only wounded her, when instantly clubbing the musket, he belaboured her on the sides, snout, head, &c. till the stock of the gun was shivered, and the barrel wrenched and twisted in an extraordinary manner. After a sustained combat, in which the bear tore his clothes to pieces and scratched him severely, he took an opportunity (when, from the bleeding of her wounds and weakness, she began to flag,) to run away for assistance. On returning with his master they killed the old bear and both her cubs."*

A friend of the author's, while traversing a wood near Fort Snelling, on the Missouri, saw a she-bear accompanied by two cubs, (about the size of puppies at a month old) a short distance before him. The cubs immediately ascended a tree, and the dam, raising herself on her hind legs, sat erect at its foot in order to protect them: the rifle, discharged with a fatal aim, laid the parent lifeless on the earth. The hunter then approached and stirred the body with the butt of his gun, on which the little cubs hastily descended the tree and attacked him with great ear-

---

* Vide, Cyclop. Am. Ed. vol. iv.

nestness, attempting to bite his legs and feet, which their youth and want of strength prevented them from injuring. When he retired to a short distance, they returned to the dead body of their dam, and by various caresses and playful movements endeavoured to rouse her from that sleep which "knows no waking."

Black bears are still numerous in the wooded and thinly settled parts of Pennsylvania, as well as in most of the other states of the Union, and where their favourite food is plenteous they grow to a great size, and afford a large quantity of oil. BARTRAM relates that he was present at the cutting up of one which weighed five or six hundred pounds, and his hide was apparently as large as that of an ox of six or seven hundred weight.

The food of this animal is principally grapes, plums, whortle-berries, persimmons, bramble and other berries; they are also particularly fond of the acorns of the live oak, on which they grow excessively fat in Florida, &c. In attempting to procure these acorns they subject themselves to great perils, for after climbing these enormous oak-trees, they push themselves along the limbs towards the extreme branches, and with their fore-paws bend the twigs within reach, thus exposing themselves to severe and even fatal accidents in case of a fall. They are also very fond of the different kinds of nuts and esculent roots, and often ramble to great distances from their dens in search of whortle-berries, mulberries, and indeed all sweet flavoured and spicy fruits: birds, small quadrupeds, insects, and eggs, are also devoured by them whenever they can be obtained. They are occasionally very injurious to the frontier settlers, by their incursions in search of potatoes and young corn, both of which are favourite articles of food; their claws enable them to do great mischief in potato grounds, as they can dig up a large number in a very short time, and where the bears are numerous their ravages are occasionally very extensive.

In the north, the flesh of the black bear is fittest for the table after the middle of July, when the berries begin to ripen, though some berries impart a very disagreeable flavour to their flesh. They remain in good condition until the following January or February; late in the spring they are much emaciated, and their flesh is dry and disagreeable in consequence of their long fasting through the season of their torpidity. Their flesh is also rendered rank and disagreeable by feeding on herring spawn, which they seek and devour with greediness, whenever it is to be obtained. The southern Indians kill great numbers of these bears at all seasons of the year, but no inducement can be offered to prevent them from singeing off the hair of all that are in good condition for eating, as the flesh of the bear is as much spoiled by skinning as pork would be; the skins these people bring the traders are consequently only such as are obtained from bears that are too poor to be eaten.

In the vicinity of Hudson's Bay the black bear has been observed to feed entirely on water-insects during the month of June, when the berries are not ripe. These insects, of different species, are found in astonishing quantities in some of the lakes, where, being driven by gales of wind in the bays, and pressed together in vast multitudes, they die, and cause an intolerable stench by their putrefaction, as they lie in some places two or three feet deep.* The bear swims with his mouth open, and thus gathers the insects on the surface of the water: when the stomach of the animal is opened, at this season, it is found to be filled with them, and emits a very disagreeable stench. They are even believed to feed on those which die and are washed on shore. The flesh of the animal is spoiled by this diet, though individuals killed at a distance from the water are agreeably flavored at the same time of year.

The black bear is in fact very indiscriminate in his feeding; and though suited by nature for the almost exclusive consumption of vegetable food, yet refuses scarcely any thing when pressed by hunger. He is moreover voracious as well as indiscriminate in satisfying his appetite, and frequently gorges until his stomach loathes and rejects its contents. He seeks, with great assiduity, for the larvæ or grub-worms of various insects, and exerts a surprising degree of strength in turning over large trunks of fallen trees, which, whenever sufficiently decayed to admit of it, he tears to pieces in search of worms.

During the season when the logger-head turtles land in vast multitudes from the lagoons at the south, for the purpose of laying, the black bears come in droves to feast on their eggs, which they dig out of

---

* See Hearnes' Journey, p. 371, 8vo. ed.

the sand very expeditiously, and they are so attentive to their business, that the turtle has seldom left the place for a quarter of an hour, before the bear arrives to feast on her eggs.

While Major LONG's party were passing through the country west of the Missouri, they often saw black bears, which they observed most commonly to feed on grapes, plumbs, dog-wood berries, &c.; but they were also frequently seen disputing with the wolves and buzzards for a share of the carcasses of animals abandoned by the hunters, or of such as had perished by disease. When the bear seizes a living animal, he does not, as most other beasts do, first put it to death, but tears it to pieces and devours it, without being delayed by its screams or struggles, and may be actually said to swallow it alive.

DUPRATZ, who has been properly considered an intelligent and veracious historian, relates the following circumstance, in his History of Louisiana, to prove that the black bear is by no means carnivorous. We must in this case believe that he was misinformed, or mistaken, since this bear is well known to feed on flesh, even where the provocation is much less than that he relates.

"The black bears," says he, "appeared in Louisiana during the winter, being driven from the northern regions by the snow which covered the earth, and prevented them from obtaining their food. They fed on fruits, acorns, and roots, and were most fond of honey and milk, which, when obtained, they would sooner be killed than relinquish. In spite of the prejudice which supposes this bear to be carnivorous, I maintain, with all the inhabitants of this province and the surrounding country, that this is not the fact. It has never happened that they have devoured a man, notwithstanding their numbers and the extreme hunger they sometimes suffer; they even do not eat flesh when it is thrown in their way. During the time that I resided at Natchez, there was a very severe winter in the northern regions, and the black bears came south in great numbers; they were so numerous that they were all starving and very poor. Their great hunger drove them out of the woods bordering the river; they were seen at night entering yards which were not well closed, and where fresh meat was exposed, yet they left it untouched, and ate nothing but such grain as they could find. It was certainly on such occasions that their carnivorous disposition should have been displayed. They never killed animals to devour them, and so little carnivorous are they, that they abandon the snow-covered countries, where they could kill men and animals at pleasure, to wander so far to the south, in search of fruits and roots that a carnivorous animal would not touch." Sweet creatures!—either the Abbé Dupratz was sadly misinformed, or the disposition of the black bear has astonishingly deteriorated since the year 1755, as those of the present day, so far from refusing meat left in their way, will break into enclosures in search of it, and if opportunity offers, when pressed by hunger, they do not scruple to kill pork for their own use.

The following instance occurred in the western part of the state of New York, in the year 1824. The back window of a farm-house was forced open one night, and a considerable quantity of pork carried off. The proprietor, without suspecting the nature of the plunderer, placed a loaded musket opposite the window, having a string so adjusted that the gun would be discharged by any thing attempting to enter the room through the window. During the night the report of the gun was heard, and in the morning the body of a large black bear was found at a short distance from the spot where he had received his death wound.

The usual residence of the black bear is in the most remote and secluded parts of the forest, where his den is either in the hollow of some decayed tree, or in a cavern formed among the rocks. To this place he retires when his hunger is appeased, and in the winter he lies coiled up there during the long period of his torpidity. The female of the black bear, during the period of gestation, which commences in the month of October, and continues for about one hundred and twelve days, leads a retired and concealed life,—for we have not a single instance on record of a pregnant bear being killed either by white men or Indians, though the mother and very young cubs are frequently destroyed. During an extremely hard winter the inhabitants of the borders of James' river, Virginia, killed several hundred bears, among which two only were females, and those not with young.

In the northern parts of this continent, the subterraneous retreats of the black bear may be readily discovered by the mist which uniformly hangs about the entrance of the den, as the animal's heat and breathing prevent the mouth of the cave from being entirely closed, however deep the snow may be. As the black bear usually retires to his winter quarters before any quantity of snow has fallen, and does not again venture abroad, if undisturbed, until the end of March or beginning of April, he must consequently spend at least four months in a state of torpidity, and without obtaining food. It is therefore not surprising that, although the bear goes into his winter quarters in a state of excessive fatness, he should come out in the spring of the year extremely emaciated.

The northern Indians occasionally destroy the bear by blocking up the mouth of the cave with logs of wood, and then, breaking open the top of it, kill the animal with a spear or gun; yet this method is considered both cowardly and wasteful, as the bear can neither escape nor offer the slightest injury to his disturbers. Sometimes they throw a noose round his neck, draw him up to the top of the hole, and kill him with a hatchet.

The black bear is occasionally captured in large and strong steel-traps, well secured by a chain to a neighbouring tree, and laid in a path over which a freshly-killed carcass has been drawn along,—or he is taken in a noose suspended from a strong sapling. A common mode of hunting this animal is to follow him with two or three well trained dogs. When he finds that he is pursued, he generally pushes directly forward for eight or ten miles, or farther, if not overtaken; as the dogs come up with him their repeated attacks cause him to turn for the purpose of striking at them, and if they do not dexterously avoid his blows they will be killed, as he strikes with very great force. To avoid the vexation produced by the dogs, he mounts a tree, ascending for twenty or thirty feet, but is allowed very little rest, for the hunter now approaching, he throws himself to the earth, and hurries onwards, being still pursued and worried by the dogs. Again he is obliged to take refuge in a tree, and sometimes climbs as near as possible to the top, endeavouring to conceal himself among the foliage. The hunter now strikes against the trunk of the tree, as if engaged in cutting it down; the poor bear soon betrays his hiding place, and slipping to the end of the longest branch, gathers his body up, and drops from a vast height to the ground, whence he often appears to rebound for several feet, and then runs off as actively as he can. At length, worn out by frequently repeated exertions to escape, he is finally shot, while attempting to screen himself by aid of the trunk of a tree, or while employed in resisting the attacks of the dogs.

Among other modes of killing the black bear the Indians employ a trap composed of logs, which, when the animal attempts to remove the bait, either falls on his body and kills him outright, or secures him until he is put to death by the proprietor of the snare. Our enterprising countryman, SCHOOLCRAFT, relates an instance of his having seen one thus caught. "The animal sat up on his fore-paws facing us, the hinder paws being pressed to the ground by a heavy weight of logs, which had been arranged in such a manner as to allow the bear to creep under, and by seizing the bait he had sprung the trap and could not extricate himself, although with his fore-paws he had demolished a part of the works. After viewing him for some time a ball was fired through his head, but it did not kill him. The bear kept his position and seemed to growl in defiance. A second ball was aimed at the heart, and took effect, but he did not resign the contest immediately, and was at last despatched with an axe. As soon as the bear fell, one of the Indians walked up, and addressing him by the name of *muck-wah*, shook him by the paw, with a smiling countenance, as if he had met with an old acquaintance, saying, in the Indian language, he was sorry they had been under the necessity of killing him, and hoped the offence would be forgiven, especially as the *che-moek-o-men* (white men) had fired one of the balls."

The Indians consider this bear as one of the noblest objects of the chase, and they always manifest the highest degree of exultation when they are successful in killing one. Every part of the animal is valuable to them, even to its intestines and claws; the latter are bored at the base and strung on deer sinews to be worn as ornaments. The flesh is considered most delicious food, and the fore-paws as an exquisite dainty.

The fat of the bear is accumulated in different parts of the body to an excessive degree towards autumn, after the animal has been plentifully supplied with food; the oil obtained by liquefying it, is a well known popular remedy against baldness, as well as for rubbing stiff or rheumatic joints. The fat obtained from the paws is most highly prized, either because it is most difficult to procure in any quantity, or because it is really finer than that obtained from the body generally. It is very certain that few, or indeed perhaps none, of the animal oils are finer when properly prepared than that of the bear, and hence, in any case where the external application of oil is thought to be proper, bear's oil will be preferable to any other; but that it possesses many other virtues except those depending on its tenuity, we are not prepared to admit.

The black bear, like all the species of this genus, is very tenacious of life, and seldom falls unless shot through the brain or heart. An experienced hunter never advances on a bear that has fallen, without first stopping to load his rifle, as the beast frequently recovers to a considerable degree, and would then be a most dangerous adversary. The skull of the bear appears actually to be almost impenetrable, and a rifle-ball, fired at a distance of ninety six yards, has been flattened against it, without appearing to do any material injury to the bone. The best place to direct blows against the bear is upon his snout; when struck elsewhere, his dense woolly coat, thick hide, and robust muscles, render manual violence almost entirely unavailing.

When the bear is merely wounded, it is very dangerous to attempt to kill him with such a weapon as a knife or tomahawk, or indeed any thing which may bring one within his reach. In this way hunters and others have paid very dearly for their rashness, and barely escaped with their lives; the following instance may serve as an example of the danger of such an enterprise:

"Mr. Mayborne, who resides in Ovid township, Cayuga county, between the Seneca and Cayuga lakes, in the state of New York, went one afternoon through the woods in search of his horses, taking with him his rifle and the only load of ammunition he had in the house. On his return home, about an hour before dusk, he perceived a very large bear crossing the path, on which he instantly fired, and the bear fell, but immediately recovering his legs, made for a deep ravine a short way onwards. Here he tracked him awhile by the blood, but night coming on, and expecting to find him dead in the morning, he returned home. A little before day-break the next morning, taking a pitchfork and hatchet, and his son, a boy of ten or eleven years of age with him, he proceeded to the place in quest of the animal. The glen or ravine into which he had disappeared the evening before, was eighty or ninety feet from the top of the bank to the brook below: down this precipice a stream of three or four yards in breadth is pitched in one unbroken sheet, and, forming a circular basin or pool, winds away among the thick underwood. After reconnoitering every probable place of retreat, he at length discovered the bear, who had made his way up the other side of the ravine, as far as the rocks would admit, and sat under a projecting cliff, stedfastly eyeing the motions of his enemy. Mayborne, desiring his boy to remain where he was, took the pitchfork, and descending to the bottom, determined from necessity to attack him from below. The bear kept his position until the man approached within six or seven feet, when on the instant, instead of being able to make a stab with the pitchfork, he found himself grappled by the bear, and both together rolled towards the pond, at least twenty or twenty-five feet, the bear biting on his left arm, and hugging him almost to suffocation. By great exertion he thrust his right arm partly down his throat, and in that manner endeavoured to strangle him, but was once more hurled headlong down through the bushes, a greater distance than before, into the water.— Here, finding the bear gaining on him, he made one desperate effort and drew the animal's head partly under water, and repeating his exertions, at last weakened him so much, that calling to his boy, who stood on the other side in a state little short of distraction for the fate of his father, to bring him the hatchet, he sunk the edge of it by repeated blows into the brain of the bear. This man, although robust and muscular, was scarcely able to crawl home, where he lay for nearly three weeks, the flesh of his arm being much crushed, and his breast severely mangled. The bear weighed upwards of four hundred pounds."

The black bear, in common with other species of the genus, endeavours to suffocate an adversary by violently hugging and compressing its chest. A man might end such a strgggle in a few instants, if one hand be sufficiently at liberty to grasp the throat of the animal with the thumb and fingers, externally, just at the root of the tongue, as a slight degree of compression there will generally suffice to produce a spasm of the glottis, that will soon suffocate it beyond the power of offering resistance or doing injury.

The black bear differs from other species of the genus by having the nose and forehead nearly on the same line, though the forehead is slightly prominent. This projection of the front is less at the upper part than in the brown bear of Europe.* The palms of the hands and soles of the feet are very short, and the whole body is covered with long, shining, straight black hair, which is by no means harsh to the touch. The sides of the face are marked with fawn colour, and a small spot of the same exists in some individuals in front of the eye; others have the muzzle of a clear light yellow, with a white line commencing on the root of the nose and reaching to each side of the angle of the mouth. This continues over the cheek to a large white space, mixed with a slight fawn colour, covering the whole of the throat, whence a narrow line descends upon the breast. It was this variety which Geoffroy called *Ours Gulaire*. The yellow bear of Carolina is also a variety of the black or American bear. Capt. FRANKLIN saw adults of this species in the vicinity of Cumberland house which were red, and remarked that the cubs of these red bears were black; while the cubs of the black individuals were as frequently of a red colour.

---

* We are informed by Capt. J. LE CONTE, who, as a naturalist, ranks deservedly high, that the black bear is distinguished with still greater certainty from the brown or European bear by having one more molar tooth than that animal.

1. Black Bear. 2. Grizzly Bear.

Species II.—*The Grizzly Bear.*

Ursus Horribilis.—Ord.

Ursus Horribilis. Say. Long's Exped. to the Rocky Mountains. vol. ii.
Ursus Cinereus. Desm. Mam. p 164.
Ursus Ferox. The Grizzly, White, Variegated and Brown Bear of Lewis & Clark.*

This bear, justly considered as the most dreadful and dangerous of North American quadrupeds, is the despotic and sanguinary monarch of the wilds over which he ranges. Gigantic in size, and terrific in aspect, he unites to a ferociously blood-thirsty disposition a surpassing strength of limb, which gives him undisputed supremacy over every other quadruped tenant of the wilderness, and causes man himself to tremble at his approach, though possessed of defensive weapons unknown to any but the human race. To the Indians the very name of the Grizzly Bear is dreadful, and the killing of one is esteemed

---

* In the English translation of Cuvier's Régne Animal, the translator has given to this bear the name of *Ursus Candescens*, proposed by Major Hamilton Smith, which must of necessity be rejected, along with all others, except the one we have adopted, because no person has a right to change a name given by the original describer of an animal, according to the rules now well established, and almost universally accepted by naturalists, to prevent the confusion resulting from changes of nomenclature, which are frequently proposed with no better reason than the mere pleasure of an individual. Say, who first systematically described this species, adopted the name proposed for it by Ord, which cannot now, with any shadow of propriety, be changed.

equal to a great victory:—the white hunters are almost always willing to avoid an encounter with so powerful an adversary, and seldom or never wantonly provoke his anger.

This formidable bear unhesitatingly pursues and attacks men or animals, when excited by hunger or passion, and slaughters indiscriminately every creature whose speed or artifice is not sufficient to place them beyond his reach. The Bison, whose size and imposing appearance might seem to be a sufficient protection, does not always elude his grasp, as the grizzly bear is strong enough to overpower this animal, and drag its carcass to a convenient place to be deposited and devoured at leisure.

However singular it may appear that an animal endowed with such a fondness for destruction and blood, can exist altogether on vegetable food, it is a fact that the grizzly bear, no less than all other species belonging to the same genus, is capable of subsisting exclusively on roots and fruits: this may be inferred from the peculiarities of their system of dentition. It is by no means surprising that hunters and travellers should suppose the grizzly bear to be almost wholly carnivorous, seeing that he displays such an unappeasable ferocity of disposition, and so uniform an eagerness to destroy the life of any animal that falls within his power.

This bear at present inhabits the country adjacent to the eastern side of the Rocky Mountains, where it frequents the plains, or resides in the copses of wood which skirt along the margin of water courses. There is some reason to believe that the grizzly bear once inhabited the Atlantic regions of the United States, if we may be allowed to form any inference from traditions existing among the Delaware Indians, relative to the Big Naked Bear which formerly existed on the banks of the Hudson. The venerable HECKEWELDER informs us that Indian mothers used to frighten their children into quietness by speaking to them of this animal.

Notwithstanding it was mentioned a long time since by LA HONTAN and other writers, it has been but recently established as a distinct species in the works of systematic Zoologists. SAY was the first to give a full description of it, in the well known work we have quoted at the head of this article.

Two cubs of the grizzly bear were some time since kept alive in the menagery of PEALE's (now the Philadelphia) Museum. When first received they were quite small, but speedily gave indications of that ferocity for which this species is so remarkable. As they increased in size they became exceedingly dangerous, seizing and tearing to pieces every animal they could lay hold of, and expressing extreme eagerness to get at those accidentally brought within sight of their cage, by grasping the iron bars with their paws and shaking them violently, to the great terror of spectators, who felt insecure while witnessing such displays of their strength. In one instance an unfortunate monkey was walking over the top of their cage, when the end of the chain which hung from his waist dropped through within reach of the bears; they immediately seized it, dragged the screaming animal through the narrow aperture, tore him limb from limb, and devoured his mangled carcass almost instantaneously. At another time a small monkey thrust his arm through an opening in the bear cage to reach after some object; one of them immediately seized him, and, with a sudden jerk, tore the whole arm and shoulder-blade from the body, and devoured it before any one could interfere. They were still cubs, and very little more than half grown, when their ferocity became so alarming as to excite continual apprehension lest they should escape, and they were killed in order to prevent such an event.

To the venerable founder of the Philadelphia Museum, CHARLES WILLSON PEALE, to whom American students of natural history are under the most lasting obligations for his zeal and liberality, we are indebted for the following letter, written by the gallant and lamented PIKE, relative to the two grizzly bears above mentioned:

Washington, Feb. 3d, 1808.

SIR,—I had the honor of receiving your note last evening, and in reply to the inquiries of Mr. PEALE, can only give the following notes:

The bears were taken by an Indian in the mountains which divide the large western branches of the Rio Del Norte, and some small rivers which discharge their waters into the east side of the gulf of California, near the dividing line between the provinces of Biscay and Senora. We happened at the time to be marching along the foot of those mountains, and fell in with the Indian who had them, when I conceived the idea of bringing them to the United States, for your Excellency, although then more than 1600 miles from our frontier post, (Natchitoches) purchased them of the savage, and for three or four days made my men carry them in their laps, on horseback. As they would eat nothing but milk, they were in danger of starving. I then had a cage prepared for both, which was carried on a mule, lashed between two packs, but always ordered them to be let out the moment we halted, and not shut up again until we were prepared to march. By this treatment they became extremely docile when at liberty, following my men (whom they learned to distinguish from the Spanish dragoons by their always feeding them and encamping with them,) like dogs through our camps, the small villages and forts where we halted. When well supplied with sustenance they would play like young puppies with each other and the soldiers; but the instant they were shut up and placed on the mule they became cross, as the jostling of the animal knocked them against each other, and they were sometimes left exposed to the scorching heat of a vertical sun for a day without food or a drop of water, in which case they would worry and tear each other, until nature was exhausted, and they could neither fight nor howl any longer. They will be one year old on the first of next month, (March, 1808) and, as I am informed, they frequently arrive at the weight of eight hundred pounds.

Whilst in the mountains we sometimes discovered them at a distance, but in no instances were we ever able to come up with one, which we eagerly sought, and *that* being the most inclement season of the year, induces me to believe they seldom or never attack a man unprovoked, but defend themselves courageously;* an instance of this kind occurred in New Mexico, whilst I sojourned in that province;—three of the natives attacked a bear with their lances, two of whom he killed, and wounded the third, before he fell the victim.

With sentiments of the
highest respect and esteem,
Your obedient servant,
Z. M. PIKE.

His Excellency, Thomas Jefferson,
 President of the United States.

The grizzly bear is remarkably tenacious of life, and on many occasions numerous rifle-balls have been fired into the body of an individual without much apparent injury. Instances are related by the travellers who have explored the countries in the vicinity of the Rocky Mountains, of from ten to fourteen balls having been discharged into the body of one of these bears before it expired. In confirmation of these statements we shall here introduce some sketches from narratives given in the journals of Lewis and Clark, and Long's Expedition to the Rocky Mountains.

One evening the men in the hindmost of one of Lewis and Clark's canoes perceived one of these

---

* It is very possible that the bears thus seen, may have been common black bears; the reader will find in the sequel abundant proofs that it is not difficult to come up with the grizzly bear, and that this animal does not often wait to be attacked.

THE GRIZZLY BEAR. 137

bears lying in the open ground about three hundred paces from the river, and six of them, who were all good hunters, went to attack him. Concealing themselves by a small eminence, they were able to approach within forty paces unperceived; four of the hunters now fired, and each lodged a ball in his body, two of which passed directly through the lungs. The bear sprang up and ran furiously with open mouth upon them; two of the hunters, who had reserved their fire, gave him two additional wounds, and one breaking his shoulder-blade, somewhat retarded his motions. Before they could again load their guns, he came so close on them, that they were obliged to run towards the river, and before they had gained it the bear had almost overtaken them. Two men jumped into the canoe; the other four separated, and concealing themselves among the willows, fired as fast as they could load their pieces. Several times the bear was struck, but each shot seemed only to direct his fury towards the hunter; at last he pursued them so closely that they threw aside their guns and pouches, and jumped from a perpendicular bank, twenty feet high, into the river. The bear sprang after them, and was very near the hindmost man, when one of the hunters on the shore shot him through the head, and finally killed him. When they dragged him on shore, they found that eight balls had passed through his body in different directions.

On another occasion the same enterprising travellers met with the largest bear of this species they had ever seen; when they fired he did not attempt to attack, but fled with a tremendous roar, and such was his tenacity of life, that although five balls had passed through the lungs, and five other wounds were inflicted, he swam more than half across the river to a sand bar, and survived more than twenty minutes. This individual weighed five or six hundred pounds at least, and measured eight feet seven inches and a-half from the nose to the extremity of the hind-feet, five feet ten inches and a-half around the breast, three feet eleven inches around the middle of the fore-leg, and his claws were four inches and three-eighths long.

In fact, the chance of killing the grizzly bear by a single shot is very small, unless the ball penetrates the brain, or passes through the heart. This is very difficult to effect, since the form of the skull, and the strong muscles on the side of the head, protect the brain against every injury except a very truly aimed shot, and the thick coat of hair, the strong muscles and ribs, make it nearly as difficult to lodge a ball fairly in the heart.

Governor CLINTON, in the notes to his discourse delivered before the Literary and Philosophical Society of New York, says, "that Dixon, an Indian trader told a friend of his, that this animal had been seen *fourteen feet long;* that notwithstanding its ferocity, it had been occasionally domesticated, and that an Indian belonging to a tribe on the head waters of the Mississippi, had one in a reclaimed state, which he sportively directed to go into a canoe belonging to another tribe of Indians, then returning from a visit: the bear obeyed, and was struck by an Indian. Being considered as one of the family, this was deemed an insult, resented accordingly, and produced a war between these nations."

Mr. JOHN DOUGHERTY, a very experienced and respectable hunter, who accompanied Major LONG's party during their expedition to the Rocky Mountains, several times very narrowly escaped from the grizzly bear. Once, while hunting with another person on one of the upper tributaries of the Missouri, he heard the report of his companion's rifle, and when he looked round, beheld him at a short distance endeavouring to escape from one of these bears, which he had wounded as it was coming towards him. Dougherty, forgetful of every thing but the preservation of his friend, hastened to call off the attention of the bear, and arrived in rifle-shot distance just in time to effect his generous purpose. He discharged his ball at the animal, and was obliged in his turn to fly; his friend relieved from immediate danger, prepared for another attack by charging his rifle, with which he again wounded the bear, and saved Mr. D. from further peril. Neither received any injury from this encounter, in which the bear was at length killed.

On one occasion several hunters were chased by a grizzly bear, who rapidly gained upon them. A boy of the party, who could not run so fast as his companions, perceiving the bear very near him, fell with his face toward the ground. The bear reared up on his hind-feet, stood for a moment, and then bounded over him in pursuit of the more distant fugitives.

Mr. DOUGHERTY, the hunter before mentioned, relates the following instance of the great muscular strength of the grizzly bear: Having killed a Bison, and left the carcass for the purpose of procuring assistance to skin and cut it up, he was very much surprised on his return to find that it had been draged off, whole, to a considerable distance, by a grizzly bear, and was then placed in a pit, which the animal had dug with his claws for its reception.

This bear strikes a very violent blow with its forepaws, and the claws inflict dreadful wounds. One of the cubs before mentioned as belonging to the Philadelphia Museum, struck the other a blow over part of its back and shoulder, which produced a large wound like a sabre cut. It is stated in Long's Expedition, that a hunter received a blow from the fore-paw of a grizzly bear, which destroyed his eye and crushed his cheek bone.

The grizzly bear is unable to climb trees like other bears; he is much more intimidated by the voice than the aspect of man, and on some occasions, when advancing to attack an individual, he has turned and retired merely in consequence of the screams extorted by fear. The degree of ferocity exhibited by the grizzly bear appears to be considerably influenced by the plenty or scarcity of food in the region it inhabits.

Anterior to the time of Lewis and Clark's expedition, nothing very satisfactory was known relative to this bear, and it was not until the publication of the journal of Long's Expedition to the Rocky Mountains, that a correct scientific description was given by that distinguished naturalist, SAY.

It may be with certainty distinguished from all the known species of this genus, by its elongated claws, and the rectilinear or slightly arched figure of its facial profile. Its general appearance may be compared with the Alpine bear of Europe, (U. Arctos) especially with the Norwegian variety. The Alpine bear has not the elongated claws, and the facial space is deeply indented between the eyes.—This bear is also a climber; the grizzly bear is not.

On the front of the grizzly bear the hair is short, and between and anterior to the eyes it is very much so. On the rest of the body it is long and very thickly set, being blacker and coarser on the legs, feet, shoulders, throat, behind the thighs, and beneath the belly; on the snout it is paler. The ears are short and rounded, the forehead somewhat convex or arcuated, and the line of the profile continues on the snout without any indentation between the eyes.

The eyes are quite small, and have no remarkable supplemental lid. The iris is of a light reddish brown, or burnt Sienna colour. The muffle of the nostrils is black, and the sinus very distinct and profound. The lips are capable of being extended anteriorly, especially the upper one, which has on it a few more rigid hairs or bristles than the lower lip. The tail, which is very short, is concealed by the hair. The length of the hair gradually diminishes on the legs, but is still ample in quantity on the upper part of the foot.

The claws on the fore-feet are slender and elongated, and the fingers have five sub-oval naked tubercles, separated from the palm, each other, and the base of the claws, by dense hair. The anterior half of the palm is naked, and is of an oval figure transversely,—the base of the palm has a rounded naked tubercle encircled by hair.

The soles of the (hind) feet are naked, and the

nails are more curved and not so long as those on the fore-paws. The nails are not in the least diminished at tip, but they grow sharper at that part only by lessening from beneath.*

The colour of the grizzly bear varies very considerably, according to age and its particular state of pelage. Hence they have been described as brown, white, and variegated, by Lewis and Clark, although evidently of the same species, judging by all the other characters. The colour of the young animal approaches more nearly to the brown bear of Europe than any other; in advanced life the colour is that peculiar mixture of white, brown, and black, which has procured for this bear the appropriate designation of "Grizzly."

The following are the dimensions of the specimens preserved in the Philadelphia Museum, as given by Say.

Length from the tip of the nose to the origin of the tail,	5 ft. 2 in.
The tail exclusive of the hair at tip,	$1\frac{3}{4}$
From the anterior base of the ear to the tip of the nose,	6
Orbit of the eye,	$\frac{3}{4}$
Between the eyes,	$4\frac{2}{5}$
Ears from their superior base,	3
Longest claw of the fore-foot,	$4\frac{1}{5}$
Shortest,	$2\frac{3}{4}$
Longest claw of the hind-foot,	3
Shortest,	$1\frac{3}{4}$

---

* Ursi Horribilis testes saccis duobus distinctis, etiam ad quatuor pollicum segregatis, pendunt.

1

2

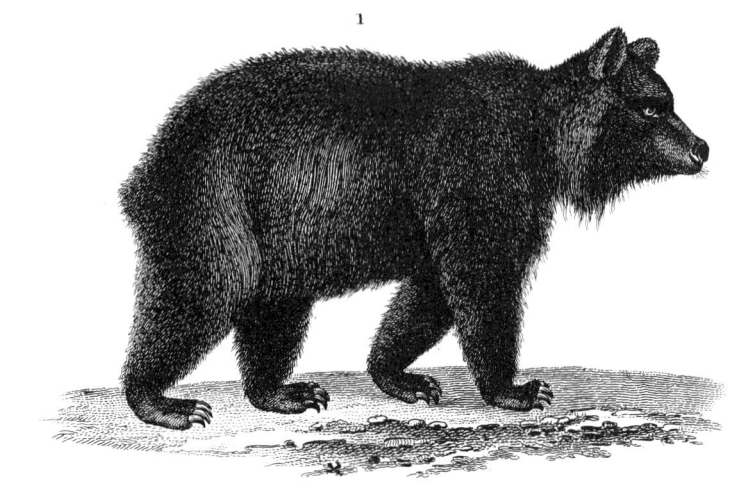

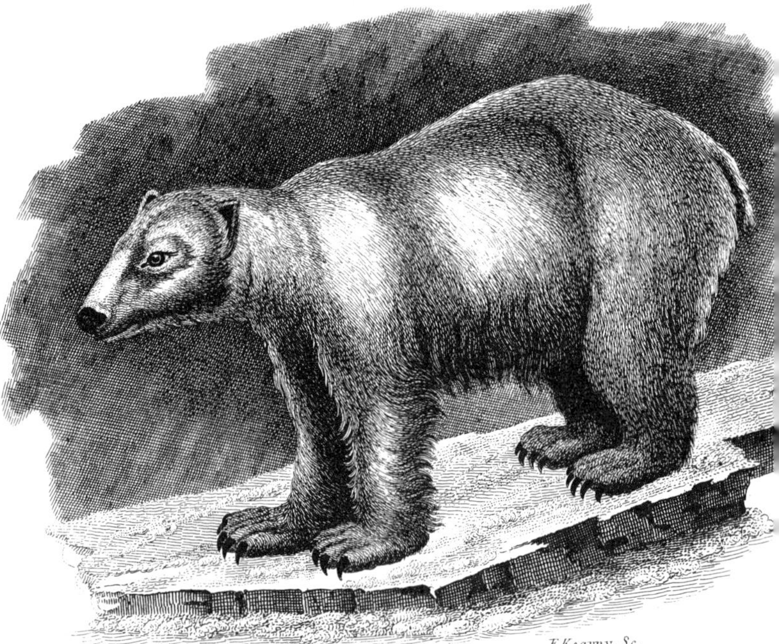

C. A. Lesueur. Del   F. Kearny. Sc.

1. *Brown Bear.* 2. *Polar Bear.*

Hair at the tip of the tail,	4½ in.
Length of the hair on the top of the head,	1¾ to 2
Beneath the ears,	2½ to 3½
On the neck above,	3
On the shoulders above,	4½
On the throat,	4
On the belly and behind the fore legs the longest hairs are	6

These measurements are taken from two individuals which were by no means full grown, as may be perceived by comparing them with the measurements heretofore cited from Lewis and Clark.— They will serve, however, to give a fairer idea of the proportions of this animal than any which have been previously given, as they are so much more detailed and very carefully made.

Species III.—*The Polar Bear.*

Ursus Maritimus.—Lin.

Ursus Albus; Briss. Régne Anim. p. 260. Sp. 2.
Ours Blanc; Buff. Supp. tom. 3. pl. 34.
Ours Blanc; Desm. Mam. p. 16. Sp. 257.
The Polar Bear; Pen. Syn. quad. p. 192. tab. 20. fig. 1. Pallas, spicil. Zool. XIV. tab. 1.

In the desolate regions of the north, where unrelenting winter reigns in full appanage of horrors during the greater part of the year, and even the stormy ocean itself is long imprisoned by " thick ribbed ice," the Polar Bear finds his most congenial abode. There, prowling over the frozen wastes, he satiates his hunger on the carcasses of whales deserted by the adventurous fishermen. or seizes on such marine animals as come up to bask in open air; and when occasion calls, he fearlessly plunges into the sea in pursuit of his prey, as if the deep were his native and familiar element: To most other animals extreme cold is distressing and injurious; to him it is welcome and delightful: to him the glistening ice-bank or snow-wreathed shore, canopied by louring and tempestuous clouds, are far more inviting and agreeable, than verdant hills or sunny skies.

Being endowed with extremely acute senses, great strength, and a savagely ferocious disposition withal, it is not surprising that this animal is dreaded as the most formidable quadruped of the region he inhabits. Notwithstanding his great size and apparent heaviness, he is very active, and though his ordinary gait may appear clumsy, when excited by rage or hunger, his speed on the ice far exceeds that of the swiftest man.

When on an extensive ice-field, the polar bear is often observed to ascend the knobs or hummocks, for the purpose of reconnoitering, or he stands with head erect to snuff the tainted air, which informs him where to find the whale carrion at astonishing distances. This substance, so unpleasant and disgusting to human sense, is a luxurious banquet to the bear, and a piece of it thrown on a fire will allure him from a distance of several miles.

A considerable part of the Polar bear's food is supplied by seals, but very probably he suffers long fasts and extreme hunger, owing to the peculiar vigilance of these creatures; occasionally he is much reduced by being carried out to sea on a small island of ice, where he may be forced to remain for a week or more without an opportunity of procuring food. In this situation they have been seen on ice-islands two hundred miles distant from land, and sometimes they are drifted to the shores of Iceland, or Norway, where they are so ravenous as to destroy all the animals they find. Most commonly such invaders are soon destroyed, as the natives collect in large numbers and commence an immediate pursuit, but frequently do not succeed in killing them, before many of their flocks are thinned. An individual polar bear has occasionally been carried on the ice as far south as Newfoundland, but this circumstance very rarely occurs.

This animal swims excellently, and advances at the rate of three miles an hour. During the summer season he principally resides on the ice-islands, and leaves one to visit another, however great be the distance. If interrupted while in the water, he dives and changes his course; but he neither dives very often, nor does he remain under water for a long time. Captain Ross saw a polar bear swimming midway in Melville Sound, where the shores were full forty miles apart, and no ice was in sight large enough for him to have rested on. The best time for attacking him is when he is in the water; on ice or land he has so many advantages that the aggressor is always in danger. Even in the water he has frequently proved a formidable antagonist, has boarded and taken possession of a small boat, forcing the occupants to seek safety by leaping overboard. Instances are related in which this animal has climbed up the sides of small vessels, and been with difficulty repelled from the deck.

Generally the polar bear retreats from man; but when pursued and attacked he always resents the aggression, and turns furiously on his enemy. When struck at with a lance, he is very apt to seize and bite the staff in two, or wrest it from the hands. Should a ball be fired at him, without taking effect in the head or heart, his rage is increased, and he seeks revenge with augmented fury. It has been remarked that, when wounded and able to make his escape, he applies snow to the wound, as if aware that cold would check the flow of blood.

A great majority of the fatal accidents following engagements with the polar bear, have resulted from imprudently attacking the animal on the ice. Scoresby, in his interesting narrative of a voyage to Greenland, relates an instance of this kind. " A few years ago, when one of the Davis's Strait whalers was closely beset among the ice at the ' south west,' or on the coast of Labrador, a bear that had been for some time seen near the ship, at length became so bold as to approach alongside, probably tempted by the offal of the provision thrown overboard by the cook. At this time the people were all at dinner, no one being required to keep the deck in the then immoveable condition of the ship. A hardy fellow who first looked out, perceiving the bear so near, imprudently jumped upon the ice, armed only with a handspike, with a view, it is supposed, of gaining all the honour of the exploit of securing so fierce a visitor by himself. But the bear, regardless of such weapons, and sharpened probably by hunger, disarmed his antagonist, and seizing him by the back with his powerful jaws, carried him off with such celerity, that on his dismayed comrades rising from their meal and looking abroad, he was so far beyond their reach as to defy their pursuit."

"A circumstance, communicated to me by Capt. Munroe of the Neptune, of rather a humorous nature as to the result, arose out of an equally imprudent attack made on a bear, in the Greenland fishery of 1820, by a seaman employed in one of the Hull whalers. The ship was moored to a piece of ice, on which, at a considerable distance, a large bear was observed prowling about for prey. One of the ship's company, emboldened by an artificial courage, derived from the free use of rum, which in his economy he had stored for special occasions, undertook to pursue and attack the bear that was within view. Armed only with a whale-lance, he resolutely, and against all persuasion, set out on his adventurous exploit. A fatiguing journey of about half a league, over a yielding surface of snow and rugged hummocks, brought him within a few yards of the enemy, which, to his surprise, undauntedly faced him, and seemed to invite him to the combat. His courage being by this time greatly subdued, partly by evaporation of the stimulus, and partly by the undismayed and even threatening aspect of the bear, he levelled his lance, in an attitude suited either for offensive or defensive action, and stopped. The bear also stood still; in vain the adventurer tried to rally courage to make the attack; his enemy was too formidable, and his appearance too imposing. In vain also he shouted, advanced his lance, and made feints of attack: the enemy, either not understanding or despising such unmanliness, obstinately stood his ground. Already the limbs of the sailor began to quiver; but the fear of ridicule from his messmates had its influence, and he yet scarcely dared to retreat. Bruin, however, possessing less reflection, or being regardless of consequences, began, with audacious boldness, to advance. His nigh approach and unshaken step subdued the spark of bravery and that dread of ridicule that had hitherto upheld our adventurer; he turned and fled. But now was the time of danger; the sailor's flight encouraged the bear in turn to pursue, and being better practised in snow-travelling, and better provided for it, he rapidly gained upon the fugitive. The whale-lance, his only defence, encumbering him in his retreat, he threw it down, and kept on. This fortunately excited the bear's attention; he stopped, pawed it, bit it, and then renewed the chase. Again he was at the heels of the panting seaman, who, conscious of the favourable effects of the lance, dropped one of his mittens; the stratagem succeeded, and while Bruin again stopped to examine it, the fugitive, improving the interval, made considerable progress a-head. Still the bear resumed the pursuit with a most provoking perseverance, except when arrested by another mitten, and, finally, by a hat, which he tore to shreds between his fore-teeth and paws, and would, no doubt, soon have made the incautious adventurer his victim, who was now rapidly losing strength, but for the prompt and well-timed assistance of his shipmates—who, observing that the affair had assumed a dangerous aspect, sallied out to his rescue. The little phalanx opened him a passage, and then closed to receive the bold assailant. Though now beyond the reach of his adversary, the dismayed fugitive continued onwards, impelled by his fears, and never relaxed his exertions, until he fairly reached the shelter of his ship. The bear once more came to a stand, and for a moment seemed to survey his enemies with all the consideration of an experienced general; when, finding them too numerous for a hope of success, he very wisely wheeled about, and succeeded in making a safe and honourable retreat.*"

The polar bear is stated to be generally four or five feet high, from seven to eight feet long, and nearly the same in circumference. Individuals have frequently been met with of much greater size; Barentz killed one in Cherie Island, whose skin measured thirteen feet.† The weight is generally from six to eight hundred pounds. The hair of the body is long, and of a yellowish white colour, and is very shaggy about the inside of the legs. The paws are seven inches or more in breadth, with claws two inches long. In some individuals, the canine teeth

---

* Scoresby's Greenland Voyage.

† DESMAREST states in a note that the largest individuals of this species which have been observed, are not more than six feet seven inches long. This does not agree with the accounts given by many northern voyagers: we have selected Captain Ross's measurements, (not because the individual from which they were taken is the largest that has been seen, but) because his scientific character is so generally and advantageously known. It would have been very easy to have selected measurements of larger specimens, from other sources.

have been found an inch and a-half long, exclusive of the portion imbedded in the jaw: the strength of the jaws is very great, and enables the animal to inflict dreadful injury when he bites.

The following measurements are from an individual, killed during Capt. Ross's voyage, in the vicinity of Prince William's Sound:—

	ft.	in.
Length, from the snout to the tail,	6	8
to the shoulder-blade,	2	10
Circumference near the fore-legs,	6	
of the neck,	3	2
Breadth of the fore-paw,		10
of the hind-foot,		$8\frac{1}{2}$
Circumference of the hind-leg,	1	10
of the fore-leg,	1	8
of the snout, before the eyes,	1	8
Length of the snout to the occiput,	1	6
Height to the fore-shoulder,	4	
Fore-claws,		$2\frac{1}{2}$
Hind-claws,		$1\frac{3}{4}$
Tail,		4

Weight of the animal, after losing thirty pounds of blood, $1131\frac{1}{2}$ pounds.

We have stated that the polar bear preys on seals, fish, and the carcasses of whales; it also preys on birds, and their eggs, and not unfrequently destroys young whales and walruses: it is also said to disinter human bodies, and devour them with great greediness. Occasionally they break into the huts of the Greenlanders, attracted by the smell of seal's flesh, on which these people almost exclusively subsist. Yet we are credibly informed, that, when their accustomed food is to be obtained in sufficient quantity, they neither show much disposition to attack men, nor cattle, however accessible these may be.

In the morse or walrus, this bear has an enemy of great power and fierceness, with which he has at times dreadful combats, most generally terminating in the defeat of the bear, as the walrus is armed with long tusks, capable of giving deadly wounds. The whale is also a perpetual enemy of the polar bear, chasing him from the waters it frequents, and killing him by blows with its tail. Notwithstanding, the bear succeeds in catching and feasting on many of the young whales.

The dwelling place of the polar bear on shore, is by no means well ascertained, but is most probably in caves, or in some well concealed situation; it has been stated, that they reside, during winter, in excavations made in the permanent ice,—but Fabricius, from personal observation, declares the statement to be incorrect. Certainly this animal does not often go to any great distance from the sea, on which he is almost exclusively dependent for food. Hence the flesh of the polar bear is generally fishy and rank, though it is said to be whitish, and similar to mutton. Captain Cook's people always preferred it to the flesh of the walrus or morse, yet they never considered it a very desirable food, except when none other was to be obtained. The fat resembles tallow, becoming as clear as whale-oil after liquefaction, and free from disagreeable smell; the oil obtained from the feet has been used medicinally, but except in fineness, has no qualities which the oil of other parts does not possess.

One of the most singular facts relative to the polar bear is, that its liver is to a great degree poisonous, a circumstance unknown in almost every other animal. Three of BARENTZ's sailors were very much injured by eating of it; and Capt. Ross, in his late Arctic voyage, verified the observation by experiment. The principle, which imparts this noxious quality to the liver, is as yet undiscovered; we know of no article of diet used by the animal, to which it can be attributed, and even if we did, this would not account for the deleteriousness of the liver, while all other parts of the body remain free from any injurious property.

The skin of the polar bear, dressed with the hair on, forms very substantial mats for carriages, or hall floors. The Greenlanders sometimes take it off without ripping up, and inverting the skin, form a very warm sack, which serves the purpose of a bed, the person getting into it in order to sleep comfortably. It cannot well be dressed at any other than the winter season, on account of its great greasiness when freshly removed from the animal. The nations residing in the vicinity of Hudson's Bay dress it in the following manner: they first stretch it out on a smooth patch of snow, and stake it down, where it soon becomes stiffly frozen. While in this condition the women scrape off all the fat till they come to the very roots of the hair. It is occasionally permitted to remain in that situation for a considerable time, and when taken up it is suspended in the open air. When the frost is very intense, it dries most perfectly; with a little more scraping it becomes entirely dry and supple, both skin and hair being beautifully white. Not withstanding that this bear is so large and powerful, his skin is both light and spongy.

The time of the year at which the sexes seek each other is not positively known, but it is most probably in the month of July, or of August. HEARNE, who is an excellent authority, relates that he has seen them killed during this season, when the males exhibited an extreme degree of attachment to their companions. After a female was killed, the male placed his fore paws over her, and allowed himself to be shot rather than relinquish her dead body.

The pregnant females during winter seek shelter near the skirt of the woods, were they excavate dens in the deepest snow-drifts, and remain there in a state of torpid inaction, without food, from the latter part of December or January till about the end of March; then they relinquish their dens to seek food on the sea shore, accompanied by their cubs, which are usually two in number. The size of the cubs is very small; when they first leave the cave with the mother they are not larger than rabbits; yet we have seen that the weight of the full grown animal sometimes exceeds a thousand pounds. HEARNE states that he has seen them not larger than a white fox, and their foot-prints on the snow not larger than a crown piece, when the impression of their dam's foot measured upwards of fourteen inches long by nine in breadth. This length and breadth appear excessive, and were probably rather more than the actual size of the foot itself, as the impression of the hair projecting over the feet would give an appearance in the snow which might lead to an incorrect notion of the size of the animal,

The enterprising observer above mentioned is of opinion that these animals breed when very young, or at least when half grown, as he has killed young females " not larger than a London calf," having milk in their teats; " whereas one of the full-grown ones are heavier than the largest of our common oxen. Indeed, I was once at the killing of one, when one of its hind feet, being cut off at the ancle, weighed fifty-four pounds."

The female polar bear is as rugged in her appearance, and as savagely ferocious in disposition, as her mate; yet to her offspring she displays a tenderness of affection which strongly contrasts with her fierce and sanguinary temper. When her cubs are exposed, danger has no existence to her, and nothing but death can compel her to desist from struggling desperately to defend or save them. The death of her offspring is with great difficulty acknowledged by the parent; when they are shot by her side the poor beast solicits their attention by every fond artifice, and endeavours to awaken them from their unnatural sleep: she offers them food, licks their wounds, caresses and moans over them in such a manner as to evince a degree of feeling which could scarcely be anticipated from so rude and terrible a quadruped.

Numerous instances of this fondness of attachment have been observed, and some of them attended with most singular displays of sagacity on the part of the mother. The following circumstance is related in SCORESBY's account of the Arctic Regions, and is entitled to the fullest credence, because coming from so competent and excellent an observer:

THE POLAR BEAR.

"A she-bear, with her two cubs, were pursued on the ice by some of the men, and were so closely approached, as to alarm the mother for the safety of her offspring. Finding that they could not advance with the desired speed, she used various artifices to urge them forward, but without success. Determined to save them, if possible, she ran to one of the cubs, placed her nose under it, and threw it forward as far as possible; then going to the other, she performed the same action, and repeated it frequently, until she had thus conveyed them to a considerable distance. The young bears seemed perfectly conscious of their mother's intention, for as soon as they recovered their feet, after being thrown forward, they immediately ran on in the proper direction, and when the mother came up to renew the effort, the little rogues uniformly placed themselves across her path, that they might receive the full advantage of the force exerted for their safety."

The most affecting instance on record of the maternal affection exhibited by this bear, is related in one of the Polar Voyages; it conveys so excellent an idea of this creature's strong feeling of parental love, that we should deem the history of the animal imperfect, were such an illustration omitted.

"Early in the morning, the man at the mast-head gave notice that three bears were making their way very fast over the ice, and directing their course towards the ship. They had probably been invited by the blubber of a sea-horse, which the men had set on fire, and which was burning on the ice at the time of their approach. They proved to be a she-bear and her two cubs; but the cubs were nearly as large as the dam. They ran eagerly to the fire, and drew out from the flames part of the flesh of the sea-horse, which remained unconsumed, and ate it voraciously. The crew from the ship threw great pieces of the flesh, which they had still left, upon the ice, which the old bear carried away singly, laid every piece before her cubs, and dividing them, gave each a share, reserving but a small portion to herself. As she was carrying away the last piece, they levelled their muskets at the cubs, and shot them both dead; and in her retreat, they wounded the dam, but not mortally.

"It would have drawn tears of pity from any but unfeeling minds, to have marked the affectionate concern manifested by this poor beast, in the last moments of her expiring young. Though she was sorely wounded, and could but just crawl to the place where they lay, she carried the lump of flesh she had fetched away, as she had done the others before, tore it in pieces, and laid it down before them; and when she saw they refused to eat, she laid her paws first upon one, and then upon the other, and endeavoured to raise them up. All this while it was piteous to hear her moan. When she found she could not stir them, she went off; and when at some distance, looked back and moaned; and that not availing to entice them away, she returned, and smelling around them, began to lick their wounds. She went off a second time, as before; and having crawled a few paces looked again behind her, and for some time stood moaning. But still her cubs not rising to follow her, she returned to them again, and with signs of inexpressible fondness, went round first one and then the other, pawing them, and moaning. Finding at last that they were cold and lifeless, she raised her head towards the ship, and growled her resentment at the murderers; which they returned with a volley of musket balls. She fell between her cubs, and died licking their wounds."

The sagacity of the polar bear is well known to the whale fishers, who often find all their ingenuity insufficient to entrap him, as the following instance may serve to show. A noose, baited with a piece of "*kreng,*" or whale carcass, was placed at a proper distance from the ship, which soon attracted the attention of a large bear. In attempting to secure the bait, the animal by some movement drew the noose, so as to catch him by one of his fore-paws. Apparently unconcerned by this circumstance, and conscious of knowing how to free himself from restraint, he quietly loosened the slip-knot with the other paw, and leisurely walked off to enjoy his morsel. The trap was again baited, and the bear once more approached to obtain his favourite food, but, grown wise by experience, he carefully avoided the rope, and carried off the bait, to the mortification of the captain, who wished to obtain his skin. The whaler, resolved to baffle the address of the bear, re-arranged his noose once more, carefully burying the rope at a considerable depth in the snow: but his precautions were unavailing; the bear cautiously examined the vicinity, scented the ground with attention, detected the situation of the rope, dug it up and threw it out of his way; then securing his prize, he once more triumphantly withdrew to enjoy it.[*]

---

[*] See Scoresby's Arctic Regions, vol. i. whence several of these anecdotes are sketched.

Captain Scoresby shot a she-bear and took her two cubs alive, as they did not offer to leave the body of their mother, and he kept them on board of his ship, until they were tame enough to be allowed to go about the deck. On one occasion a cub, tied by the neck with a long rope, was allowed to go out of the ship, when he immediately swam to the ice, and as soon as he attained it, made a violent effort to break the rope by running at full speed until he put the rope as suddenly on the stretch as possible. Failing in his first attempt, he went back far enough to slacken the cord, and again renewed his race, in order, if possible, to break it. Convinced by these experiments, that it was a hopeless attempt, he lay down, sullenly growling his vexation. Another artifice resorted to by this animal was still more singular; passing a chasm or fissure in the ice, about eighteen inches or two feet wide, and three or four feet deep, the slack (or bight) of his rope dropped into it; young Bruin returned, and going down head foremost into the chasm, he hung by the edges, holding on with one hind-foot on each side of it, and tried with both his fore-paws to loosen the rope and slip it off his head, as if he was aware that in this position he would be assisted by the weight of the portion which had dropped lower into the cavity.

The polar bear, like the other species of this genus, is able to live exclusively on vegetable food, as has been repeatedly proved by experiment on those brought to Europe. One which was exhibited in France, ate six pounds of bread a-day, and was altogether fed with this substance. It appears that the carnivorous habits of this animal, are greatly de pendent on the circumstances of its situation, for being placed where vegetation is exceedingly scanty, if it even exists at all, and surrounded by seals, fish, &c. there can be no choice; notwithstanding, the animal is provided by nature with proper organs for the mastication and digestion of vegetable food.

The polar bear in captivity seems to suffer much from heat, which renders his confinement very uncomfortable, as is expressed by his restlessness and roaring. This is in some degree quieted by repeatedly throwing buckets of cold water over his body, which is always grateful and refreshing.

As far north as navigators have yet advanced, polar bears have been found, but their numbers evidently diminish where seals are scarce, while they are very numerous where seals are found in greatest abundance.—Near the east coast of Greenland they have been seen in large flocks, at a distance resembling sheep more than beasts of prey. On the shores of the Arctic Ocean, Spitzbergen, Greenland, and Nova Zembla, from the river Ob in Siberia, to the mouths of the Jenesei and Lena, and in the vicinity of Hudson's Bay, they are found in various degrees of abundance.

The polar bear is peculiarly distinguished from other species of this genus by the length of the body, compared with its height, by the length of the neck, the smallness of the external ears, and length of the soles of the feet; which, according to CUVIER, are one-sixth of the whole length of the animal. In the fineness and length of its pelage it also differs materially from the other species. The forehead and muzzle of the polar bear are nearly on the same line, or flat; while in the European or brown bear, they are separated by a deep depression. In the black, or American bear, the profile is rather an arched line, and in the grizzly bear it is slightly convex between the eyes. The forehead of the polar bear is flat; the European bear has it rounded. The polar bear has the head narrow and the muzzle large; the brown bear has the head large and the muzzle narrow.*

---

* The following measurements of the polar bear are given by Capt. Lyon, in the excellent and interesting narrative of his Arctic Voyage in company with Captain Parry.

Length—From the snout to the insertion of the tail, 8 ft. $7\frac{1}{2}$ in.—the head only 1 ft. 6 in.—from the eye to the ear, 10 in.—from the nose to the centre of the eye, 8 in.— of the ear alone, $4\frac{1}{2}$ in.—the tail from root to tip, 5 in.—fore claws, $5\frac{1}{2}$ in.—hinder claws, $1\frac{1}{2}$ in.—canine teeth, $2\frac{1}{2}$ in.

Girth—Round the body, 7 ft. 11 in.—neck, 3 ft. $4\frac{1}{2}$ in.—fore leg, 2 ft. 3 in.—hind leg, 3 ft. 3 in.—round the snout, 1 ft. $9\frac{1}{2}$ in.—round the forehead, 2 ft. 1 inch.

Breadth—Paws 10 in.—between the ears, 1 ft. 3 in.— canine teeth, 3 in.—[*Weight*, 1600 *lb.*]

Capt. Lyon, in consequence of having seen a polar bear prowling about during the coldest part of the year, infers that naturalists are mistaken in thinking that this animal becomes torpid during winter. We do not feel authorised to draw a similar conclusion from Capt. L.'s observation; especially as the habits of the *genus* in this respect are well known, and because the usual food of the polar bear must be extremely difficult to obtain, if it be at all accessible to the animal, during the severest part of the winter.

CHAPTER VIII.

GENUS IX. RACCOON; *Procyon*, STORR. C.

Germ. Waschthier.   *Fr.* Raton.

GENERIC CHARACTERS.

THE head is short and triangular, having the nose to project beyond the lower jaw. The tongue is smooth; the eyes not large; the ears short and oval; the body short and rather slender. The teats are six in number, and situated on the belly. The feet are five-toed, and provided with large and strong nails: the soles of the (posterior) feet are naked, but the animal does not always place their whole length on the ground in walking. The tail is long and pointed, but not prehensile. The habits of the genus are given with the description of the species belonging to it.

Dental System.

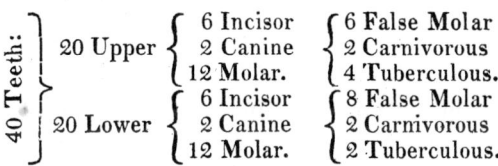

IN THE UPPER JAW we find three incisors, the two central being smaller than the external, which is slightly separated from them, longer and more coni cal. The canine tooth is more slender and trenchant than that of the dog. There are three false molars; the first touching the canine is rudimental. the second is regular, but smaller and more delicate than the third, and remarkable by the thickness of its base, and the rudiment of a tubercle. The carnivorous tooth, on its exterior surface, still presents the three characteristic divisions peculiar to this aspect of all carnivorous teeth in the upper jaw, but the internal and anterior tubercle has a considerable development, and a second tubercle arises behind this on the posterior edge, which changes it into a true tuberculous tooth. The tuberculous tooth which succeeds to the carnivorous, also presents on its external face two divisions, or the two tubercles which are observed in the analogous tooth of dogs; but like the carnivorous, it is thickened, and has on its interior (after the two external tubercles) three other tubercles placed on the same line, and separated from the first by a deep depression; in short, a fourth tubercle shows itself on the internal border of this tooth, at its posterior part, so as apparently to be a mere division of the third internal tubercle. The last upper tuberculous tooth, one-third smaller than the foregoing, and much smaller on its internal than on its external surface, seems to present the same number of tubercles, but those of the middle of the crown, instead of standing on the same line, are placed in a triangle on account of the narrowing of the part they occupy.

In the lower jaw the incisors are all similar to each other, and the external closely approaches the canine tooth. The latter is long, inclining forwards, and slightly recurved at the point; its posterior face being concave. The false molars are four in number; the first, placed at the base of the canine, is rudimental; the three others increase progressively to the last, which is thickened and extended at its posterior part. The carnivorous tooth, at its anterior part, is composed of three principal tubercles, disposed in triangles, and a small point is seen at the base of the first tubercle, and its posterior part is composed of two thick and blunt tubercles. The tuberculous tooth, nearly as large as the carnivorous, appears to be nothing but the latter reversed; that is, it has anteriorly two tubercles, one on its external and another on its internal edge, and posteriorly three tubercles disposed in a triangle.

In their reciprocal position, the relations of these teeth consist in the interlocking of their tubercles, with the intervals left between those of the other jaw.

Species I.—*The Raccoon.*

Procyon Lotor.—L.*

Ursus Lotor; Lin. Erxl. Bodd.
Vulpes Americana; Charleton.
Le Raton; Buff. Hist. Nat. 8, pl. 43.
Procyon Lotor; Cuv. Reg. An. p. 143. Sabine app. p. 649.
Coati Brasiliensium; Klein.
Mapach, etc. Mexicanorum.

There are few parts of the American continent in which the Raccoon has not, at some period, been

---

* The specific name "Lotor" was given by Linne. The removal of this species from the genus *Ursus*, by Storr, to form his *genus* Procyon, will not justify the appropriation of the *species* to the latter naturalist.

found native, from the borders of Nootka Sound to the forests of Mexico, and still more southern regions. Yet the Count de BUFFON asserts, that this animal was originally from South America, and is most numerous in hot climates, without giving any fact on which his opinion is founded, or supporting his declaration by the observations of other naturalists. Sonnini properly observes, that neither Frezier, Ulloa, nor Molina, who have given descriptions of the animals of Peru, Brazil and Chili, make any mention of the raccoon; and, in his own long and numerous journeys in Guiana, he never found one among the great number of quadrupeds which hold undisturbed possession of the vast forests, by which that interesting region is overshadowed.

But the most positive proofs of their existence, in the northern parts of this continent, are to be found in the journals of the most respectable observers. By Dampier, they were seen near the southern point of California, in the 22° of N. latitude; Bartram found them on the isle of St. Simon, near the coast of Georgia, in 30° of N. latitude; and the celebrated Capt. Cook saw them in considerable numbers at Nootka and Prince William's Sound Most probably, had this enterprising voyager landed still farther north, he would have discovered the raccoon there, as the natives of Prince William's Sound were in a great degree, clothed with skins of this animal.

Were we to form an opinion of this animal's character solely from external appearances, the mingled expression of sagacity and innocence exhibited in his aspect, his personal neatness and gentle movements, might all incline us to believe that he possessed a guileless and placable disposition. But in this, as in most other cases, where judgments are formed without sufficient examination, we should be in error, and find, that to the capricious mischievousness of the monkey, the raccoon adds a blood-thirsty and vindictive spirit, peculiarly his own. In the wild state, this sanguinary appetite frequently leads to his own destruction, which his nocturnal habits might otherwise avert; but as he slaughters the tenants of the poultry-yard with indiscriminate ferocity, the vengeance of the plundered farmer speedily retaliates on him the death so liberally dealt among the feathered victims. This destructive propensity of the raccoon is more remarkable, when we observe that his teeth are not unsuited for eating fruits. When he destroys wild or domesticated birds, he puts to death a great number, without consuming any part of them, except the head, or the blood which is sucked from the neck.

Being peculiarly fond of sweet substances, the raccoon is occasionally very destructive to plantations of sugar-cane,* and of Indian corn. While the ear of the Indian corn is still young, soft and tender, or "in the milk," it is very sweet, and is then eagerly sought by the raccoons; troops of them frequently enter fields of maize, and in one night commit extensive depredations, both by the quantity of grain they consume, and from the number of stalks they break down by their weight.

Sir Hans Sloane; Natural History of Jamaica.

The raccoon is an excellent climber, and his strong sharp claws effectually secure him from being shaken off the branches of trees. In fact, so tenaciously does this animal hold to any surface upon which it can make an impression with its claws, that it requires a considerable exertion of a man's strength to drag him off; and as long as even a single foot remains attached, he continues to cling with great force. I have had frequent occasion to pull a raccoon from the top of a board-fence, where there was no projection which he could seize by; yet, such was the power and obstinacy with which the points of his claws were stuck into the board, as repeatedly to oblige me to desist for fear of tearing his skin, or otherwise doing him injury by the violence necessary to detach his hold.

The conical form of the head, and the very pointed and flexible character of the muzzle or snout, are of great importance in aiding the raccoon to examine every vacuity and crevice to which he gains access; nor does he neglect any opportunity of using his natural advantages, but explores every nook and cranny, with the most persevering diligence and attention, greedily feeding on spiders, worms, or other insects which are discovered by the scrutiny. Where the opening is too small to give admittance to his nose, he employs his fore-paws, and shifts his position or turns his paws sidewise, in order to facilitate their introduction and effect his purpose. This disposition to feed on the grubs or larvæ of insects must render this animal of considerable utility in forest lands, in consequence of the great numbers of injurious and destructive insects he consumes. He is also said to catch frogs with considerable address, by slily creeping up, and then springing on them, so as to grasp them with both paws.

The circumstance which has procured for the raccoon the specific name of "Lotor," or the Washer, is very remarkable and interesting: this is, the habit of plunging its food into water, as if for the purpose of soaking or cleansing it. To account for this disposition, some naturalists have supposed that the raccoon is not as liberally supplied with salivary organs as other animals, and is therefore obliged to prepare its food by softening it in water. The raccoon, however, does not invariably wait to subject his food to this preparation, but frequently devours it in the condition he receives it, although it may be nothing but dry bread, and clean water be within a few steps of where he stands.

Water seems to be essential to their comfort, if not of absolute necessity for the preparation of their food. I have had for some time, and at the moment of writing this have yet, a male and female raccoon in the yard. Their greatest delight appears to be dabbling in water, of which a large tub is always kept nearly full for their use. They are frequently seen sitting on the edge of this tub, very busily engaged in playing with a piece of broken china, glass, or a small cake of ice. When they have any substance which sinks, they both paddle with their fore-feet with great eagerness, until it is caught, and then it is held by one, with both paws, and rubbed between them; or a struggle ensues for the possession of it, and when it is dropped the same sport is renewed. The coldest weather in winter does not in the least deter them from thus dabbling in the water for amusement; nor has this action much reference to their feeding, as it is performed at any time, even directly after feeding till satiated. I have frequently broken the ice on the surface of their tub, late at night, in the very coldest winter weather, and they have both left their sleeping place with much alacrity, to stand paddling the fragments of ice about, with their fore-legs in the water nearly up to the breast. Indeed, these animals have never evinced the slightest dislike to cold, or suffered in any degree therefrom: they have in all weathers slept in a flour-barrel thrown on its side, with one end entirely open, and without any material of which to make a bed. They show no repugnance to being sprinkled or dashed with water, and voluntarily remain exposed to the rain or snow, which wets them thoroughly, notwithstanding their long hair, which being almost erect, is not well suited to turn the rain. These raccoons are very fond of each other, and express the greatest delight on meeting after having been separated for a short time, by various movements, and by hugging and rolling one another about on the ground.

My raccoons are, at the time of writing this, more than a year old, and have been in captivity for six or eight months. They are very frolicsome and amusing, and show no disposition to bite or injure any one, except when accidentally trodden on. They are equally free from any disposition to injure children, as has been observed of other individuals. We frequently turn them loose in the parlour and they appear to be highly delighted, romping with each other and the children, without doing any injury even to the youngest. Their alleged disposition to hurt children especially, may probably be fairly explained by the fact above mentioned, that they always attempt to bite when suddenly hurt, and few children touch animals without pinching or hurting them. They exhibit this spirit of retaliation, not only to man, but when they accidentally hurt themselves against an inanimate body; I have many times been amused to observe the expression of spite with which one of them has sprung at and bit the leg of a chair or table, after knocking himself against it so as to hurt some part of his body.

These animals may be tamed while young, but as they grow to maturity most generally become fierce and even dangerous. I have had one so tame as to follow a servant about through the house or streets, though entirely at liberty; this was quite young when obtained, and grew so fond of human society as to complain very loudly, by a sort of chirping or whining noise, when left alone. Nothing can possibly exceed the domesticated raccoon in restless and mischievous curiosity, if suffered to go about the house. Every chink is ransacked, every article of furniture explored, and the neglect of servants to secure closet-doors, is sure to be followed by extensive mischief, the evil being almost uniformly augmented by the alarm caused to the author of it, whose ill directed efforts to escape from supposed peril, increase at the same time the noise and the destruction.

To complete the history of the raccoon in a state

of captivity, we shall insert here the greater part of a letter written by Mr. BLANQUART DE SALINES to Count de BUFFON, on the correctness of which full reliance may be placed.

" My raccoon was always kept chained before he came into my possession, and in this captivity he seemed sufficiently gentle, though not caressing; all the inmates of the house paid him the same attention, but he received them differently; treatment he would submit to from one person, invariably offended him when offered by another. When his chain was occasionally broken, liberty rendered him insolent; he took possession of his apartment, suffering no one to approach him, and was, with difficulty, again confined. During his stay with me, his confinement was frequently suspended; without losing sight of him, I allowed him to walk about with his chain on, and he expressed his gratitude by various movements. It was otherwise when he escaped by his own efforts: he would then ramble for three or four days together over the neighbouring roofs, and only descend at night into the yards, enter the hen-roosts and destroy the poultry, especially the Guinea-fowls, eating nothing but their heads. His chain did not render him less sanguinary, though it made him more circumspect; he then employed stratagem, allowing the poultry to familiarize themselves with him by partaking of his food, nor was it until he had induced them to feel in perfect security that he would seize a fowl and tear it to pieces: he also killed kittens in the same manner.

" If the raccoon be not very grateful for favours received, he is singularly sensible of bad treatment:

a servant one day struck him some blows with a stick, and often afterwards vainly endeavoured to conciliate him, by offering eggs and shrimps, of which the animal was very fond. At the approach of this servant he became enraged, and with sparkling eyes would spring towards him, making violent outcries; under such circumstances he would accept of nothing until his enemy had withdrawn. The voice of the raccoon, when enraged, is very singular, sometimes resembling the whistling of a curlew, and at others the hoarse barking of an old dog. When struck by any one, or attacked by an animal stronger than himself, he offered no resistance; like the hedgehog, he hid his head and paws, by rolling his body in form of a ball, and would have suffered death in that position. I have observed that he never left hay nor straw in his bed, preferring to sleep on the boards; when litter was given, he threw it away immediately. He did not seem very sensible to cold, and passed two out of three winters exposed to all the rigours of the season, and did well, notwithstanding he was frequently covered with snow. I do not think he was solicitous to receive warmth; during some frosts I gave him separately warm water and water almost frozen, to soak his food in, and he always preferred the latter. He was at liberty to sleep in the stable, but often preferred passing the night in the open yard."

Captivity and domestication produce great changes in the habits of this animal, as he learns to spend nearly the whole of the daytime in active exercise, and sleeps during the greater part of the night.—When inclined to sleep, the raccoon forms itself into a sort of ball, by sitting on its hind legs and doubling the head under the body, so as to apply the forehead to the ground; the tail is then brought forward so as to conceal the feet and face on one side, and the true figure of the animal is no longer discernible. In this position the raccoon sleeps profoundly, and is not startled readily, nor by slight sounds.

The size of the raccoon varies with the age and sex of the individual. A full grown male may be stated to have the body a foot, or a few inches more, in length; the highest part of the back is about a foot from the ground, while the highest part of the shoulder is ten inches. The head is about five, and the tail rather more than eight inches long. The female is larger than the male in every respect, at least such is the fact in relation to the raccoons now in my possession, which, however, have not yet attained their full growth. They are of the same age, and the female is strongly distinguished from the male by the black markings on all parts of the body being more purely black, and the fur and hair longer, thicker, and more glossy than that of the male; these peculiarities, in addition to her greater size, uniformly lead strangers to suppose this individual to be the male, instead of the female. The pelage of the male is not only less purely black at the extremities of the hairs, but there is a much greater intermixture of fawn-coloured hair than in the female, giving more of a rusty appearance to the whole surface of his body. A young raccoon of thirty days old is about the size of a common cat of a year old, though the greater length of its legs, and the bushiness of its pelage, make it at first sight appear much larger.

The general colour of the body is a blackish gray, which is paler on the under part of the body, and has over considerable part of the neck, back and sides, some fawn or light rust-coloured hair intermixed. The general gray colour is owing to the manner in which the hairs are alternately ringed with black and dingy white. The tail is very thickly covered with hair, and is marked by five or six black rings around it, on a yellowish white ground.

The head, which is about five inches long, is very triangular, and from its pointed snout reminds us of the aspect of the fox: the snout terminates in a smooth and shining black membrane, through which the nostrils open, having the slit to rise slightly at the sides. The nose is prolonged considerably beyond the upper jaw, and this, together with its great flexibility, gives the animal great advantages in exploring little crevices and crannies for insects, &c. The pupils of the eyes are round; the ears are oval, or rather elliptic, and of a yellowish white colour on their extremities and anterior edges. The face is whitish in front, but there is a black patch surrounding the eye, that descends entirely to the lower jaw, over the posterior part of which it is diffused, and a black line running from the top of the head down the middle of the face, ending below the eyes. The rest of the hair between the eyes, the ears, and eyebrows, is almost entirely white, and directed downwards. The hair on the muzzle is usually very short; on the feet also, and on one half of the legs; the short hair of the feet and legs is of a dirty whitish colour. The whiskers on the upper lip are long and strong.

All the feet have five toes each, terminated by strong curved and pointed claws; and each foot is furnished with five thick and very elastic tubercles beneath. The first toe or thumb of the fore-foot is the shortest of all; the little or external finger is next in length, and then the fore-finger: the remaining two are equal. The first tubercle, which is a very strong one, is situated near the wrist; the second is at the base of the little finger; the third at the root of the inner finger or thumb; the fourth opposite the second digit, and the fifth opposite the two longest. The hind feet are throughout similar, except that the first tubercle is farther distant from the heel.

The raccoon has a gland on each side of the anus, which secretes a strong scented fluid; but this does not impart an unpleasant smell to the animal. Its liver has five lobes, and is provided with a large gall-bladder; the bowels have no cœcum, and the stomach, which is situated entirely on the left side, is elongated and small, compared with the size of the animal.

The pelage of the raccoon is subject to considerable variations of colour at different periods of life, and in different individuals. The rings on the tail and the patches around the eyes are, however, uniform and constant. The tail of the raccoon is not affected by the coldest weather; hence this quadruped is never known to gnaw his tail, as has been observed of animals closely allied to it in configuration and habits.*

---

* This is especially the case with the *coati* or coatamundi of South America, and it has been considered very wonder-

THE RACCOON. 175

The fur of the raccoon forms an article of considerable value in commerce, as it is largely employed in the fabrication of hats. Vast numbers of raccoon skins are collected by the different fur companies; and we occasionally see in our furrier shops, skins which must have belonged to individuals of much larger size than those from which the measurements have been hitherto taken.

Raccoons are found throughout the whole of North America, and they still continue to be numerous in many of the well peopled parts of the United States. Occasionally their numbers are so much increased as to render them very troublesome to the farmers in the low and wooded parts of Maryland, bordering on the Chesapeake Bay. Their season of sexual intercourse begins in the first week of March: the female usually produces two or three cubs at a litter; her den is then made in some hollow tree or very secure situation.*

---

ful that the animal should *eat its own tail*, which certainly *appears* to be the fact. The extreme length of its tail, in which the blood circulates but feebly, exposes it to the influence of the cold or frost; and the exceedingly tormenting irritation produced thereby, leads the animal to gnaw and scratch the tail to relieve the excessive itching. The disease spreads, and the anguish induces the coatamundi to gnaw more furiously, and eventually his life is destroyed by the extension of the inflammation and irritation to the spine, &c.

* Os peni inest, leviter versus glandem curvatum; testiculi et caput penis, tempore amoris incipiente, notabiliora pendentioraque deveniunt. Fœminam contra terram vel aliquod durum, frequentissime genitalia fricare notavi; profecto et marem aliquando, simili modo, sese diligenter agitare vidi.

CHAPTER IX.

GENUS X. BADGER; *Meles*, BRISS.*

Gr. Μελις.
Lat. Taxus, Meles.
Ital. Tasso.
Swed. Graf-Svin.

Fr. Blaireau.
Sp. Tassugo, Texon.
Ger. Tachs, Dachs, Dar.
Scot. Brock, Tod, Pate, &c.

GENERIC CHARACTERS.

THE head is conical and the muzzle elongated; the ears are rounded and the eyes small. The body is robust and the limbs comparatively short, the digits being all covered by the integuments, as far as to the roots of the claws, which on the fore-feet are long, and admirably adapted for burrowing. The teats are six in number, two of them are placed on the lower part of the chest, two on the belly, and

---

* DESMAREST has incorrectly quoted LINNÉ as having appropriated the name "*Taxus*" to this genus. In this, as in almost every other inaccuracy, the French naturalist has been servilely copied in this country.

LINNÉ made *Taxus* a species of his genus *Ursus*, [see ed. 6, genus 10, sp. 1.] and it occupies a place in the same genus in Gmelin's edition of the Systema Naturæ, p. 102, vol. i. The term *Taxus* is preoccupied in Botany; we have adopted the name given to this genus by BRISSON, which has the advantage of being the same as that used by Varro and other Latin writers for the only species of badger then known. This name of the genus, in some of the books, is inaccurately attributed to STORR. [See *Ranzani*, Elem. di Zoologia, ii. p. 249.]

Drawn by C. A. Lesueur.   Eng.d by G. B. Ellis.

1. Raccoon.   2. American Badger.

THE BADGER.

two in the groin. The most remarkable character which distinguishes this genus, is a pouch situated beneath the tail, whence an unctuous and fetid substance is discharged.

Dental System.

36 Teeth:	16 Upper	6 Incisive 2 Canine 8 Molar.	4 False Molars 2 Carnivorous 2 Tuberculous.
	20 Lower	6 Incisive 2 Canine 12 Molar.	8 False Molars 2 Carnivorous 2 Tuberculous.

The dental system of the badger is very analogous to that belonging to the genus *mephitis*, or skunk, hereafter to be described; differing from it, however, by some modifications of the carnivorous and superior tuberculous teeth.

IN THE UPPER JAW the incisors and canine teeth are similar to those of the martens, which we shall treat of in another part of this work. The false molars have the regular forms of such teeth. The carnivorous tooth, (remarkable by its small size, owing to the diminution of its posterior part, which gives it when externally viewed the appearance of a false molar,) is composed at the internal part of a base furnished with three small tubercles separated by a perceptible depression. The tuberculous tooth is disproportionately large, and is as broad as it is long; having on its external edge three tubercles, on its internal edge a serrated crest, and on its middle another crest, separated into two principal parts by a slight groove.

IN THE LOWER JAW the incisors and canine teeth offer nothing remarkable. Of the four false molars, the first is rudimental and has but one root, the three others have the regular forms of such teeth. The carnivorous, at its anterior part, is composed of three tubercles, (as in the skunk and others,) but its posterior part, besides two tubercles, has a spur terminating in a serrated crest. The tuberculous is a small rounded tooth marked on its surface by some irregular projections and depressions.

In their reciprocal position, the principal character consists, as we have seen, in the lower carnivorous and upper tuberculous teeth; thus the relations existing between these teeth are more extended. The two first tubercles of the superior carnivorous are in relation with the posterior edge of the opposite carnivorous; (this is the carnivorous portion of this dental system;) the extremity of the first of these two tubercles fills the hollow separating the three small tubercles of the enlarged base, which belongs to the inner face of the upper carnivorous. The whole remainder of the inferior carnivorous is opposed to two-thirds of the superior tuberculous tooth. The last third corresponds with the lower tuberculous tooth.

From this dental system it is evident that the badger is an animal partly frugivorous, and the arrangement for triturating vegetable substances much exceeds that for the mastication of flesh.

Species I.—*American Badger.*

Meles Labradoria.—Sabine.

Ursus Labradorius: L. Gmel.
Carcajou: Buff. Sup. 3, p. 49.
Braro: Lewis and Clark, vol. ii. p. 177.
Meles Labradoria: Sabine, App. to Franklin's Journey, p. 649.

The American Badger has been for a long time known to naturalists, though it is but recently established as a species distinct from the badger of Europe. By some of the European naturalists and compilers, our badger has been considered as a mere variety of the European species, while by others it has been regarded as entirely the same; the fact was never fairly decided, until the publication of Sabine's Appendix above cited. Say had, however, arrived at the same conclusion, and applied nearly the same name to it in the the journal of Long's Expedition to the Rocky Mountains, previous to the appearance of Sabine's observations.

" Schreber was the first author who considered the American to be a distinct species, and Gmelin adopted this conclusion in his edition of Linné, though he was led to give the incorrect specific character of " palmis tetradactylis," in consequence of Buffon's statement, that the carcajou had but four toes on its anterior feet."*

Nature has destined this animal to a subterraneous and solitary mode of life, which, together with its timid disposition and nocturnal habits, throw great

---

* Sabine.

difficulties in our way while endeavouring to ascertain its peculiarities. It is entirely inoffensive, and apparently feeble, but if denied the advantages of swiftness of motion or great size, it has not been left entirely destitute of the means of providing for its own safety. The long claws on its fore feet are admirably adapted for removing the earth, and the celerity with which it can escape from danger, by burrowing, is really surprising. It is altogether fruitless to attempt to secure the animal by digging after it, as its progress is too rapid, and the depth to which it descends too great. It is only by artifice that the badger can be brought from its retreat: this is effected by the aid of dogs, smoke, &c. and when driven to the last extremity, the strength of its jaws and the sharpness of its teeth, enables the animal to inflict the severest injury on its persecutors. The body of the badger is thick and heavy, and its movements on the ground slow and creeping; there is little appearance of vivacity or intelligence in its aspect, yet it does not exhibit any appearance of dulness or stupidity. It is in fact endowed with exactly the degree of understanding which is suited to its peculiar condition: having the proper instruments for securing itself from ordinary enemies, as well as strength and courage enough to defend itself when pressed, but little sagacity is necessary to enable it to obtain the requisite food, or to continue its kind. Neither should we indulge in reflections similar to those found in many books of natural history, and believe that the life of this animal is gloomy or wretched. To men it may appear gloomy or dreadful to live under ground, or to steal forth under cover of the night in search of food; but this is the only mode of life the badger is susceptible of enjoying, and the only kind of action he is capable of.

The burrows of the badger are deep and extensive, and several individuals have been found inhabiting one excavation. Within his subterraneous retreat he passes the day in sleep, and it is not until night veils all objects in shade that he comes forth to seek his subsistence. Then, fruits of different sorts, frogs, insects, and most probably any small animals to be procured, constitute his food.

The badger has its young in summer, and generally two, three, or four, at a litter, which are occasionally brought out to the mouth of the burrow to enjoy the sunshine. The young become capable of procreating when two years old, and the period of their lives is extended to ten, twelve, or fifteen years. If taken when young, the badger is easily tamed, soon becoming quite familiar and obedient.

The American badger is a pretty little animal, and its aspect is not unlike that of some small pug-faced dogs. It is found most frequently on the plains adjacent to the Missouri and its tributaries, as well as on those near the Columbia river. It is not uniformly found in the open country; Lewis and Clark sometimes observed them in the woods.

This animal is about two feet five inches long, including the tail, which measures three inches, and its body appears long in proportion to its thickness. The fore and hind legs are short, but remarkably muscular, the fore paws are provided with the long claws peculiar to this genus. which gives them the means of burying themselves with great celerity even in a hard soil. The neck is short and the mouth wide; the eyes are black and small; the ears short, wide, and appearing as if a portion had been cut off them. The whiskers are arranged in four points on each side near the nose, and on the jaws close to the opening of the mouth. The hairs are much shorter on the sides and rump than on other parts of the body, which imparts an appearance of flatness, especially when the badger rests upon its belly.— The length of the hair is upwards of three inches, especially upon the rump, whence it extends so far towards the extremity of the tail as to conceal it entirely, and gives to the whole of the posterior parts of the body, " the appearance of a right angled triangle, of which the tail forms an acute angle." Intermixed with the hair we find a small quantity of coarse pale reddish yellow fur.

The American badger differs from the European by generally being smaller and more slenderly formed; its head is full as long, but not so pointed towards the nose; neither is the profile at all similar to the badger of Europe. In the European animal, the outline drawn from the forehead to the nose is quite straight, while in the American there is a considerable depression on a line with the eyes.

There is also a very striking difference between the markings of these animals. In the American badger there is a narrow white line running from between the eyes towards the back; the remainder of the superior part of the head is brown, the under jaw and whole of the throat are white. A semicircular brown spot is seen between the ears and the light coloured part of the cheeks. Above the eyes the white marking extends triangularly for a short distance, and below it runs in a line with the eyes towards the fore part of the mouth; yet the whole eye is within the dark colour of the upper part of the head, and this colour runs at the corner of the eye, with an acute angle, into the white.

The badger of Europe has three broad white marks, one on each side and one on the top of the head, between which there are two broad black lines, including the eyes and ears. All the parts under the throat and jaw are black. The hairs on the upper part of the body and sides of the American badger are fine, long and grayish; in the European the hairs on the same parts are darker, longer, and coarser. In the American, the under are lighter than the upper parts; in the European they are darker. In our animal the legs are of a dark brown; in the European quite black. Notwithstanding the European badger is generally the largest, its dark coloured nails are smaller than those of the American, which are of a light horn colour. The tail of the American badger is shorter than that of the European.*

The American badger weighs from fourteen to eighteen pounds.

---

* For the details of difference in the markings of the two species, we are indebted to Capt. SABINE's Appendix, above quoted. We have no specimen of the European badger in the collection of the Philadelphia Museum.

CHAPTER X.

Genus XI. Glutton; *Gulo.* Storr.

Ger. Vielfrass; Rosomak. *Russ.* Rosamaka; Rosamak.
Fr. Glouton. *Swed.* Järff; Filfras; Snop.

GENERIC CHARACTERS.

The head is but moderately elongated, and the body is long in proportion to its distance from the ground. The ears are very short and rounded; the tail is intermediate in length to the tails of the badger and raccoon. There is a simple fold of the integuments below the tail, instead of the pouch observed in the badger, to which animal the movements of those belonging to the present genus bear some resemblance. The feet are five-toed, the toes being distinctly separated and armed with hooked claws. The soles of the (posterior) feet are capable of being applied either wholly or partially to the ground.

Dental System.

36 or 38 Teeth:
- 16 or 18 Upper { 6 Incisive, 2 Canine, 8 or 10 Molar. } { 6 or 8 False Molars, 2 Carnivorous, 2 Tuberculous.
- 18 or 20 Lower { 6 Incisive, 2 Canine, 10 or 12 Molar. } { 6 or 8 False Molars, 2 Carnivorous, 2 Tuberculous.

The dental system of this genus offers nothing peculiar in the different sorts of teeth to distinguish them from those hereafter to be described as proper

1. Wolverene. 2. Fisher.

to the mustelæ or martens, between which and this genus there is a very close affinity. This genus may be considered as intermediate to the true plantigrade and digitigrade animals.

Species I.—*The Wolverene.*

Gulo Luscus. L.*

Gulo Luscus: Sab. app. p. 650.
Ursus Gulo: Pal. Spicil. 14, 125, pl. 2. Schreb. pl. 144.
Glouton: Buff. Sup. 3. pl. 48.
Meles Gulo: Bod. Elench. Anim. p. 80, sp. 5.
Ursus Luscus: L. Gmel.
Ursus Freti Hudsoni: Briss. quad. 188.
Quickhatch, or *Wolverene:* Edw. av. 2. p. 103. pl. 108.
Wolverene; Penn. quad. p. 195, no. 140, pl. 20, fig. 2.
Gulo Arcticus: Desm. Mammal. p. 174.
Quickhatch: Catesby's Carolina, app. xxx.
Carcajou: [so called by the Canadians.]

The Wolverene has served as a fruitful theme for exaggeration and fiction, which has continued the longer in proportion to the remoteness of the animal, and the difficulty of ascertaining its real manners. It is true that ferocity and destructiveness are among its most striking characteristics, and it is known to feed ravenously and fully when it has secured its prey, yet in none of these respects is the wolverene different from numerous other animals, nor is it at

---

* The specific name of "*Luscus*" was bestowed on this species by Linné, who arranged it in his genus *Ursus*, in consequence of its plantigrade character. Though it is now removed to Storr's genus *Gulo*, we believe it improper to withdraw the credit of the specific title from Linne since, notwithstanding the change stated, his specific name must always be continued.

all the prodigy that book-makers have heretofore represented it to be.

The wolverene inhabits the northern parts of America generally, quite to the Arctic Sea, and it is probable that its visits extend beyond the continent towards the Pole, as a skull of this animal was found on Melville Island by Capt. Parry. It is an inhabitant alike of the woods and barren grounds, and is capable of enduring the severest cold. The motions of the wolverene are necessarily slow, and its gait heavy, but the acuteness of its sight and power of smelling are an ample compensation; as they are seldom or never killed without being found fat, there is good reason for believing that they rarely suffer much from hunger. This animal is surprisingly strong, and an overmatch for any quadruped near its own size;—indeed its sharp claws and teeth enable it to offer a very effectual resistance even to the bear.

The strength of the wolverene, joined to its great gust for animal food, causes much trouble to hunters and travellers who attempt to secure provisions by burying them in the snow, or protect them by coverings of boughs and trunks of trees. It is almost impossible to prevent this creature from finding access to such places of deposit either by strength or stratagem, and destroying the stock on which the voyager may have counted for his future subsistence and safety. To the hunters the wolverene is also very injurious, by robbing their traps of the animals which are taken in them, before the arrival of the owners. The wolverene is fierce and dauntless, and has been seen to take away from the wolf the carcass of a deer, and when itself engaged in feeding, has refused to move though warned of the approach of an armed hunter, who shot it while standing as if prepared to maintain its prize.

It is stated in all the books of natural history, that this animal is in the habit of ascending trees for the purpose of leaping down upon the necks of rein-deer and other similar animals; and that it has sagacity enough to carry with it into the top of the tree some of the moss of which the deer are fondest, and drop this immediately under it, so as to secure the intended victim, by placing it in the most favourable position for being leaped on. When the deer approaches to pick up this moss, the watchful glutton is said to drop from his perch upon the neck of the animal, drive his crooked claws into the flesh, fasten himself firmly, and from some deep wound to drink the blood of the unfortunate deer, until exhaustion and death is produced. Such relations are so frequently repeated of this animal that they have long ago ceased to be doubted, and it may seem like supererogatory scepticism to doubt on the subject at present. Thus much, however, it is due to truth to state, that we have examined with some interest the authorities originating such accounts of the sagacity or instinct of the wolverene, and have not been able to find any thing more satisfactory than mere assertions relative to the European glutton. It is not objected here that these assertions are unfounded, but they are gratuitous, at least as applied to the wolverene or American glutton, since HEARNE and other travellers residing in the regions where this animal is most abundant, make no mention of any such thing concerning it. The necessity of scepticism relative to the habits of the wolverene becomes the more obvious, when it is recollected how much of what was formerly believed as unquestionable, has been proved to be fable, resting on nothing better than the fancy of Olaus Magnus.*

The regions inhabited by this animal are supplied most abundantly with small quadrupeds, and with birds as well as with the larger animals, so that it is quite probable that without any very great exercise of ingenuity it is capable of procuring a plentiful subsistence. When taken captive and retained in confinement, its disposition does not seem by any means untameable, nor is its voracity especially remarkable.

Nothing however is better ascertained, than that the wolverene is one of the most destructive animals found in the northern part of this continent. It destroys great numbers of young foxes during summer, while they are small, discovering their burrows by its keen scent, and, if necessary, enlarging the cavity so as to gain access to the bottom of the den, where the mother and cubs are speedily destroyed. —The wolverene is not less inimical or destructive to the beaver than other animals, though the habits of the beaver expose it less to this sanguinary quad-

---

* Hoc animal voracissimum est, reperto namque cadavere tantum vorat ut violento cibo corpus instar tympani extendatur: *inventaque angustia inter arbores se stringit ut violentius egerat*: sicque extenuatum revertitur ad cadaver et ad summum usque repletur, *iterumque se stringit* angustia priore!" Ol. Mag. Hist. de Gent. Septentrion.

ruped, which is generally successful in securing his prey only when the beaver is caught at any distance from the margin of the water.

The fur of the wolverene is of considerable value, and the natives of the northern parts of Asia highly esteem the skin of the glutton for making or ornamenting their robes. The skin of the wolverene is not so highly valued by the Indians, nor by the fur traders. The animal does not breed in sufficient numbers to furnish any great collection to the fur traders, and but few skins are sent by the companies to the merchants.

The wolverene is about two feet two inches long from the end of the nose to the origin of the tail, and the latter is about eight inches in length, if the hair on its extremity be included; without measuring the hair, the length of the tail is but four inches. The fore legs are upwards of eleven inches long, and the hind legs one foot. The face is blackish as high as the eyebrows, and between these and the ears we find a space of a whitish and brownish tint. The ears are covered with harsh hair, the lower jaw and inside of both fore legs are spotted with white; the upper part of the back, thighs, and the under part of the belly are brown, or brownish black. The sides are of a fine chesnut colour, from the shoulders to the beginning of the tail. There is a white spot over the navel; the parts of generation are reddish. The short hair of this animal is whitish. The eyes are small and black.

There is a small tubercle at the under part of each digit, and four others under the palm, forming a semicircle with another posterior tubercle; the hind feet have a similar arrangement, but have no tubercles at the heel, which are slightly raised from the ground in walking.* There is considerable difference in the markings of the skins brought from Hudson's Bay, some being darker than others, and some having the band of lighter hairs, which runs along the sides and over the back and tail, very obscure. In numerous individuals greater variations may be found, owing to the circumstances of age, state of pelage, &c.

---

* See Desmarest, Mammalogie, p. 174.

CHAPTER XI.

Tribe II.—Digitigrada; *Digitigrade Animals.*

The animals belonging to this tribe are characterized by moving on the extremities of their digits, and being endowed with a greater degree of agility than that possessed by the plantigrade tribe.

The first subdivision of this second tribe has but one tuberculous behind the upper carnivorous tooth. The length of their bodies, and the shortness of their legs, enable them to pass through very small openings with facility, and their vermiform appearance has procured for them the general appellation of vermin. They are small and weak creatures, but are extremely sanguinary and destructive, living in a great degree on the blood of their victims. They have no cœcum, and do not become torpid during the winter.

Genus XII.—*Mustela;* L. Marten.

generic characters.

A slender vermiform and very long body, the back of which is convex when the animal is in a state of repose. The head is small, oval, and apparently flattened above, having short and rounded external ears. The limbs are very short and five toed, the digits being armed with very sharp and crooked nails; the tail is of a middling length. There are some small glands on each side of the anus, which secrete a very fetid fluid, of a powerful musky and unpleasant odour. The teats are situated upon the belly.

Dental System.

34 or 38 Teeth:			
	16 or 18 Upper	6 Incisive 2 Canine 8 or 10 Molar.	4 or 6 False Molars 2 Carnivorous 2 Tuberculous.
	18 or 20 Lower	6 Incisive 2 Canine 10 or 12 Molars.	6 or 8 False Molars 2 Carnivorous 2 Tuberculous.

IN THE UPPER JAW we find three incisive, one canine, two or three false molars, one carnivorous and one tuberculous tooth. The incisive and canine offer nothing remarkable, except that their internal lobe is very small. The first false molar is a very small tooth with a single root; the crown of this is terminated by a very blunt point, and the tooth is rudimental. The two succeeding teeth have several roots, thin from without inwards, broad from before backwards, and very pointed. The first is rather smaller than the second, and they are both regular. The carnivorous does not differ materially from that of the cat, except that the internal tubercle is more distinct, and the middle part larger and more acute.

IN THE LOWER JAW there are three incisive teeth, one canine, three or four false molars, one carnivorous and one tuberculous tooth. The false molars come immediately after the base of the canine.—The first is rudimental and has but a single root, the three following have two roots, and the form of regular false molars, and are placed somewhat obliquely in the jaw. The carnivorous tooth resembles that of the cat, except in the spur, which is developed at its posterior part. The tuberculous is small and round, its crown being terminated by three small points.

In their reciprocal position, the relations of these teeth are nearly the same as those we have heretofore examined. All the difference is, that the anterior part of the lower tuberculous rests against the posterior portion and internal part of the upper tuberculous tooth; the great development of the internal tubercle of the inferior carnivorous, corresponds to the superior tuberculous tooth.

Species I.—*The Ermine Weasel.* *

Mustela Erminea.—L.†

Mustela Candida sive animal Ermineum recentiorum; Ray. Syn. quad. p. 198.
Hermellanus; Charlet. Exer. p. 20.
L'Hermine; Buff. t. 7, p. 29, fig. 2.
The Ermine; Penn. quad. ii. p. 35.
Stoat Ermine; Sabine, App. p. 652.

Among the small quadrupeds inhabiting this continent, few are to be found equalling the ermine in beauty—perhaps none that excel it in the qualities of courage, graceful celerity of movement, and un-

---

* *Germ.* Hermelin, Hermelinwiesel; *Dutch*, Hermyn; *Swed.* Sekat; *Norw.* Roskat; *Russ.* Gornostai; *Fr.* Rosselet, Hermine; *Ital.* Armellino, Ermellino.

† The beautiful animal generally known throughout North America as the *weasel*, or *common weasel*, and considered by naturalists as the common weasel of Europe (*m. vulgaris.*)

tiring activity. Its whole aspect inspires the beholder with an idea of its character, which is well supported by its actions. The long and slender body, bright and piercing eyes, keen teeth and sharp claws, clearly show that, however diminutive the animal may appear, it is destined by nature to destroy other creatures more numerous and less powerful than those of its own race; this length and slenderness of body is accompanied by a peculiar degree of flexibility, and by a strength of limb, which, in so small an animal, may be fairly esteemed surprising. There is scarcely an opening through which its prey can enter, where the ermine cannot follow, and having once gained access, its instinctive destructiveness is only allayed when no other victim remains to be slaughtered.

In the northern parts of this continent, and the northern portions of Asia, the ermine is found in the greatest abundance; yet it is by no means limited to northern regions, since it is found throughout a vast expanse of country, reaching from the highest northern latitudes to the middle states of the Union. In the middle and eastern states it is most generally known as the *weasel;* farther north it is called *stoat* in its summer, and *ermine* in its winter pelage of pure white.

The habits of the ermine weasel are very analogous to those of the common weasel of Europe, and as its

---

has, by recent examination and comparison, been proved to be the ermine in summer pelage. For this interesting observation we are indebted to the researches of that assiduous cultivator of Natural Science, CHARLES L. BONAPARTE

THE ERMINE WEASEL.

general configuration is so nearly similar, it is not surprising that this animal should have been confounded with the European species. This weasel frequents the barns and outhouses of plantations, and its retreat is generally well secured beneath the floors or rafters, amid accumulations of timber or stone, or in similar situations. Mice, and various other depredators on the granary, are the special objects of its pursuit, and the rapid multiplication of many of these devourers of grain could scarcely be sufficiently restrained, were it not that the ermine is capable of tracing them throughout their labyrinths, and possesses the disposition to destroy all that come within its reach. If the efforts of this weasel were confined to the destruction of these little depredators, we might consider it as the best friend to the husbandman; but occasionally a contribution is levied on the hen-roost, and the morning's light exhibits an universal slaughter of the poultry, whose throats are cut, or heads eaten off. It is scarcely possible to prevent such occurrences when these animals are resident in the vicinity, as they can gain access where few other creatures can enter; then their swiftness of motion and keen bite soon render the escape of their victims impossible.

Still it must be acknowledged that there are many situations in which the services of this little animal may be esteemed a positive good; for such is the fecundity of many of the depredators on the grain, that nothing short of the destruction of the whole crop would ensue, were it not that the weasel is continually thinning their ranks and killing greater numbers than are required for its mere subsistence.

The disposition which makes this weasel so useful under ordinary circumstances, forbids an attempt to increase its usefulness by domestication, for the purpose of freeing our houses from mice, &c. Notwithstanding it might be so far tamed as to take up its residence about our dwellings, it would be exceedingly dangerous to expose the lives of the inmates to the blood-thirstyness of this quadruped, which is rendered doubly dangerous from the circumstance of seeking its prey during the hours devoted by man to sleep.

The ermine is very common in the vicinity of Hudson's Bay, yet it is found in greater abundance on barren grounds or open plains than in the woods, which in all probability is owing to the greater number of mice that frequent the former situations.

While pursuing their prey, ermines are said to resemble little hounds running upon a trail; their tails are carried horizontally, while with eager haste and most agile movements they follow their prey by the scent. Except when in their summer dress, it is very difficult to distinguish their actions, as in winter their pure white pelage is so nearly the colour of the snow, as to render it almost impossible to see them. When the ermine is hunted and closely pursued, like other species of this genus, it has the faculty of ejecting from a peculiar glandular apparatus a fluid of a powerful musky odour; this, though it may serve to retard the pursuit of some of its enemies, is too harmless a resource to save the ermine from the hands of man.

There is but little probability of taming the ermine unless it be captured very young, and even then the period of its mildness would pass away with its early youth. When caught in a trap and subsequently kept in a cage, it exhibits every sign of the most unappeasable disposition to kill or injure every being it is able to master. Various attempts have been made to domesticate the ermine, but all without success, and frequently the restlessness and impatience of the animal has appeared to increase with the duration of its imprisonment.

The following interesting account of an attempt to tame the ermine is from the pen of Capt. R. LYON, of the British navy, (to whom we owe many of the facts here stated,) whose excellent observations, conveyed in a delightful style, while they impart the purest satisfaction, always awaken our regret, that his opportunities of studying from living nature were not more ample, in order that our instruction and pleasure might have been extended in the same degree.

"In the night my servant caught in a small trap a very beautiful ermine, and I had soon a convenient cage made, for perhaps the first of these animals which was ever caught on board a ship four hundred yards from the land. He was a fierce little fellow, and the instant he obtained day-light in his dwelling, he flew to the bars and shook them with the greatest fury, uttering a shrill passionate cry, and emitting the strong musky smell which I formerly noticed. No threats nor teazing could induce him to retire to the sleeping place, and whenever he did so of his own accord, the slightest rubbing on the bars was sufficient to bring him out to the attacks of his tormentors. He soon took food from the hand, but not until he had used every exertion to reach and bite the fingers which conveyed it; this boldness gave me hopes of being able to keep my little captive alive through the winter; but he was killed by an accident in a few days."*

We have mentioned that in the eastern and middle parts of the United States the ermine weasel frequents out-houses, stone-heaps, piles of timber, &c. and though capable of following its prey into small holes, does not burrow in the earth. Captain Lyon had an opportunity of observing a singular kind of burrow made by this animal in the snow, resembling the elevations of the soil produced by the passage of a mole. These galleries in the snow were serpentine in their direction, and in the neighbourhood of the hole or residence the circles were multiplied, as if to render the approach more intricate.

The ermine weasel, in its summer pelage, is of a light ferruginous or chesnut-brown colour over the whole of the head; this colour extends in a rounded spot below the angle of the jaw; the whole back, sides, and half of the tail next the body being of the same colour. The other portion of the tail is blackish, becoming gradually darker as it approaches the extremity, where it is quite black, and the hairs terminate in a point resembling that of a camel's hair pencil. The external and anterior half of the fore legs are of the same colour as the upper part of the body, and there are three small spots of white over the base of the toes of the right foot, and one on the left,

---

* Lyon's Narrative, &c.

over the first or shortest digit, in the specimen before me.

The under part of the animal is nearly of a pure white, beginning at the extremity of the under jaw and spreading broadly as it passes over the throat, where it forms a point on each side, almost reaching to the base of the ear. The white then narrows slightly in descending the neck, spreads broadly upon the breast, and then suddenly growing narrower passes down the inner and posterior part of the fore legs. Thence it passes along the belly, where it is again narrowed, and then spreading out widely at the groin, it terminates at the upper and anterior part of the thigh, becoming visible for a short distance on its outside.

The fur in summer is short, soft and silky to the touch, not varying perceptibly in length except on the snout, where it is quite short, and covering the digits of the fore and hind feet, where it is rather longer than on the other parts, and conceals the nails entirely. On the tail the hairs are longer and coarser than on the rest of the body, though still soft.

The ermine weasel, in its winter pelage, is of a pure white over the whole head, body and limbs; half of the tail to its extremity only retaining its black colour. This white colour is so pure in the northern regions as to render it almost impossible to distinguish these animals upon the snow, when the ends of their tails are not in sight. The whiteness is not always thus pure, but the fur is slightly tinted with pale yellow on the tip.

The ear of the ermine weasel is broad at its basis, and the orifice leading to the internal ear large: the ears are not covered with fur on their posterior surface, but by a very short down. On the superior and anterior part of the external ear, there is some hair of considerable length growing from that part of the ear which would correspond with the helix and anti-helix of the human ear, and almost covering the concha. The eyes of this animal, are small and black, yet prominent, clear and lustrous.

The fur of the ermine becomes longer, thicker, and finer in winter than in summer; this effect seems to be a general consequence of rigorous seasons on all animals, without reference to the permanence or mutability of their colouring.*

Species II.—*The Pine Marten.*

Mustela Martes; L.

Mustela Martes; L. Gmel. p. 95.
Mustela Abietum; Raii. quad. p. 200.
Martes Gutture Luteo; Agric. An. Subter. 485.
La Marte; Buff. vii. 185, pl. 22.
Baum Marter; Klein, quad. p. 64.
Yellow Breasted Marten; Penn. quad. ii. p. 42, No. 244.
Pine Weasel; Penn. quad. ii. p. 42, No. 244.
Pine Marten; Sabine, Zool. App. to Franklin's Exp. p. 651.

The pine forests of the northern parts of this continent are inhabited by vast numbers of these martens, and the common name by which the species is known is derived from the preference they show for the pines, in whose lofty tops they reside. This species is also found in Northern Asia and in Europe; its fur is highly esteemed wherever to be ob-

* See Sabine, p. 653.

1 Sable. 2 Weasel. 3 Pine Weasel.

tained, and the numbers of the pine marten may be fairly inferred from the vast amount of skins annually collected. In the year 1743 the Hudson's Bay company sold 12,370 good, and 2,367 damaged skins of the pine marten; in the same year the French sent from Canada 30,325; if to these be added the numbers consumed in America, and those rejected by the traders when finally arranging their packages, we must believe that this species is one of the most numerous belonging to our continent.

The pine marten very closely resembles the common European marten, (M. Foina,) and the two species may be readily mistaken for each other when inattentively observed. But they may be distinguished with facility by remarking that the head of the pine marten is not so long as that of the European, neither is its body as large. The pine marten has longer legs, finer, thicker and more glossy fur, and the throat is marked by a broad yellow or orange coloured spot, while the same part in the European marten is white.

The pine marten resembles the ermine weasel in habits and disposition, destroying great numbers of small quadrupeds and birds, but shows no disposition to approach the habitations of man. The pine marten frequently has its den in the hollows of trees, but very commonly takes possession of the nest of some industrious squirrel, which it enlarges to suit its own convenience, after putting the builder to death. It *is said* to feed occasionally on fruits, berries, honey, &c., but, with the exception of the last substance, we should have strong doubts of the statement; when very much pressed by hunger, this marten may feed on " fruits, berries, &c." but it is too exclusively carnivorous to use such food if animal matter is to be obtained.

The pine marten is of a brilliant fulvous brown colour over the whole of the body, with the exception of the throat and anterior part of the breast, which are of a yellow or orange hue, varying in depth in different individuals. The general colour of the fur is owing to the intermixture of two sorts of hair, one of which is longer and coarser, of an ash colour near the body, then of a clear fawn colour, and ending at the tip with a brown mixed with bright red: the other is a soft, fine, downy hair, slightly coloured with white and pale brownish yellow. The end of the snout is of a blackish brown colour, and the legs and tail are of the same, having little or none of the yellow brown. The margins and internal surface of the ears are covered by a whitish yellow fur.

The pine marten is most frequently obtained in its summer pelage, which is neither so brilliant, nor of the silky fineness which it possesses in winter. The colour becomes paler, particularly about the head; the yellowness on the throat cannot be easily distinguished from that on the rest of the body, as it changes to a dingy white, which runs into and becomes blended with the lighter brown of the surrounding parts.* The fur of this marten is extensively used in the manufacture of hats, and is most generally preferred for ornamenting and increasing the warmth of winter dresses.

Sabine.

The length of the body of this animal is about eighteen, and that of the tail about ten inches.—There is a very marked difference in the size of the sexes, the male being one-third larger than the female. Their season of sexual intercourse, according to Linné, is the month of February, and in December the female brings forth seven or eight young; hence there is not much probability of the species being speedily extinguished. notwithstanding the vast numbers annually killed for the sake of their skins. A little care on the part of the hunters to avoid destroying animals during their breeding seasons, would in all cases tend to secure the permanence of their sources of profit.*

Species III.—*Pennant's Marten.*

(*Commonly called the* Fisher.)

Mustela Pennanti; Erxl.

Mustela Pennanti; Erxl. Syst. Mam. p. 470, sp. 10.
Mustela Melanorrhynca; Bodd. Elench. an. p. 88, sp. 13.
Mustela Piscator; of various authors.
Fisher Weesel; Penn. Hist. Quad. Ed. 3. p. 50. No. 246. Arct. Zool i. p. 94.
Fisher; Sabine, App. to Franklin's Exped. p. 652.
Wejack; Hearne, Journey, &c. 8vo. ed. p. 378.

The impropriety of giving to animals names that may mislead the inexperienced, is clearly shown in the case of the present species. As it is commonly

---

* The names of northern Indian girls are chiefly taken from some part or property of a marten; such as the white marten, the black marten, the summer marten, the marten's

called the *fisher*, most persons directly infer that the animal subsists on fish, and hence resembles the otter and other quadrupeds in its fondness for an aquatic mode of life. Neither of these conclusions is correct. Like the pine marten and various kindred species, this animal subsists by preying on various small quadrupeds, birds, eggs, &c. and so far from being addicted to the water, Hearne states that it manifests as much repugnance to that fluid as a domestic cat. That it will eat fish when they are thrown on shore there is little doubt, as almost all the carnivorous animals are delighted with such food; but we have no proof that this marten is in the habit of *fishing* for itself.

Since the common name is injudicious, by inducing erroneous notions of the habits and manners of the animal, and has no connexion with any distinctive character, we have preferred a translation of the scientific name given to this species by ERXLEBEN, which, if equally inexpressive of any peculiarity, does not produce any false or incorrect opinions.

Pennant's marten is found in various parts of North America, from the state of Pennsylvania to as far north as the Great Slave Lake, where it was seen by Capt. Franklin. Its habits are stated to be very similar to those of the pine marten, climbing trees, catching mice, rabbits, and partridges, with as much

---

head, the marten's foot, the marten's heart, the marten's tail, &c. Matonabbee had eight wives, and they were all called Martens. *Hearne*, p. 94. This *variety* of names may serve to remind the reader of Dandie Dinmont's celebrated family of Pepper and Mustard terriers.

facility as that animal. Hearne informs us that this species is easily domesticated, becomes fond of tea-leaves, is very playful, and has a pleasant musky smell. We may correctly infer that this species is not very scarce or uncommon, if we remark the numbers of them which are collected by the fur traders. Pennant says that five hundred and eighty skins were sent in one year from the states of New York and Pennsylvania, and Sabine remarks that the Hudson's Bay company sent eighteen hundred skins to England in one year.

The length of this marten is from twenty-four to thirty inches without the tail, which is from thirteen to seventeen inches long. The snout is pointed, and the fur near the nose is brown, in some individuals approaching to black: the ears are broad, short, and rounded, having a dusky fur on the outside, which appears lighter on their tips: the throat is brown, with a few white tipped hairs: the belly and legs are of a dark brown. The feet are very broad and covered with hair, which conceals the sharp, strong, white claws. All the fur on the superior part of the body is dark at the base, yellowish above, then tipped with black. The fur on the head is short, but gradually increases in length towards the tail, and its colour changes, losing much of the yellowish, and assuming a chestnut hue. The tail is full, bushy, black and lustrous, being smallest at the end.*

---

* Penis hujus, *P. Lotori* simillimum est.

Species IV.—*The Mink.*

Mustela Lutreola; L.

Viverra Lutreola; Pallas. Spicil. Zool. xiv. p. 46. t. iii. p. 462.
Lutra Minor: Erxl. Mamm. p. 451, No. 3. Schreb. Saeugth, iii. p. 462.
Lesser Otter; Penn. Quad. ii. p. 51. No. 249, Arct. Zool. i. p. 89. No. 28.
Jackash of the fur traders; Hearne. Journey, &c. 8vo. ed. p. 377.
Mænk, of the Swedish colonists in America.

This animal is found throughout a great extent of country from Carolina to Hudson's Bay, and in its habits and appearance so much resembles the otter, as to have acquired the common name above quoted from Pennant.

The favourite haunts of this species are the banks of streams, especially in the vicinity of mill-seats or farm-houses, where it inhabits holes near the water, or in the ruins of old walls, &c. Its food in a great degree consists of frogs and fish, but it frequently invades the poultry yards and commits as extensive ravages as any of its kindred species, cutting off the heads and sucking the blood of the fowls in a similar manner. Rats, mice, and other small animals, also fall victims to the mink, and when this animal takes up its residence about wharves or bridges, it does great service by the destruction of such vermin. Lawson, in his history of Carolina, says that the mink is very destructive to the tortoise, scraping their eggs out of the sand and destroying them, and adds that it feeds upon the fresh water muscles, the shells of which are often found in considerable quantity about the mouth of its hole.

The mink is an excellent swimmer and diver, and can remain much longer under water than the musk rat. When provoked this animal ejects from its anal glands a fetid liquor, which is exceedingly unpleasant.

Hearne states that like the larger otter they are frequently found in winter several miles from any water, and are often caught in traps set for martens. They are supposed to prey on mice and partridges like the marten; but when near the rivers and creeks they generally feed on fish. They vary very much in size and colour. "They are very easily domesticated, and in a short time become so familiar that it is scarcely possible to keep them from climbing up one's legs and body, and never feel themselves happier than when sitting on the shoulder of their master. They sleep very much during the day, but prowl about and feed in the night; they are very fierce when at their meals, not suffering those to whom they are most attached to take it from them. I have kept several of them, but their overfondness made them troublesome, as they were always in the way, and their so frequently emitting a disagreeable smell rendered them quite disgusting.*"

In addition to the latter circumstances, it may be suggested that an animal naturally so much addicted to destruction and so blood-thirsty, might be a dangerous pet in case it were not regularly supplied with food, especially in a house in which there might be small children to tease or excite its anger.

Octavo edition. p. 377-8.

This animal is about twenty inches in length from head to tail, and the tail is four inches long. The ears are rounded, the top of the head in some individuals hoary, in others tawny; the hair of the body is of two sorts, the short being of a tawny colour and the longer of a dusky hue. The feet are broad, webbed, and covered with hair. The tail is dusky and ends in a point. The fur is principally used by the hatters.*

Species V.—*The Sable.*

Mustela Zibellina; L.

Martes Zibellina; Briss. Quad. p. 248.
Zibeline; Buff. Hist. Nat. xii. p. 309.
Sable Weesel; Penn. Quad. ii. p. 43, Arct. Zool. 1, p. 90. *A specimen in winter pelage in the Philadelphia Museum.*

In Siberia, Kamtschatka, and the Kurile islands, this species is very common, while in North America it is almost unknown; more we believe because

---

* From a careful comparison of the descriptions given by the systematic writers, and an examination of numerous skins of the animals, we are inclined to believe that the Pekan, (*M. Canadensis,*) and Vison, (*M. Vison,*) are both nothing more than mere varieties of the Mustela Lutreola, which we have seen having all the markings by which they would distinguish the species above mentioned. Nothing but a reference to living nature can decide the doubt, and this reference is not at present in our power. A new species has been proposed on no better foundation than an overstuffed and faded skin of the common mink, belonging to the Philadelphia Museum.

the northern parts of this continent have not been sufficiently explored, than owing to the absence of the animal. PENNANT, in his Arctic Zoology, states, on the authority of PALLAS, that this animal " extends across the whole continent the skins being frequently found among the furs which the Americans traffic with among the inhabitants of the Tschutscki noss." The skin from which the first named author described the species was from Canada, and " was of the bleached or worst kind."

Captains Lewis and Clark, while on their expedition, obtained one of these sables, and the specimen was prepared and mounted for the Philadelphia Museum, in which it may now be seen. It is in that state of pelage which Pennant speaks of above, the fur being " bleached," or rather the individual is in its winter dress, the colour being of a dingy white, or white so tipped with brownish red as to give something of a faintly reddish hue to the whole animal, except the tail, which is distinctly brownish, becoming darker towards the extremity. The person who prepared this individual was led to believe it an albino from the colour of the fur, and hence has inaccurately given it pink-coloured or albino eyes, but, with this exception, the specimen is still in good condition.

The habits of the sable very closely resemble those of the martens we have before described, but we have so little positive information relative to the species as it exists on this continent, that we do not feel at liberty to enter into any details merely inferred from what is known of the European species. The skins of sables are esteemed among the most precious peltries, and yield large sums annually to the Russian government.

The length of the animal from the nose to the tail is about twenty inches; the tail, including the hair at the extremity, is eight inches long, half of which length is due to the hair. The general colour, when in winter dress, is an obscure fulvous or tawny hue. The head and ears are whitish and broad, somewhat triangular in shape, and rather pointed at top. The feet are very large and covered with hair on the upper and under surfaces, which conceal the nails.

CHAPTER XII.

GENUS XIII. SKUNK; *Mephitis,* C.

Fr. Chinche; Enfant du diable; Bête puante.
Germ. Stinkthier.

GENERIC CHARACTERS.

THE head is conical and small, having a somewhat blunt snout, and small rounded ears. The feet have five toes, those on the fore-feet being large, strong, and suited for burrowing.

Dental System.

32 Teeth:	14 Upper	6 Incisive 2 Canine 6 Molar.	2 False Molars 2 Carnivorous 2 Tuberculous.
	18 Lower	6 Incisive 2 Canine 10 Molar.	6 False Molars 2 Carnivorous 2 Tuberculous.

IN THE UPPER JAW the incisors and canines are exactly similar to those of the martens. There are two false molars, one of which is very small and rudimental, and the other regular, with two roots and one point. The carnivorous tooth is remarkable for the great developement of the internal tubercle, which adds much to its thickness, and gives it a triangular form; and the tuberculous tooth is also peculiar in its dimensions, which are nearly the same from the anterior to the posterior edge as from the internal to the external. In the martens this tooth is only extended in the latter direction, and its slightly salient and rounded tubercles are not distinctly marked. In the present genus these tubercles are very strong and angular, which makes this really a triturating tooth; there are four principal tubercles separated by depressions of some depth, but their extremely irregular figure sets description at defiance.

In the lower jaw the incisors and canines are similar to those of the martens, without exception, and the same is the fact relative to the three false molars: the first is much smaller than the other, which has the form and proportions of the regular false molar. The carnivorous is divided into two nearly equal parts by a rather deep cavity: the anterior is formed by three pointed tubercles arranged triangularly, and the posterior by a spur ending in two sharp and rather slender tubercles, separated by a deep depression. The tuberculous tooth is similar to that of the martens.

In their reciprocal position, the peculiar characters of the carnivorous and tuberculous teeth cause the only difference between the relations of these teeth and those we have remarked to exist in the martens. The great internal tubercle of the upper carnivorous tooth, fills the space between the three triangularly ranged tubercles of that belonging to the lower jaw, and the spur of the latter is in relation with the anterior half of the great superior tuberculous tooth, the posterior part of which corresponds with the inferior tubercle.

The genus mephitis is therefore much less carnivorous than the marten and wolverene, on account

1. Skunk   2. Ermine   3. Mink

of the thickening of the cutting teeth, and more frugivorous in consequence of the enlargement of the tuberculous teeth.

Species I.—*The Skunk.*

Mephitis Americana; Desm.

Viverra Mephitis; Gmel. [L.] Syst. Nat. p. 88, No. 13.
Chinche; Buff. Hist. Nat. tom. 33, pl. xx. fig. 2.
Enfant du diable; Charlev. Nouv. France v. 196.
Skunk Weesel. Penn. Quad. ii. p. 65, No. 263.

Pedestrians, called by business or pleasure to ramble through the country during the morning or evening twilight, occasionally see a small and pretty animal a short distance before them in the path, scampering forward without appearing much alarmed, and advancing in a zig-zag or somewhat serpentine direction. Experienced persons generally delay long enough to allow this unwelcome fellow-traveller to withdraw from the path: but it often happens that a view of the animal arouses the ardour of the observer, who in his fondness for sport thinks not of any result but that of securing a prize. It would be more prudent to rest content with pelting this quadruped from a safe distance, or to drive it away by shouting loudly; but almost all inexperienced persons, the first time such an opportunity occurs, rush forward with intent to run the animal down. This appears to be an easy task; in a few moments it is almost overtaken; a few more strides and the victim may be grasped by its long and waving tail—but that tail is now suddenly curled over the back, its pace is slackened. and in one instant the condition of things is entirely reversed;—the lately triumphant pursuer is eagerly flying from his intended prize, involved in an atmosphere of stench, gasping for breath, or blinded and smarting with pain, if his approach were sufficiently close to allow of his being struck in the eyes by the pestilent fluid of the skunk. Should the attack on this creature be led by a dog, and he be close at hand when the disgusting discharge is made, he runs with tail between his legs howling away, and by thrusting his nose into the soil as he retreats, tries to escape from the horrible effluvium which renders the air in the immediate vicinity of the skunk too stifling to be endured. Thus is an animal, possessed of very trifling strength and no peculiar sagacity, protected by the hand of nature against the most powerful and destructive enemies. A few glands secrete a most noisome and intolerably stinking fluid, and this scattered with peculiar force upon the body of its enemies, or even in the air, is sufficient to disarm the violence of most quadrupeds, and induce man himself rather to avoid than to seek an encounter.

The organs by which this fluid is formed are placed near the termination of the digestive tube, and the ducts from the glands open into the rectum, by the aid of whose muscles the fluid is ejected with astonishing force, and is aimed with great accuracy, rarely missing the object if discharged while within the proper distance. The faculty this animal possesses of annoying its enemies by the discharge of the fluid just mentioned, causes it rather to be shunned than hunted, which the value of its skin would otherwise be sure to occasion.

THE SKUNK.

The skunk inhabits the whole of North America. and is also found throughout a considerable part of the southern portion of this continent. As the coloured markings vary exceedingly in different individuals, it is not surprising that naturalists have made several species of this animal, though without any foundation in nature Of several specimens in the Philadelphia Museum there is not one corresponding with the other in colour; neither have we ever seen two exactly alike. Sometimes they are of a uniform dark brown colour, having a white spot on the top of the head; sometimes they have two white stripes, commencing from a white patch on the back of the head and neck, passing outwards as far as the middle of the back, while only the tip of the tail is pure white; again, other individuals are found with white and black rays on the back and sides, and the tail in great part white, as the skunk is represented in the ordinary figures. All the species proposed by systematic writers are reducible to one, the subject of this article, *Mephitis Americana*, or American skunk.

The fetor produced by the skunk is especially characterized by all who have experienced it as suffocating or stifling, which is owing to its peculiar concentration. The predominant odour is that of muskiness, but in so condensed and aggravated a form as to render it almost insupportable, even at a considerable distance from the spot where it is first discharged. A very good idea may be formed of this stench by breaking and smelling a leaf or stalk of the plant called skunk cabbage, (the *Dracontium fetidum*, or *pothos fetidum*) resembling it in every respect except in strength, which perhaps no artificial accumulation of this vile scent could ever equal.

The fluid ejected by the skunk is not merely offensive by its stench, but also in consequence of its highly stimulating and acrimonious qualities. When any of it is thrown into the eyes, it is productive of very violent and dangerous inflammation; we must suppose that this peculiar acrimony, rather than any mere offensiveness of odour, is the cause of the marked repugnance evinced by dogs, as these animals show not the slightest sign of uneasiness from the presence of the most nauseous and putrid effluvia from animal or vegetable substances, yet run howling and trying to thrust their noses into the ground after having been exposed to this pungent perfume from the skunk.

In its extreme volatility it bears a considerable resemblance to true musk. The smallest drop is sufficient to render a garment detestable to the wearer and his companions for a great duration of time, and without any perceptible diminution of intensity. Washing, smoking, baking and burying articles of dress, and in fact every effort short of destroying the materials of which they are made, seem to be equally inefficient for its removal.—This scent is not only thus enduring when the fluid is sprinkled upon clothing, but the spot where the animal is killed, or where the matter was ejected, retains it for a great lapse of time. " When I was at Cumberland House, (says Hearne, p. 378) in the fall of 1774, some Indians that were tenting on the plantation killed two of these animals, and made a feast of them, when the spot where they were singed and gutted was so impregnated with the nauseous smell which they emit, that after a whole winter had elapsed, and the snow had thawed away in the spring, the smell was still intolerable." A friend informed the author of this work, that he had plainly perceived the odour of the fluid ejected by this animal from across the Hudson river, near Albany; we have no doubt of its being possible to smell it at a much greater distance when the wind blows from the spot where the effluvium is thrown out.*

However singular the fancy may appear, we are assured by Catesby that he has seen one of these animals tamed as a pet, and following its owner like a little dog, without offering to offend any one by its peculiar odour, which it has the power of dispensing at will. When it is recollected that on any provocation or threatened injury, the skunk immediately fires upon his disturber, it will be conceded that such

---

* Professor Kalm gives the following anecdotes in his travels:—"In 1749 one of these animals came near the farm-house in which I lodged; it was winter, and during the night the dogs were aroused and pursued it; in a moment so fetid an odour was diffused, that, being in bed, I thought I should have been suffocated. About the end of the same year, another slipped into our cellar, but did not make the least unpleasant smell, because it only diffuses this when hunted or disturbed. A woman who discovered it at night by its shining eyes killed it, and at the instant it filled the cellar with such a stench, that not only was the woman sick during several days, but the bread, meat and other provisions kept in that cellar were so infected as to be entirely spoiled, and required to be thrown away."

a pet must require a very cautious management, for to startle it suddenly, or injure it accidentally, would expose both friends and enemies to a shower of "liquid sweets," which all " the odours from the spicy shore of Araby the blest" could not correct.

If the skunk be killed while unsuspicious of the approach of danger, or before time has been allowed for the discharge of his artillery of perfume, the animal is not in any way disagreeable, and may be approached closely, or even eaten without the least unpleasantness, if the glands be carefully taken out. Its flesh, when the odorous parts have been carefully removed, is said to be well flavoured, and resembles that of a pig considerably. It is eaten by the Indians, and occasionally by hunters, with much relish.

The skunk is most generally found in the forests or their immediate vicinity, having its den either in the hollow of an old tree or stump, or an excavation in the ground. It feeds upon the young and eggs of birds, and on small quadrupeds, wild fruits, &c. Occasionally the skunk gains access to the poultry-yard, where it does much mischief by breaking and sucking the eggs, or by killing the fowls. When resident in the vicinity of farm-houses, it remains for a long time without giving notice of its presence by emitting its offensive fluid, which proves how ridiculous is the notion that the urine of this animal is the source of its disgusting fetor; for, as Hearne justly observes, were this the fact, the whole country it inhabits would be rendered almost insupportable to every other creature.

We have already stated that the colour of the hair is very various in different individuals of this species at different seasons and periods of life. Very commonly it is of a blackish brown over the whole of the body, except on the top of the head, or immediately between the ears, where there is a white spot, and the tip of the tail, also, is white. Some individuals have a slight white mark on the breast. The hairs of the tail are long and bushy, and, with the exception of their tips, are of a dark brown colour. But, as heretofore stated, scarcely two of them are coloured precisely in the same way. The length of a full grown skunk is about eighteen inches, and the tail about seven, the long hair at the extremity making nearly one-half of this length.*

---

* For a very long time the offensive fluid of this animal was almost universally believed to be its urine, with which it was said first to wet its bushy tail, and then, by a vigorous flourish, to scatter the perfume far and wide against its disturbers. It is not surprising that common observers should mistake the action intended to withdraw the tail from the course of the discharge for the manner in which the fluid is scattered.

We are informed by our friend, Mr. T. KEARNEY, that on one occasion, while going to visit a trap before day-light, he disturbed a skunk which was running along the path at some distance in advance of him, and was much surprised to observe that the course of the fluid discharged was rendered perfectly visible by a distinct *phosphorescent* light; the odour left no doubt of the animal whence it proceeded.

CHAPTER XIII.

Genus XIV. Otter; *Lutra,* Briss.

Ger. Fischotter; Flussotter; Fischdieb, &c.   *Ital.* Lontra.
Fr. Loutre.   *Swed.* Utter.

GENERIC CHARACTERS.

The animals belonging to this genus are characterized by a broad and flat head, which terminates in a blunt snout; small eyes and very short rounded ears. The whiskers are large; the tongue, though not very rough, is papillous. The body is larger in proportion than that of the marten. The very short legs are terminated by five toes, which have their phalanges united by a web or membrane, and are armed with sharp, not retractile claws. The pelage is composed of two sorts of hair, one of which is silky, soft, thick and short, the other long, scattered and rather bristly. The teats are placed upon the belly; on each side of the anus is situated an orifice leading to a small sack containing a fetid matter. These animals are excellent swimmers, reside in holes along the banks of fresh water streams, and feed almost entirely upon fish.

Dental System.

36 Teeth:
- 18 Upper
  - 6 Incisive
  - 2 Canine
  - 10 Molar.
    - 6 False Molars
    - 2 Carnivorous
    - 2 Tuberculous.
- 18 Lower
  - 6 Incisive
  - 2 Canine
  - 10 Molar.
    - 6 False Molars
    - 2 Carnivorous
    - 2 Tuberculous.

THE OTTER.

In the upper jaw the incisor and canine teeth are exactly similar to what we have seen in the marten, the glutton and skunk. The false molars are three in number, the first is very small and rudimental, the second slightly larger than the first, but much smaller than the third, is regularly conformed like all normal false molars. The carnivorous tooth is principally remarkable for the extent and form of its internal tubercle. It is no longer a salient point reposing upon a very large base as in the skunk, but a broad surface terminated at the inner side by a circular line, and bordered at this part by a continuous and salient spine. The tuberculous tooth has the dimensions and form of the same tooth in the marten; it is also more extended from the out to the inside than from before backward, and the inequalities which divide its surface differ in nothing from what we have observed in the animal just referred to.

In the lower jaw the incisors and canines have nothing to distinguish them from the dental system of the skunk; and the same is true of the false molar, the carnivorous and tuberculous teeth.

In their reciprocal position it results from the differences we have pointed out between the skunk and otter, that in the latter no tubercle fills the space left by the tubercles arranged in a triangle on the inferior carnivorous tooth. The first of these tubercles is at the anterior part of the tooth, and opposed to the hollowed centre of the broad surface, bordered by a spine which replaces in these animals the tubercle found in the same tooth of the skunk. The two other tubercles fill the void remaining between the carnivorous and opposite tuberculous tooth, and this last presents almost the whole of its crown to the posterior spur of the lower carnivorous tooth. Nothing is opposed to the inferior tuberculous tooth but the posterior edge of the analogous tooth of the upper jaw.

It is well known that the otters subsist principally upon fish; they also may be fed with flesh, and may be accustomed without difficulty to use vegetable food. Nevertheless, it would be difficult to determine by the teeth, whether they are more carnivorous than the skunk; because, while they appear to have carnivorous teeth which separate them farther from the marten than the carnivorous teeth of the skunk remove that animal, they have, on the other hand, less extensive tuberculous teeth than those of the latter genus.

Species I.—*The American Otter.*

Lutra Brasiliensis; Ray.

Mustela Lutra Brasiliensis: Gmel. L. p. 93.
Lutra Brasiliensis: Ray. Syn. Quad. p. 91.
Lutra Brasiliensis: Briss. Quad. p. 278.
Saricovienne de la Guyane: Buff. Supp. 6. p. 287.
Loutre d'Amerique: C. Rég. Anim. i. p. 151. tom. iv. fig. 3.
Lutra Canadensis: Sabine, App. to Franklin's Exp. p. 654.

Though the Brasilian Otter has long been regarded as a distinct species, the North American or Canadian Otter was almost uniformly confounded with its European analogue, until the distinctive characters were accurately pointed out by Sabine in the work above quoted. Cuvier considered the North

THE AMERICAN OTTER. 223

American as identical with the Brasilian species, and in this correct opinion he has been followed by most of the naturalists.

This otter inhabits South, as well as various parts of North America, along the fresh water streams and lakes, as far north as to the Copper Mine river. In the southern, middle and eastern states of the Union, they are comparatively scarce, but in the western states they are in many places still found in considerable numbers. On the tributaries of the Missouri they are very common; but it is in the Hudson's Bay possessions that these animals are obtained in the greatest abundance, and supply the traders with the largest number of their valuable skins. Seventeen thousand three hundred otter skins have been sent to England in one year by the Hudson's Bay company.

Nature appears to have intended the otter for one among her efficient checks upon the increase of the finny tribes, and every peculiarity of its conformation seems to have this great object in view. The length of body, short and flat head, abbreviated ears, dense and close fur, flattened tail, and disproportionately short legs with webbed feet, all conspire to facilitate the otter's movements through the water. In the crystal depths of the river, few fish can elude this swiftly moving and destructive animal, which unites to the qualities enabling him to swim with fish-like celerity and ease, the peculiar sagaciousness of a class of beings far superior in the intellectual scale to the proper tenants of the flood. In vain does the pike scud before this pursuer, and spring into the air in eagerness to escape; or the trout dart with the velocity of thought from shelter to shelter; in vain does the strong and supple eel seek the protection of the shelving bank or the tangled ooze in the bed of the stream; the otter supplies by perseverance what may be wanting in swiftness, and by cunning what may be deficient in strength, and his affrighted victims, though they may for a short time delay, cannot avert their fate. When once his prey is seized, a single effort of his powerful jaws is sufficient to render its struggles unavailing; one crush with his teeth breaks the spine of the fish behind the dorsal fin, and deprives it of the ability to direct its motions, even if it still retain the least power to move.

The residence of the otter is a burrow or excavation in the bank of a stream or river, and the entrance to this retreat is under water; at some distance from the river an air-hole is generally to be found opening in the midst of a bush or other place of concealment. The burrow is frequently to be traced for a considerable distance, and in numerous instances leads to the widely spreading roots of large trees, underneath which the otter finds a secure and comfortable abode. The winter residence is generally chosen in the vicinity of falls or rapids where the water is least liable to be closed from the severity of the cold, and where the otter may find the readiest access to the fish upon which his subsistence depends. Otters have been seen during the coldest parts of winter at very considerable distances from their usual haunts, or from any known open water, as well as upon the ice of large lakes, a circumstance that appears the more singular as this animal is not known to kill game on land at this season. When the otter is in the woods where the snow is light and deep, it dives if pursued, and moves with considerable rapidity under the snow. But its route is always betrayed by the rising of the superincumbent mass, and numbers of them are occasionally killed with clubs by the Indians, while thus endeavouring to make their escape. The old otters, however, are often able to disappoint their pursuers by force, if not by address, for they turn upon them with great fury and ferocity, and so desperate are the wounds inflicted by their teeth, that few individuals are willing to encounter the severity of their bite. The Indians have various methods of killing the otter, one of which is that of concealing themselves near the haunts of the animal on moon-light nights, and shooting them when they come forth for the purpose of feeding or sporting. A common mode of taking them is by sinking a steel-trap near the mouth of their burrow, over which the animal must pass in entering or leaving the den.

We have alluded to the sporting of the otter, and may now remark that its disposition in this respect is singular and interesting. Their favourite sport is *sliding,* and for this purpose in winter the highest ridge of snow is selected, to the top of which the otters scramble, where, lying on the belly, with the fore-feet bent backwards, they give themselves an impulse with their hind-legs, and swiftly glide head-foremost down the declivity, sometimes for the distance of twenty yards. This sport they continue, apparently with the keenest enjoyment, until fatigue or hunger induces them to desist.

In the summer this amusement is obtained by selecting a spot where the river-bank is sloping, has a clayey soil, and the water at its base is of a considerable depth. The otters then remove from the surface, for the breadth of several feet, the sticks, roots, stones and other obstructions, and render the surface as level as possible. They climb up the bank at a less precipitous spot, and starting from the top slip with velocity over the inclining ground, and plump into the water to a depth proportioned to their weight and rapidity of motion. After a few slides and plunges the surface of the clay becomes very smooth and slippery, and the rapid succession of the sliders show how much these animals are delighted by the game, as well as how capable they are of performing actions, which have no other object than that of pleasure or diversion. This amusement is so congenial to the frolic spirit of boyhood, that in vicinities where otter slides are found, youngsters while bathing sometimes take possession of one, and sitting at the top glide thence with great glee into the water, in imitation of the disports of the otter.* But not recollecting that the skin of the otter is protected by a thick and fine fur against the

---

* "We had an opportunity of seeing on the ice of Boyer Creek a considerable number of the tracks or paths of otters; they were the more readily distinguishable from there being snow of but little depth on the ice, and they appeared as if the animal was accustomed to slide in his movements on the ice, as there were in the first place the impressions of two feet, then a long mark clear of snow for the distance of a yard or more, then the impressions of the feet of the animal, after which the sliding mark, and so on alternately. These paths were numerous, and passed between

effects of friction, the poor lads find, on relinquishing their play, that, notwithstanding the apparent smoothness of the slide, the fine sand mingled with the clay has robbed them of a broad surface of cuticle, the loss of which experimentally convinces them, before they can limp home, that an otter slide, in the end, is not altogether well suited for the recreation of human bathers.

The American otter is about five feet in length, including the tail, the length of which is eighteen inches. The colour of the whole of the body, (except the chin and throat, which are dusky white) is a glossy brown. The fur throughout is dense and fine.

The differences between this species and the European otter are thus pointed out by Capt. Sabine. " The neck of the American otter is elongated, not short, and the head narrow and long in comparison with the short broad visage of the European species; the ears are consequently much closer together than in the latter animal. The tail is more pointed and shorter, being considerably *less* than one half of the length of the body, whilst the tail of the European otter is *more* than half the length of its body."

The fur of the otter is much valued by the hatters and other consumers of peltries, and as the animal is hunted at all times without any regard to the preservation or increase of the species, it must ultimately become as rare in North America as the kindred species has long since become in Europe.

---

the bank and a situation where a hole had been in the ice, now frozen over."—Long's Exped. to the Rocky Mountains, vol. i. p. 455.

SPECIES II.—*The Sea Otter.*

Lutra Marina; ERXL.

Lutra Marina, Kalan: Nov. Com. Petrop. ii. p. 367.
Castor Marin: Hist. Kamtschatka, p. 444.
Mustela Lutris: L. GMEL. p. 92.
See Biber, oder See Otter: STELLER, Kamtschatka.

This animal is very interesting on account of its habits and manners, the peculiar beauty, fineness and value of its fur, and from the singular circumstance of the comparatively narrow limits within which the species is restricted. The sea otter spends the greater part of its life in the ocean, where it is abundantly supplied with food, and to all appearance there is nothing to prevent it from roving wherever inclination or curiosity invites; yet the species is exclusively resident within the 49th and 60th degrees of north latitude, and from the 126th degree west longitude to the 150th east from London, on the north-eastern coasts and seas of North America, and on the shores of Kamtschatka and of the islands lying between the two continents.

The sea otter when full grown is about the size of a large mastiff, and weighs from seventy to eighty pounds. In general appearance there is a considerable degree of resemblance between this animal and the seal, especially in the flat and webbed feet of the hinder extremities. It is always found on the coast, or in the immediate vicinity of the salt water, and in tempestuous weather seeks shelter among the weeds which are collected in great quantities in many parts of the seas it inhabits. Its food is vari-

1. Sea Otter. 2. American Otter.

ous, but principally cuttle-fish, lobsters and other fish.

The sea otter, like most other animals which are plentifully supplied with food, is entirely harmless and inoffensive in its manners, and might be charged with stupidity, according to a common mode of judging animals, as it neither offers to defend itself nor to injure those who attack it. But as it runs very swiftly and swims with equal celerity it frequently escapes, and after having gone some distance turns back to look at its pursuers. In doing this it holds a fore paw over its eyes, much in the manner we see done by persons who in a strong sunshine are desirous to observe a distant object accurately. It has been inferred that the sight of this animal is imperfect; its sense of smelling however is said to be very acute.

The female sea otter brings forth on land after a pregnancy of eight or nine months, and but one at a birth; the extreme tenderness and attachment she displays for her young are much celebrated. Before the cub can swim she is observed to play with and fondle it in various ways; sometimes she carries it about, holding it carefully in her mouth; at other times is observed to play with it by throwing it up into the air and catching it between her fore-paws; and before the young otter has learned to swim, she carries it in her arms, swimming about on her back. Nothing but death can induce the parent to relinquish her offspring, and instances are related of the sea otter pining and dying in the vicinity of the spot where her young has died or been killed. The young continues with the dam until it becomes old enough to seek a mate, and after pairing, the male and female are very constant to each other, their union not being disturbed by the wanderings of either party. Sea otters swim on their backs, their sides, and sometimes as if placed upright in the water, and they are frequently observed in this attitude, as if embracing and caressing each other, or engaged in play.

The method of capturing the sea otter is commonly that of placing a net among the sea weed; occasionally they are hunted with two boats, by means of which the animal is soon tired down, being unable to remain a very long time under water: they are also killed by clubs or spears when they are found asleep upon the rocks. The flesh is eaten by the hunters, but its qualities are very differently represented; some have stated it to be tender, juicy, and flavored like young lamb, and by others it is declared to be hard, insipid, and tough as leather; this variance we suppose to arise from the different ages of the individuals eaten, the young being doubtless tender and delicate, and the old being tough and unsavory.

The length of the sea otter, (including the tail, which is ten or twelve inches long,) is about five feet, and the whole of the body appears to vary but little in thickness. The upper jaw is long and considerably broader than the lower; the ears are not an inch long and are pointed, thick and fleshy, stand erect, and are covered with short hair. The fore legs are very short, thick and covered with fur, having five broad toes covered with short hair, and joined by a membrane. The hind feet, which we have stated to bear much resemblance to those of the

THE SEA OTTER.

seal, have a strong and larger membrane between the toes, which is somewhat like the shark skin called shagreen; on the outside of the external toe there is a skin skirting it, analogous to what is seen in some water fowl.

The colour of the sea otter when in full season is perfectly black, at other times of a uniform dark brown. When the fur is opened it is of a lighter colour than on the surface, and intermixed with the fur are some longer, black and shining hairs, which greatly increase the beauty of the pelage. In some individuals the fur about the ears, nose and eyes is of a lighter colour, sometimes brown. The young are sometimes of a cream colour, with white about the nose, eyes and forehead. The fur of the young animal is not equal to that of the adult, which is unrivalled for richness and beauty. The prices at which they have been sold in China appear enormous, from seventy to a hundred rubles having been given for a single skin.

CHAPTER XIV.

Genus XV. Dog; *Canis,* L. C.

Gr. Κυων.
Ger Hund.
Fr. Chien.
Russ. Pes, Sobaka.
Dutch, Hond.
Ital. Cane.
Span. Perro.
Port. Caô.

GENERIC CHARACTERS.

The face is prolonged and the naked glandulous part of the nose rounded; the cheeks are somewhat elevated, the tongue smooth, and the ears (in the wild animals) erect and pointed. The teats are in the greatest number of instances placed in part on the belly and in part on the chest. On the fore feet there are five, and on the hind feet four toes, armed with strong, slightly curved nails, which are not retractile.

Frederick Cuvier remarks that the animals of this genus may be considered in point of carnivorous regimen, as standing between the glutton and skunk, but nearer the former than the latter. We have seen that the appetite for animal food, or the disposition to feed upon flesh, diminishes both in proportion to the increased numbers of tubercles, and the augmented thickness of the carnivorous teeth, which in undergoing this change proportionally loose their trenchant character.

THE DOG.

Dental System.

42 Teeth:	20 Upper	6 Incisive 2 Canine 12 Molar.	6 False Molar 2 Carnivorous 4 Tuberculous.
	22 Lower	6 Incisive 2 Canine 14 Molar.	8 False Molar 2 Carnivorous 4 Tuberculous.

In the upper jaw the number, proportion and respective situation of the incisive teeth resemble those of the marten, but in their forms they have characters peculiar to themselves; they are *trilobed*, having a principal middle lobe and two smaller ones on each side; their internal surface is not divided by a transverse groove, but is bordered by a crest or spine, which rises upon the edges of the two small lobes, and by reuniting at the commencement of the root, forms an angle varying in acuteness in different individuals. The canine teeth also resemble those of the marten, and the false molars exhibit an equal similarity, except that they are separated by a vacant space from the canine, and the two last have their posterior part prolonged into a very evident spur, formed by a particular lobe separated from the principal one by a groove. The carnivorous is entirely like the analogous tooth of the marten: its principal part is divided into two lobes, an anterior, which is larger and more pointed, and a posterior, which is more trenchant, and presents on its internal surface anteriorly only a very small blunt or rounded tubercle, according to the species. The first tuberculous tooth is very large, and its external is larger than the internal part, by which it is distinguished from that belonging to the marten. On its external face it has two pointed tubercles, bordered externally by a spine; in its middle there are two small eminences which appear to be connected with the exterior spine, and between these and the tubercle on the external face there is a broad and deep depression; finally, the internal face, which is rounded, has on it a spine, which marks its outline, and terminates in a groove that separates it posteriorly from the eminences before mentioned. Between these eminences and this last spine is found a second well marked depression. The second tuberculous tooth resembles the one we have described, only that it is more than one-third smaller.

IN THE LOWER JAW the incisors are merely bilobed, and the lobe nearest to the canine is one-half larger than the other. The canine does not at all differ from that of the marten. After a vacant space come four false molars; the first is merely rudimental, and the three others, which have all the characters of these teeth, do not differ from each other, except that they increase slightly in size, from the second to the fourth, and in having their posterior part divided by two serrations. The anterior part of the carnivorous resembles the analogous tooth of the cat; its edge is trenchant and divided in its middle by a groove into two parts, but the anterior is less elevated than the other, and we find at its base interiorly, and rather at the back part, the little pointed tubercle which we have mentioned in describing the dentition of the marten. Its posterior part is a spur, which is principally composed of two obtuse tubercles, one on the outside and the other on the inside. The first tuberculous is twice as long as it is broad. Its anterior part is composed of two tubercles, one below and the other on the outside, the posterior consists of a spur edged by an irregular spine. The second tuberculous tooth is very small, circular, and composed of two small tubercles, and is surrounded, especially on the inside, by a small spine.

In their reciprocal position the relations of these teeth, as to the incisor, canine and false molars, are such as we have heretofore described. The internal tubercle of the superior carnivorous fills the space which separates the fourth false molar and the inferior carnivorous. The external face of the anterior part of the latter is found in relation with the internal part of the posterior surface of the opposite tooth, and the spur of the first tuberculous fills the vacuities of the opposite tuberculous tooth, and these in turn occupy the vacuities on the spur of the lower carnivorous. The first part of the lower carnivorous fills the vacuity which exists between the two superior tubercles, and the second pair of tubercles of this first tooth is in opposition with the second lower tuberculous, which appears to be merely rudimental and without function.

The species belonging to this genus are endowed with very acute senses, especially of sight, smell, and hearing. Their food varies accordingly to circumstances, and is composed wholly or in part of animal matter, either recently killed, or in a putrid state.*

---

* Illis omnibus penis nodosus est, quo pacto, cohærent copula junctis. Secundum Clariss. Linnei verba. "gravida

Were we desired to propose a creature fit to be an emblem of incorruptible fidelity, unwavering friendship, forbearing and enduring affection, combined with all that renders gratitude commendable, and honesty of high value, we know not one more worthy to be thus distinguished than the dog, which, under all circumstances of adverse or prosperous fortune, adheres with untiring and zealous vigilance to the cause of his master, being ever ready to lay down his own life in defence of him he has chosen to serve and obey. Without the co-operation of this highly gifted quadruped, how could man have opposed the noxious animals by which his path was beset, or his dwelling surrounded? In those dreary regions of the earth, where the face of nature wears a veil of almost perpetual gloom, and the wretched wanderers of the human family are forced to maintain a perpetual struggle against the combined severities of cold and famine, how much more abjectly miserable would their condition be, had not nature endowed this animal with the disposition to seek the society of man, despite of all the sufferings incident to his poverty, and all the injuries inflicted through his barbarity or neglect.

We have to regret, however, that the good mankind derives from the faithful services of this animal, is closely connected with the possibility of receiving from it the most terrible of evils, as if nature could not operate without balancing or antagonizing every

---

lxiii. diebus parit, sæpe iv. ad viii; masculis patri similibus, femineis matri."

thing by its opposite. It is the dog kind that from time to time inflicts on our race a malady, perhaps the most agonizing and horrific to which humanity is liable, from whose aspect or endurance the stoutest hearts and best regulated minds shrink away in terror. Against this dreadful disease the resources of medicine have hitherto proved inadequate and unavailing, and though we continue to hope that the augmentation of power which medicine is daily acquiring through the zeal of its cultivators may ultimately triumph over this afflicting disorder, experience teaches us how much cause there is to fear that many must still perish before a remedy can be found.*

An examination of dogs bred in long civilized countries will speedily convince us of the impossibility of arriving, through them, at any satisfactory conclusion as to the source or origin of the species, because the influence of domestication and intermixture of races has produced changes and varieties so numerous and interminable as to set at defiance the most scrutinizing and indefatigable research. But if we visit the regions where man has long remained almost entirely stationary, and is still under the go-

---

* A dog which has bitten any person, and is supposed to be mad, should *not* be immediately killed, but securely confined, in order that the madness may be positively ascertained. There is no reliance to be placed on any other preventive treatment than the immediate and perfect excision of the part bitten, or the removal of the poison by sucking the wound forcibly and for a considerable time. Expe-

vernance of few other feelings than those relating to his animal wants, we shall find the dog, though long domesticated, continuing to exhibit as much of peculiar and specific character as if yet in the wild state.

The inhabitants of the northern parts of this continent, and of the north of Asia, have for ages beyond the memory of man employed dogs as beasts of burthen, or for draught: the dogs thus used by the Eskimaux, as well as those kept about the lodges of our southern and western Indians, retain so much of the external appearance and general carriage of the wild animal, as to leave no question of their descent from the same stock as the wolf residing in the vicinity, and do not appear to be distantly removed from that species, however long they may have been in the service of man.

Notwithstanding the endless varieties produced by the causes first referred to, and the fact that between the dog and wolf there are striking external differences, yet they actually constitute but one *natural* group, of which the individuals should be regarded as *varieties* rather than *species*, because they may all be indiscriminately bred together, in such a manner as to result in new and prolific races, bearing but slight resemblance to the parent individuals, and exhibiting new modifications of external characters, of mental qualities, and of internal configuration.

---

rience leads us to believe that the excision of the bitten part at any time, before the general symptoms of the disease have appeared, will avert the disease.

This is the best, and perhaps the only sufficient test of the appropriateness of specific subdivisions, formed in what is called systematic natural history, and affords the most satisfactory proof of the *natural* alliance of various portions of the animal kingdom. Wherever we find one race of animals capable of having its peculiarities entirely obliterated by intermixture with another, and the altered offspring is as prolific as the parent stock, we may feel certain that such races are but varieties of a natural family, however they may be arranged in arbitrary nomenclatural catalogues. We have heretofore adverted to the fact, that there is a law of nature, shown by the prolonged experience of ages to be invariable, which is, that although two species of the same genus may produce offspring partaking in great degree of the qualities of both parents, yet that offspring is sterile and unproductive, thus opposing an impassable barrier to the confusion of species which would inevitably result, were these mule or hybrid beings capable of continuing their race. The converse of the proposition is equally true, animals of the *same* species, however dissimilar in external character, habits, or manners, are capable of breeding with varieties of their own species in illimitable progression, and every successive crossing of breed may result in new modifications of form and in improved physical and intellectual conditions. In domestic economy this has been carried to the greatest extent in the improvement of animals raised for various purposes of utility or luxury, especially in the instance of horses, cattle, sheep, swine, poultry, &c.

In proof of the *specific* identity of the wolf and dog we may here introduce the following very interesting account of a tamed wolf, given by FREDERICK CUVIER, one of the most distinguished scientific naturalists of the present age, and perhaps second to no man, but his illustrious brother, the acknowledged Coryphæus of Zoologists throughout the world.

This wolf was brought up like a young dog, and became familiar with every person whom he was in the habit of seeing, followed his master everywhere, appeared to suffer much from his absence, obeyed his voice, always exhibited the most entire submission to his commands, and in fact differed in no respect from the tamest domestic dog. When his master set out upon his travels, this wolf was presented to the menagerie at Paris, and there, shut up in his den, he remained for many weeks without showing any signs of gaiety, and almost refusing to eat. From this state of depression he gradually recovered, attached himself to his keepers, and had apparently forgotten all his former attachments when his master returned after an absence of eighteen months. At the first word which he uttered, although he was surrounded by a crowd, and could not be seen by the animal, the wolf instantly recognized him and expressed his joy by his movements and cries, and when set at liberty exhibited all the fondness that is commonly expressed by a dog after a separation of a few days, almost overwhelming his old master with caresses. In a short time he was left as before in the menagerie, and again evinced the deepest distress at the absence of his master, but in

OF THE DOG.

time resumed his vivacity. After an absence of three years his master visited him in the evening, when the den was shut up, and there was no possibility that the wolf could discover him by sight; but the moment he heard the voice, he recognised, and responded to it by cries expressive of the most impatient anxiety to approach his master. As soon as the door was opened his cries were redoubled, he rushed forward, placed his two fore-feet on the shoulders of his old friend, licked every part of his face, and threatened his keepers with his teeth for venturing to approach, though an instant before he had testified to them the warmest affection. When his master was under the necessity of again leaving him, the poor wolf became sad and immoveable, refusing all sustenance; he pined away, his hairs bristled up, and in eight days' time he was scarcely to be known by those who had previously seen him. It was expected that he would die; however, he eventually recovered, resumed his healthy condition and glossy coat, but his keepers dared no longer to approach him; he would not endure the caresses of any one, and answered strangers only by threats of injury.*

It is nevertheless true, that occasional and apparent exceptions to the great law of nature heretofore stat-

---

* In the year 1800, a full grown she-wolf that had been caught in a trap was brought to the Royal Menagerie. This animal became so tame as to live among the dogs, and produced with them repeatedly. She barked as they did at all strangers, and so far laid aside her carnivorous disposition as to offer no injury to the poultry, to which she had free access.

ed are on record; but granting them the fullest force, they are few and mere exceptions, which, by their singularity and rareness, confirm rather than disturb the stability of the general rule. If the united experience of all observers throughout the world, during a period of three or four thousand years, cannot produce more than two, three, or four instances of exception, of which number few or none are adduced in any other than a "questionable shape," we need nothing better to show that this law is as old as nature herself, and must continue to operate in the same manner until time shall be merged in eternity.

We shall not enter into any farther discussion of the circumstances which may have led to the production of many of the permanent varieties to be found among animals, of which the (so called) species are capable of producing prolific offspring together. We know that, in their wild state, similarity of instinct and mode of life is sufficient cause for the association of races with individuals having similar peculiarities, and this preserves the characters of the race from change or deterioration during an indefinite lapse of time; but circumstances occasionally place animals, apparently of different species, together at times when they are under the influence of an imperious necessity, and the result is the commencement of a new race, which, by subsequent association and multiplication with each other, may present such a permanence of external characters as to induce observers to consider them as specifically distinct from all others.

The breed of dogs produced from the wolf and varieties of the domestic dog, during a considerable succession of generations, is characterized by a marked predominance of the qualities inherited from their savage progenitors, in the keen and vivid expression of the eye, ferocity of disposition, and severity of their bite. It is impossible, however, without a fuller and more authentic history of the dog from the earliest day to the present time than we possess, to decide by what combinations all the different varieties have been produced. That such changes are still going on among these animals daily observation proves; and we have ocular demonstration that the most curious of these varieties may propagate its kind with others, which, if external characters were to be relied on, might be considered radically and specifically distinct. The domestic dog and wild fox breed together, but there is every reason to believe that the offspring are true *mules*, or incapable of continuing their kind.

Species I.—*The Domestic Dog.*

Canis Familiaris; L.

The dogs of the Eskimaux and other aboriginals of this continent differ much in size and colour, yet they are all of a breed, apparently intermediate to the wolf and the fox. They have sharp noses, pointed and erect ears, and long bushy tails, which are most commonly carried over the back and curled toward the left side. They are strong, fierce, and courageous, biting with so much vigour and keenness, that the smallest of them, when he can get into a corner, is able to keep several large European dogs at bay.

These dogs serve the Eskimaux, and the traders to the snow covered countries of the North, instead of horses and other animals, and though at first view they may appear inadequate to the task of drawing heavy burthens for great distances, yet on a closer examination we shall perceive that they are better adapted for such service in those regions than any other quadruped. In the first place, the ground during great part of the year is deeply buried in snow, and the slight vegetation is beyond the reach of any animal not taught by necessity how to remove the snow with its feet, even if a sufficiency of food were then to be obtained. To employ in such countries the animals commonly used for draught, it would be necessary to make up the lading chiefly of provender for their support; and their own weight would speedily fatigue and disable them, on account of their repeatedly sinking into the snow, on which the crust is generally too thin to bear a large animal. But the dog is not only light and strong, and can draw a sledge or slide carrying a considerable weight, but, in consequence of his carnivorous regimen, can live on the offal of the provision on which his driver subsists; to this may be added the fact, that the facility of procuring flesh is far greater in those countries than almost any other substance. We need not feel surprised that the European and American traders should have uniformly adopted the modes of travelling used by the aborigines, or have substituted their simple conveyances for the more artificial though less appropriate European vehicles.

In the country north of Hudson's Bay these animals are employed by the Indians and traders to draw their furs and other baggage, though the Indians are frequently too indolent to be at the trouble of making sledges, and then the women are under the necessity of placing a part of their load upon the dogs, which they do by lashing their tents, kettles, and other utensils, on the backs of the animals, as is usually done with pack-horses. In the fall of the year and beginning of winter the squaws sew the skins of deer legs together in the shape of long portmanteaus, which, when hauled on the snow as the hair lies, soon become quite slippery, and answer very well instead of sledges while travelling over the open country.

The sledges made by the northern Indians vary in size, accordingly as they are designed for women or the dogs. Occasionally they are used as long as twelve or fourteen feet, and from fifteen to sixteen inches wide; but the common length is eight or nine feet, and the breadth from twelve to fourteen inches.

These sledges are made of boards, which are not more than a quarter of an inch thick, and from five to six inches wide. As the Indians have no other tool with which to fashion them boards than a common knife, slightly turned up at the point, they sew the boards together with thongs made of parchment deer skin; and on the upper side several cross bars of wood are lashed, which serves to increase the strength of the sledge, and secure the ground-lashing, to which the load is always fastened by small strips or thongs of leather.

The fore part or head of the sledge is turned up, so as to form a semicircle of at least fifteen or twenty inches in diameter, which prevents it from plunging into light snow, and enables it to slide over the snow-drifts and solid projections. The dogs are harnessed to these vehicles by traces of raw hide, which are fastened to leathern collars round their necks. They are generally harnessed two abreast, and the whole number attached to a sledge is commonly five; the foremost or leader being always, if possible, a well broken and long trained dog.*

These faithful and invaluable drudges are treated with shameful neglect and cruelty both by the Indians and the scarcely less savage Voyageurs or Engagées employed in the fur trade. They not only flog them unmercifully, and vent all their ill humour upon these suffering quadrupeds, but overload, overdrive, and starve them most barbarously. In consequence, these poor animals are almost always in wretched plight, and their dispositions rendered so ferocious and irritable as to keep them, whenever they halt on a march, continually worrying and teasing each other. The consequences are sometimes severe to the Voyageurs, for occasionally a team of dogs, when unobserved, takes off at full speed in chase of some animal with a view of running it down, and either dash with their sledge so furiously against stones, trees, &c. as to break it and lose the

---

* See Hearne's Journey to the Northern Ocean.

lading, or they escape so far as to be entirely lost to their unworthy drivers.

The sledges of the Eskimaux* are exceedingly various in their forms, and in the materials of which they are composed, scarcely two of them resembling each other. The best of these vehicles are made from the jaw-bones of the whale, sawed to about two inches in thickness, and from six inches to a foot in depth. These bones form the runners, and are shod with a plank of the same substance: pieces of bone, wood, or deer's horn, are lashed across within a few inches of each other, to connect the side pieces, and these yield to any great strain which the sledge receives. The upper part of the sledge is generally twenty inches broad, but as the runners lean inwards, the breadth is somewhat greater at the bottom. The weight of these bone sledges is very great; one of moderate length, or not more than ten or twelve feet long, weighed 217 lbs.

In the coldest part of the winter the skin of the walrus is very frequently used for runners, and when frozen stiff answers very well, being more than an inch in thickness and exceedingly tough. The Eskimaux have another ingenious contrivance to make runners, which is to case earth and moss in seal skin, and by pouring in some water a hard bolster is easily formed. The sticks, bones, &c. are placed across the upper part of the runners, as before mentioned, and the surface that is to move upon the snow is

---

* All the facts here stated relative to the Eskimaux sledges and dogs are from the " Private Journal of Capt. G. F. Lyon, R. N."

covered with ice by mixing snow with fresh water and allowing it to freeze on the runner, which enables the sledge to slide along with great facility. An Eskimaux who owns but one or two dogs, sometimes drives them in a little tray made of a rough piece of walrus hide, or a flat slab of ice hollowed to resemble a bowl. Boys are often seen to divert themselves by yoking several dogs to a small piece of seal skin, sitting on which they hold by the traces. They then dash off at full speed, and the one who sustains the greatest number of bumps before he relinquishes his hold is the cleverest fellow.

The following spirited description of Eskimaux sledge travelling, we give in the words of Capt. Lyon, whom we are always happy to have an opportunity of quoting. "Our eleven dogs were large and even majestic looking animals, and an old one of peculiar sagacity was placed at their head by having a longer trace, so as to lead them through the driest places, these animals having such a dread of water as to receive a severe beating before they will swim a foot. The leader was instant in obeying the voice of the driver, who never beat, but repeatedly called to him by name. When the dogs slacked their pace, the sight of a seal or bird was sufficient to put them instantly to their full speed, and even though none of these might be seen on the ice, the cry of a seal, a bear, a bird, &c. was enough to give play to the legs and voices of the whole pack. It was a beautiful sight to observe the two sledges racing at full speed to the same object, the dogs and men in full cry, and the vehicles splashing through the holes of water with the velocity and spirit of rival stage coaches.

There is something of the spirit of professed whips in these wild racers; for young men delight in passing each other's sledges, and jockeying the hinder one by crossing the path. In passing on different routes, the right hand is always yielded, and should an inexperienced driver endeavour to take the left, he would have some difficulty in persuading his team to do so. The only unpleasant circumstance attending these races is, that a poor dog is sometimes entangled and thrown down, when the sledge with perhaps a heavy load is unavoidably drawn over his body. The driver sits on the fore part of the vehicle, whence he jumps when requisite to pull it clear of any impediments which may lie in the way, and he also guides it by pressing either foot upon the ice. The voice and long whip answer all purposes of reins, and the dogs can be made to turn a corner as dexterously as horses, though not in such an orderly manner, since they are constantly fighting, and I do not recollect to have seen one receive a flogging without instantly wreaking his passion on the ears of his neighbours. The cries of the men are not more melodious than those of the animals, and their wild looks and gestures when animated, give them the appearance of devils driving wolves before them. Our dogs had eaten nothing for forty-eight hours, and could not have gone over less than seventy miles of ground; yet they returned to all appearance as fresh and active as when they first set out."

The Eskimaux dog bears a considerable resemblance to the English shepherd dog, but is much stronger and broader across the breast, in consequence of the severe labour he is constantly engaged in. A fine dog about equals a Newfoundland dog in size, but the nose is broad like that of the mastiff. Both in winter and summer the hair is very long, but there is a soft downy under covering in cold weather, which is not found in the warm season.

As soon as these dogs can walk they are put into harness, and are soon taught to pull, by their continual efforts to regain their liberty, or to go in search of their dams. When two months old, sometimes eight or ten little ones are harnessed to the sledge along with an old steady animal, where, by dint of repeated and severe beatings, they are at length educated for service. Each dog has his name, which when it is angrily called out, has an immediate effect on the animal. The whip has a lash of seemingly immoderate length, it being from eighteen to twenty-four feet long, and the handle but one foot. The dogs are guided or stopped by throwing this lash on one side or the other of the leader, and by peaking certain words. They are taught to lie down by throwing the whip gently over their backs, and will remain in this position during whole hours.

The weight these animals are capable of moving over the ice or snow is really surprising. Capt. Lyon has seen a walrus drawn along by three or four of them. He found by several experiments that three dogs could draw a man on a sledge weighing one hundred pounds, at the rate of one mile in six minutes; and in evidence of the strength of a well grown dog, his leader drew one hundred and ninety-six pounds singly, and to the same distance in eight minutes. On another occasion, seven of his dogs ran a mile in four minutes and thirty seconds, drawing a heavy sledge full of men. Afterwards nine dogs drew one thousand six hundred and eleven pounds one mile in nine minutes! The sledge was on wooden runners, neither shod nor iced; had they been iced, Capt. L. thinks forty pounds might have been added for each dog.

The Eskimaux dogs, notwithstanding the intensity of cold to which they are subjected, have no particular breeding season, but their females bring forth their young at any season, seldom having more than five at a litter. These dogs also sleep in the open air, during the coldest weather, without inconvenience. They are very poorly fed, thin, gaunt and savage looking animals; they are remarked to be more insolent and irritable when well supplied with food than in their ordinary half starved condition.

We have examined the observations of the late Professor Barton, relative to the native American or Indian dog, with much interest, and should be very happy to believe that he has thrown light upon the origin of this variety of the species. The facts he adduces are too few, and of too negative a character, to lead to any other conclusion than that the domesticated dog of the Indians bears a closer resemblance to the wolf, or to " a half breed" between the wolf and the fox, than to any other animal. The *conjectures* of the learned Professor, however ingenious, we do not feel at liberty to receive, except as an evidence of the vividness of his imagination. Fortunately for mankind the reign of imagination is rapidly passing, if it have not already entirely passed away, and the conjectures of the highest authorities are " void, and of none effect" when opposed by facts, and subjected to the scrutiny of common sense.

It is with much pleasure, however, that we witness the feeling with which Dr. Barton replies to the illiberal observations of celebrated European authors, concerning the character of the aboriginals of this continent. As we fully coincide with him on this subject, we give his remarks at full length.

" It is well known how much ingenuity, eloquence and science, have within the last fifty years been employed to respresent the Americans as the degenerated or imperfectly organized children of the earth. To complete the large volume of calumny against these poor people, even the manner in which they treat their dogs is not suffered to pass unnoticed by the historians of the new world. ' Prior to their intercourse with the people of Europe,' says the eloquent Dr. Robertson, ' the North Americans had some tame dogs which accompanied them in their hunting excursions, and served them with all the ardour and fidelity peculiar to the species. But instead of that fond attachment which the hunter naturally feels towards those useful companions of his toils, they requite their services with neglect, seldom feed, and never caress them.'

" It would, I believe, (adds Dr. Barton) be a much easier task to prove that Dr. Robertson was unqualified to write the history of America; to prove that the Indian-Americans are not the inferiors of the people of the old world, in the measure of their intellectual endowments; and to show that more than one-half of the charges which have been brought against these people, are charges resulting from ignorance, or from systematic zeal, than to prove that the Indians are peculiarly entitled to the character of kind and tender dog-masters. After some attention to this subject, I must candidly confess that I possess not the materials for a satisfying defence of the Indian.* The charges which have been brought against him by the writers whom I have mentioned, will be convictive. But why, in this inquiry, if the historian will condescend to mention the fact and interweave it with his eloquence, should he forget the hardships of the savage life? Where the master labours under a scarcity of food, his servants, the animals which depend upon him for their subsistence, must share in the hardships and the evils of his state. The miserable condition of the Indian dogs, is a necessary result of the miserable condition of the Indians themselves.†"

Captains Lewis and Clark, while on their celebrated expedition, were, on several occasions, under the necessity of subsisting their men upon the flesh of dogs purchased from the Indians. Their prejudices against such food were soon overcome, and they quickly learned to prefer it to the poor fare of

---

* In our opinion no defence is necessary; the treatment of the untutored savage to his dog, is comparatively as good as the treatment of many illiterate Europeans to their own children. The Indian may be cruel and neglectful; more *brutal* than the lower classes of many parts of Europe he cannot be.

† Medical and Physical Journal, vol. i. p. 27.

the Indians, who held the "dog-eaters" in sovereign contempt.*

The Dog of Newfoundland is remarkable for its sagacity, size, strength and beauty, and in external characters differs almost entirely from the Eskimaux and Indian dog. The Newfoundland dog has a broader and more expressive visage, and a blunter nose than either of the dogs yet mentioned, the orbits of his eyes have more prominent superciliary ridges, the ears are broad, soft and pendulous, and the whole body is more robust, and covered with long, soft and glossy hair. On the tail the hair is still longer than on the body and forms a handsome brush, which appears to greater advantage when the animal is in motion, as it is then carried slightly curved upwards at its extremity. The Newfoundland dog is very fond of the water, and swims with great ease; the Eskimaux dog is by no means disposed to enter the water, most probably from experience of the effects of extreme cold on coming out again. The Newfoundland dog is employed with great advantage by the settlers to draw heavily laden sledges, and

---

* "The dog is unusually small, about the size of an ordinary cur; he is usually particoloured, amongst which the black, brown, white and brindle, are the colours most predominant; the head is long, the nose pointed, the eyes small, the ears erect and pointed like those of the wolf: the hair is short and smooth, except on the tail, where it is long and straight like that of the ordinary cur dog. The natives never eat the flesh of this animal, and he appears to be in no way serviceable to them, except in hunting the elk."—Lewis and Clark, ii. p. 165.

1. Newfoundland Dog. 2. Esquimaux Dog.

is an invaluable servant. Of his peculiar sagacity numerous instances are on record, and almost every one has made observations enough in relation to this dog, to render any repetition of such instances unnecessary in this place. He breeds with all the known varieties of the domestic dog, and also with the common wolf.*

Species II.—*The Common Wolf.*

Canis Lupus; L.

When the aboriginal Americans first gave place to European adventurers, and the forests which had flourished for ages undisturbed, began to fall before the unsparing axe, the vicinity of the settler's lonely cabin resounded with the nightly howling of wolves, attracted by the refuse provision usually to be found there, or by a disposition to prey upon the domestic animals. During winter, when food was most difficult to be procured, packs of these famished and ferocious creatures were ever

---

* Scientia naturali multum versato et fide digno viro Sabine, se canem Terræ-novæ cum lupa coire frequenter vidisse dictum est; ex quo nobis ejusdem speciei esse ambos, minime dubitare licet. Tamen, si sciolo aliquo, canem sue etiam coire aliquando tentare, objectum sit; respondendum est, neminiunquam visum cognitumve canem *vere copulari*, aut *speiem* vel suam, vel abnormem ex sue procreare. Genitalia lupi canisque nullo modo inter sese differunt: dentes, viscera, victus, vivendique modus eadem sunt; alter ex altero catulos gignere potest.

at hand, to run down and destroy any domestic animal found wandering beyond the enclosures, which their individual or combined efforts could overcome, and the boldest house-dog could not venture far from the door of his master without incurring the risk of being killed and devoured. The common wolf was then to be found in considerable numbers throughout a great extent, if not the whole of North America; at present it is only known as a resident of the remote wooded and mountainous districts where man has not fixed his abode, nor laid bare the bosom of the earth to the enlivening radiance of the sun.

The common wolf of America is considered to be the same species as the wolf of Europe, and in regard to habits and manners gives every evidence of such an identity. Like all the wild animals of the dog kind, they unite in packs to hunt down animals which individually they could not master, and during their sexual season, engage in the most furious combats with each other for the possession of the females.

The common wolf is possessed of great strength and fierceness, and is what is generally called a cruel animal, tearing the throat of his victim, drinking its blood, and rending it open for the purpose of devouring its entrails. The great strength of its jaws enables the wolf to carry off with facility an animal nearly as large as itself, and makes its bite exceedingly severe and dangerous. Aged or wounded individuals, as well as the hinds and fawns of the deer, sheep, lambs, calves and pigs, are killed by these wolves, and the horse is said to be the only domestic animal which can resist them with success. They gorge with much greediness upon all sorts of carrion

which they can discover at great distances; and where such provision is to be obtained in great plenty, they become very fat and lose their ferocity to a singular degree.

When this wolf has been caught in a trap, and is approached by man, it is remarked to be exceedingly cowardly, and occasionally suffers itself to be beaten without offering the slightest resistance. If a dog be set upon a wolf thus captured, the assault is patiently endured so long as his master is present; but as soon as the wolf is freed from the restraint imposed by the presence of his captor, he springs upon and throttles the dog, which, if not speedily assisted, pays the forfeit of his presumption and temerity with his life. When kept in close confinement and fed upon vegetable matter, the common wolf becomes tame and harmless, but is very shy, restless and timid, expressing the greatest alarm at the approach of a stranger, and striving to escape from observation. The voice of this wolf is a prolonged and melancholy howl, which, when uttered by numerous individuals at once, is discordant and frightful. The period of gestation, &c. in this species is in every respect analogous to that of the common dog.

In the regions west of Hudson's Bay wolves are often seen, both in the woods and on the plains, though their numbers are inconsiderable, and it is not common to see more than three or four in a pack. They appear to be very fearful of the human race, but are destructive to the Indian dogs, and frequently succeed in killing such as are heavily laden, and unable to keep up with the rest. The males are not so swift as the females; and they seem to lead a forlorn life during the winter, being seldom seen in pairs until the commencement of spring. They bring forth their young in burrows, and though it might well be inferred that they are fiercer at those times than under ordinary circumstances, yet Hearne states that he has frequently seen the Indians take the young ones from the dens and play with them. They never hurt the young wolves, but always replace them in the dens, sometimes painting the faces of these whelps with vermilion or red ochre.*

In the highest northern latitudes which have yet been explored, the wolves are very numerous and exceedingly audacious. They are generally to be found at no great distance from the huts of the Eskimaux, and follow these people from place to place, being apparently very much dependent upon them for food, during the coldest season of the year. They are frequently seen in packs of twelve or more, prowling about at a short distance from the Eskimaux huts, lying in wait for the domestic dog, which they are successful in killing if he wanders so far as to be out of reach of assistance from his master. In one instance two of them rushed upon a fine Newfoundland dog belonging to Capt. Lyon, in the day time, and would have killed him but for the timely interference of his master. Capt. Lyon relates the following singular instance of the cunning of one of these wolves which had been caught in a trap, and, after being to all appearance dead, was dragged on board

---

* See Hearne, p. 362-3. 8vo. ed.

of the ship:—" The eyes, however, were observed to wink whenever any object was placed near them; some precautions were therefore considered necessary, and the legs being tied, the animal was hoisted up with his head downwards. He then, to our surprise, made a vigorous spring at those near him, and afterwards repeatedly turned himself upwards, so as to reach the rope by which he was suspended, endeavouring to gnaw it asunder, and making angry snaps at the persons who prevented him. Several heavy blows were struck on the back of his neck, and a bayonet was thrust through him, yet above a quarter of an hour elapsed before he died, having completely convinced us that for the future we should not too easily trust to the appearance of death in animals of this description."

Animals exposed to so much suffering from hunger, we may readily believe, are in no way exclusive in their preference of food, and these wolves may be said to feed on every creature they can master, or on the remains of any animal left by the natives.

The common wolf is about four feet and a-half in length, including the tail, which is rather more than a foot long. The height, before, is two feet three inches; behind, it is two feet four inches. The tail is bushy and bending downwards, having upon it hairs upwards of five inches in length.

The general colour of this wolf is reddish brown, intermixed with ferruginous and black; but a great variety is to be observed in the colouring of the wolf as found in the northern, middle and southern regions, exhibiting various gradations from grizzly white to pure black.* The wolf found in Pennsylvania has more of the reddish brown colour, the hair being tipped with black, but especially so over the fore shoulders and sides.

Species III.—*The Prairie or Barking Wolf.*

Canis Latrans; Say.

Canis Latrans: Say. Long's Exp. to the Rocky Mountains, vol. i. p 168. A well prepared specimen in the Philadelphia Museum.

This wolf frequents the prairies or natural meadows of the west, where troops or packs containing a considerable number of individuals are frequently seen following in the train of a herd of buffalo or deer, for the purpose of preying on such as may die from disease, or in consequence of wounds inflicted by the hunters. At night they also approach the encampments of travellers, whom they sometimes follow for the sake of the carcasses of animals which are relinquished, and by their discordant howlings, close to the tents, effectually banish sleep from those who are unaccustomed to their noise. According to Say's observation they are more numerous than any of the other wolves which are found in North America.

---

* " The wolves of Florida are larger than a dog, and are perfectly black, except the females, which have a white spot on the breast, but they are not so large as the wolves of Canada and Pennsylvania, which are of a yellowish brown colour." Bartram's Travels in Florida, p. 199.

THE BARKING WOLF. 261

The barking wolf closely resembles the domestic dog of the Indians in appearance, and is remarkably active and intelligent. Like the common wolf, the individuals of this species frequently unite to run down deer, or a buffalo calf which has been separated from the herd, though it requires the fullest exercise of all their speed, sagacity and strength, to succeed in this chase. They are very often exposed to great distress from want of food, and in this state of famine are under the necessity of filling their stomachs with wild plums, or other fruits no less indigestible, in order to allay in some degree the inordinate sensations of hunger.

This wolf barks in such a manner as to resemble the domestic dog very distinctly; the first two or three notes are not to be distinguished from those produced by a small terrier, but differs from that dog by adding to these sounds a lengthened scream. On account of this habit of *barking*, Say has given the specific name of "latrans" to this wolf, which we prefer to translate for a trivial name, instead of using that of "*prairie* wolf" which is equally applicable to other species.

In confirmation of the sagacity of this wolf we shall quote from Say, to whom we owe all that has yet been made known on this species, some anecdotes respecting it. "Mr. [Titian] Peale constructed and tried various kinds of traps to take them, one of which was of the description called a "live trap," a shallow box reversed and supported at one end by the well known kind of trapsticks usually called the "figure four," which elevated the front of the trap upwards of three feet above its slab flooring; the trap was about six feet long, and nearly the same in breadth, and was plentifully baited with offal. Notwithstanding this arrangement a wolf actually burrowed under the flooring, and pulled down the bait through the crevices of the floor: tracks of different size were observed about the trap. This procedure would seem to be the result of a faculty beyond mere instinct."

"This trap proving useless, another was constructed in a different part of the country, formed like a large cage, through which the animals might enter but not return; this was equally unsuccessful; the wolves attempted in vain to get at the bait, as they would not enter by the rout prepared for them. A large double "steel trap" was next tried; this was profusely baited, and the whole with the exception of the bait was carefully concealed beneath the fallen leaves. This was also unsuccessful. Tracks of the anticipated victims were next day observed to be impressed in numbers on the earth near the spot, but still the trap with its seductive charge remained untouched. The bait was then removed from the trap and suspended over it from the branch of a tree; several pieces of meat were also suspended in a similar manner from trees in the vicinity. The following morning the bait over the trap alone remained. Supposing that their exquisite sense of smell warned them of the position of the trap, it was removed and again covered with leaves, and the baits being disposed as before, the leaves to a considerable distance around were burned; and the trap remained perfectly concealed by ashes; still the bait over the trap was avoided."* It was not until a log-trap was used that an individual of this species was caught. This log trap is made by raising one log above another at one end by means of an upright stick, which rests upon a rounded horizontal trigger on the lower log.

The barking wolf is about three feet and a-half in length, of which the tail forms thirteen and a-half inches, exclusive of the hair at its extremity. The ears are four inches long from the top of the head, and the distance from the anterior canthus of the eye to the end of the snout is three inches and three-fourths.

The general colour of the barking wolf is cinereous, or gray intermingled with black and dull fulvous or cinnamon above. The hair is of a dusky lead colour at base; of a dull cinnamon in the middle of its length, and gray or black at tip, being of greater length along the middle of the back than on other parts of the body. The ears are erect and rounded at tip, having the hair on the back part of a cinnamon colour, and dark plumbeous at base, while that on the inside is gray. The eyelids are edged with black; the superior eyelashes are black beneath, and on the superior surface of their tips. The supplemental lid is margined with black brown before, and edged with black brown behind. The iris is yellow, and the pupil of a black blue; upon the lach-

---

* See Journal of Long's Expedition to the Rocky Mountains, vol. i. p. 168.

rymal sac is a spot of black brown. The face is cinnamon coloured, tinted with grayish on the nose; the lips are white edged with black, and have three series of black bristles. The colour of the head between the ears is an intermixture of gray and dull cinnamon; the colour of the sides is paler than that of the back, and faintly banded with black above the legs, which are cinnamon coloured on the outside, and more distinctly so on the posterior hair. The tail is straight, broad and bushy, of a gray colour, mingled with cinnamon above and black at the tip. The extremity of the trunk of the tail reaches to the projection of the os calcis when the leg is extended.

Different individuals exhibit very considerable variations in colouring, and hence many of the minute markings given by Mr. Say, in the above description, may not be found applicable. He states himself that other specimens which he saw differed much from his first description; one " was destitute of the cinnamon colour, except on the snout where it was but slightly apparent; the general colour was therefore gray, with an intermixture of black, in remote spots and lines, varying in position and figure with the direction of the hair." Perhaps no two individuals could be found exhibiting throughout precisely the same arrangement of coloured markings.

Species IV.—*The Dusky Wolf.*

Canis Nubilus; Say.

Canis Nubilus; Say. Long's Exp. to the Rocky Mountains, vol. i. p. 169,
A well prepared specimen in the Philadelphia Museum.

This wolf is more robust in form, and fiercer and more formidable in appearance, than either of those we have described. It is found in the Missouri country, frequenting the same districts as the prairie or barking wolf, but is by no means in equal numbers. It is distinguished from the common wolf by the length of its ears and tail, while it is separated from the barking wolf by its greatly superior size, difference of colour, &c. But little is known of the peculiar habits of this species, though there is no reason for believing that they differ much from those of its kindred species or varieties. It is remarkable for diffusing a strong and disagreeable odour.

The general colour of this wolf is dusky, the hair being ash-coloured at base, then brownish black, then gray, and next black. The proportion of black upon the hairs is so considerable as to impart a much darker colour to the whole animal than is found in the darkest of the barking wolves, but in the general effect a mottled appearance is produced by the gray of the hairs combining with the black of the tips. The gray colour predominates on the lower part of the sides. The ears are short and of a deep brownish black, with a patch of gray hair on the anterior side within. The muzzle is blackish above, and the superior lips, anterior to the canine teeth, are gray; the same colour extends from the tip of the lower jaw, in a narrow line backwards, nearly to the commencement of the neck. On the under part of the body the colour is dusky ferruginous, with long grayish hairs between the thighs, and with a large white spot on the breast. On the neck the ferruginous colour is very much narrowed, but it becomes broader on the lower part of the cheeks. The legs are of a brownish black, having but a slight admixture of gray hairs, except on the front ridge of the thighs, and the lower edging of the toes, where the gray predominates. The tail is short and fusiform, slightly tinged with rust colour; ar the base and at the tip it is black on the upper surface; the end of the tail itself does not quite reach the os calcis. The longer hairs of the back, particularly over the shoulders, resemble a short spare mane.

Length from the tip of the nose to the origin of the tail, 4 feet $3\frac{3}{4}$ inches —Length of the trunk of the tail, 1 foot 1 inch.—Ear, from the anterior angle to the tip, $3\frac{3}{4}$ inches.—From the anterior angle of the ear to the posterior canthus of the eye, $4\frac{3}{4}$ inches.—From the anterior canthus of the eye to the middle of the tip of the nose, $5\frac{1}{2}$ inches.—Between the anterior angles of the ears rather more than 3 inches.*

---

* The descriptions of this and the preceding species are given nearly in Mr. Say's words, from the work cited at the head of this article.

Species V.—*The Black Wolf.*

Canis Lycaon; L.

Canis Lycaon; Shreb. Saeugthiere, pl. 89.
Loup noir; Buff. 9. pl 41.
Black Wolf; Say, in Long's Exp. to the Rocky Mountains, p. 102, vol

The Black Wolf has been found in the Missouri country, and the British possessions in North America, but not in such numbers as either of the foregoing species. It is not yet satisfactorily established that this wolf is precisely of the same species as the European black wolf, whose scientific name is applied to it. Desmarest is much inclined to thi*t* it a new species, and it is most probable that a close comparison of the two will show differences between them supporting his opinion.

In general appearance, and in the relative proportions of the different parts of its body, this wolf resembles the common wolf, (*C. Lupus*) but in size it is intermediate to the fox and common wolf. The colour of the animal is its most remarkable characteristic, as it is entirely black, without the slightest admixture of any other colour.

In passing through the Missouri country, where this wolf is quite common, Mr. Say had an opportunity of examining a young black wolf, which was chained near the door at a settler's hut. This individual was one of five which had been taken from the same den; it had become familiar with the family, but was shy towards strangers. It was commonly fed on bread and milk; when fed on meat it became ferocious, and attempted to bite the children.

SPECIES VI.—*The Isatis or Arctic Fox.**

Canis Lagopus; L.

Le Renard Bleu; BUFF. 13, p. 272.
C. Lagopus, Arctic Fox; SABINE, Zool. App. p. 858
Arctic Fox; PENN. Arct. Zool. 1, p. 48, No. 10.

In the highest northern latitudes which have yet been explored this fox is found in great abundance, and does not come lower than a few degrees south of the polar circle. Vast numbers are annually killed in the Hudson's Bay possessions for the sake of their skins, which at the proper season are covered with a thick and fine coat of hair. This species is also found on all the Arctic islands, and in the northern parts of Asia.

The Arctic Fox is a pretty and an intelligent animal, bearing a considerable resemblance in figure to the common fox, though its head is not so pointed, and it is of smaller size. In its winter dress the soft woolly white hair gives it some similarity to a small shock dog. Its eyes are of a clear light hazel colour, and are bright and piercing in expression: the ears are thickly clothed with hair, short and apparently doubled in, or cropped on the edges. A projecting ruff of hair descends from behind the ears and surrounds the lower part of the face, producing a very pleasing appearance. The legs are

---

* CUVIER forms a *subgenus* of the foxes, on account of their longer and more bushy tail, more pointed muzzle, nocturnal pupils, less slanting superior incisive teeth, fetid odour and habit of burrowing.

1. Black Wolf. 2. Arctic Fox.

THE ARCTIC FOX. 269

strong and muscular; the large feet are armed with strong claws, and the toes are covered with hair of considerable length above and below; from this character the scientific name of the animal is made, signifying the hare-footed fox.

It has been stated that the Arctic fox* visits the Hudson's Bay settlements only once in five or seven years. Hearne† remarks on this subject, that there is not one year in twenty in which they are not taken in greater or less numbers at Churchill, and he has known not less than from two to four hundred to be caught each year, for three successive years, within thirty miles of the fort. The Arctic foxes always come along the coast from the north, generally making their appearance at Churchill towards the middle of October; their skins are not in the best condition until November. In the interval they are never molested, and are permitted to feed in the vicinity of the fort until they become almost tame. The large numbers of these animals which visit Churchill river in some years do not come together, as they could not find subsistence enough for one-fourth of them while on the way. When they approach the fort the whale carcasses lying along the shore, and the offal left after boiling out the oil, furnishes them abundantly with food, and detains them about the fort until, by frequent reinforcements from the north, they are increased in such vast numbers as almost to be beyond credibility.

---

* Pennant from Graham.
† P. 363. All the facts relative to this species in the vicinity of Hudson's Bay are from Hearne.

They are caught in traps, or killed by guns set for the purpose by the Hudson's Bay people, principally during the month of November, though some few are found to remain during the winter. Upwards of forty have been killed in one night within half a mile of Prince of Wales' fort; this, however, is rarely done, except on the first or second night of the season. The main body of the survivors cross Churchill river near its mouth after it is frozen over, and move to the southward, in some years assembling in considerable numbers at York fort and Severn river. It is not known that any of them ever again return to the north.

Their breeding places have not been positively determined, but Hearne states without doubt that they breed during the summer season in every part of the coast they frequent. He has known them to breed near Churchill, and along the west coast of Hudson's Bay, particularly at Cape Eskimaux, Whalecove, and also on Marble Island. They bring forth from three to five whelps at a litter, which when young are almost entirely of a sooty black colour; as the autumn advances the belly, sides and tail, change to a light ash, while the back, legs, part of the face and tip of the tail, become of a lead colour. As the winter commences they grow perfectly white, changing colour last on the ridge of the back and the tip of the tail, on which a few dark hairs are to be found almost at all times.*

The Arctic fox preys upon various small quadru-

---

* Penis calamo communi similis magnitudine; testes (qui vix conspicui) amygdalis æquant.

peds, such as hares, marmots, lemings, &c. and upon partridges and other birds, as well as their eggs, the carcasses of fish left on shore; and, when driven by necessity, they eat indiscriminately whatever may promise to allay their hunger.* The Arctic foxes which are killed at a sufficient distance from the sea coast are said to be very fit for the table, their flesh being as well flavored as the rabbit on which they feed. But near the sea they devour great quantities of putrid whale flesh and similar matters, which impart a disgusting rankness to their flesh.

These foxes are taken with great facility in traps, and, unlike the wolves, seem to have no idea of the design of such contrivances. They are also occasionally shot while they are feeding together in considerable numbers. This shooting is most successful on moonlight nights, as they lie hid during the day time in the holes among the rocks, or under the hollow ice above the high water mark.

It is singular that these animals will prey upon each other when they find individuals killed, wounded, or caught in traps, as readily as upon other animals, a fact which altogether invalidates the old saying that "dog will not prey upon dog." Hearne informs us that he has known upwards of a hundred and twenty foxes which were caught in traps to be

---

* "As a proof of what foxes will eat to satisfy hunger, I may mention having examined the stomach of one which contained a mass of rope-yarn and line of the size of the doubled fist; amongst which some pieces of sinnet or plaited stuff were above six inches in length." Lyon's Private Journal, p. 109.

destroyed and eaten by their comrades during a single winter, within half a mile of the fort.

We shall conclude this account of the Arctic fox by introducing the observations Capt. G. F. LYON* made on different individuals which he kept on board his ship during the tedious winter spent by the expedition under Capt. PARRY at the North Georgian islands.

"The Arctic fox is an extremely cleanly animal, being very careful not to dirt those places in which he eats or sleeps. No unpleasant smell is to be perceived even in a male, which is a remarkable circumstance. To come unawares upon one of these creatures is impossible, for even in an apparently sound sleep they open their eyes at the slightest noise which is made near them, although they pay no attention to sounds at a short distance. The general time of rest is during the daylight, in which they appear listless and inactive; but the night no sooner sets in than all their faculties are awakened; they commence their gambols and continue in unceasing and rapid motion until the morning. While hunting for food they are mute, but when in captivity or irritated, they utter a short growl like that of a young puppy; it is a singular fact, that their bark is so modulated as to give an idea that the animal is at a distance, although at the very moment he lies at your feet. It strikes me that nature has gifted these creatures with this kind of ventriloquism in order to deceive their prey as to the distance they

---

* The name of this excellent observer is inaccurately given as Capt. R. Lyon in p. 197 of this work.

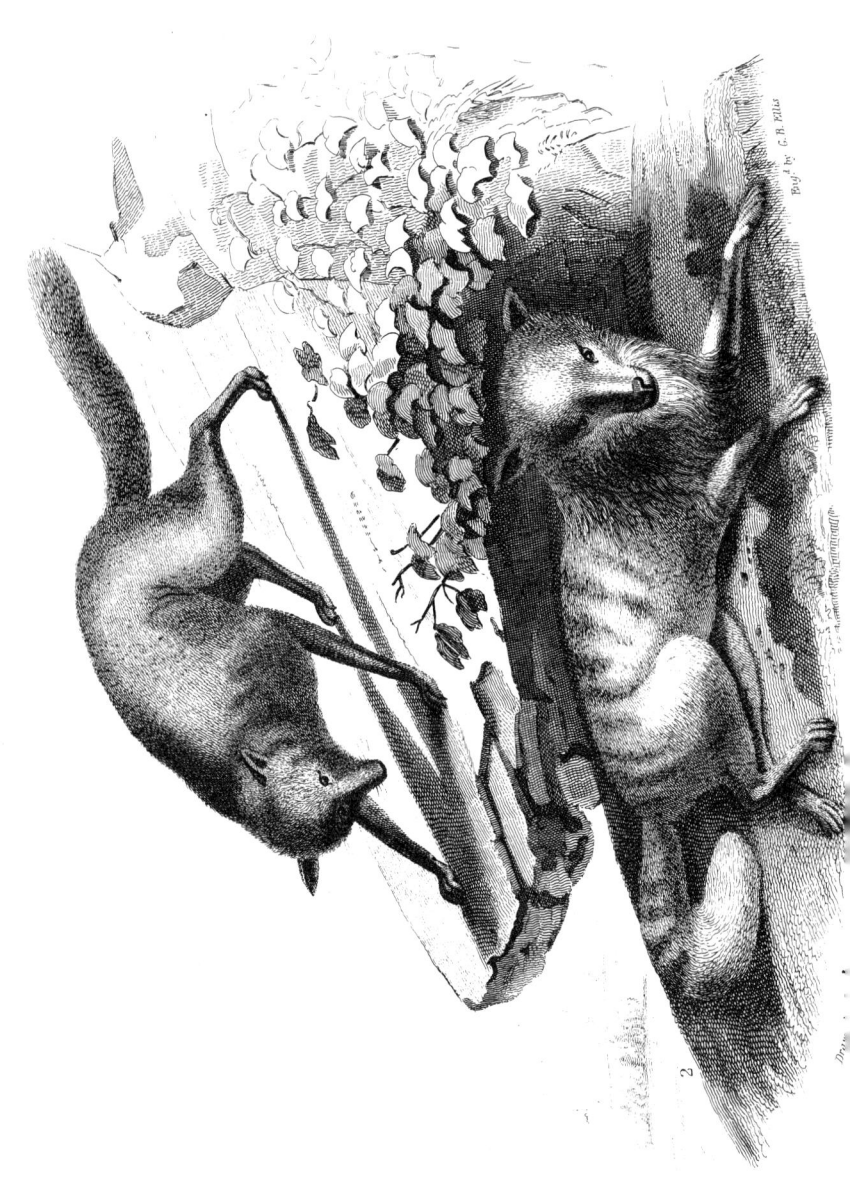

are from them. Although the rage of a newly caught fox is quite ungovernable, yet it very rarely happened that on two being put together they quarrelled. A confinement of even a few hours sufficed to quiet these creatures; and some instances occurred of their being perfectly tame, although timid, from the first moment of their captivity. On the other hand there were some which after months of coaxing never became more tractable. These we supposed were old ones.*

"Their first impulse on receiving food is to hide it as soon as possible, even though suffering from hunger, and having no fellow prisoners of whose honesty they are doubtful. In this case snow is of great assistance, as being easily piled over their stores and then forcibly pressed down by the nose.† I frequently observed my dog-fox, when no snow was obtainable, gather his chain into his mouth, and in that manner carefully coil it so as to hide the meat. On moving away, satisfied with his operations, he of course has drawn it after him again, and sometimes with great patience repeated his labours five or six times, until in a passion he has been constrained to eat his food without its having been rendered luxurious by previous concealment. Snow is the sub-

---

* "If taken young they are easily domesticated in some degree, but I never saw one that was fond of being caressed; and they are always impatient of confinement." Hearne, p. 366.

† We have seen the domestic dog covering up food with earth in the same manner, first pushing the dirt forward and then pressing it down with the end of the nose.

stitute for water to these creatures, and on a large lump being given them, they break it in pieces with their feet and roll on it with great delight. When the snow was lightly scattered on the decks they did not lick it up as dogs are accustomed to do, but by repeatedly pressing it with their nose, collected a small lump at its extremity, and then drew it into the mouth with the assistance of the tongue.

"On the 19th of December, I was so fortunate as to catch one of these foxes in a trap. He was small and not perfectly white, but his temper was so remarkable that I could not resolve to kill him, but confined him on deck in a small hutch with a scope of chain. The little animal astonished us very much by his extraordinary sagacity, for, during the first day, finding himself much tormented by being drawn out repeatedly by his chain, he at length, whenever he retreated to his hut, took this carefully up in his mouth and drew it so completely after him, that no one who valued his fingers would endeavour to take hold of the end attached to the staple."

Species VII.—*The Black or Silver Fox.*

Canis Argentatus.

Renard Noir ou Renard Argenté: Geoff. Coll. du Mus. Charlevoix, Nouv. France, 3. p. 123.
Renard Argenté: F. Cuvier, Mam. Lithog. livr. 5.
Canis Argentatus: Sabine, Zool. App. p. 657.

Were it not for the copious and beautiful black fur of this species, there would be scarcely a characteristic by which to distinguish it from the common fox, (*C. Cinereo-argenteus*) to which in all other respects it is strikingly similar. In form and proportions, as well as in gait and expression of countenance, in sports, mode of feeding and exhibition of anger, there is nothing which may not be equally applied to this and the species above mentioned.

The colour of this fox, as its name implies, is a richly lustrous black, having a small quantity of white mingled with it in different proportions on different parts of the body. Individuals vary considerably in this respect; some have no white except at the extremity of the tail, or a few white hairs scattered along the middle of the breast, or else at irregular intervals a sprinkling of solitary white hairs along the sides, which by contrast add to the intensity and brilliance of the black. This last colour is produced by the longer silken hair which forms the great mass of the pelage, and is occasionally tipped with white; there is a grayish silken hair that constitutes the immediate covering of the skin. Over the whole body and tail the hair is long and tufted, on the paws it is short, and on the face still shorter, the colour of the latter being mostly whitish on the fore part; the eyes are of a yellowish tint.*

This fox resembles the kindred species in the unpleasant odour it diffuses, which is in a considerable

---

* "The silver fox is an animal very rare even in the country he inhabits. We have seen nothing but the skins of this animal, and those in the possession of the natives of the woody country below the Columbia falls, which makes us conjecture it to be an inhabitant of that country exclusively. It has a long deep lead coloured fur for a foil, intermixed with long hairs either of a black or white colour.

degree owing to its urine, as well as to a peculiar glandular secretion. After having satisfied its hunger, it hides the remainder of the food, and when disturbed, expresses its anger by growling like a dog. A comparison of the American black fox with the black fox of Europe, may hereafter show differences sufficient to authorize it to be considered as distinct from that species. But until better opportunities of examination are afforded we shall hazard no opinion on the subject. The black fox is found throughout the northern parts of America, and is also obtained in the north of Asia, where it is considered among the richest and most valuable of furs.*

Species VIII.—*The Red Fox*.

Canis Fulvus; Desm.

Renard de Virginia: De Beauvois, Bullet. de Soc. Philom.
Red Fox: Lewis & Clark, vol. ii. p. 159.
Canis Fulvus: Sabine, Zool. App. p. 656.

This beautiful fox is found throughout North America, and is the species which frequently has

---

at the lower part and invariably white at the tip, forming a most beautiful silver gray." Lewis & Clark, ii. p. 169.

The *black* fox of Lewis & Clark is Pennant's marten, (*M. Pennanti*,) improperly called *fisher*, described at p. 203 of this work.

* Colle pelli si fanno pelliccie bellissime, e ricercatissime ordinariamente il valore di una pelle corrisponde a tanti scudi, quanti ne può la medesima contenere. Le più belle fra le testè indicate pellicie si pagano a Costantinopoli sino 50 mila piastre. *Ranzani*, tomo 2°, parte 2°.

been thought identical with the common fox of Europe, to which it bears a resemblance sufficiently striking to mislead an incidental observer. But by the fineness of its fur, the liveliness of colour, length of limbs and slenderness of body, as well as the form of its skull, it is obviously distinguished.

Red foxes are very numerous in the middle and southern states of the Union, and are every where notorious depredators on the poultry-yards. Their haunts are most commonly in exceedingly dense thickets of young pine, where they can scarcely be followed even by dogs.

Like all his kindred species, the red fox is distinguished by the possession of keen senses and great sagacity or craftiness, which enables him almost to bid defiance to traps, while his strength and swiftness of foot render it extremely difficult to capture him in the chase. Once fairly roused by the hounds, this animal dashes off with great speed, and soon far outstrips pursuit, and did he not lose the advantage of his celerity by remitting his efforts, might soon render the exertions of the sportsman nugatory. But the persevering hounds again and again drive him to his utmost speed, and eventually wear him down, though not until a wide extent of country has been traversed, and huntsmen, horses and dogs have suffered severely from fatigue.

The general colour of this fox when in full summer pelage is bright ferruginous on the head, back and sides, but less brilliant towards the tail. Beneath the chin it is white, while the throat and neck are a dark gray, which colour is continued along the anterior part of the belly in a narrower stripe that passes along the breast. The under parts of the body towards the tail are very pale red; and the anterior parts of the fore legs and feet, as well as the fronts of the inferior part of the hind legs, are black. The tail is very bushy but less ferruginous than the body, the hairs being mostly terminated with black, which is more obvious towards the extremity than at the origin of the member, giving the whole a dark appearance. A few of the hairs are lighter at the end of the tail, but not sufficiently to allow us to state that it is tipped with white.

In summer the fur of the red fox is long, fine, brilliant in colour, and lustrous over the whole body. In winter its length and denseness is considerably increased. The red fox is nearly two feet long and about eighteen inches high: the tail is about sixteen inches long. The peltry is of considerable value, and employed in various ways by the manufacturers.*

---

* We subjoin in this note a description of the *Canis Decussatus*, Geoff. given by Sabine in the work above quoted, as the species is entirely doubtful, and may prove to be a variety of the black fox, (*C. Argentatus*) or a *mule* produced between the fox last named and the red fox. "The cross fox, in comparison with the red, is shorter on its legs, and has a larger and longer body, being altogether a stronger animal. The front of the head is gray, composed of black and white hairs, the latter predominating on the forehead; the ears are large, covered with short, soft, black fur behind, and within by long yellowish hairs; the back of the neck and shoulders are pale, ferruginous, crossed with dark stripes, one extending from the head to the back, the other passing the first at right angles. The rest of the back is gray, composed of black fur tipped with white; the feet are white beneath the

THE RED FOX.

The red fox when caught young may be domesticated to a considerable degree, but it is rendered extremely unpleasant by the fetor of its urine, which very strongly resembles the abominable odour of the skunk, (*Mephitis Americana.*) We have lately had the pleasure of seeing a female red fox in the possession of Dr. BETTON, of Germantown; this animal is very interesting by its playfulness and vivacity. It lives in the same cage, and in perfect harmony with a raccoon, (*Procyon Lotor;*) shows no fear nor enmity towards the dogs of the farm, but always exhibits the greatest delight on being allowed to play with them. All the gestures and movements of this fox are exceedingly similar to those of a small dog, but are performed with remarkable quickness.

A very young whelp of this fox was some time ago brought to the Philadelphia Museum in company with its foster mother, a common cat, which had adopted and appeared to be very fond of it. She continued to nurse the little fox for several weeks, expressing much affectionate solicitude when he

---

under part of the tail, and the adjacent parts of the body are pale yellow; the gray character of the back extends to the upper part of the tail, at the commencement; the rest of the tail is dark above and lighter beneath, being tipped with white. The character of the fur is thick and long. The specimen when set up will stand about fourteen inches high; it is two feet four inches in length, and the tail, which is thick and bushy, is sixteen inches long." The colours are very variable in different individuals, some being very nearly of the colour of the red fox, while others more closely approach the black or silver fox.

wandered from her, notwithstanding the frequent ungrateful bites inflicted by her vicious foundling. How long this singular relation might have continued, or to what result it would have led, is unknown. The fox strayed too far from his cautious nurse, fell from the platform of a tall staircase to the ground, and was killed: the poor cat evinced as much sorrow for her loss as if it had been really her own offspring.

Species IX.—*The Gray Fox.*

Canis Cinereo-Argentatus; Gmel.

Renard Gris: Briss. Quad. p. 41.
Agourachay: Azara, Quad. du Paraguay, i. p. 317.
Canis Cinereo-Argentatus: Sabine, Zool. App. p. 657.
Fulvous Necked Fox: Shaw, Zool. Miscel.

The gray fox is very common throughout this country, and is found more immediately in the vicinity of human habitations than either of the other species. It is pursued by our sportsmen with more pleasure than the red fox, because it does not immediately forsake its haunts and run for miles in one direction, but, after various doublings, is generally killed near the place whence it first started.

The gray fox, like all the species we describe, exhibits considerable differences of colour at different ages and in different states of pelage. The length of the head and body is about twenty-four, and of the tail eleven inches. The general colour of the animal is grizzly, becoming gradually darker from the fore shoulders to the posterior

1. Black Fox. 2. Grey Fox.

parts of the back, produced by the intermixture of fulvous hairs with those constituting the mass of the pelage, which are thus coloured; near the body the hair is rather plumbeous, then yellowish, then white, and then uniformly tipped with lustrous black. The front, from the top of the head to the edge of the orbits, is gray, while the rest of the face, from the internal angle of the eye to within half an inch of the extremity of the snout, is blackish; at the extremity on each side of the granulated black tip of the nose it is of a yellowish white. A fine line of black tipped hairs extends upwards and outwards, from half an inch below the internal angle of the eyes until it is intersected by a similar black line about half an inch beyond the external angle of the eye, thus forming a very acute triangle, whose basis is on the side of the face. This blackish gray triangle, joined to the peculiar sharpness of the face, and the line produced by the black whiskers on the sides of the nose, singularly increase the appearance of slyness and cunning expressed in the physiognomy of this animal. The face below this triangle is white, and the latter colour is continued semicircularly upon the upper part of the throat.

The under jaw is blackish, this colour extending along the line of the mouth, and passing about half an inch beyond the junction of the lips at the angle. The inner surface of the ears is clothed with short light yellowish hair; their tips on the outside are blackish gray, and the whole of the rest of their posterior surface is yellow, which colour descends encircling the neck, and is the only colour on the anterior parts, with the exception of a white spot on the breast. The inferior parts of the body are white, tinted slightly in some individuals with faint reddish brown. The tail is thick and bushy, and the fur on the upper side is pale yellow, slightly tipped with black; the under part is rust coloured, and the end entirely black.

Species X.—*The Swift Fox.**

Canis Velox; Say.

Burrowing Fox: Lewis & Clark, ii. p. 351.
Canis Velox: Say, Long's Exp. to the Rocky Mountains, i. p. 486.

This interesting species inhabits the open plains which extend from the base of the Rocky Mountains towards the Missouri River, and forms its dwelling by burrowing in the soil. It is smaller than any other fox we have described, and is not known to frequent forest countries.

The most remarkable circumstance peculiar to this fox is its extraordinary swiftness, which all who have seen it agree in declaring to surpass that of any other animal with which we are at present acquainted. The fleetest antelope or deer, when running at full speed, is passed by this little fox with

---

* We prefer to translate Say's specific name for the trivial appellation of this fox, rather than adopt the common name of *burrowing* fox, since all the foxes burrow more or less, and the surpassing swiftness of this animal is a much more distinctive attribute of the species.

the greatest ease, and such is the celerity of its motion that it is compared by the celebrated travellers above quoted to the flight of a bird along the ground, rather than the course of a quadruped. Other observers have stated that when in full speed over the plain, the effect produced on the eye makes the animal resemble a line drawn rapidly along the surface, so impossible is it to distinguish any of the parts of its body on account of the surprising velocity of its motion.

Unfortunately for us the notes taken by SAY of the external characters of this animal were lost, and he was obliged to make known the species from nothing but the head and part of the neck of one individual, and the cranium of another. He gives a full account of the peculiarities of the skull of this animal in order to prove its specific distinction from the red fox, (*C. Fulvus*) which we do not think it necessary to repeat in detail, but shall append to this slight notice of the animal his description of the head and skull which were preserved.

" The entire length, from the insertion of the superior incisors to the tip of the occipital crest, is rather more than four inches and three-tenths; the least distance between the orbital cavities nine-tenths of an inch; between the tips of the orbital processes less than one inch and a-tenth; between the insertions of the lateral muscles at the junction of the frontal and parietal bones, half an inch. The greatest breadth of this space on the parietal bones thirteen-twentieths of an inch.

"The hair is fine, dense and soft. The head above is fulvous, verging on ferruginous intermixed with gray, the fur being of the first mentioned colour, and the hair whitish at base, then black, then gray, then brown. The ridge of the nose is somewhat paler, and a more brownish line passes from the eye to the nostrils, (as in the *C. Corsac.*) The margin of the upper lip is white; the orbits are gray; the ears behind are paler than the top of the head, intermixed with black hairs, and the margin, excepting at tip, white; the inner side is broadly margined with white hairs; the space behind the ears is destitute of the intermixture of hairs; the neck above has longer hairs, of which the black and gray portions are more conspicuous; beneath the head is pure white. The body is slender and the tail rather long, cylindrical and black."

CHAPTER XV.

Genus XVI. Cat; *Felis*, L.

Gr. Αιλουροσ. *Lat.* Felis, Catus, Cattus.
Ger. Katze. *Ital.* Gatto; *Span.* Gato.
Fr. Chat. *Russ.* Kot; fem. Kotscha.

GENERIC CHARACTERS.

The head is rounded, having a short face and large eyes, of which the pupils open either circularly or vertically. The tongue is covered with horny prickles or papillæ, which have their points turned backwards. The body is long, compared with the length of the legs; the teats are either four in number, and situated on the belly, or three or four on the chest and four on the belly. The anterior feet have five toes, not joined by an intervening membrane; the toes on the posterior feet are generally four in number; all the toes are armed with sharp hooked claws, which the animals have the power of projecting or retracting at will. The tail varies in length according to the species.

Dental System.

In the upper jaw the incisors are ranged side by side in a straight line. The two first are of equal size, wedge-shaped and transversely grooved on their internal face; the third is twice as large as these two, pointed and grooved on its internal face. A vacant space separates the last incisor from the canine, which is very large, conical, slightly hooked, rounded on its internal and external face, and angular at its anterior and posterior edges. The first false molar succeeds the canine, and is a small, very obtuse tooth, with a single root. A vacant space separates that tooth from the succeeding one, or second false molar, which we regard as having the regular form; it is very broad, has several roots, broad from before backwards, compressed from within outwards, trenchant, and presenting nearly the form of a right angle; its edges are divided by two grooves, or rather by two serrations, which augment the trenchant power. The carnivorous, which has at least three roots, succeeds the false molar immediately; it is a third larger than that tooth from before backwards, and is in this direction divided into three parts; the first is a small tubercle with cutting edges; the second or middle one presents a large tubercle trenchant on its edges, of a right angled figure; the third is terminated by a nearly straight line, only a little inflected in its middle, and with cutting edges. At the internal face of this tooth, and at the base of the small obtuse tubercle, is a still smaller tubercle, which is connected by a salient point to the middle tubercle. Finally, the tuberculous molar is a very small tooth, very narrow from before backward, broader from the external to the internal side.

rounded, and having one or two roots; this tooth concealed at the base of the carnivorous tooth is entirely rudimental.

IN THE LOWER JAW the first incisive is somewhat smaller than the second, and this than the third; they are shaped like blunt wedges, and show a slight groove from before backwards, narrower on the side next to the canine than to the opposite side. The canine, which immediately follows the incisors, is strong, conical, and more hooked than that of the opposite jaw, rounded on its anterior and exterior, and angular at its internal face and exterior edge. A large vacancy separates this tooth from the first false molar, which is broad from before backward, slender from the inner to the outside, with trenchant edges, and having the right angled figure, whose edges are divided by a groove as in the upper jaw. The succeeding false molar does not differ except in being larger and having an additional groove on its posterior edge; both of these teeth are normal. The carnivorous tooth is, like the preceding, compressed from within outwards, having cutting edges; but it is divided into two nearly equal parts by a deep groove in its middle, still more evident on its internal than its external face.

In their reciprocal position the incisors are opposed crown to crown, whence in old animals the grooves of which we have spoken wear away, and as these teeth are alternate, that is, the middle of these teeth in one jaw corresponds with the interval which separates the two opposite incisors, they wear unequally and become pointed instead of remaining in a straight line. The anterior edge of the upper canine is in relation with the posterior and exterior edge of the lower canine. The superior false molar only fills the void between the canine and the first inferior false molar. The posterior edge of the last named tooth acts against the anterior edge of the opposite false molar; the latter, by its posterior edge, acts upon the internal and anterior face and internal tubercle of the opposite carnivorous tooth. The whole breadth of this tooth is opposed to the external face of the lower carnivorous, which only touches the tuberculous tooth by its basis, or by that part of it which at its posterior part is nearest to the roots.

It will be seen by the number, form and disposition of these teeth, that the jaws of the cat are very short, and that the teeth being situated near the power which moves the jaws, they can act with great force, more especially as the points of articulation of the jaws, or condyles, are on a line with the teeth. The cat is absolutely carnivorous, eating nothing but flesh, and preferring much that of animals recently killed; they do not eat bones, except such as are not very hard, or when strongly pressed by hunger.

Nature having destined the animals of this genus to subsist exclusively on the flesh of other creatures, has endowed them with insatiably blood-thirsty dispositions, and furnished them with the most effective means of destruction. Their muscular strength, especially that moving the jaw, is exceedingly great, and gives their keen lacerating teeth, and strong sharp edged, pointed claws, terrific efficacy in the infliction of wounds, while their peculiarly flexible bodies and agility of motion enables them to spring with vast force upon their destined victims. They are sly, insidious and ferocious, approaching their prey by stealth or stratagem, and secure it by a sudden bound, or, failing in their first attempt, sneak off as if ashamed, but in reality because they are not well adapted to succeed in running it down. Most of them climb with great facility, the larger species especially bounding with a few efforts into the tops of lofty trees; all of them (with the exception of one species not found in this country) have their horrid claws entirely concealed in a sheath, and protected from wearing or becoming blunted, unless protruded at the will of the animal when about to strike its prey, or to aid in ascending trees, &c.

Few creatures exist which are destitute of peculiar beauties, either of adaptation or ornament, and it may appear singular to such as examine but slightly, that nature should have been lavish of adornment to a race of animals so generally injurious and hateful, or that the species most remarkable for invincible ferocity and destructive habits should be most eminently beautiful. From the dread monarch of the desert, whose limbs combine every attribute of vigour, and whose tawny mane adds terrific grandeur to his aspect of savage defiance, down to the domestic cat, whose smooth glossy fur and demure countenance might induce a belief in her peculiar innocence, we may observe every degree of beauty connected with fine proportions, graceful forms and agile movements, clothed with skins most richly variegated; an unequivocal evidence that external ornament is no test of disposition, and that mere beauty should not persuade us that the possessor is therefore excellent.

All these animals are called *cruel*, and their disposition to destroy the lives and drink the blood of other creatures growing out of a physical necessity, has been termed *cruelty*. Such expressions applied to animals are entirely incorrect, because it implies a power of discrimination which we do not believe them to possess. It would be exceedingly cruel in a human being to play with the terrors of an animal about to be put to death, and allow it to make numerous fruitless efforts to escape; yet this act is perfectly natural to the cat, and not more *cruel* than the most ordinary movement, inasmuch as the cat can have no idea of the suffering inflicted.

We must not rashly conclude that these animals are an evil unattended by any utility or good. They are designed by nature to occupy regions where animal life is most likely to increase in undue proportion, and it is their province to keep this increase from becoming excessive. It is in the sultry deserts of Africa and the vast plains of Asia, that the gaunt lion raises his hollow roar, threatening destruction; it is in the jungles of India that the untameable tiger lurks for his prey, and in the dense and remote forests of our own country that the fierce and vindictive cougar utters his startling scream. In all cases their haunts are in the vicinity of numerous herds of animals, whose superabundant increase they are engaged in restraining. They live far retired from the habitations of man, are princi-

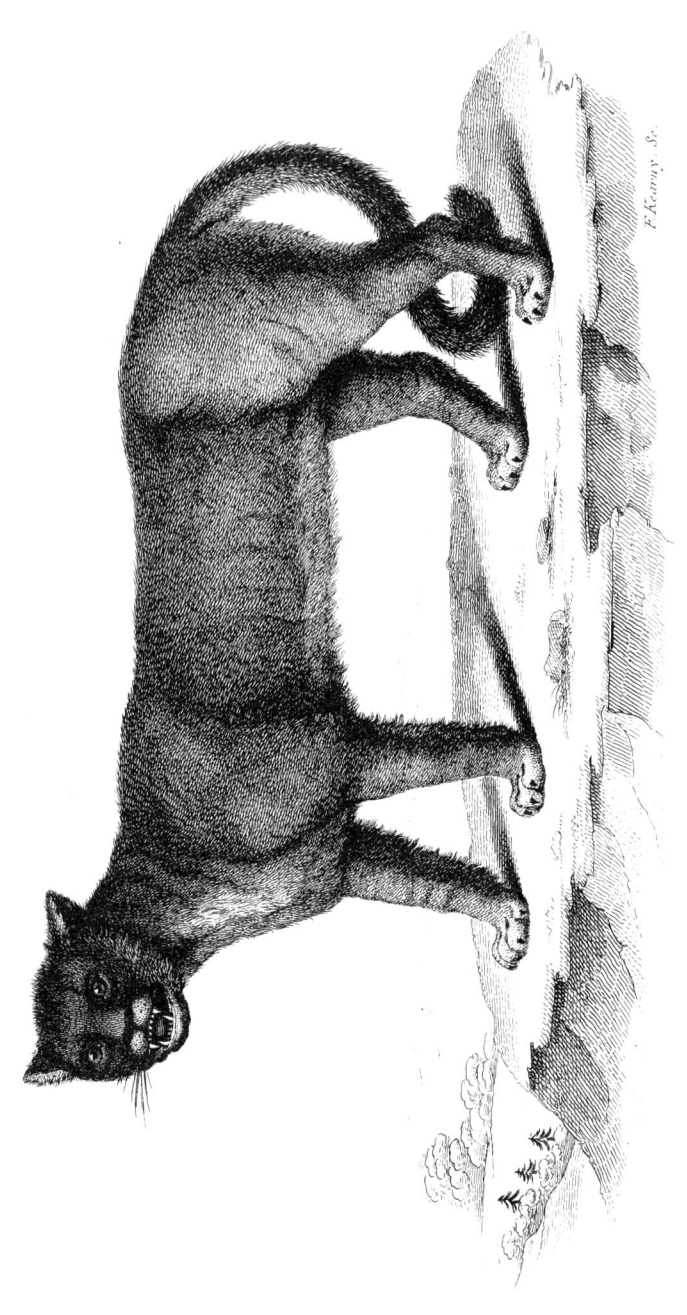

pally solitary and nocturnal, and appear to diminish in direct proportion to the advances of the human race, and the extension of cultivation. Hence we find that nature has not in these animals entailed a curse on mankind, which are so circumstanced, that their number may be diminished and their race even finally extinguished, as their services become less required.*

Species 1.—*The Cougar.*

(*Improperly called* Panther.†)

Felis Concolor; L.

Felis Concolor et discolor; L. Gmel. Syst. Nat. i. p. 79, Sp. 9-12.
Le Cougar; Buff. Quad. vol. 9, tab. 59; *Cougar de Pennsylvanie et Cougar Noir;* Buff. Supp. 3. p. 41, 42.
Pouma; Garcilasso, liv. 8. chap. 18.
Gouzara; De Azzara, Quad. du Paraguay, vol. i. p. 133.
Felis Concolor et discolor; Temminck, Monog. de Mam. livrais 4, p. 134.

The Cougar is the largest animal of the cat kind found in North America, and has occasionally re-

---

* For the gratification of our readers, we subjoin a note on this genus in the words of Linne, not only the most learned, but one of the most eloquent naturalists whose writings we possess.

"Felis genus sobrium, scandit facile arbores, cursu celer, noctu potissimum speculatur, casu delabitur in pedes contusioni vix obnoxium; glans penis ipsis muricatus retrorsum; retractis unguibus acutissimis incedunt, iisque sæviunt; inhiando sanguinem hauriunt; visa præda caudam movent; vegetabilibus ultro non vescunt."

† There is a great deal of confusion in the books relative to the true *panther,* which Temminck has decided to be the

ceived the name of American lion, from the similarity of its proportions and colour to the lion of the old world. It is very little inferior in size, and not at all in the qualities of magnanimity, clemency, and generosity which have been so lavishly yet so falsely attributed to the " king of beasts," whom we might believe to have been distinguished as royal in derision of some of the human species, who despotically rule over their fellow creatures by virtue of the " right divine" of power.* The cougar

---

Felis Pardus of Linné. The name employed by our country people generally, is "*Painter*," evidently a corruption of Panther. We have deemed it advisable to use the term *cougar* for a trivial name, to avoid the confusion which would be occasioned by that of *panther*, more especially as the former is more generally known and used than that of *puma*, applied by some European naturalists to this animal.

* The " lordly lion" conceals himself near the places where deer and other animals come to drink, and springs upon them from his ambush, like the veriest tom-cat; having feeble sight, and being unfit for the chase, he follows the wild dogs and chacals, which run down buffaloes, antelopes, &c. and when they have succeeded, drives them off and gorges to repletion: as he relinquishes the carcass when satiated he is called *generous*; as he does not attack and devour men—when not hungry, he is considered *magnanimous;* he retires slowly, facing his enemies, being unable to run with speed, and is celebrated for his noble spirit, and as he does not kill the wild dogs and other small animals because—it is not in his power to catch them, he is then called *clement;* while in virtue of his great strength, dreadful claws, horrid teeth and awful roar, he is considered as altogether *royal.* Yet this king of quadrupeds has not half the moral excellence of a poodle dog, nor a thousandth part of the dignity of character possessed by the elephant. He is,

may be stated to be about one-third less in size than the lion, and has no mane nor tuft at the extremity of the tail, which is about half the length of the body and head.

The skin of this animal is clothed with a soft and close hair over the limbs and body, of a brownish yellow colour, or a mixture of red and blackish, with occasional patches of a rather deep reddish tint, which are only remarkable in certain lights, and disappear entirely with the advancing age of the individual. A dark red is spread over all the upper parts, produced by the tips of the hair, which is black at base, and this colour is deeper upon the back, the head and upper part of the tail, than upon the sides. The belly is pale red; the breast, inside of the thighs and legs, are of a reddish white, and the lower jaw and throat entirely white. The ear is whitish internally and black externally, with the exception of the small external lobule, which is reddish gray. The head has a great many gray hairs upon it; the whiskers are white and rise in a blackish space; the end of the tail is black. The round spots varying in size, are more or less distinct according to the age of the individual, and are found in greatest number on the thighs, where they are not

---

moreover, no match for the great tiger of Asia, which, in ferocity, savage daring, audacious destructiveness, unconquerable and inappeasable hatred to mankind, is infinitely more *royal*, and a more consistent emblem of a great number of human *kings* who have aided in various ages and countries to retard the progress of improvement and the march of mind.

so close together. The sexes are not distinguishable by the colour of the pelage.*

The cougar was at an early period distributed in considerable number over the whole of the warm and temperate regions of this continent, and is still found, though by no means abundantly, in the southern, middle and north-western parts of the Union, becoming, however, gradually more rare as the population increases and cultivation is extended. It is a savage and destructive animal, yet timid and cautious: in ferocity it is quite equal to most of its kindred species, and kills numbers of small animals for the sake of drinking their blood, and when pressed by hunger attacks large quadrupeds, though not always with success. When the cougar seizes a sheep or calf, it is by the throat, and then flinging the victim over its back it dashes off with great ease and celerity to devour at leisure; deer, hogs, sheep and calves, are destroyed by the cougar whenever they are within reach, and occasionally one or two of these animals have committed extensive ravages among the stock of the frontier settlers. They climb or rather spring up large trees with surprising facility and vigour, and in that way are enabled, by dropping

---

* We have chosen to draw up this description from the excellent monography of TEMMINCK, on account of the admirable opportunities he has enjoyed and profited by, of visiting all the rich collections of Europe, especially of Paris, where there is a full series of cougars preserved in the museum. In this city there are very few specimens, and of these a still smaller number fit to furnish materials for a description of extensive application.

suddenly upon deer and other quadrupeds, to secure prey which it would be impossible for them to overtake.

In the day time the cougar is seldom seen, but its peculiar cry frequently thrills the experienced traveller with horror, while camping in the forest at night, or he is startled to hear the cautious approaches of the animal, stealing step by step towards him over the crackling brush and leaves, in expectation of springing on an unguarded or sleeping victim, whom nothing but a rapid flight can save. That the cougar will attack animals of large size and great strength is well known to those who have resided where this beast is found, in proof of which we may here insert a letter received from a scientific friend, who, during twenty-five or thirty years, has resided in the state of Ohio, and made the study of natural history his chief amusement.

Lexington, Ky. March 26, 1826.

DEAR FRIEND,—About the first of December, 1819, I visited Shane's prairie with the view of procuring specimens of every quadruped to be obtained in that district of country. I made early inquiry of SHANE* as to what animals might be

---

* " We travelled for twelve miles over a swampy country through which the St. Mary flows, after which we struck a dry plain, known by the name of Shane's prairie, and at eighteen miles from Fort St. Mary, we crossed the river at a settlement called Shaneville; both the prairie and the settlement, (which consists of but one family) owe their appellation to an interpreter, who is a half bred Indian; his

collected, and whether it was possible a cougar could be found. On the subject of the cougar he expressed some doubt, but believed that one was still lurking somewhere between Shane's crossing of St. Mary's river and Fort Wayne. This individual, he was of opinion, was the only one which for many years past had resided within his vicinity. He related various anecdotes of the alarm caused to travellers by this animal, especially at the twelve mile swamp, then a common camping ground on the Fort Wayne road. The following is the most interesting:

About the close of the late war, a merchant of Piqua named Herse, received a considerable sum of money in small bills, which made it appear of still greater magnitude to several suspicious looking persons who were present when it was received. Mr. Herse being unarmed, was apprehensive that an attempt would be made to rob him at the camping ground, and expressed his apprehensions to a single

---

father was a Canadian and his mother an Ottawa. He was employed as an interpreter and spy by General HARRISON during his western campaigns, and is considered as having acquitted himself of his duties faithfully; on the conclusion of the war he was rewarded by the grant of half a section of land, (320 acres) which he has divided into town lots; he resides within a short distance of Shanesville on part of his grant. No man is better known in this part of the country than Shane; his influence among the Indians is great, and he enjoys a high degree of popularity among the whites, founded upon the uniformly good character which he maintained during the war, and upon the unbounded confidence reposed in him, by General Harrison.—*Long's Expedition to the Source of St. Peters' River.* By W. H. KEATING. A. M. vol. i. p. 76.

fellow traveller who was also unprovided with arms. In consequence, they resolved not to go to the camping ground, but to pass the night in the woods without fire; there, turning their horses loose, they lay down in their blankets on the leaves. In the night they were aroused by hearing the horses snort, as they are apt to do on the approach of Indians, and shortly after they were heard to make several bounds through the woods, as if some one had unsuccessfully attempted to catch them. After some time had elapsed they both distinctly heard what they supposed to be a man crawling towards them on his hands and feet, as they could hear first one hand cautiously extended and pressed very gently on the leaves, to avoid making a noise, then the other, and finally the other limbs in like manner and with equal care. When they believed that this felonious visitor was within about ten feet of them, they touched each other, sprang up simultaneously, and rushed to some distance through the woods, where they crouched and remained without further disturbance. A short time after they heard the horses snorting and bounding furiously through the woods, but they did not venture to arise until broad day-light, being still ignorant of the character of their enemy.

When sufficiently light to see, by climbing a sapling they discovered the horses at a considerable distance on the prairie. On approaching them it was at once evident that their disturber had been nothing less than a cougar. It had sprung upon the horses, and so lacerated with its claws and teeth their flanks and buttocks, that with the greatest difficulty were they able to drive the poor creatures before them to Shane's. Several other instances of annoyance to travellers had happened at the same place, and Shane believed by the same cougar.

I now offered, through Shane, a reward of ten dollars to any of the Indians who would bring in this animal, and a few evenings after, on returning from a day's hunting, I found an Indian waiting with the body of the cougar, which he had killed about two hours before. This Indian found its track about fourteen miles from Shane's, and tracked it to within about two miles of Shane's house, where he was on the point of abandoning the chase. At this moment he heard the bushes rustle, and turning he saw the beast, which had sprung against the body of a tree to observe its pursuer. He instantly fired, and shot him through the heart, as I found on dissecting the animal. The Indian dragged the body about a quarter of a mile on the snow, but finding it too heavy, came to Shane's and obtained a sledge, on which he brought it in.

The following measurements were made by myself on the spot:—Length from the tip of the nose to the root of the tail, 4 feet 5 inches.—Length of the tail, 2 feet 4 inches.—Height before and behind, (toes extended) 2 feet 7 inches.—Circumference immediately before the fore legs, 2 feet 7 inches.—Around the body just before the hind legs, 2 feet 6 inches.—Of the neck, close to the head, 1 foot 5 inches.—Of the wrist, $7\frac{1}{2}$ inches.—Length of the fore leg to the body, $11\frac{1}{2}$ inches.—Of the hind leg in the same way, 20 inches.—Of the head, just before the ears, $20\frac{1}{2}$ inches.—The eyes were brown.

ROBERT BEST, M. D.

This fine specimen was carefully prepared, and is now in the Western Museum at Cincinnati, Ohio. We have been furnished with a drawing of this animal from the pencil of Mr. C. Corwine, a respectable and promising artist, resident in that city, through the kindness of Daniel Drake, M. D. a gentleman who is not less distinguished by his profound acquaintance with his profession, than for the zeal and liberality with which he devotes himself to the advancement of natural science.

In the remote and thinly settled parts of Pennsylvania the cougar is still occasionally found, and the following relation of the manner in which two of these animals were recently killed, will be read with interest from the singularity of the attendant circumstances. We have the account through our friend Isaac Hays, M. D. direct from Mr. John Mitchell, the respectable gentleman who killed them, and presented their skins and heads to the Philadelphia Academy of Natural Sciences.

About five miles from Philipsburg, Centre County, Mr. Mitchell, on the 8th of December, 1825, shot at a buck (*Cervus Virginianus*) and wounded him in the shoulder. He followed the animal for some time, and at length perceived him at the distance of about forty yards, lying with his heels upwards, and a cougar holding him by the throat. The hunter discharged his rifle at the cougar and shot him through the heart, when this animal relinquished the buck, advanced four or five yards, and fell lifeless. Having again charged his rifle, and believing the panther to be dead, Mr. Mitchell, turning to wards the wounded buck, was surprised to see another cougar in the act of pulling down the head, and as it now appeared, the buck had been held down by the throat by both cougars at the moment the first was killed. The body of the buck was between the hunter and the second cougar, nothing but the head of which was visible. At this Mr. Mitchell levelled his rifle, and the ball entered it at the angle of the eye. The beast remained still for a few minutes, and then, for the first time, relinquished his hold of the buck and walked over it towards the hunter, who fired his rifle a second time and shot him through in the neighbourhood of the heart. At this moment the buck recovered his legs, stumbled over the body of the cougar, finally extricated himself, and ran off. A third discharge of the rifle pierced the cougar with another ball, yet he still remained on his feet, and it was not until the rifle was again charged and a fourth ball driven through the back part of the under jaw, that the animal fell and expired. What is most singular, is that the male should not have relinquished his hold of the buck when the female was killed, but continued in the same position until the ball entered his own head near the eye.* The buck ran near a mile before

---

* "Major Smith witnessed an extraordinary instance of the abstracted ferocity of this animal, when engaged with its food. A puma [cougar] which had been taken and confined, was ordered to be shot, which was done immediately after the animal had received its food. The first ball went through its body, and the only notice he took of it was by

he was finally overtaken and killed. During his walk home, loaded with the trophies of his success, Mr. Mitchell killed another buck, having, during an absence of four days, killed two cougars and four bucks.

The following account of the destruction of a large cougar, which is still preserved in the New York Museum, was given by the late Mr. Scudder. Two hunters, accompanied by two dogs, went out in quest of game near the Kaatskill mountains. At the foot of a large hill, they agreed to go round it in opposite directions, and when either discharged his rifle the other was to hasten towards him to aid in securing the game. Soon after parting, the report of a rifle was heard by one of them, who, hastening towards the spot, after some search, found nothing but the dog, dreadfully lacerated and dead. He now became much alarmed for the fate of his companion, and while anxiously looking around, was horror struck by the harsh growl of a cougar, which he perceived on a large limb of a tree, crouching upon the body of his friend, and apparently meditating an attack on himself. Instantly he levelled his rifle at the beast, and was so fortunate as to wound it mortally, when it fell to the ground along with the body of his slaughtered companion. His dog then rushed upon the wounded cougar, which with one blow of its paw laid the poor animal dead by its side. The surviving

---

a shrill growl, doubling his efforts to devour his food, which he actually continued to swallow with quantities of his own blood till he fell."—*Griffith's Translation of Cuvier, p.* 438.

hunter now left the spot, and quickly returned with several other persons, when they found the lifeless cougar extended near the dead bodies of the hunter and the faithful dogs.*

Species II.—*The Northern Lynx.*

Felis Canadensis; Geoff.

Felis Lynx: L. Syst. Nat. p. 83.
Le Lynx du Canada: Buff. Hist. Nat. Suppl. iii. p. 44.
Lynx du Canada: C. Ossem. Fossiles, Nouv. ed. iv. p. 443.
Gatto Lince: Ranzani, ii. p. 309. sp. 8.
Felis Canadensis: Geoff. Catal. des Mammif. p. 120.
Felis Canadensis: Sabine, Zool. App. to Franklin's Exped. p. 659.
Felis Borealis: Temminck, Monog. de Mammal. livr. 4e. p. 111.
 [Loup-Cervier; Lynx de Suède of the Furriers.]

The researches of the justly distinguished Temminck have reduced the catalogue of lynxes or wild cats inhabiting North America to two species, of which one is common to both continents, and one proper to the American. The species now to be described is found only in the northern regions of both continents.

---

* These incidents may remind many of our readers of the spirited and highly interesting account given in Cooper's "*Pioneers of the Susquehanna,*" of a combat between a female cougar and a mastiff, which may be referred to with pleasure and advantage, on account of its verisimilitude; a merit by no means common in such works, and more especially in relation to American animals. Nothing can well be productive of deeper disgust to one who has any knowledge of natural history, than the inappropriate and ridiculous re-

1 Canada Lynx. 2 Wild Cat.

THE NORTHERN LYNX.

The northern lynx* is a fierce and subtle creature, exhibiting most of the traits of character which distinguish animals of the cat kind. To the smaller quadrupeds, such as rabbits, hares, lemmings, &c. it is exceedingly destructive, never leaving the vicinities they frequent until their numbers are altogether destroyed, or exceedingly thinned. But the ravages of the northern lynx are not confined to such small game; it drops from the branches of trees on the necks of deer, and clinging firmly with its sharp hooked claws, ceases not to tear at the throat and drink the blood of the animal until it sinks exhausted and expires. It attacks sheep and calves in the same manner, and preys upon wild turkies and other birds, which it is capable of surprising, even on the tops of the highest trees.

The northern lynx is found in great abundance in the country south-west of the Hudson's Bay settlements, but it is scarce to the north of Churchill river. Large packages of the skins of this species are exported by the Hudson's Bay company to Europe annually; in one year nine thousand were sent. The fur is highly esteemed by the dealers in peltry.

The northern lynx is fearful of man, offers very little resistance when attacked, and is easily killed by a smart blow over the back. This animal is

---

lations of the habits and appearance of animals frequently introduced into popular books; when given correctly or consistently with the true character of the animals, references to nature always impart pleasure, and add to the durability of the work they are intended to adorn.

* The description of the *Felis Rufa*, bay lynx, or wild cat, will be given in the appendix.

not often found to approach closely to the European settlements, but frequents the plains and woods where the animals on which it subsists are obtained in greatest abundance. The flesh of this lynx is considered good food by hunters, being fat, white, and flavoured like the hare, on which it principally feeds.

The northern lynx has a large body and strong legs, and measures about three feet from the tip of its nose to the end of its tail, which is about six or seven inches long, and black for half its length towards the extremity. The head is thick and round, and the ears sharp and tipped with a tuft of black hair. There are four or five small undulating bands on the cheeks, and the labial whiskers are white. The animal is about sixteen inches high.

The general colour of the northern lynx is deep reddish, marked on the flanks with small oblong spots of a reddish brown, with small round spots of the same colour on the limbs. The ears are black externally, but covered by an angular space of shining ash colour; the eyes are surrounded by a whitish circle to a black longitudinal mark above them, running from each side towards the front. The back is never marked by a black band along its middle.

In summer dress the pelage is short, the hair being brown at the base and of a bright red at the point. In winter the hairs are longer and all their points are whitish; the silky hairs, which are most numerous and long in winter, render the colour of the animal ash or whitish, which in summer gives place to the more decided red, marked with brown spots.*

---

* Temminck, Monog. de Mammal. livr. 4e. p. 111.

CHAPTER XVI.

Family IV.—Carnivora Amphibia; *Amphibious Flesh-Eaters.*

The peculiarities of conformation observable in these animals clearly demonstrate that they are destined to pass the greater part of their lives in the water. Their bodies are elongated, tapering from the anterior part of the chest to the posterior extremities, very flexible and powerfully muscular. The members are so modified as to be enclosed by the skin of the body, allowing only their flexible extremities to project; the hind legs presenting backwards, and placed in such a manner as to correspond to the tail of a fish, both in use and position. The heads of these animals are either entirely destitute of projecting ears, and have a small slit, which is closed by muscular action to prevent the entrance of water, or they have very small triangular ears, little more than perceptible. The nostrils are provided with a peculiar muscular apparatus, by which their orifices are perfectly closed at the will of the animal, effectually excluding water from the lungs during submersion. Their power of swimming is still further augmented by their coat of close smooth hair, which uniformly has the points presenting towards the posterior extremities. The feet are very imperfectly adapted for walking.

Genus XVII. Seal; *Phoca*, C.

Gr. Φοκη.
Ger Robbe; Seehund.
Fr. Veau Marin, Phoque.
Dutch, Rob; Zeehond.
Den. Sœlhund, Sœl.
Swed. Själ.
Ital. Foca.

GENERIC CHARACTERS.

The seals, like other mammiferous animals, are provided with four limbs, yet nothing but their extremities appear externally, being closely covered up by the integument of the trunk of the body, the fore limbs to the wrist, the hinder to the heel. The digits of the fore feet are successively shorter from the thumb, which is the longest. The posterior feet have the lateral digits either longer than the intermediate, or the whole nearly of an equal length. On the upper lip are strong erectile whiskers; the tongue is smooth and bifid at tip. The stomach is simple, the cœcum short, the digestive tube long and nearly equal in size. The nose is closed by the action of muscles at the extremity of the nostrils. The heart is formed in all respects analogous to terrestrial warm blooded animals, but their blood is very black and abundant. They have a large venous sinus in their liver, which (Cuvier thinks) may aid them in diving, by rendering respiration less necessary to the circulation of their blood.

Dental System.

We have seen, in describing the dental systems of insectivorous and carnivorous animals, what strong resemblance exists between the molars of the former and the tuberculous molars of the latter, being alike in their forms and destinations, composed of the same tubercles, and arranged according to the same relations, but only a little more obtuse in the carnivorous than in the insectivorous animal, and in all better suited for crushing than cutting.

We shall now see in the seals of our first division all the molars assume the form of the regular false molars, varying in their degree of slenderness and trenchant character, with deeper or more numerous serrations on their edges, and having several roots; and in those of the second division, we shall find them take in thickening a more or less conical form, which seems so much the more to make the transition from these teeth to those of some cetaceous animals, as each of them appears to have but a single root.

These are the only two general forms under which we find the molars of seals; but the divisions they characterize, and which may be considered as suborders or families, are divided into several groups by other considerations, and among these by the incisive teeth, which differ in number in different species. In this respect the seals, having teeth with several roots, form three divisions: 1st, those having six superior and four inferior incisors, among which we find the common seal:—2nd, with four superior and four inferior, (as in the *P. Monachus*):—3d, having four superior and two inferior, (the only example of which is the *P. Mitrata.**)

---

* Sent from New York to Paris, and thus named by Mr. Milbert.

Seals having several Roots to their Teeth.

34 Teeth:	18 Upper	6 Incisive 2 Canine 10 Molar.
	16 Lower	4 Incisive 2 Canine 10 Molar.

IN THE UPPER JAW the first incisor is rather smaller than the second, and that, half the size of the third; all are hooked, terminating in a point resembling canine teeth in form, especially the last one. The canine follows after a vacant space; it is strong, uniformly rounded, except on its inner surface, where there are slight longitudinal ribs, separated at the base and united at their points. The first molar, situated at the base of the canine, is one-half smaller than the others, rounded, terminated by a point, around which are placed some other very small points irregularly disposed. The four which follow and resemble each other have the forms of false molars, but their posterior cutting edge is separated by two grooves into two serrations, the first very deep and the second slighter. These grooves are not so distinctly marked upon the last of these teeth.

IN THE LOWER JAW the first incisor is much smaller than the second, and they both partake slightly of the canine form. The canines resemble those of the other jaw, as do the molars, except that there are one or two grooves, and by consequence one or two serrations on the anterior cutting edge of those in the lower jaw.

In their reciprocal position the incisors and canine teeth of both jaws have the same relations as in the carnivorous animals, and the molars also resemble, in this respect, the false molars of these quadrupeds. They are alternate, and do not pass each other so as to cut like scissor-blades, but the trenchant surfaces of the opposite teeth are applied immediately against each other, dividing the food by direct compression.

The common seal (*Phoca Vitulina*) furnished Mr. F. Cuvier with this type of dentition.

A comparatively short time has elapsed since the animals of this genus served as fruitful themes to declaiming theorists, and gave ample scope to their ingenuity in explaining the supposed relations existing between them and other parts of the animal kingdom. From the ponderous volumes of Aldrovandus and Gesner, down to the fascinating eloquence of the inaccurate Buffon, these beings have been considered as a sort of anomaly, bearing the same relation to fish as that in which the bat was supposed to stand to birds; they have been invested by ignorant observers with various imaginary attributes, which have been frequently perpetuated through the heedlessness and prejudices of the learned. Happily for us, the absurdities involving this part of our subject have gradually and finally disappeared before the increasing light of science, leaving no food for wondering credulity, but developing innumerable objects for enlightened admiration, in the study of the beautiful modifications of structure by which the great Author of nature has enabled breathing and warm blooded quadrupeds to dwell in " the vasty deep" without in the slightest degree depriving them of the intelligence or other characteristics of their order.

The natural history of the seal has been known during a great lapse of time, and what is more singular, is as correctly given by Aristotle as by Buffon, with all his advantages. This difference may be readily accounted for by observing that Aristotle states the facts which he had ascertained, without endeavouring to suit them to any preconceived opinion. De Buffon, believing in the absurd notion that animals capable of living for a considerable time under water had an opening between the right and left auricles of the heart, insists upon the existence of such a communication in the seal, even in opposition to positive demonstration. The prejudicial influence of error, when favoured by great men, is very clearly seen in the instance of the French naturalist, for the mere expression of his opinion was sufficient to induce M. La Verniere, who had dissected a seal and disproved the existence of the opening in the heart, to discredit the testimony of his own senses,* a degree of complaisance which we may hope will meet with few imitators at the present day.

---

* Je ne sais si le changement d'habitudes que cet animal avoit contractèes auroit pu former une membrane de cette structure; mais *il me suffit*, monsieur, *que vous en affirmiez la possibilité*, pour être de votre sentiment."—*Rep. de M. de la Vernière;* BUFF. 34. p. 47.

OF THE SEAL.

Seals are found on the sea coasts throughout the world in various degrees of abundance, and some species are peculiar to certain latitudes. They are most numerous very far to the north, where they almost exclusively furnish the Eskimaux resident in those chill regions with food, clothing, and implements made from their bones, &c.

Seals are viviparous, bringing forth and suckling their young on land; they are polygamous and gregarious, living in large families together, and exhibiting curious traits of character which will be described when treating of the species. They swim with admirable facility, remain for a considerable time under water, and derive their food entirely from the sea. They are very fond of sunning themselves upon the sea-beach or on ice-banks, scrambling upon them by aid of their flippers or fore feet. On land their movements are awkward and heavy, but not so slow as we might suppose from their appearance: to this motion of the seal the term " walloping" has been aptly applied. They are vigilant, intelligent, and tenacious of life. It was from imperfectly made observations on these animals, that the ancient fictions of sea-nymphs, mermaids, sirens and tritons were founded.*

---

* We subjoin a translation of Aristotle's account of the seal, that the reader may judge of the accuracy of this very ancient and truly illustrious naturalist, who flourished upwards of three hundred years before the christian era. The parts of his statement which are incorrect or doubtful are italicised.

"The seal is an amphibious animal; it does not inhale

De Buffon proposed to divide this genus into two parts or subgenera, founded on a character which to him appeared very striking and natural, the absence or presence of external ears.* This division was subsequently adopted by Peron, who formed his subgenera *phoca* and *otaria* for the reception of the species thus distinguished. The same arrangement is followed by Cuvier and Mr. Desmarest, but is rejected by Frederick Cuvier for the following reasons:—He thinks there is nothing sufficiently absolute in this character, and that the common seal, which is considered to belong to *phoca* and not to *otaria*.

---

water, but on the contrary breathes the air; sleeps and brings forth its young on land as if it were a terrestrial animal, but couples on the margin of the sea. It passes the greater part of its life in the water, and there obtains its food; absolutely viviparous internally and externally, the female bringing forth living animals enveloped in a chorion; she has milk like a sheep. The young are one, two, or *at most three* in number. The teats are two in number, which the young suck like other quadrupeds. *The seal brings forth like the human race at all seasons of the year;* nevertheless most frequently in the season when the kids are dropped. When the young are about eleven days old the mother leads them several times a day to the sea, to accustom them gradually thereto; but as their feet are not yet able to sustain them, *they allow themselves to slide along without walking.* The seal can easily *draw up its body and double it on itself*,* because it is fleshy, supple, and *the bones are cartilaginous*. Its great quantity of flesh renders it difficult to kill, if it be not struck on the side of the head.† Its voice resembles that of the bull.

* Buffon, par Sonnini, tom. 34e. p. 3.

* This is entirely true of *the neck* of the seal.
† "Αν μη τισ παταξη παοα τον κροταφον." Κεφ. F.

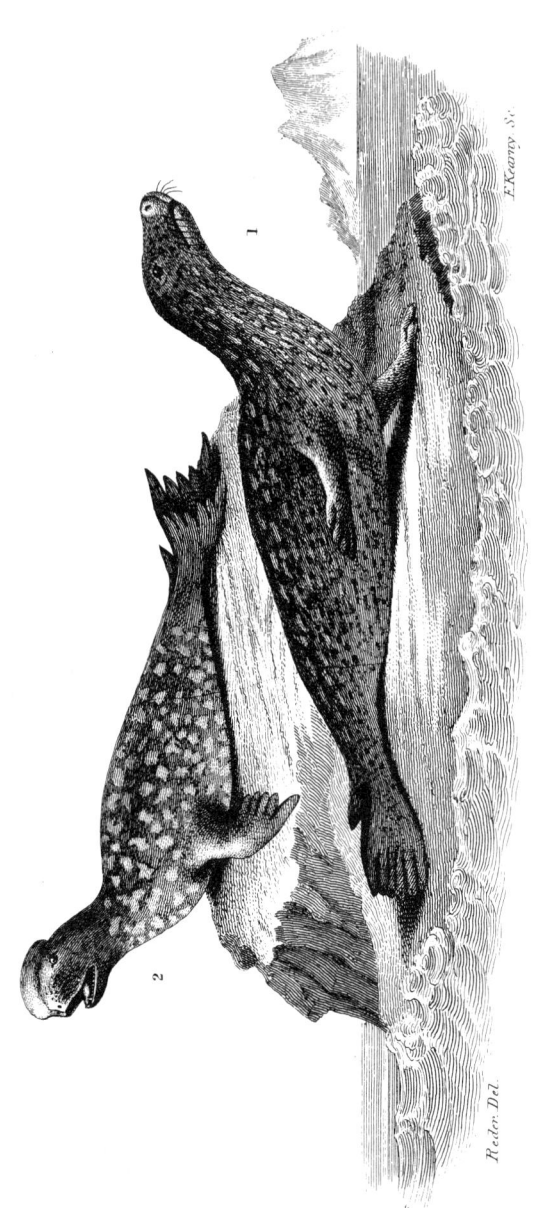

1. Common Seal. — 2. Hooded Seal.

has nevertheless a perfectly formed though a very small external ear. According to his views, the character drawn from the incisor teeth, first employed by Mr. Blainville for the divisions of an inferior order, is preferable to that commonly adopted, after BUFFON and PERON.*

SPECIES I.—*The Common Seal.*

Phoca Vitulina; L.

Vitulus Oceani: ROND. 453.
Kassigiak: CRANTZ, Hist. of Greenland, i. 123.
Le Phoque Commun: BUFF. p. 34.

Our impressions relative to this animal will be very opposite, according as we see it for the first

---

* "Sans examiner la valeur physiologique de ce caractére, je ferai seulement remarquer qu'il n'a rien d'assez absolu, et que le phoque commun qui est considéré comme un phoque et non point comme un otarie, a une conque externe très petite il est vrai, mais très nettement formée. Quoiqu'il en soit cette division a généralement été suivie, et les derniers travaux des zoologistes la reproduisent. Le point de vue sous lequel j'envisage les phoques, ne me permet pas de m'y conformer; je crois même que les caractéres pris des incisives, et que M. Blainville a le premier employé pour les divisions d'un ordre inférieur, est préférable á celui qui est tiré de l'oreille, quoiqu'il ne produise point encore de reunions naturelles d'espéces comme nous aurons occasion de le faire remarquer. Ce sont donc les divisions formées par les dents que je suivrai dans les détails ou je vais entrer."
—*Des Dents Des Mammif.* p. 117

time upon the land or in the water. Viewed while basking on shore, its peculiar form and seeming helplessness lead us to misjudge its strength and activity, while its motions are so clumsy and awkward that it excites a degree of compassion for its apparent deformity and imperfection. But, beheld in that fluid to which every peculiarity of its conformation is so admirably adapted, we relinquish all ideas of its imperfectness, with surprise that we could for a moment have indulged in them. Its countenance enlivened by large, dark, and lustrous eyes, which vigilantly regard surrounding objects, is remarkably expressive of intelligence. All its actions indicate the exertion of great strength, which is every moment displayed in movements of most graceful activity, whether while surmounting with head erect the foaming crests of the billow, and cleaving the waves with wonderful swiftness in chase of the finny tenants of the deep, wheeling in easy circles by gentle flexures of the body while sporting with its frolicsome companions, or diving profoundly to elude the pursuit of an eager enemy.

The common seal frequents the sea-coasts perhaps throughout the world, but is most numerous in high northern latitudes, and furnishes the inhabitants of those frigid regions with nearly all their necessaries and luxuries. In such situations, the Eskimaux are denied the opportunity of deriving their subsistence from animals which depend upon the vegetable kingdom for nutriment, but to compensate for this disadvantage, the seas which wash their ice-bound shores are thronged with seals and walruses, supplying to them the place of flocks and herds, without

THE COMMON SEAL.

requiring from those fed and clothed by them any provision for their maintenance.

This seal has a round head, which at the fore part bears considerable resemblance to that of an otter, though the whole aspect is not unlike that of some varieties of the dog, whence the name of sea-dog, sea-wolf, &c. has been applied to different species of seal. The extremity of the snout or muzzle is flat and broad; the posterior part of the head is very large and without bony projections; the upper lip is peculiar, moveable and extensible, garnished with long, unequally thick, strong whiskers, which are capable of being erected, or thrown forward by the action of a peculiar muscle. The seal has no external ear, but instead of it a very small tubercle on the anterior edge of the opening to the tympanum, which is placed considerably posterior to the orifice. Over the eyes, which are much nearer to the ears than the nose, there are seven or eight bristles similar to the whiskers, but smaller. The fore limbs are short, and the feet have five digits, joined together by a membrane, having thick, long, black nails projecting from their extremities: these nails are longer on the hind than they are on the fore feet.

The general colour of the seal is of a yellowish gray, varied or spotted with brown or blackish in different degrees, according to the age of the animal. On the head and back the colour is generally darkest, while on the flanks and belly it is pale. In advanced age the colour is generally whiter. The hair in this species differs remarkably from what we find in others, being close and not presenting entirely backward. The hairs are individually stiff, harsh, flat and pointed, yet slender, dry and shining; they are blackish brown until near the point, and then yellowish gray.

From the organization of the common seal, as well as on account of the medium in which the greater part of its life is spent, we should not be induced to believe that any of its senses are remarkably acute. Its powers of vision appear to be considerable, though it sees much better in a moderate than in a strong light; its sense of smelling cannot be exercised to much advantage while the animal is under water, as at that time the nostrils are perfectly closed by muscular action. From the manner in which the whole external surface excepting the end of the nose is covered with hair, the sense of touch would appear to be slight, and the small size of the ears, as well as the manner in which they are generally immersed, lead to a belief that this sense also is not very acute.

Notwithstanding all these apparent defects, the seal is susceptible of a remarkable degree of education, learns to distinguish his feeder, to perform various actions when commanded, both in the water and on land, and acquires fondness enough for the society of domestic animals kept with him to attempt following them, in spite of his awkward and disadvantageous movements. The brain of the seal is very large, when compared with that of various other quadrupeds of less remarkable intelligence, and it is fair to infer that to this circumstance its intellectual superiority is attributable.

THE COMMON SEAL. 317

The manner in which the seal feeds is very interesting; when fish are thrown into a tub where several of these animals are kept together, they eagerly spring to a considerable distance in the air, raising half their bodies out of water and elongating their necks to the utmost. In most instances the fish is swallowed directly, without the slightest chewing, the swallowing being facilitated by the elevation of the head and straightening of the neck. If the fish be caught by the tail, it is immediately disabled by being crushed between the teeth, and is then turned head foremost and swallowed without chewing. When a fish too broad to be thus bolted is given, then the seal chews it, rather with a view of compressing it sufficiently to allow of its passage into the stomach than for the purpose of comminution. While engaged in feeding, the aspect of the seal is very different from what it is when the animal is quiescent. The upper lip is thickened and projected forwards, the bristles or whiskers fiercely erected, and the nostrils dilated and closed with force. They also feed while under water and swallow with as much ease as in the air, but in a different manner. Under water it opens the mouth but partially, and lowers the under jaw, while it separates the lips at the extremity, apparently drawing in the prey by suction. It is not yet ascertained in what manner the animal avoids the ingurgitation of water at the moment of thus swallowing.

In a state of captivity the seal expresses little or nothing of fearfulness, and does not avoid either man or animals, except when very closely approached. They are not inclined to bite or injure persons examining them, so long as no attempt is made to touch them with the hand, or otherwise disturb them, but if thus annoyed they snap fiercely, and also strike with their flippers or fore feet. Their characteristic vigilance never appear to forsake them a moment. Three of these animals, two adults and one nearly full grown, were exhibited in this city last season, in a large box or tub containing water. Though they appeared to be very sleepy at times, and unwilling to be disturbed, yet every half minute the eyes were slowly opened, and it was almost impossible to succeed in touching them without their being first alarmed. These were caught in a bay on the coast of Massachusetts, by aid of a net; when first captured they fought desperately, and were with great difficulty secured. After being a few months in captivity they became quite harmless, and the younger one learned to perform several tricks, placing himself in different attitudes at the command of the keeper. They all died during the winter, in consequence of too much exposure to cold, in a small quantity of nearly fresh water, together with a suspension of all their natural habits. One of these animals occasionally made a noise resembling the loud snorting of a horse, and all of them were in the habit of drawing the head under water every few minutes, as if to moisten the eyes.

Frederick Cuvier has published, in his splendid work on mammiferous animals, some very interesting observations on two young seals kept in the menagerie at Paris, from which we shall here introduce a sketch. These individuals showed no fear in the presence of men or other animals, never at tempting to escape or withdraw themselves, unless to avoid being trodden on, and then merely removing to a short distance. One of them would occasionally threaten with its voice and strike with its paw, but would never bite unless extremely provoked. They were very voracious, yet showed no ill temper when their food was taken from them, and some young dogs, to which one of the seals was attached, would snatch the fish from his mouth just as he was about to swallow it, without the seal showing any sign of anger. When both the seals were suffered to eat together, they usually fought with their paws, and the strongest drove the other away.

One of these seals was at first very shy, and retreated when any one attempted to caress him, yet in a few days he became quite tame and confident of the kindness of those who approached him.— When shut up with two little dogs that used to mount upon his back and playfully bark and bite at him, he soon entered into the spirit of their actions, and took pleasure in their frolics, striking them gently with his paw rather to encourage than repress them. When the dogs ran off he would follow them, though the ground was covered with stones and mud. During cold weather he would lie in close company with the dogs for the sake of their mutual warmth. The other seal evinced a strong degree of attachment to the keeper, recognising him at a considerable distance, and using many expressive gestures and looks to solicit his attention and obtain food, the idea of which was no doubt associated with the presence of the keeper. These seals barked commonly in the evening, or on a change of weather, though with a much feebler voice than that of the dog; their anger was exhibited by a kind of hissing noise.*

The common seal brings forth two young in autumn, and suckles them on shore until they are six or seven weeks old, when they are gradually accustomed by their parents to frequent the sea. At this period they are generally of a whitish or light fawn colour, covered with soft or woolly hair, and when in distress or hurt have a sort of whining voice. Seals are mostly associated in families consisting of a few males and a large number of females and young ones.† They are fond of landing on the sea-beach, ledges of rocks or ice-banks, for the purpose of basking in the sun, and in fine weather prefer being on the ice to remaining in the water;

---

* "A young seal, which was given by the master of a whaler to the officers of the Alexander, one of the ships on the former voyage, became so entirely domesticated and attached to the ship that it was frequently put into the sea and suffered to swim at perfect liberty, and when tired would return of itself to the boats and be taken in."—*Sabine*, p. 191.

† The principal part of the materials used in preparing the rest of our account of the seals and walrus is obtained from the valuable writings of CRANTZ, SCORESBY, PARRY and LYON. A very considerable number of other respectable authorities have been carefully examined, and their observations compared, with a view of correcting and enlarging the natural history of these useful and interesting animals. Much still remains to be desired to render this account complete; yet we may be allowed to hope that the reader will find a more ample collection of facts in the present instance than has yet been presented at any one view.

sometimes indeed they are very averse to take to the water when they have been out of it long enough to become perfectly dry.

When on their passage from one place to another they swim in very large flocks or shoals, and become visible to the mariner every few minutes, as they are obliged to come to the surface to breathe; this is generally done by the whole company nearly at the same time, when they spring up so as to raise their heads, necks, and even their whole bodies, out of the water. From the peculiar vivacity of their movements and general sportiveness of the company, such a shoal of these animals has obtained from the sailors the designation of a "seal's wedding."

The seal is peculiarly vigilant, and whenever a herd of them visit the shore some are always on the look out, and a seal when alone is observed very frequently to raise its head for the purpose of discovering the approach of enemies. Should they be on a large field of ice, they are always careful to secure a retreat by lying near the edge of it, or keeping a hole in the ice always open before them. The old ones are exceedingly vigilant and distrustful, the largest crowd of them immediately disperse at the approach of a boat, and few or none of them are taken, while the young ones, which are not so cautious, frequently fall victims to their inexperience by suffering the hunters to approach. The food of the common seal is fish, crabs and birds, which last it contrives to secure by rising under them and seizing their feet before they can be aware of its approach. Feeding on much the same food as some whales, the latter are not found where seals are very numerous. In the spring of the year the seals are fattest, and yield several gallons of blubber, small ones affording four or five gallons of oil.

In the high northern latitudes, during the winter season, the common seal is found many miles from any open water, and makes a very circular hole through the ice, even when it is several feet in thickness, and there comes up to breathe. This opening is continually kept clear, and allows the entrance of the seal's body, the top being permitted partially to freeze over. These breathing places bear a considerable resemblance to mole-hills, and have a small crack through their upper part.

Since the whale fishery has in some degree declined in productiveness, seal hunting has risen in importance to Europeans and Americans, some ships being now sent almost expressly for the purpose of procuring the oil and skins of these animals, which are of extensive importance in commerce and manufactures. We shall make mention of the methods used by the *sealers* to take these creatures, before we refer to the implements and hunting of the Eskimaux, to whom the seal is more important than bread to other people, inasmuch as they depend on it almost for every thing.

Seals are sometimes enticed to the surface by music, or the whistling of an individual who is prepared to shoot them, and this proves that they can hear far better when under water than we might be inclined to believe from a mere glance at their external ears. When they hear this sound they come to the surface, elongate their necks to the utmost extent, and expose them fully to the aim of the hunter. They are most effectually secured however by firing duck or other shot, which blinds them, so that they may be approached and despatched; when killed at once by a single bullet they most commonly sink. Another mode of killing the seal is to go to the caves on shore, into which herds of seals occasionally enter. When the sealers are properly placed they raise a simultaneous shout, at which the affrighted animals rush out in great confusion, and are despatched with wonderful quickness by a single blow on the nose, struck with a club. They are very tenacious of life when struck or wounded on any other part of the body.

The best situation for *sealing* in the Arctic Seas is stated by SCORESBY to be in the vicinity of Jan Mayen's Island, and the best season the months of March and April. When the boats arrive at the ice, the sealers immediately attack the animals with clubs and stun them by a single blow over the nose, which mode enables one person to destroy a large number of seals; when they are seen on pieces of drift-ice they are hunted by means of boats, each boat pursuing a different herd; should the seals attempt to leave the ice before the arrival of the boat, the sealers shout as loudly as possible, and produce such amazement in the seals by this uproar as to delay their flight till the boat arrives and the work of destruction is begun. Where the seals are very numerous the sealers stop not to flay those they have killed, but set off to another ice-field to kill more, merely leaving one man behind to take off the skins and fat. When the condition of the ice forbids the use of boats, the hunter is obliged to pursue the seals over it, jumping from piece to piece, until they succeed in taking one, which he then stops to flay and *flense,* or to remove the skin and fat. This sometimes is a horrible business, since many of the seals are merely stunned, and occasionally recover after they have been flayed and flensed. In this condition, too shockingly mangled for description, they have been seen to make battle and even to swim off.

The number of seals destroyed in a single season by the regular *sealers* may well excite surprise; one ship has been known to obtain a cargo of four or five thousand skins and upwards of a hundred tons of oil. Whale ships have accidentally fallen in with and secured two or three thousand of these animals during the month of April. The sealing business is, however, very hazardous when conducted on the borders of the Spitzbergen ice. Many ships with all their crews are lost by the sudden and tremendous storms occurring in those seas, where the dangers are vastly multiplied by the driving of immense bodies of ice. In one storm that occurred in the year 1774 no less than five seal ships were destroyed in a few hours, and six hundred valuable seamen perished.

Seal oil when properly prepared is pure and fine, and may be employed for all purposes to which whale oil is adapted. The skins of these animals are extensively consumed in various manufactures, especially in trunk making, saddlery, &c. The leather made from seal skin is perhaps the most pleasant material which can be worn in boots, on account of its lightness, softness and pliability, but it is too porous to be worn during the winter season, or in wet weather.

The Eskimaux hunt the seal in various modes, according to circumstances. When the breathing place already described is discovered, the hunter raises near it a small wall about four feet high, of slabs of snow, to shelter himself from the wind, and sits under the lee of his snow shelter, having deposited his spear, lines and other implements upon several little forked sticks set up in the snow, in order to avoid making the slightest noise in moving them when wanted. The most curious precaution, taken with a similar intention, is that of tying his own knees together with a thong to prevent any rustling of his dress, which would alarm the seal. In this situation the Eskimaux will frequently sit for many hours, when the thermometer is below zero, attentively listening to ascertain whether the animal is working below.

When he thinks the hole is almost completed, he carefully raises his spear, to which the line is previously tied, and the moment the breathing of the seal is distinctly heard, the ice being then of course very thin, he strikes the spear into him with both hands, and cuts away the ice with his knife to repeat his blow. At other times, having enlarged the breathing place, he takes his position behind the shelter, and the animal, when he next comes to the hole, rises fearlessly out of the water, exposing his head and shoulders, and repeats this action with increased confidence. As he is not in haste to dive again, the hunter now starts up suddenly and drives his spear forcibly into him. Another method adopted consists in covering the breathing hole with light snow, and making an opening through the top of it with the spear handle about as large as the mouth of a bottle. The hunter then withdraws the spear and takes his place behind his snow-screen, listening vigilantly until he hears the seal breathing beneath the snow, when he silently rises and plunges his weapon through the snow-covering into the body of the seal. The moment the seal is struck, the hunter endeavours to catch the line behind one leg to act as a strong check; and as an additional security, a hitch is taken round the ring finger, which is sometimes either dreadfully lacerated or entirely torn off by the violent struggles of a large seal. The animal is then stabbed until dead; the hole being enlarged, it is drawn out on the ice where it speedily freezes and is in condition to be drawn home.

When seals are seen on the edge of the ice next the open sea, the seal hunters dispose themselves in a single file, so as to conceal their number, and appear as few as possible when viewed from the point towards which they are moving. In this manner they creep cautiously towards the edge of the ice. When nearly close enough to throw the spear they all crouch low, and remain in this position for a quarter of an hour, during which time they get all their implements in in readiness for immediate service. Then when the seals are intercepted from view, they creep forward, gaining a few paces at a time, until they approach close enough to throw the spear, which is done suddenly and with full force. This mode of hunting the seal is occasionally attended by fatal consequences to the poor Eskimaux, especially if the ice be of recent formation, as large cakes are at times detached by the force of the tide, and swept out to sea without allowing the slightest opportunity for escape. When a seal is observed to come up through a hole in the ice, a hunter sets off for the purpose of approaching it by stratagem. He crawls through the snow on his hands and knees, taking care to remain stationary whenever the seal raises his head, and advancing as soon as the animal allows it again to rest on the snow. When stationary as well as when moving forward, the hunter imitates the actions of a seal with singular fidelity, and improves every opportunity of approaching, until he comes near enough to strike his spear into the body of his victim and secure his prize.

The implements used by the Eskimaux in the vicinity of Winter Island in their seal hunting are principally the spear and the knife. The spear is called *akliak* or *oonak*, has an ivory point, which is not fastened to the handle, but attached to it by a line, and when struck into the animal is immediately liberated from the handle, which is provided with a float or bladder. A few of the spears are made of a single narwal's horn of solid ivory, about four feet long, well rounded and polished. The bone of whale ribs is used for spear handles, or wood, when it can be procured, but the ivory of the sea-unicorn or narwal is more easily procured at Igloolik or Winter Island than either of these substances. When engaged in sealing on the ice the spear and knife are generally the only weapons used.

While hunting on the ice the Eskimaux use a long bone feeler, both for the purpose of sounding the cracks through which seals are thought to breathe, as well as for ascertaining the safety of the road. An instrument is occasionally used by them, when watching a breathing hole, serving the same purpose as the float of a fishing line. This is an exceedingly delicate rod of ivory, about a foot long and as thick as a fine knitting needle, having a small knob at the lower end not larger than a pin's head, and a very fine piece of sinew to the upper extremity, by which it is loosely attached to the side of the hole. This small object is not observed by the seal on rising in the hole, and as he raises it with his nose, the watchful Eskimaux strikes the unsuspecting animal as he approaches the surface. Whenever the seal is secured, the orifices made by the spear and knife are carefully closed by small pins of bone or ivory, to prevent the loss of the blood, which is highly prized by the Eskimaux.

The seal is generally very fat, as his supply of food is abundant, and the amount of blood contained in his body is far greater than would be inferred from comparing him with other animals. The flesh is of a very dark red colour, and rather soft; that of the young animal is thought to be quite good by Europeans, but the Eskimaux are extremely fond of it at every age and under all circumstances.

The common seal or *Neitiek* is the only seal the women are allowed to cut up among the Eskimaux residing near Winter Island. Before the knife is used on the animal they pour into its mouth a little water as it lies on its back, and with a little lampblack and oil taken from the under part of the lamp they touch each flipper and the middle of the belly. The object of this ceremony is unknown, but from the seriousness with which it is performed, seems to be one of high importance. The first operation in the division of the seal consists in cutting the animal into two parts, laying open the cavity of the belly. The boys then come eagerly forward to have a small piece of membrane or bladder stuck upon their foreheads, of which they are very proud, as it is to make them expert seal hunters. The intestines are then removed; next the blood is carefully collected and put into the cooking pot over the fire. The head and flippers are then separated from the carcass and the ribs divided.

The loose scraps are then put into the pot for immediate use, except such as the *lady* butchers occasionally cram into their own mouths with great relish, or distribute to the bystanders, who are all eager to catch such precious favours. The little children, old enough to make their way between the legs of their friends, also present themselves to the attention of those who are engaged in this division of food, and are highly delighted when their efforts are rewarded by lumps of raw meat which are thrust into their mouths. The poor dogs seem to be the only visitants not allowed to participate at this preliminary banquet, as their attempts to subtract some small portion is always rewarded by heavy blows with the knife-handle. During the whole operation the surrounding friends are eagerly engaged in chewing portions of the raw intestines, which they sometimes allow to become frozen, and then snap off with the same kind of glee as is displayed by our youngsters in munching molasses-candy. One of the women from each of the different huts attends with her cooking pot to receive her portion of the flesh. When nothing is left but the blubber attached to the skin, it is removed, and the two portions of the latter are rolled up and laid by with the store of flesh and blubber.

Then the feasting may be truly said to begin, and the voracious Eskimaux seem determined to indemnify themselves for all the privations to which they have ever been subjected. They gorge until they are absolutely stupified, and frequently are only saved from death by the occurrence of copious bleeding from the nose. One good trait, however, has been observed among them, and that is worthy of record; however hungry they may be, the children are supplied before any of the grown persons touch the food.

The descriptions given by various eye-witnesses of their filthy feasts, are such as almost to turn the stomach of a reader, though several of these accounts are given with great spirit and truth. We shall here introduce Capt. Lyon's picture of one of these entertainments in his own words, referring those who may wish to form a more extended acquaintance with the habits of the Eskimaux to his highly entertaining volume.

"On the 16th I was rejoiced to find the seal hunters had been successful; blood, blubber, entrails, skins and flesh, lying sociably intermixed in savoury heaps. Abundant smoking messes were in preparation, and even the dogs looked happy as they uninterruptedly licked the faces of the children, who were covered with blood and grease from the chin to the eyes. Universal merriment prevailed, and such men and children as could bear more food stood lounging round the women, who sat sucking their fingers and cooking as fast as possible. While the messes were preparing the children solaced themselves by eating such parts of the raw uncleaned entrails as their young teeth could tear, and those morsels which proved too tough were delivered over to their mothers, who soon reduced them to a proper size and consistency for their tender offspring.

"At the distribution of the contents of one of the pots I was complimented with a fine piece of half stewed seal's flesh, from which the kind donor, a most unsavoury looking old lady, with the most obliging politeness, had first licked the gravy and dirt, and bitten it all round in order to ascertain the most tender part on which I should make the first attack. My refusal of this delicacy did not offend, and we had much laughing on the subject, particularly when the old woman, with well feigned disgust and many wry faces, contrived to finish it herself. In my rambles on this day of plenty I found, beyond a doubt, that the women do not eat with the men, but waiting till they are first satisfied, then enjoy a feast by themselves. In the meantime, however, the females who superintend the cooking have the privilege of licking the gravy from the lumps of meat as they are taken out, and before they are presented to their husbands. Both sexes eat in the same manner, although not in equal proportions, the females very seldom, and the men very frequently stuffing until they become quite stupified. A lump of meat being given to the nearest person, he first sucks it all round and then pushes as much as he can into his mouth, cutting it from the larger piece close to his lips, to the great danger of them and his nose. The meat then passes round until it is consumed, and the person before whom it stops is entitled to the first bite of the next morsel. In this manner a meal continues a long time, as each eats or rather bolts several pounds, and the pots are in consequence frequently replenished. In the intermediate time the convives suck their fingers or indulge in a few lumps of delicate raw blubber. The swallows of the Eskimaux are of such a marvellous capacity that a piece of flesh of the size of an orange very rarely receives half a dozen bites before it is bolted, and that without any apparent exertion. The rich soup of the meat is handed round at the close of the repast, and each takes a sup in turn until it is finished, when the pot is passed to the good woman of the house, who licks it carefully clean and then prepares to make a mess for herself. On all occasions the children are stuffed almost to suffocation. The meals being finished, every one scrapes the grease, &c. from his face into his mouth, and the fingers are then cleaned by sucking."*

The food of the Eskimaux is cooked by the aid of a lamp which is supplied with seal oil, the wick being composed of moss. On this lamp are they also dependant for warmth in their huts, which are made of snow, as well as for their supply of water to drink, which, during a great part of the year, is only to be obtained by melting snow. A scarcity of seals, therefore, is accompanied by a series of ills, the hut is deprived of light and warmth, and the sufferings of famine are increased by the torment of thirst, against which they have no other resource under such circumstances except to eat the snow, which affords but

---

* Private Journal of Capt. G. F. Lyon, p. 141.

a partial relief. In judging of the filth and voracity of these poor creatures, we must ever bear in mind the circumstances which during so much of the time render water almost unattainable, except to quench their thirst, as well as the frequent and severe starvation to which they are subjected.

The Eskimaux apply the skins of seals to various purposes, amongst which the most important is the construction of their boats. The small boat to carry but one person is called *kayak*, and has been aptly compared in shape to a weaver's shuttle, having the head and stern equally sharp. There is an opening or hole in which the rower sits, having a rim or projection to which a part of the dress may be fastened in such a manner as entirely to exclude the water. The weight of the whole does not exceed fifty or sixty pounds, so that the boat may be readily carried by the owner on his head, and from the peculiar form of the rim without applying his hands.

The Eskimaux are very proud of their boats; they place a warm skin in the bottom to sit upon, and the position of the paddler is with the legs extended and the feet pointed forwards. Whenever any weight is to be raised, or the stowage of the boat to be changed, two kayaks lie together, and the paddles of each being laid across, a steady double boat is formed. When not paddling the occupant must preserve a very nice balance, and a tremulous motion is always to be observed in the boat. The Eskimaux in the vicinity of Winter Island have not the art of regaining the upright position when overturned by a dexterous use of the paddle. An inflated seal bladder is a constant appendage to the canoe equipage; the weapons are kept in their places on the upper surface of the boats by small lines of whale-bone, tightly stretched across so as to receive the points or handles of the spears beneath them. The stem or stern of the boat is frequently stowed with flesh, birds or eggs; a seal, notwithstanding its roundness and liability to roll, is so carefully balanced on the boat as seldom to require being tied on. When going before the wind while a smart swell is running, the kayak requires the nicest management, as the slightest inattention would expose the broadside to the sea and be followed by immediate peril to this frail vessel. The extreme velocity with which the *kayak* is impelled, and the dexterity with which it is turned and guided, render it a very interesting object.*

The Eskimaux use another boat made of seal-skin,

---

* " A flat piece of wood runs along each side of the frame, and is in fact the only piece of any strength in the kayak. Its depth in the centre is four or five inches, and its thickness about three-fourths of an inch; it tapers to a point at the commencement of the stem and stern projections. Sixty-four ribs are fastened to the gunwale piece, and seven slight rods run the whole length of the bottom and outside the ribs. The bottom is rounded and has no keel: twenty-two little beams, or cross pieces (made of ground-willow or small whale-bones) keep the frame on a stretch above; and one strong button runs along the centre from stem to stern, being of course discontinued at the seat part.

" *Length.*—Body, 19 feet; stem projection, 3 feet 2 inches; stern projection, 2 feet 10 inches; total, 25 feet. Abaft the hole, 8 feet; before it, 9 feet 7 inches. *Height.*—Rim in front, 10 inches; behind, 1¾ inches; breadth at centre, 1 foot 9 inches; depth at the same place, 10 inches. Circumference of the rim, 5 feet 1 inch."—*Lyon's Private Journal, p.* 321.

THE COMMON SEAL. 335

which is larger and destined to carry luggage, or to transport their families. This is called *umiak* or *oomiak*, and is made nearly square at the head and stern. Its frame is made of whale-bone or wood, and the bottom is flat. The seal skins with which the frame is covered are deprived of the hair, and are at all times transparent, but especially so when wet. Each of these boats has five or six seats or thwarts, placed as in ours, and is moved by two very clumsy oars with flat blades, which are used by the women, and steered with a similar oar by another. They vary much in size, having the sides very flat and about three feet high. Sometimes they are as large as twenty-five feet in length by eight in breadth, and are capable of containing women, boys and small children, to the number of twenty-one persons.

To those who wish to become well acquainted with the details of Eskimaux manners and ingenuity, the works of Crantz, Ross, Parry and Lyon, will afford a fund of the most satisfactory information, more especially as in these works will be continually experienced the force of that charm which always accompanies statements made by zealous and well qualified observers.*

---

* An interesting paper, by the celebrated Otho Fabricius, published in the fifth volume of the Royal Danish Society's Transactions, gives a complete account and representation of the weapons and implements used by the Greenland Eskimaux in seal hunting. These implements are very similar to those used by the American Eskimaux, and what is more interesting, are called by the same names

When Lewis and Clark wintered on the Columbia river they found the seals very abundant on the coast, and believed that several species frequented the shores which they had no opportunity of examining. The common seal was observed in the river as far up as the great falls.*

Species II.—*The Hooded Seal.*

Phoca Cristata; L.

Phoca Leonina: Fabricius, Faun. Grænl.
Klap-Myssen: Egede. Greenl. p. 84. (Eng. trans.)
Neitersoak; Klapmütz: Crantz, Greenl.
Hooded seal: Penn. Syn. No. 268, p. 342.
Phoca Cristata: De Kay, Ann. of the Lyceum of N. Y. vol. i. p. 94.

The hooded seal is most commonly found on the shores of Greenland, of Davis's Strait, and occasionally of Newfoundland. Recently an individual of this species has been captured in a small creek emptying into Long Island Sound, at East Chester, about fourteen miles from the city of New York.

The species is very obviously distinguished by the singular appendage it has on the head, formed by an extension of the skin of the front, which com-

---

by both people. This identity of language we have heretofore referred to as being of the highest importance in aiding us to form an opinion as to the original peopling of this continent.

* Lewis and Clark, vol. ii. p. 172.

municates with the nostrils, and can be inflated or elevated and depressed at the pleasure of the animal. The size of this hood, which extends from the end of the snout to five inches behind the eyes, is twelve inches, and its height nine. Through the anterior part of this hood the nostrils open, each two inches in diameter, and when the hood is undistended the cartilaginous partition of the nose may be felt from the outside, rising about six inches at its greatest elevation.

Internally the hood is strongly muscular, with numerous muscular fibres surrounding the orifices of the nostrils, and most probably corresponding to the sphincters which close the nostrils of other seals to prevent the entrance of water. Externally it is covered with short, bright brown hairs, and is slightly sprinkled transversely in many places.— Where the skin of the hood joins the common integuments of the head a few strong hairs are found, which are considered to resemble those which in other animals of this genus pertain to the eye. Over the sides of this appendage, as well as the cheek, we find twenty-five or thirty strong bristly whiskers arranged in rows and converging forwards. These are black and small in the upper rows; whitish, flattened and very stout in the lower, being about five inches long, all directed downwards, and when minutely examined appear to have a series of short alternate bevels on each edge, but no spiral turns.

Such is the curious structure with which the head of this seal is provided, and the question immediately occurs, what can be the use of so peculiar a contrivance? To this question no satisfactory answer has yet been given, nor have we any thing better than conjecture to offer. As it projects over the eyes at the inner angle when depressed, the opinion was long since advanced, that the hood was intended to protect those delicate organs from the sand and other substances thrown into the air by the violent storms common on the shores frequented by the animal. But none of the other seals residing under circumstances of similar exposure, have any such protection, and their eyes are equally liable to injury from the same causes. The fishermen consider it a reservoir for air, to be used while the seal is under water. Dr. De Kay thinks that its great bulk when distended would prevent the animal from descending into or moving with facility beneath the water. If we compare the extent of this appendage when distended with the bulk and weight of the animal, which is estimated at about five or six hundred pounds, this objection does not appear to us sufficient to invalidate the probability of the last mentioned use of the hood. The opinion suggested by Dr. De Kay is, that the hood is subsidiary to the sense of smell, which he concludes to be peculiarly necessary to this animal, nature having left it unprovided with efficient weapons of offence or defence. It would be a very easy matter to speculate upon this subject, and remark that the use last suggested for the hood is by no means incompatible with that generally attributed to it by the fishermen, and that the offensive and defensive weapons of this seal are as efficient as those possessed by most other species which have no such subsidiary appendage to the nose. To this might be added the probable correctness of what has been often stated, that this hood is a merely sexual distinction, &c. but we believe that all speculation not based on an actual and frequently repeated examination of the animal in its living state, and in its proper haunts, can lead to nothing better than multiplication of words. When the individual described by Dr. DE KAY was attacked, he inflated the hood and uttered a bellowing noise, until killed by repeated discharges of a musket.

The hooded seal is seven feet long from the centre of the chin, or symphysis of the lower jaw, to the root of the tail, which is six inches and a-half long, and three broad at its base. The body is cylindrical, gradually decreasing to the tail, which is flat and tapering to a point, the whole skin being covered with flat hairs about an inch in length. The general colour is gray and dark brown, distributed in irregular patches, the grayish appearance arising from very short hairs beneath the white. The head, when the hood is undistended, appears small compared with the body, and the eyes are large, of a dull greenish hue, and distant six inches and a-half from the extremity of the upper jaw. The orifices of the ears are distinct, situated about two inches and a-half behind the eyes, without any rudiment of external cartilage or concha.

The fore paws or *flippers* resemble those of the common seal, but appear small in proportion to the size of the animal, are of a uniform dark brown colour, except near the body, where they assume the common mottled appearance, and are twenty inches distant from the end of the jaw; their length is fifteen inches. Each digit is furnished with a strong, compressed channeled claw, the exterior of which is largest; at the base these are dark coloured; their tips are light horn colour. The hind paws are of the same length as the flippers, and lunated at their extremities, which are fifteen inches broad when expanded. They have five depressed claws or horny laminæ, of which the external are largest, all placed at some distance within the webbed extremities.

To the above description, which is drawn up from that given in Dr. DE KAY'S excellent paper quoted at the head of this article, we shall subjoin at full length his remarks on the peculiar dentition of this animal.

"Teeth, thirty in number; above, four incisors, two canine and ten jaw teeth; below, two incisors, two canine and ten jaw teeth. The incisors above are cylindrical and approximated; the two inner are small; the exterior much larger, and nearly half the size of the neighbouring canine. The canine are considerably larger than those of the lower jaw, and more incurved. The incisors of the lower jaw are very small and cylindrical; the jaw teeth above and below are small, distant, and have each a cutting edge; on the posterior part of this edge a notch or transverse indentation is visible. The first is placed at some distance from the canine, and is much smaller than the others.

"In the foregoing description the remarkable peculiarities presented by the teeth cannot escape notice. The incisors resemble the canine so much in form that their actual position alone can serve to point out their nature. Pennant, in the Arctic Zoology, describes four above and four below, being led into this error by confounding the laniary teeth with the incisors. The molares, or what may with more propriety be designated as jaw teeth, are very small in proportion to the size of the animal, hardly exceeding those of a child of five years old. The whole number of teeth in this tribe varies from thirty to thirty-six. It is a curious coincidence that the different species distinguished by a great development of the hood, or appendage to the head, are equally remarkable for the same number of teeth. Thus the *P. Leonina*, Gm. (*proboscidea*, Peron.) and the *P. Cristata*, Gm. have but two incisors below, indicating a natural division in this partially known family."*

We have mentioned (p. 307,) a seal sent to Paris by Mr. Milbert under the name of *Phoca Mitrata*, which corresponds, in relation to the teeth, very exactly with the above. On this Dr. De Kay remarks, " it is possible that this may have been brought here from the north by a whaling vessel; should this prove to be the case, I should incline to believe it absolutely identical with the *Phoca Cristata* described above."

---

* Dr. De Kay's paper in the annals of the Lyceum, is followed by a very interesting detail of the appearances observed on dissection of this seal by Drs. Ludlow and King, which may be referred to by the reader with much advantage.

Species III.— *The Great Seal.*

Phoca Barbata; Mull.

Urksuk utsuk: Crantz, Greenl. (Eng. trans.) i. 125.
Lakktak: Hist. Kamtschatka, 420.
Urskuk: Egede, Dict. Greenl. Hafn. 1750.
Ogiuke: Parry's Voyage; *Oghioo:* Lyon's Journal, 830.
Phoca Barbata: O. Fabricius, Faun. Grænl. p. 15.
Phoca Major: Parsons, Phil. Trans. 47. p. 121.
Grand Phoque: Buff. p. 34.
Great Seal: Penn. Arct. Zool. i. 185. Sp. 73.

This seal, which grows to the size of ten or twelve feet, is found in the Greenland seas, and on the northern extremities of this continent. It most commonly rests upon the floating ice, and when it comes up on the fixed ice it is through holes near the outer edge of the field.

Its skin is about half an inch in thickness and covered with black hair, which in summer is almost entirely shed, leaving the animal bare. The whiskers of the great seal are long, pellucid and white, having the points softer than the other part, and curled. The middle digits of the fore-feet are longer than the others, which in relative length are like the fingers of the human hand.

The great seal breeds in the month of March, having a single cub, usually upon the ice among the islands; it approaches the land more closely at that season than at any other.

The great seal resembles the common seal in habits as well as in general appearance, being distinguished from it readily by its great size and large beard-like whiskers. The adults of this species swim slowly; their peculiar timidity and watchfulness render it difficult to approach them, so that very little has been observed relative to their peculiar habits.

At some seasons the great seal is found to be remarkably fat, and its flesh is said to be very similar to veal. The Greenland Eskimaux cut its skins into thongs, and twist them into ropes, to be used in their whaling, and for various other purposes.

Species IV.—*The Harp Seal.*

Phoca Grænlandica; Mull.

Phoca Grænlandica: O. Fab. Faun. Grænl. p. 11
Svart-side: Egede, Nat. Hist. Greenl. pl. 3.
Attarsoak: Crantz, i. p. 124. (Eng. trans.)
Harp Seal: Penn. Arct. Zool. i. p. 190. Sp. 77.

The harp seal measures from six to nine feet in length from the tip of the nose to the end of the tail, which is from five to seven inches long. In circumference at the thickest part of the body it is from four to six feet. It has a round head and high forehead, with a short nose, large black eyes, and whiskers disposed in ten rows of hairs.

Crantz informs us that this seal when full grown is almost entirely of a white gray colour, having a black figure on its back like two half-moons, with their horns uniformly directed towards each other. No seal varies its colour so much as this, and the Greenland Eskimaux change its name with these variations of colour. The fœtus, which is white and woolly, they call *iblau;* in the first year it is cream coloured and the name is *attarak;* in the second year it is gray and then called *atteitsiak;* in the third year *aglektok* or painted; in the fourth *milektok* or spotted; and in the fifth year, when it has attained its adult age and the half-moon mark, it is called *attarsoak*. At this period the Russians call them *krylatka* or winged, on account of the half-moon marks.

The harp seal is quite common in the Greenland seas, where it frequents the deep bays, migrates twice a year, going in the month of March and returning in May, and again in June to return in September. It is also found near the shores of Newfoundland. The breeding season begins in July, and the female has one cub near the end of March or the beginning of April, which she suckles on fragments of ice remote from land.

The harp seal is very incautious, and shows much of the frisky or frolicsome disposition of the common seal. It is occasionally seen swimming in various attitudes, and whirling about as if for sport. This species lives in great herds that swim apparently under direction of a leader, who watches over the safety of the whole. They do not frequent the field or fixed ice, but the *floes* or large drifting ice. They are said greatly to dread the Physeter Microps, which forces them to seek safety on shore. The Greenlanders also drive them on shore by surrounding and pursuing them with loud noises whenever they come to the surface to breathe.

The harp seal has a large quantity of blubber, which yields a greater proportion of pure oil than is obtained from any other seal. Its skin is used by the Eskimaux to cover their tents and boats. The skins of the young make excellent boot-leather.

Species V.—*The Fetid Seal.*

Phoca Fetida; Mull.

Neitsek: Crantz, Greenland, i. p. 124. (Eng. trans.)
Phoca Fetida: O. Fabr. Faun. Grænl. p. 13.
Phoca Hispida: Schreb. Saeugth. tab. 86.
Phoque Neitsoak: Buff. xxxiv.

The fetid seal when full grown is about four feet and a-half long, and its skin is covered with a brownish or dingy white hair, spotted above and whitish beneath, composed of stiff bristles intermixed with a softer material. The hair does not lie smooth, but is rough and similar to that of a pig. The old animals are remarkably fetid, and this nauseous odour taints their flesh and fat equally.

The head of the fetid seal is short and rounded, about a third of its length being formed by the snout. The whiskers are pale, pointed, and compressed, having all their border undulated; the smallest of them are black. The eyes are small, the iris brown. The feet, ears, tongue and tail are similar to those of the common seal. The heels of the hind feet are scarcely apparent on account of the fatness of the animal.* The colouring varies very much according to the age. When the skins are wrought into clothing by the Greenlanders, the rough side is generally turned inwards.

The fetid seal frequents the fixed ice near frozen lands, and never relinquishes its haunts when old. It

---

* Desmarest, Mammalogie, p. 246.

has holes in this ice for the purpose of fishing, and is solitary in its habits, pairs being rarely seen together. It is not a timid animal, and is occasionally preyed upon by the eagle, being taken while asleep upon the surface. The flesh is not esteemed as food even by the Eskimaux, though they employ the different parts of its body as they do those of other seals.

Species VI.—*The Ursine Seal.*

Phoca Ursina; L.*

Ursus Marinus: Steller, Nov. Com. Petropol. ii. 231.
L'ours Marin: Briss. Quad. 166; Schreb. Saeugth. 122
Ours Marin: Buff. xxxiv. p. 94. (Ed. Sonnini.)
Chat Marin: Kracheninikow, Hist. du Kamtschatka.
Ursine Seal: Penn. Quad. ii. 281. No. 485.
Otaria Ursina: Desm. Mammal. 240, Sp. 381.

The ursine seal is a large animal, being when full grown eight feet in length by five in circumference,

---

* This species belongs to Peron's subgenus *otaria*, and to F. Cuvier's second division, or seals having *single roots to their teeth*. The dental formula is the following:

36 Teeth: { 20 Upper { 6 Incisive / 2 Canine / 12 Molar.
         { 16 Lower { 4 Incisive / 2 Canine / 10 Molar.

These teeth interlock when the mouth is closed, and in their form and arrangement differ very considerably from those of the other seals. Their roots are remarkable for having a narrowing immediately below the crown; they then swell out strongly, and are elongated so as to form a cone

and weighing about eight hundred pounds. The female is much smaller than the male, but otherwise they are very similar to each other in appearance. The anterior part of the body is very thick, the posterior slender and tapering to the tail. The head is rounded, rising suddenly from the nose, which projects like that of a pug-dog; the eyes are large and prominent, and the ears conical and pointed. The whiskers are very long and white, and the lips thick.

The ursine seal differs very materially from most other seals in having the anterior limbs entirely at liberty, or not enveloped by the integument of the body. But the wrist, bones of the palm and digits, are covered with a naked skin, which is smooth on the superior and wrinkled on the inferior surface. The thumb is the longest of the digits, which decrease successively to the external or little one. All of them have a small nail. The posterior extremities are about twenty-two inches in length, articulated like those of other seals, but they can on account of their length be used by the animal to scratch the head. They have five toes, the internal of which is as long as the three next it, and the last is smallest of all. Each toe is united to the others by a broad web, which gives a breadth of twelve inches to the hinder feet when spread out.

The general colour of the ursine seal is black,

---

twice as long as the crown of the tooth. The same dental formula is applicable to the *Phoca Jubata*. See Dents des Mammif. p. 22.

which in old individuals becomes of a dull gray. The hair is long and stiff, having a soft down of a bay colour intermixed. The colour of the females differs as much from that of the male as her size. Sometimes they are ash coloured, and sometimes of a reddish brown.

This species is principally found on the islands which lie between America and Kamtschatka, and like the sea otter are there only seen between the 50th and 60th degrees of latitude. Ursine seals have also been killed on the shores of New Zealand, Staten Land, New Georgia and the Falkland islands.

They arrive at the islands between America and Kamtschatka in the month of June, and remain until September. When they first arrive they are excessively fat and lazy, moving very seldom, and sometimes remaining for several days near one spot, without being at the trouble of seeking food.*

They lie upon the shores in vast herds, but are separated into distinct families, each male having a seraglio of from eight to thirty or more females, over which he watches with incessant jealousy. The family, with the young and half grown individuals, sometimes amount to a hundred or more.

---

* Tempus illis coeundi, ad solis occasum est; mas feminaque ad horam ante, se mare immergunt. leniter natant, unaque littus petunt. Femina prius ascendit, se resupinat et mari amore flagranti cedit. Ille tam magno ardore, quam suo pondere, feminam in arena lutove præter caput et pedes sepelit: tunc temporis, sic ad voluptatem intentus et sui oblitus est, ut cuique illum accedere et impune tangere liceat.

THE URSINE SEAL.

The old seals which are deserted by the females live by themselves, and are very fierce, irascible and quarrelsome, every intruder upon their resting places is immediately attacked, and they will incur any danger rather than resign their accustomed seats. At the approach of a disturber of their own species they relinquish their indolence and attack him, and should the two in their struggles disturb another, this third mingles in the fray, and thus at times the war extends throughout the whole flock on shore.

The younger males are excessively provoked at any attempt made by their neighbours to entice away one of their wives, and furious contests are the result of such interference with their families. After the battle, however, the females go quietly over to the conqueror, and become a part of his establishment. They inflict very severe wounds upon each other during their combats, and when they cease to fight plunge into the sea, in order to wash off the blood with which they may be stained.

The males are quite fond of their offspring, but cruelly tyrannical to the females. When any one attempts to catch one of their cubs the male opposes the aggressor, while the female tries to secure the cub by carrying it off in her mouth. But should she unfortunately drop it, the male attacks her and beats her dreadfully against the stones. When she recovers she crawls towards his feet with signs of great submission. Should the young be carried off the male then appears to be much afflicted. The female has commonly but one cub, which she brings forth in the month of January. The cubs are quite fierce, and bark and bite at the sailors passing them.*

The ursine seal is a very swift swimmer, moving at the rate of seven or eight miles an hour, and is able to remain under water a much longer time than the common seal. It is very tenacious of life, and survives dreadful wounds for a considerable time. The old ones have a very disagreeable odour, which taints their flesh and fat. The flesh of the female and young is pleasant to the taste and similar to that of a lamb.

---

* See Foster's Voyage, ii. 429, 514.

1. Harp Seal. 2. Walrus.

CHAPTER XVII.

Genus XVIII. Morse; *Trichecus;* L.

These animals resemble the seals in their elongated and conical bodies, and the construction of their anterior extremities. They have a round head, no external ears, and small eyes. The orifices of the nostrils are far distant from the upper lips. The posterior feet are horizontally placed, have five digits, of which the two external are the longest, all being provided with small incurvated nails and united by a membrane. The most striking peculiarity of the genus is the tusks or prolonged canine teeth, which descend from the superior maxillary bone and project far below the lower jaw, serving the animal as offensive weapons as well as instruments to aid in climbing on ice-banks, &c.

Dental System.

22 Teeth:
- 14 Upper { 4 Incisive / 2 Canine / 8 Molar.
- 8 Lower { 0 Incisive / 0 Canine / 8 Molar.

In the upper jaw the first incisor, (separated by a vacant space of some extent from its fellow in one species, and very close in another,) when it first emerges from the socket, is a very small, conical and hooked tooth, and being rudimental is soon worn out and disappears. The second, which is much larger than the first, is cylindrical and obliquely cut from the outer to the inside of the jaw. The canine is a very large *tusk* which is directed downwards, being curved towards the body; it is rounded on its external face, marked by a longitudinal groove on its internal surface, and rises from the maxillary bone as high as the nostril. The first molar, separated by a vacant space from the second incisive, and much larger than that tooth, is, like it, cut obliquely, but the oblique surface is slightly hollowed. The second molar, twice as large as the preceding, is cut in the same direction, but has two depressions or hollows at this part, one anterior and the other posterior, separated by an obtuse prominence, and narrow at its summit; the third strongly resembles the second, and the fourth is merely a rudimental tooth which falls out by age. All these teeth have but one very strong conical root, and they are formed entirely of a very hard compact substance analogous to that of the tusks.

In the lower jaw it appears that in early life the first tooth falls out, for which reason we have not counted it among the others. The four molars appear to have the same form, and are more extended from before backwards than from right to left, and the surface of their crowns is slightly convex. The last is somewhat smaller than the others, which are of equal size. These teeth are of the same nature as those of the upper jaw.

In their reciprocal position the first molars are opposed crown to crown; the second are alternate. These teeth are described from several heads, which appear to have belonged to two distinct species, judging from the proportion of some of their parts, and not merely by the extent of the tusks. The differences observed in the latter led Shaw to conclude that there are two species of walrus, which, however, has not yet been ascertained.*

---

* Introductory to this dental system F. Cuvier has the following remarks:—"We have seen that the seals are related by their dental systems on the one hand to the terrestrial carnivorous, and on the other to the cetaceous animals. The morse, whose organs of motion are very similar to those of the seal, is widely different from that animal in respect to the teeth. In this particular the present genus has a system altogether singular, as it is not better adapted for bruising vegetable than for cutting animal matter. We might say that the teeth of these amphibia are especially destined to crush hard materials, because by their structure and relation to each other they act like a pestle against a mortar. They form one of those insulated groups which break the necessarily continuous series of classifications, and may be connected almost indifferently, according to the point of view under which we consider them, with either of the branches of the general system we admit. We should have placed them next to the seals, which would leave a large void between them and the frugivorous marsupial animals, but induced by certain analogies, which have also some foundation, we are led to treat of them after the ruminant quadrupeds. Furthermore, we should remark, that in these animals we begin to see the number of teeth vary in the individuals, because those which are rudimental disappear according to the age of the animal We may say that these organs diminish in importance, and should be considered rather in relation to their number than to their form and structure."

Fully assenting to the general correctness of the views of this distinguished naturalist, we do not deem the analogy above stated of sufficient force to authorize us to swerve

Species I.—*The Walrus.*

Fricheus Rosmarus; L.

Juak: Crantz, i. 125. (Eng. transl.)
Wallross: Marten's Spitzbergen, 78.
Equus Marinus, &c.: Ray, Quad. 191.
Odobenus: Briss. Quad. 48.
Le Morse: Buff. xxxiv. p. 158, Sonn.
Aywek of the Eskimaux: Lyon's Journal, p. 329.

This large and unwieldy creature bears a stronger resemblance to the seal than to any other quadruped, but is strikingly distinguished by the proportions of its body and its elephant-like tusks. Vast herds of this species formerly frequented the shores of the islands scattered between America and Asia, the coasts of Davis's Straits and those of Hudson's Bay, in latitude 62°. They have been found as far south as the Magdalen Islands in the Gulf of St. Lawrence, between latitude 47° and 48°. At present they are not met with in very great numbers, except on the icy shores of Spitzbergen and the remotest northern borders of this continent.

The walrus attains the size of an ox, being when full grown from twelve to fifteen feet in length, and from eight to ten in circumference. The head is oval, short, small, and flat in front, having the eyes

---

from the arrangement of his brother, which places the morse next to the seal. The resemblance of form, structure, modes of life and action, existing between these animals, render it not only more useful but more natural, and therefore more correct to view them in succession, rather than to separate them to so great a distance as has been done in the " Dents Des Mammifères."

set in deep sockets so as to be moved forwards or retracted at pleasure. The flat portion of the face is set with very strong bristles, which are pellucid, a span long, as thick as a straw, and twisted like a three plied rope. The orifices of the ears are very small, placed far back on the head and destitute of external cartilages. The nostrils open on the upper part of the snout; through these the walrus is accustomed to blow the water in a manner similar to that of the whale. The fore feet are from two to two and a-half feet long, and when extended are fifteen or eighteen inches broad, the digits being connected by a membrane forming a sort of webbed hand. The palms of old individuals have the cuticle on them a-quarter of an inch in thickness, in consequence of the friction to which they are subject in clambering up the rocks, &c. The hind feet are from two to two and a-half feet in length, and their breadth when fully extended is from two and a-half to three feet. Each toe is terminated by a small nail.

The ivory tusks, or prolonged canine teeth measure from ten to twenty inches, exclusive of the portion which is imbedded in the jaw, of seven inches and upwards in length. Tusks of the walrus have been seen thirty-six inches long, and weighing from five to ten pounds. The circumference of one twenty-seven inches in length is about eight inches at base. The inside is more compact and of finer grain than the ivory of the elephant; in the centre the colour is somewhat brownish, but otherwise it is pure white, though speedily growing yellow from exposure to the air. They are slightly hollowed where they arise from the skull, are somewhat notched, and not entirely round. At the base they are about three inches distant from each other, and at the point about nine inches.

The skin of the walrus is generally about an inch thick, but is thicker on the neck than any other part of the body. The hair is short and of a yellowish brown colour, and the whole surface of the skin is marked by numerous chaps and wrinkles.

This thick skin is used by the Eskimaux for various purposes, such as the fabrication of cordage and coverings for tents, &c. It is used by the whale-fishers instead of mats, for protecting the yards and rigging of ships from being injured by friction. By tanning it is converted into a thick porous or spongy leather, by no means so serviceable as the raw hide. Previous to the establishment of the whale-fishery near Spitzbergen the walrus was considered of some importance, and voyages were made expressly to obtain them, for the sake of their ivory and oil. Since the whaling business has become so successful the walrus is allowed to escape unmolested, except by the Eskimaux, who feed upon its flesh with an eagerness only second to that with which they devour the seal.

On land the walrus is a slow and clumsy animal, but in the water its motions are sufficiently quick and easy. The head of a young walrus without tusks, when observed from some distance above water, bears considerable resemblance to the human face, and has been occasionally mistaken by persons unaccustomed to their appearance for that of a man.

The walrus is a fearless, but when undisturbed an inoffensive animal; it is monogamous, and displays great attachment to its mate and young. The season in which the sexes seek each other is about the month of June; the female brings forth her cub early in the spring. When attacked the walrus is both fierce and formidable, more especially if in company with its young. Under such circumstances they become very furious, attempting to destroy their enemies by rising and hooking their tusks over the sides of the boat, in order to sink it. Frequently the violence of their blows is sufficient to stave in the planks of small boats. In speaking of an instance in which an attack was made on a herd of walruses, Capt. LYON remarks,—"Mr. Sherer described the fury of the wounded animals as being quite outrageous, but those which were unhurt quickly forsook their suffering companions. The beast which sank the boat struck his tusks repeatedly through her bottom, and she filled immediately. Had she been alone not a soul of her crew could have been saved, for there was no ice within three miles, and to swim would have been impossible in such cold water."

The same author gives in another place the following account of a battle with some of these animals:—"On some stream-ice near us were several herds of walruses basking in the sun. They allowed us even to land on the pieces of ice on which they lay, before they commenced their cumbrous retreat, facing us with open mouth. We killed one, but he sunk before we could get the boat to him, and wounded several others, when seeing the Fury's boat had been more successful, we went to assist in towing her. On our way we met a male and female, attended by their cub, and soon wounded the two old ones. They fought us, however, with desperation, and would not retreat. The female on being killed was secured alongside, but the male, even when shot in three places, and having two lances sticking in him, attacked us furiously, although each time he approached he received a bayonet to the socket. Having at last driven him near to the Fury's boat, our joint efforts despatched him, after about ten minutes struggle. This brave animal had repeatedly attempted to hook his tusks over the gunwale of the boat, had stove her slightly in three places, and left eight deep marks on her bow. The cub, which was black and without tusks, continued by its parents during the whole combat, and frequently endeavoured to mount on the back of whichever first rose to the surface. To this may be attributed the more than usual fierceness of the old ones, whose fears for their offspring prevented their own escape. The female, on being hoisted in, was considered as rather small by those who were judges. On each side she had two teats almost concealed in the belly, but they could be pulled out to the size and length of those of a sow. The stomach contained only about three pounds of pebbles and a handful of sea-weed." It has been frequently stated that the food of the walrus is sea-grass, shellfish, and not flesh.* Capt. Scoresby states that he has found, in addition to such substances, parts of young seals in the stomach of the walrus.

The flesh of the walrus is occasionally eaten by the

---

* " Fucis, corallinis, testaceis, non carne victitant." Lin. Syst. Nat. p. 59.

whale-fishers and other voyagers, but is not considered to be a very desirable food, as it is dark coloured and very coarse grained. To the crews of ships which have been long at sea and confined to salted provisions, the use of this flesh at times proves very acceptable and serviceable.

One of the Eskimaux modes of killing the walrus in summer is the following: Perceiving a large herd asleep on the floating ice, as is their custom, they paddle to some other piece near at hand which is small enough to be moved. On this they lift their canoes, and then bore holes, through which they fasten their lines. As soon as every thing is prepared they quietly paddle the cake of ice towards the herd, each hunter sitting by his own spear and line. When arrived at the place where the animals lie snoring, each man if so disposed strikes a different one, though two generally attack the same. The wounded and terrified walrus now tumbles into the water, but cannot escape from the ice to which the hunters have fastened their lines. As soon as his victim becomes tired, the hunter launches his canoe, and, lying at a safe distance, spears him to death.

We have given Capt. LYON's picture of Eskimaux feasting and gluttony when speaking of the common seal; in this place we shall introduce another sketch from the same masterly hand, which is inimitable of its kind:—"We found, on the 3d, that the party which had been adrift had killed two large walruses, which they had carried home during the early part of the night. No one therefore came to the ships, all remaining in the huts to gormandize. We found the men lying under their deer-skins, and clouds of steam rising from their naked bodies. From Kooilittuk I learnt a new Eskimaux luxury; he had eaten until he was drunk, and every moment fell asleep with a flushed and burning face and his mouth open. By his side sat Arnalooa, who was attending her cooking-pot, and at short intervals awakened her spouse, in order to cram as much as was possible of a large piece of half boiled flesh into his mouth with the assistance of her fore finger, and having filled it quite full, cut off the morsel close to his lips. This he slowly chewed, and as soon as a small vacancy became perceptible, this was filled by a lump of raw blubber. During this operation the happy man moved no part of him but his jaws, not even opening his eyes; but his extreme satisfaction was occasionally shown by a most expressive grunt whenever he enjoyed sufficient room for the passage of sound. The drippings of the savoury repast had so plentifully covered his face and neck, that I had no hesitation in determining that a man may look more like a beast by over-eating than by drinking to excess."*

The fifty-fifth volume of the Philosophical Transactions,† contains the following account of the walrus, given by Lord SHULDHAM:—" The walrus, or sea-cow as it is called by the Americans, is a native of the Magdalen Islands, St. John's and Anticosti in the Gulf of St. Lawrence. They resort very early in the spring to the former of these places,

---

* Private Journal, &c. p. 182.
† Cited in Pennant's Arctic Zoology, vol. i. 173.

which seems peculiarly adapted to the nature of the animals, abounding with clams (escallops) of a very large size, and the most convenient landing places, called *Echoueries*. Here they crawl up in great numbers, and remain sometimes for fourteen days together without food when the weather is fair, but on the first appearance of rain they retreat to the water with great precipitation. They are when out of the water very unwieldy, and move with great difficulty. They weigh from fifteen hundred to two thousand pounds, producing from one to two barrels of oil, which is boiled out of the fat lying between the skin and flesh. Immediately on their arrival the females calve, and engender in two months after, so that they carry their young about nine months. They never have more than two at a time, and seldom more than one.

"The landing places are formed principally by nature, being a gradual slope of soft rock, with which the Magdalen Islands abound, about eighty or a hundred yards wide at the water side, and spreading so as to contain near the summit a very large number of these animals. Here they are suffered to come on shore and amuse themselves for a considerable time till they acquire a degree of boldness, being at their first landing so exceedingly timid as to make it impossible for any person to approach them.

"In a few weeks they assemble in great multitudes; formerly, when undisturbed by the Americans, to the amount of seven or eight thousand. The form of the landing place not allowing them to remain contiguous to the water, the foremost are insensibly pushed above the slope. When they are arrived at a convenient distance, the hunters, being provided with a spear, sharp on one side like a knife, with which they cut their throats, take advantage of a side-wind, or a breeze blowing obliquely upon the shore, to prevent the animals from smelling them, because they have that sense in great perfection. Having landed, the hunters, with the assistance of good dogs trained for that purpose, in the night-time endeavour to separate those which are most advanced from the others, driving them different ways. This they call *making a cut;* it is generally looked upon to be a most dangerous process, it being impossible to drive them in any particular direction, and difficult to avoid them; but as the walruses which have advanced above the slope of the landing are deprived by the darkness of the night from every direction to the water, they are left wandering about and killed at leisure, those that are nearest the shore being the first victims. In this manner have been killed fifteen or sixteen hundred at a *cut.*"

Balls discharged from a musket are by no means very efficient in killing the walrus, unless aimed with care at vital parts. By shooting them with duck or other small shot, so as to blind them, they may be readily killed with lances or axes. When the walrus makes an attack upon a boat, endeavouring to mount upon the gunwale, the most successful mode of repelling it is by throwing a handful of sand into its eyes, which causes the animal to retire for a time, affording an opportunity to escape, or to make better preparations for defence.

END OF VOLUME I.

AMERICAN

Natural History,

BY

JOHN D. GODMAN, M.D.

Drawn by T. Peale Jr.   Engraved by F. Kearny

Philadelphia

H. C. CAREY & I. LEA, CHESTNUT STREET.

AMERICAN NATURAL HISTORY.

VOLUME II.

PART I.—MASTOLOGY.

BY JOHN D. GODMAN, M.D.

PROFESSOR OF NATURAL HISTORY IN THE FRANKLIN INSTITUTE OF PENNSYLVANIA; ONE OF THE PROFESSORS OF THE PHILADELPHIA MUSEUM; MEMBER OF THE AMERICAN PHILOSOPHICAL SOCIETY; OF THE PHILADELPHIA ACADEMY OF NATURAL SCIENCES, &c.

PHILADELPHIA:
H. C. CAREY & I. LEA—CHESTNUT-STREET.
R. WRIGHT, PRINTER.

1826.

Eastern District of Pennsylvania, to wit:

[L. S.] BE IT REMEMBERED, That on the third day of July, in the fiftieth year of the Independence of the United States of America, A. D. 1826, Robert Wright, of the said district, hath deposited in this office the title of a book, the right whereof he claims as proprietor, in the words following, to wit:

" American Natural History. Volume II. Part I. Mastology. By John D. Godman, M. D. professor of Natural History in the Franklin Institute of Pennsylvania; one of the Professors of the Philadelphia Museum; Member of the American Philosophical Society, of the Philadelphia Academy of Natural Sciences, &c.

In conformity to the Act of the Congress of the United States, intituled, " An Act for the encouragement of learning, by securing the copies of Maps, Charts, and Books, to the authors and proprietors of such copies, during the times therein mentioned"—And also to the Act entitled, An Act supplementary to an Act, entitled, An Act for the encouragement of learning, by securing the copies of Maps, Charts, and Books to the authors and proprietors of such copies, during the times therein mentioned, and extending the benefits thereof to the arts of designing, engraving, and etching historical and other prints."

D. CALDWELL,
Clerk of the Eastern District of Pennsylvania.

AMERICAN NATURAL HISTORY.

CHAPTER I.

FAMILY V.—MARSUPIALIA; *Pouched or Marsupial Animals.*

The beings at present considered as members of this family, offer so many peculiarities and striking differences in their construction and economy, as in strictness to constitute a distinct order of animals; or if not, the remarkable differences which exist between the species in relation to their teeth, organs of digestion, food and habits, fully justify the arrangement of them under different orders in our existing classification.

They are wonderfully unlike all other animals in relation to the production of their offspring, which are brought forth in a condition apparently imperfect or premature. The young, when they are first to be discovered in the external pouch, seem scarcely formed, are incapable of movement, exhibit but slight traces of limbs or other external organs, are found attached to the teats of the mother. and are unable to resume their hold if it be broken. They remain thus attached until they acquire size and strength enough to move about at will, and continue to take refuge in this curious retreat until they attain the size of a common rat, or are even larger. The pouch is formed by a process or elongation of the skin of the belly, and is supported by two peculiar bones which arise from the pubis, and are sustained by the abdominal muscles. What is more singular, the males of these animals also have such bones, although they have no pouch, and similar bones are observed in both sexes of species which have little or nothing of the pouch itself.*

GENUS XIX. OPOSSUM; *Didelphis;* L.

Germ. Beutelthier. *Fr.* Sarigue.

GENERIC CHARACTERS.

THE head is conical, with a pointed muzzle and lateral nostrils; rounded, nearly naked and delicate

---

* " La matrice des animaux de cette famille n'est point ouverte par un seul orifice dans le fond du vagin; mais elle communique avec ce canal par deux tubes latéraux en forme d'anse. Il parait que la naissance *prematurée** des petits tient à cette organisation singuliere. Les mâles ont le scrotum pendant en avant de la verge au contraire des autres quadrupédes."—CUVIER; Régne An. i. 170.

* We cannot avoid objecting in this place to the inaccuracy of expression occasionally indulged in by the most celebrated men, since the influence of their example under such circumstances, is as injurious as in opposite conditions it is beneficial. The *birth* of the young *is not premature* in these animals, but takes place in a perfectly regular and *mature* manner, according to their peculiar organization and nature, though it may be *apparently* premature when compared with other animals. M. DESMAREST gives as the most striking character-

Drawn by C. Burton. Engraved by W.E. Tucker.

Opossum Male & Female

with their Young.

ears. The thickness of the body is great when compared with the length of the limbs. The digits on the anterior extremities are five in number, armed with hooked claws, and all lying parallel to each other. On the posterior extremities the internal digits are not in the same range with the rest, but are opposable, or constitute proper thumbs. They are rounded at the extremity, without nails, broad and fleshy. The soles of the posterior feet are provided with large fleshy tubercles, which materially aid in grasping small objects. The females of this genus have a fold of the skin of the belly, so arranged as to form a marsupium or pouch capable of receiving the young after birth; the teats, eight in number on each side, are within this pouch, which is supported by two bones of considerable length, articulated with the pubis and connected with the muscles of the belly. The males also have similar bones, but no pouch.

Dental System.

50 Teeth:	26 Upper	10 Incisive 2 Canine 14 Molar.	6 False Molar 8 Molar
	24 Lower	8 Incisive 2 Canine 14 Molar.	6 False Molar 8 Molar.

IN THE UPPER JAW the incisors are situated at the extremity of a very elongated ellipsis. The

---

istic of this family, "*birth of the young premature*," ("*naissance des petits prematurée*") which is entirely at variance with fact and philosophy. An American translator of Desmarest has advanced still farther, and "capped the climax" of absurdity by rendering the words above quoted, "GROWTH of the young *premature!*"

first is cylindrical, hooked and longer than the four following, which resemble each other and are trenchant. To these succeed a very marked depression, and then the canine which is compressed, terminates in a point and is hooked, but with rounded edges. At its base there is a very small but normal false molar, to which succeeds a vacant space, and then two equally regular false molars, the last of which is a little larger than the preceding. The three first molars successively and gradually increase in size, and have the same forms. At first the inner base is elevated nearly as high as the prisms; these are merely distinguished by the points, which are only developed at the three angles presented by a section of them, the anterior being much smaller than the posterior; its anterior point is very small. Finally, the inner base is carried obliquely forward, in consequence of which there is left between each tooth on the inside of the jaw an angular vacuity, much larger than in the insectivorous animals, where this base is uniformly developed. The last molar only differs from the others in being truncated at its posterior part, like all the last upper molars in animals of this order.

In the lower jaw we find four incisors, obliquely inclined forwards, of a cylindrical form, and nearly equal in size. The canine, which are in no respect peculiar, follow; then come three false molars, one very small, at the base of the canine, and after a vacant space the two others, somewhat larger than the first, but the middle is the largest, and all three are normal. The four molars are composed, anteriorly, of three points, disposed in a triangle, and posteriorly, of a spur also composed of three tubercles, but less regularly disposed and less elevated than those of the anterior part: the external is the largest.

There is nothing very peculiar to be observed respecting the relative position of these teeth.

Species I.—*The Common Opossum.*

Didelphis Virginiana; Penn. Gmel. &c.

Le Manicou: Feuille, Obs. Peru, iii. 206.
Tlaquatzin: Hernand. Mexico, 330.
Opossum: Lawson's Carolina, 120. Catesby's Carolina, App. xxix.
Sarigue des Illinois: Buff. Sup. tom. vi. pl. 33. Sarigue à long poil, Ibid. pl. 34.
Didelphis Opossum: L. Gmel. Syst. Nat. i. p. 105.
Micouré premier: D'Azzara, Quad du Paragua, Trad. Fran. i. p. 244.
Virginian Opossum: Penn. Quad. ii. p. 18, No. 217, Shaw, i. part 2, pl. 107.
Didelphis Woapink: Barton, Facts and Conjectures, &c.

Centuries have elapsed since this species was first observed by European naturalists, and it has long been a frequent theme of admiration and discussion to those of America, yet it is still considered as a sort of anomaly among animals, and the peculiarities of its sexual intercourse, gestation, and parturition, are to this day involved in profound obscurity. Perhaps nothing can more clearly demonstrate the impatience of the human mind, and the reluctance with which men yield to the hard necessity of carefully observing the operations of nature, than the history of this animal. Volumes of facts and conjectures have been written on the subject, in which the proportion of *conjecture* to *fact* has been as a thousand to one, and the difficulties still remain to be surmounted. The animal is among the most common within our borders, and is annually killed or captured in large numbers; faithful investigations into the habits of a few individuals would be sufficient to settle all doubts forever, and yet these still remain to be made. Very full and interesting observations have been made at almost every other period; but the great question how the helpless offspring, weighing scarcely a grain, are conveyed into the external pouch and attached to the teat of the mother, has never been properly answered.

For obvious reasons we shall wave for the present the consideration of these particulars. In our appendix we hope to give a full description of the sexual peculiarities of this very singular animal, and may then have it in our power to remove all the obscurities from the subject, by the only true method, that of a patient and vigilant observation of nature.

The opossum is very remarkable from other peculiarities, besides those which relate to the continuation of its kind. In the first place, we have already seen that it has a very large number of teeth, and its hind feet are actually rendered hands by short, fleshy and opposable thumbs, which, together with the prominences in the palms of these posterior hands, enable the animal to take firm hold of objects which no one would think could be thus grasped. An opossum can cling by these *feet*-hands to a smooth silk handkerchief or a silk dress, with great security, and climb up by the same. In like manner he can ascend by a skein of silk, or even a few threads. The slightest projection or doubling of any material, affords him a certain mean of climbing to any desired height. Another curious and amusing peculiarity is his prehensile tail; by simply curving this at the extremity, the opossum sustains his weight and depends from a limb of a tree, or other projecting body. and hanging in full security, gathers fruit or seizes any prey within his reach; to regain his position on the limb it is only necessary to make a little stronger effort with the tail and throw his body upward at the same time.

In speaking of the more obvious peculiarities of the opossum, we may advert to the thinness and membranous character of the external ears, which may remind us in some degree of what has been heretofore said relative to the perfection of the sense of touch possessed by the bat, in consequence of the delicacy of the extended integument forming the ears and wings. The extremity of the nose of our animal is also covered by a soft, moist and delicate integument, which is no doubt very sensitive. On the sides of the nose, or rather on the upper lip, there are numerous long and strong divergent whiskers or bristles, projecting to the distance of nearly three inches; over each eye there are two long black bristles, rather softer than the others, somewhat crisped or undulated, and slightly decurved; while, on the posterior part of the cheek, and about an inch below and in front of the ear, there is a bunch of long, straight bristles, (very similar to those of a hog) six or eight in number, projecting laterally so as to form a right angle with the head. When the elongated conical form of the opossum's head is recollected, together with its nocturnal habits. we cannot avoid remarking that all these arrangements appear to have immediate reference to the safety of the animal, furnishing the means of directing its course, and warning it of the presence of bodies which otherwise might not be discovered until too late.

The mouth of the opossum is very wide when open, yet the animal does not drink by lapping, but by suction. The wideness of the mouth is rendered very remarkable when the female is approached while in company with her young. She then silently drops the lower jaw to the greatest distance it is capable of moving, retracts the angles of the lips, and shows the whole of her teeth, which thus present a formidable array. She then utters a muttering kind of snarl, but does not snap until the hand or other object be brought very close. If this be a stick or any hard or insensible body, she seldom closes her mouth on it after the first or second time, but maintains the same gaping and snarling appearance, even when it is thrust into her mouth. At the same time the young, if they have attained any size, either exhibit their signs of defiance, take refuge in the pouch of the mother, or, clinging to various parts of her body, hide their faces amidst her long hair.

The general colour of the opossum is a whitish gray. From the top of the head along the back and upper part of the sides the gray is darkest, and this colour is produced by the intermixture of coarse white hairs, upwards of three inches long, with a shorter, closer, and softer hair, which is white at base and black for about half an inch at tip. The whole pelage is of a woolly softness, and the long white hairs diverging considerably, allow the black parts to be seen.

so as to give the general gray colour already mentioned. On the face the wool is short and of a smoky white colour; that on the belly is of the same character, but longer on the fore and hind legs; the colour is nearly black from the body to the digits, which are naked beneath. The tail is thick and black for upwards of three inches at base, and is covered by small hexagonal scales, having short rigid hairs interspersed throughout its length, which are but slightly perceptible at a little distance. The opossum is generally killed for the sake of its flesh and fat. Its wool is of considerable length and fineness during the winter season, and we should suppose that in manufactures it would be equal to the sheep's wool which is wrought into coarse hats.

The opossum is a nocturnal and timid animal, depending more on cunning than strength for his safety. His motions are slow, and his walk when on the ground entirely plantigrade, which gives an appearance of clumsiness to his movements. When on the branches of trees he moves with much greater ease, and with perfect security from sudden gusts of wind; even were his weight sufficient to break the limb on which he rests, there is no danger of his falling to the earth, unless when on the lowest branch, as he can certainly catch and securely cling to the smallest intervening twigs, either with the hands or the extremity of the tail. This organ is aways employed by the animal while on the smaller branches of trees, as if to guard against such an occurrence, and it is very useful in aiding the opossum to collect his food, by enabling him to suspend himself from a branch above, while rifling a bird's nest of its eggs, or gathering fruits.

The food of the opossum varies very much according to circumstances. It preys upon birds, various small quadrupeds, eggs, and no doubt occasionally upon insects. The poultry-yards are sometimes visited, and much havoc committed by the opossum, as, like the weazel, this animal is fonder of cutting the throats and sucking the blood of a number of individuals, than of satisfying his hunger by eating the flesh of one. Among the wild fruits the persimmon (*Diospyros Virginiana*) is a great favourite, and it is generally after this fruit is in perfection that the opossum is killed by the country people for the market. At that season it is very fat, and but little difference is to be perceived between this fat and that of a young pig. The flavour of the flesh is compared to that of the roasting pig; we have in several instances seen it refused by dogs and cats, although the opossum was in fine order and but recently killed. This may have been owing to some accidental circumstance, but it was uniformly rejected by these animals, usually not very nice when raw flesh is offered.

The hunting of the opossum is a favourite sport with the country people, who frequently go out with their dogs at night, after the autumnal frosts have begun and the persimmon fruit is in its most delicious state. The opossum, as soon as he discovers the approach of his enemies, lies perfectly close to the branch, or places himself snugly in the angle where two limbs separate from each other. The dogs, how-

THE COMMON OPOSSUM.

ever, soon announce the fact of his presence by their baying, and the hunter ascending the tree discovers the branch upon which the animal is seated, and begins to shake it with great violence to alarm and cause him to relax his hold. This is soon effected, and the opossum attempting to escape to another limb is pursued immediately, and the shaking is renewed with greater violence, until at length the terrified quadruped allows himself to drop to the ground, where hunters or dogs are prepared to despatch him.

Should the hunter, as frequently happens, be unaccompanied by dogs when the opossum falls to the ground, it does not immediately make its escape, but steals slowly and quietly to a little distance, and then gathering itself into as small a compass as possible, remains as still as if dead. Should there be any quantity of grass or underwood near the tree, this apparently simple artifice is frequently sufficient to secure the animal's escape, as it is difficult by moonlight or in the shadow of the tree to distinguish it, and if the hunter has not carefully observed the spot where it fell, his labour is often in vain. This circumstance, however, is generally attended to, and the opossum derives but little benefit from his instinctive artifice.

After remaining in this apparently lifeless condition for a considerable time, or so long as any noise indicative of danger can be heard, the opossum slowly unfolds himself, and creeping as closely as possible upon the ground would fain sneak off unperceived. Upon a shout or outcry in any tone from his persecutor, he immediately renews his death-like attitude and stillness. If then approached, moved or handled, he is still seemingly dead, and might deceive any one not accustomed to his actions. This feigning is repeated as frequently as opportunity is allowed him of attempting to escape, and is known so well to the country folks as to have long since passed into a proverb. " He is playing *'possum"* is applied with great readiness by them to any one who is thought to act deceitfully, or wishes to appear what he is not.

The usual haunts of the opossum are thick forests, and their dens are generally in the hollows of decayed trees, where they pass the day asleep, and sally forth mostly after night-fall to seek for food. They are occasionally seen out during day-light, especially when they have young ones of considerable size, too large to be carried in the maternal pouch. The female then offers a very singular appearance, as she toils along with twelve or sixteen cubs nearly of the size of rats, each with a turn of his tail around the root of the mother's, and clinging on her back and sides with paws, hands and mouth. This circumstance was thought distinctive of another species, hence called *dorsigera,* but is equally true of the common or Virginian opossum. It is exceedingly curious and interesting to see the young, when the mother is at rest, take refuge in the pouch, whence one or two of them may occasionally be seen peeping out, with an air of great comfort and satisfaction. The mother in this condition, or at any time in defence of her young, will make battle, biting with much keenness and severity, for which her long canine teeth are well suited.

THE COMMON OPOSSUM.

If taken young the opossum is readily tamed and becomes very fond of human society, in a great degree relinquishes its nocturnal habits, and grows troublesome from its familiarity. We have had one thus tamed which would follow the inmates of the house with great assiduity, and complain by a whining noise when left alone. As it grew older it became mischievous from its restless curiosity, and there seemed to be no possibility of devising any contrivance effectually to secure it. The same circumstance is frequently remarked by persons who have attempted to detain them in captivity, and of all the instances which have come to our knowledge, where even a great number were apparently well secured, they have all in a short time enlarged themselves and been no more heard of. In some such instances these animals have escaped in the city, and for a long time have taken up their quarters in cellars, where their presence has never been suspected, as during the day they remain concealed. In this way it is very probable that many are still living in the city of Philadelphia, obtaining a plentiful food by their nightly labours.

In Dr. Barton's facts and conjectures on the opossum, he mentions as a circumstance worthy of curiosity, the faculty the opossum has of lying on its back. We have observed this action of the animal, but could see nothing in it very different from what is very frequently done by the dog, cat, marmot, squirrel, and various other animals, which occasionally place themselves sufficiently on the back to expose the inferior surface of the body fully; but that this action in the opossum is indicative of any peculiarity, or is the ordinary position chosen by the animal, is what we cannot state from our own observation.

The size of the full grown opossum is about twenty inches, and that of the tail twelve; the weight is about fourteen pounds. The number of young is from twelve to sixteen. There is therefore not much probability of the species becoming very scarce, especially as their nocturnal mode of life renders it by no means necessary that they should fly to very remote distances from the habitations of man.*

---

* It is amusing to read the accounts of the wonderful medical virtues which have been attributed to the tail of this animal, in some of the older writers on the natural history of our continent. The following is a good specimen of the credulity and disposition to deal in the marvellous, which was formerly thought to form an almost essential quality in the natural historian:—"The tail of this animal (says MARCGRAVE) is a singular and wonderful remedy against inflammation of the kidneys; for if it be broken, and the quantity of a drachm of the water in which it is steeped be drunk sometimes, fasting, it wonderfully cleanses the ureters, expels calculi and other obstructions, [excitat venerem, et generat lac, medetur colicis doloribus, prodest parientibus et accelerat partum, promovet menses,] and if it be chewed and placed on a part into which thorns have been thrust, it extracts them, loosens the bowels, and I believe in all New Spain there is not to be found another remedy as useful in so many cases.".—*Hist. Ker. Nat. Brasil, lib. vi. p.* 22.

The above passage may have served as a hint to the celebrated CUMBERLAND, who, in one of his amusing works, introduces a quack, soliloquising on the virtues of a dried lizard's tail in the following words:

"Thou wilt pulverize most featly," quoth he, "when I

CHAPTER II.

Order IV. Glires; L. *Gnawers.*

SECTION I.—CLAVICULATA.

Having perfect, and in some, very strong clavicles.

The animals belonging to this order have the brain nearly smooth and without convolutions; the orbits of the eyes are not separated from the temporal cavities, which are slight; the eyes are directed laterally; the zygomatic arches are delicate and curved downwards, indicating feebleness in the jaws; the fore arm can scarcely be turned, and the two bones

have thee under the pestle; but before I consign thee to the mortar and reduce thee to dust, let me ponder upon thy properties, and do nothing without forecast and circumspection. Poisonous thou can'st not be, for though I have never eaten of thy species myself, I know that others have; and if thy flesh be delicate, thy dust cannot fail to be wholesome; nay, I doubt not but it is medicinal. Thou hast other virtues, if I could but recollect them; there is something more about thee; something I have read in learned authors of the back-bone of a lizard; and thine, heaven be prais'd, I perceive is perfect and entire; but whether it is recorded as a provocative to incontinency, or as a preventative, I cannot to a certainty recollect: upon second thoughts, I suspect thou art a stimulative; as I'm a sinner, I suspect thou art of a stirring quality, for thy tail betokeneth it."

are often consolidated. Those possessing the strongest clavicles exhibit some intelligence, and use their fore feet to convey their food to the mouth. They have the posterior extremities generally higher than the anterior, by which they are rendered fitter for leaping than running. Their intestines are very long, and the stomach is simple, or but little divided; the cœcum is often larger than the stomach itself.

These animals are provided with two large cutting teeth in both jaws, separated by a vacant space from the molars, and such teeth are exclusively destined to disintegrate solid bodies by repeated efforts, by nibbling or *gnawing;* the name of the order has been derived from this circumstance in various languages. The cutting teeth are enamelled only on the anterior surface, so that as the posterior surfaces wear away first, they always preserve a beveled edge. They grow from the root as rapidly as they wear at the edge, and when an opposing tooth is broken or lost, the other grows so rapidly as to become monstrous. The lower jaw is articulated by a longitudinal condyle, and has no other horizontal movement than from behind forwards and the reverse. The molar teeth have flat crowns, with transverse projections of enamel, in opposition to the horizontal motion of the jaw.

Those which have simple lines on the crowns instead of projections, and the whole surface of the molars very plane, are more exclusively frugivorous. Those whose teeth have these projections divided into blunt tubercles, are omnivorous; the small number which have points to these teeth, attack other animals more readily, and slightly approach carnivorous animals.*

Genus XX. Beaver; *Castor*; L.

Gr. Καςεg   *Ital.* Bevero.
Fr. Biévre.   *Pol.* Bobr.
Ger. Biber.   *Swed.* Bœffwer.

GENERIC CHARACTERS.

The head is large, with a short and blunt snout, small ears and eyes, and the upper lip divided. The trunk of the body is thick, having four teats, two near the fore limbs, and two at the posterior part of the chest. The limbs are short, the anterior being somewhat larger than the posterior; all the feet have five short, free and flexible toes, which are webbed, and the posterior toes have the membrane longer and broader. The middle toe is always longest in the fore and hind feet; the thumb and little finger, on the external and internal digits, are the shortest, and equal to each other; the intermediate toes are of middling size and equal in length. All the digits are furnished with strong and slightly incurvated nails, which are fit for burrowing; those on the hind feet are rather the largest. The tail is peculiarly flattened, of an oblong, oval shape, broad and covered at base with thick fur; the remainder has a covering of scales.

See Cuvier, Régne Animal. p. 186.

Dental System.

20 Teeth:	10 Upper	2 Incisive 8 Molar.
	10 Lower	2 Incisive 8 Molar.

IN THE UPPER JAW the incisors are flat, smooth and of a very great breadth, arising from the inferior and anterior part of the maxillary bone. The molars differ slightly from each other in size, and appear all to be composed of one internal and three external grooves, which being interrupted by the wearing of the teeth, at length merely present elliptical figures. Many of these external grooves are characterized by enlarging at their extremity. The germs of these teeth show the same number of grooves as we have described from partly worn teeth.

IN THE LOWER JAW the incisors are similar to those of the upper, and not less remarkable for great size; they rise far beyond and beneath the molars, between the coronoid process and condyle. The molars present absolutely the same characters, that is, the same figures as those in the opposite jaw; excepting that the three grooves are on the inner side of the tooth, and the external has but one. The germs of these teeth have also the strongest resemblance to the figures which are seen when the teeth are partly worn away.

Species I.—*The Beaver*.

Castor Fiber; L. Erxl. &c.

Le Castor ou le biévre: Briss. Regn. An. p. 133.
Le Castor: Buff. viii. pl. 36.
Castor Fiber: Sabine, App. p. 659. Say, Long's Exped. to the Rocky Mountains, i. p. 464.

Truth, alike the object and reward of all rational inquiry, is too delicate and unobtrusive to be advantageously approached or estimated, unless the mental vision be entirely free from prejudice, and her votaries, for the sake of her unostentatious though unfading charms, forego the pride of worshipping the fantastic creatures of their own imaginations. Accessible to all who in the proper disposition seek her presence, how many ages have elapsed during which fiction has been pursued in her stead, till at length opinion gains such strength, and prejudice so deep a root, that the semblance passes into general acceptation for the substance, and what was at first the mere breath of speculation, becomes finally received and accredited as indubitable. Thankless is the office of the individual who ventures to overturn any of these *idols* of the mind;* to displace the illusions of fancy by cold reality, and disperse into thin air the fairy world which credulity first called into existence, and indolent imagination perpetuates. It must be confessed that occasionally

---

* "Excrevit autem mirium in modum, istud malum ex opinione quadam sive æstimatione inveterata verum tumida et damnosa; minui nempe mentis humanæ majestatem, si ex

this is no pleasant task; yet it is one of the duties especially incumbent on the teacher of natural history, inasmuch as the exercise of imagination is always prejudicial to the study of nature, the sober reality of which, when correctly examined, possesses an interest far transcending that of all the fugitive beauties bestowed by this deluding faculty of the mind.

Who has not heard of the wonderful sagacity of the *beaver*, or listened to the laboured accounts of its social and rational nature? Who that has read the impassioned eloquence of Buffon, to which nothing is wanting but truth in order to render it sublime, can forget the impression which his views of the economy and character of this species produced? The enchanter waves his wand and converts animals, congregated by instinct alone, and guided by no moral influence, into social, rational, intelligent beings, superior to creatures high above them in organization, and even far more exalted than vast tribes of that race which has been justly and emphatically termed " lords of creation." Alas, for all these air-drawn prospects! while we endeavour to gaze upon their beauties they fleet away and leave no trace behind.

---

perimentis et rebus particularibus sensui subjectis et in materia determinatis diu ac multum versetur; præsertim quum hujusmodi res ad inquirendum laboriosæ, ad meditandum ignobiles, ad dicendum asperæ, ad practicam illiberales; numero infinitæ et subtilitate tenues esse soleant. Itaque jam tandem huc res rediit ut via vera non tantum deserta sed etiam interclusa et obstructa sit; fastidita experientia, nedum relicta, aut male administrata.—Bacon; *Nov. Organ.*

The injury the mind receives from this source is scarcely to be appreciated, and among others, the false notions we form concerning the relative perfection and excellence of the plan of nature, may be considered as of the first magnitude. The beaver, for instance, is endowed with singular instincts, and performs actions worthy of our admiration; yet the beaver is not more sagacious than the ant or the bee, creatures far removed from it in every respect, neither are its moral qualities better than those of the common rat. Each, according to its instinct, provides for the safety and support of itself and offspring, each obeys the impulse of a power beyond its own control, and each remains through countless generations the same in point of intelligence;—untaught, incapable of teaching, and as well qualified to perform all the singular actions of its predecessors, if removed at the earliest age from its kind, as if it had grown to maturity in their midst, and aided in their operations from the time its strength became sufficient to the task.

After rejecting the exaggerated facts, as well as the numerous fictions relative to this animal, ample scope will still remain for the exercise of our admiration; for although the beaver is in no respect exclusively wonderful, yet its character and habits are such as to render it highly interesting. We shall therefore give a plain, unvarnished statement of facts obtained from the most authentic sources, and afterwards present some sketches of what, although frequently repeated in books of acknowledged authority, may be termed the *fabulous history* of the animal. This will prove serviceable as well as amusing, as it will lead the inexperienced to receive wonderful narrations of the intelligence, &c. of animals, with enough of scepticism to prevent them from being betrayed into error.

The general aspect of the beaver, at first view, would remind one of a very large rat, and seen at a little distance it might be readily mistaken for the common musk-rat. But the greater size of the beaver, the thickness and breadth of its head, and its horizontally flattened, broad and scaly tail, render it impossible to mistake it for any other creature when closely examined. In its movements, both on shore and in the water, it also closely resembles the musk-rat, having the same quick step, and swimming with great vigour and celerity either on the surface, or in the depths of the water.

In a state of captivity or insulation, the beaver is a quiet or rather stupid animal, evincing about as much intelligence as a tamed badger, or any other quadruped which can learn to distinguish its feeder, come when called, or grow familiar with the inmates of the house where it is kept. It is only in a state of nature that the beaver displays any of those singular modes of acting which have so long rendered the species celebrated: these may be summed up in a statement of the manner in which they secure a sufficient depth of water to prevent it from being frozen to the bottom, and their mode of constructing the huts in which they pass the winter.

They are not particular in the site they select for the establishment of their dwellings, but if in a lake or pond where a dam is not required, they are careful to build where the water is sufficiently deep. In standing waters, however, they have not the advantage afforded by a current for the transportation of their supplies of wood, which, when they build on a running stream, is always cut higher up than the place of their residence, and floated down.

The materials used for the construction of their dams are the trunks and branches of small birch, mulberry, willow, poplar, &c. They begin to cut down their timber for building early in the summer, but their edifices are not commenced until about the middle or latter part of August, and are not completed until the beginning of the cold season. The strength of their teeth and their perseverance in this work, may be fairly estimated by the size of the trees they cut down. Dr. Best informs us that he has seen a mulberry tree, eight inches in diameter, which had been gnawed down by the beaver. We were shown, while on the banks of the Little Miami river, several stumps of trees, which had evidently been felled by these animals, of at least five or six inches in diameter. These are cut in such a manner as to fall into the water, and then floated towards the site of the dam or dwellings. Small shrubs, &c. cut at a distance from the water, they drag with their teeth to the stream, and then launch and tow them to the place of deposite. At a short distance above a beaver-dam the number of trees which have been cut down appears truly surprising, and the regularity of the stumps which are left might lead persons unacquainted with the habits of our animal to believe that the clearing was the result of human industry.

The figure of the dam varies according to circumstances. Should the current be very gentle, the dam is carried nearly straight across; but when the stream is swiftly flowing, it is uniformly made with a considerable curve, having the convex part opposed to the current. Along with the trunks and branches of trees they intermingle mud and stones, to give greater security, and when dams have been long undisturbed and frequently repaired, they acquire great solidity, and their power of resisting the pressure of water and ice is greatly increased by the willow, birch, &c. occasionally taking root, and eventually growing up into something of a regular hedge. The materials used in constructing the dams are secured solely by the resting of the branches, &c. against the bottom, and the subsequent accumulation of mud and stones, by the force of the stream or by the industry of the beavers. In various parts of the western country, where beaver are at present entirely unknown, except by tradition, the dams constructed by their labours are still standing securely. and in many instances serve instead of bridges to the streams they obstruct. There are few states in the Union in which some remembrance of this animal is not preserved by such names as *Beaver-Dam, Beaver-Lake, Beaver-Falls, &c.*

The dwellings of the beaver are formed of the same materials as their dams, and are very rude, though strong, and adapted in size to the number of their inhabitants. These are seldom more than four old and six or eight young ones. Double that number have been occasionally found in one of the lodges, though this is by no means a very common occurrence.

When building their houses, they place most of the wood crosswise and nearly horizontally, observing no other order than that of leaving a cavity in the middle. Branches which project inward are cut off with their teeth and thrown among the rest. The houses are by no means built of sticks first and then plastered, but all the materials, sticks, mud and stones, if the latter can be procured, are mixed up together, and this composition is employed from the foundation to the summit. The mud is obtained from the adjacent banks or bottom of the stream or pond near the door of the hut. Mud and stones the beaver always carries by holding them between his fore paws and throat.

Their work is all performed at night, and with much expedition. When straw or grass is mingled with the mud used by them in building, it is an accidental circumstance, owing to the nature of the spot whence the latter was taken. As soon as any part of the material is placed where it is intended to remain, they turn round and give it a smart blow with the tail. The same sort of blow is struck by them upon the surface of the water when they are in the act of diving.

The outside of the hut is covered or plastered with mud late in the autumn, and after frost has begun to appear. By freezing it soon becomes almost as hard as stone, effectually excluding their great enemy, the wolverene, during the winter. Their habit of walking over the work frequently during its progress, has led to the absurd idea of their using the tail as a trowel. The habit of flapping with the tail is retained by them in a state of captivity, and unless it be in the acts already mentioned, appears designed to effect no particular purpose. The houses, when they have stood for some time, and been kept in repair, become so firm from the consolidation of all the materials, as to require great exertion and the use of the ice-chisel or other iron instruments to be broken open. The laborious nature of such an undertaking may easily be conceived, when it is known that the tops of the houses are generally from four to six feet thick at the apex of the cone. HEARNE relates having seen one instance in which the crown or roof of the hut was more than eight feet in thickness.

The door or hole leading into the beaver-hut is always on the side farthest from the land, and is near the foundation of the house, or at a considerable depth under water. This is the only opening into the hut.

The large houses are sometimes found to have projections of the main building thrown out, the better to support the roof, and this circumstance has led to all the stories of the different chambers or apartments in beaver-huts. But these larger edifices, so far from having several apartments, are either double or treble houses, each part having no communication with the other, except by water. Upwards of twelve such dwellings have been seen under one roof, and, excepting two or three of them, the whole of the remainder had no communication unless by water, each having its own door into the dam, which is doubtless well known to the inmates, who may have comparatively little intercourse with each other. It is a fact that the

THE BEAVER.

musk-rat is sometimes found to have taken up his abode in the huts of the beaver; the otter also occasionally intrudes his company. The latter animal, however, is a dangerous guest, for, if provision grow scarce, it is not uncommon for him to devour his host.

The northern Indians believe that the beaver always thicken the northern walls of their houses much more than the others, in order more effectually to resist the cold. In consequence of this belief, these Indians always break into the huts from the south side.

All the beavers of a community do not co-operate in the fabrication of houses for the common use of the whole. Those which are to live together in the same hut, labour together in its construction, and the only affair in which all seem to have a joint interest, and upon which they labour in concert, is the dam, as this is designed to keep a sufficient depth of water around all the habitations.

In situations where the beaver is frequently disturbed and pursued, all its singular habits are relinquished, and its mode of living changed to suit the nature of circumstances, and this occurs even in different parts of the same rivers. Instead of building dams and houses, its only residence is then in the banks of the stream, where it is now forced to make a more extensive excavation, and be content to adopt the manners of a musk-rat. More sagacity is displayed by the beaver in thus accommodating itself to circumstances, than in any other action it performs. Such is the caution which it exercises to guard against detection, that were it not for the removal of small trees, the stumps of which indicate the sort of animal by which they have been cut down, the presence of the beaver would not be suspected in the vicinity. All excursions for the sake of procuring food are made late at night, and if it pass from one hole to another during the day time, it swims so far under water as not to excite the least suspicion of the presence of such a voyager. On many parts of the Mississippi and Missouri, where the beaver formerly built houses according to the mode above described, no such works are at present to be found, although beaver are still to be trapped in those localities. The same circumstances have been remarked of the European beaver, which has been thought to belong to another species, because it does not build. This, however, as may readily be inferred from what we have just stated, is no test of difference of species.

These animals also have excavations in the adjacent banks, at rather regular distances from each other, which have been called *washes*. These excavations are so enlarged within, that the beaver can raise his head above water in order to breathe without being seen, and when disturbed at their huts, they immediately make way under water to these washes for greater security, where they are more readily taken by the hunters, as we shall presently discover.

The beaver feeds principally upon the bark of the aspen, willow, birch, poplar, and occasionally the alder, but it rarely resorts to the pine tribe, unless from severe necessity. They provide a stock of wood from the trees mentioned, during the summer season, and place it in the water opposite the entrance to their houses. They also depend in a great degree upon the large roots (of the *nuphar luteum*,) which grow at the bottom of the lakes, ponds and rivers, and may be procured at all seasons. It is remarked that these roots, although they fatten the beaver very much, impart a rank and disagreeable taste to their flesh.

The number of young produced by the beaver at a litter is from two to five. Females have been killed in which six young were found, but this occurred only in two instances out of many hundreds examined at different stages of gestation.* During the season of union, the voice of both sexes resembles a groan, the male having a much hoarser note than the female. The young beavers whine in such a manner as closely to imitate the cry of a child. Like the young of most other animals they are very playful, and their movements are peculiarly interesting, as may be seen by the following anecdote, related in the narrative of Capt. FRANKLIN's perilous journey to the shores of the Arctic Sea.— " One day a gentleman, long resident in the Hudson's bay country, espied five young beavers sporting in the water, leaping upon the trunk of a tree, pushing one another off, and playing a thousand interesting tricks. He approached softly, under cover of the bushes, and prepared to fire on the unsuspecting creatures, but a nearer approach discovered to him such a similitude betwixt their gestures and the infantile caresses of his own chil-

---

Hearne

dren, that he threw aside his gun and left them unmolested."

The beaver is a cleanly animal, and always leaves the house to attend to the calls of nature; the excrement being light, rises to the top of the water and soon separates and disappears. Thus, however great may be the number of individuals occupying the hut, no accumulation of filth of this kind occurs.

The beaver swims to considerable distances under water, but cannot remain for a long time without coming to the surface for air. They are therefore caught with greater ease, as they must either take refuge in their vaults or washes in the bank, or seek their huts again for the sake of getting breath. They usually, when disturbed, fly from the huts to these vaults, which, although not so exposed to observation as their houses, are yet discovered with sufficient ease, and allow the occupant to be more readily captured than if he had remained in the ordinary habitation.

To capture beavers residing on a small river or creek, the Indians find it necessary to stake the stream across to prevent the animals from escaping, and then they try to ascertain where the vaults or washes in the banks are situated. This can only be done by those who are very experienced in such explorations, and is thus performed:—The hunter is furnished with an ice-chisel lashed to a handle four or five feet in length; with this instrument he strikes against the ice as he goes along the edge of the banks. The sound produced by the blow informs him when he is opposite to one of these vaults. When one is discovered, a hole is cut through the ice of sufficient size to admit a full-grown beaver, and the search is continued until as many of the places of retreat are discovered as possible. During the time the most expert hunters are thus occupied, the others with the women are busy in breaking into the beaver-houses. which, as may be supposed from what has been already stated, is a task of some difficulty. The beavers, alarmed at the invasion of their dwelling, take to the water and swim with surprising swiftness to their retreats in the banks, but their entrance is betrayed to the hunters watching the holes in the ice, by the motion and discolouration of the water. The entrance is instantly closed with stakes of wood, and the beaver, instead of finding shelter in his cave, is made prisoner and destroyed. The hunter then pulls the animal out, if within reach, by the introduction of his hand and arm, or by a hook designed for this use, fastened to a long handle. Beaver-houses found in lakes or other standing waters offer an easier prey to the hunters, as there is no occasion for staking the water across.

Among the Hudson's bay Indians every hunter has the exclusive right to all the beavers caught in the washes discovered by him. Each individual on finding one places some mark, as a pole or the branch of a tree stuck up, in order to know his own. Beavers caught in any house are also the property of the discoverer, who takes care to mark his claim, as in the case of the washes.*

---

* Lewis and Clark relate an instance which fell under their observation of one beaver being caught in two traps belonging to different owners, it having one paw in each. The

The number of beavers killed in the northern parts of this country is exceedingly great, even at the present time, after the fur trade has been carried on for so many years, and the most indiscriminate warfare waged uninterruptedly against the species. In the year 1820, sixty thousand beaver skins were sold by the Hudson's bay company, which we can by no means suppose to be the whole number killed during the preceding season. If to these be added the quantities collected by the traders from the Indians of the Missouri country, we may form some idea of the immense number of these animals which exist throughout the vast regions of the north and west.

It is a subject of regret that an animal so valuable and prolific should be hunted in a manner tending so evidently to the extermination of the species, when a little care and management on the part of those interested might prevent unnecessary destruction, and increase the sources of their revenue. The old beavers are frequently killed within a short time of their littering season, and with every such death from three to six are destroyed. The young are often killed before they have attained half their growth and value, and of necessity long before they have contributed to the continuance of their species. In a few years, comparatively speaking, the beaver has been exterminated in all the Atlantic and in the

---

proprietors of the traps were engaged in a contest for the beaver, when the above named distinguished travellers arrived and settled the dispute between them by an equitable arrangement.

THE BEAVER.

western states, as far as the middle and upper waters of the Missouri; while in the Hudson's bay possessions they are becoming annually more scarce, and the race will eventually be extinguished throughout the whole continent. A few individuals may, for a time, elude the immediate violence of persecution, and like the degraded descendants of the aboriginals of our soil, be occasionally exhibited as melancholy mementos of tribes long previously whelmed in the fathomless gulf of avarice.

The Indians inhabiting the countries watered by the tributaries of the Missouri and Mississippi, take the beavers principally by trapping, and are generally supplied with steel-traps by the traders, who do not sell, but lend or hire them, in order to keep the Indians dependant upon themselves, and also to lay claim to the furs which they may procure. The name of the trader being stamped on the trap, it is equal to a certificate of enlistment, and indicates, when an Indian carries his furs to another trading establishment, that the individual wishes to avoid the payment of his debts. The business of trapping requires great experience and caution, as the senses of the beaver are very keen, and enable him to detect the recent presence of the hunter by the slightest traces. It is necessary that the hands should be washed clean before the trap is handled and baited, and that every precaution should be employed to elude the vigilance of the animal.

The bait which is used to entice the beavers is prepared from the substance called castor (*castoreum,*) obtained from the glandulous pouches of the male* animal, which contain sometimes from two to three ounces. This substance is called by the hunters *bark*-stone, and is squeezed gently into an open mouthed phial.

The contents of five or six of these castor bags are mixed with a nutmeg, twelve or fifteen cloves, and thirty grains of cinnamon, in fine powder, and then the whole is stirred up with as much whiskey as will give it the consistency of mustard prepared for the table. This mixture must be kept closely corked up, and in four or five days the odour becomes more powerful; with care it may be preserved for months without injury. Various other strong aromatics are sometimes used to increase the pungency of the odour. Some of this preparation, smeared upon the bits of wood with which the traps are baited, will entice the beaver from a great distance.

The castor, whose odour is similar to tanner's ooze, gets the name of *bark*-stone from its resemblance to finely powdered bark. The sacks containing it are about two inches in length. Behind these, and between the skin and root of the tail, are found two other oval cysts, lying together, which contain a pure strong oil of a rancid smell.

During the winter season the beaver becomes very fat, and its flesh is esteemed by the hunters to be excellent food. But those occasionally caught in the summer are very thin, and unfit for the table. They lead so wandering a life at this season, and are so much exhausted by the collection of materials

---

* Juxta preputium utroque latere existunt.

for building, or the winter's stock of provision, as well as by suckling their young, as to be generally at that time in a very poor condition. Their fur during the summer is of little value, and it is only in winter that it is to be obtained in that state which renders it so desirable to the fur-traders.

The different appearances of the fur, caused by age, season, disease, or accident, has at times led individuals to state the existence of several species of beaver in this country. No other species, however, has yet been discovered, but that whose habits we have been describing. Beavers are occasionally found nearly of a pure white, which is owing to the same cause that produces albino varieties of various animals. A specimen of the albino beaver may be seen in the Philadelphia Museum; HEARNE saw but one such specimen during a residence of twenty years in the Hudson's bay country. This was considered a great curiosity, and no other was afterwards procured there during the ten ensuing years, notwithstanding he offered a large reward to the Indians for as many of the same colour as they could procure.

The traits of character exhibited by the beaver in captivity are not very strikingly peculiar, though sufficiently interesting. It learns to obey the voice of its master, is pleased to be caressed, and cleanly in its habits. HEARNE states that he has kept various individuals about his house during his residence at Hudson's bay, and remarks, "they made not the least dirt, though they were kept in my own sitting room, where they were the constant companions of the Indian women and children, being so fond of their company that when the Indians were absent for any considerable time, the beaver discovered great signs of uneasiness, and on their return showed equal marks of pleasure by fondling on them, crawling into their laps, laying themselves on their backs, sitting erect like a squirrel, and behaving to them like children that see their parents but seldom. In general during the winter they lived on the same food as the women did, and were remarkably fond of rice and plumb-pudding. They would eat fresh venison and partridges very freely, but I never tried them with fish, though I have heard they will at times prey on them."*

Fabulous History of the Beaver.

This part of our subject is richer in materials than any other which comes within the scope of our work. We have in the beginning adverted to the grand

---

* "It is well known that our domestic poultry will eat animal food: thousands of geese that come to London market are fattened on tallow craps, and our horses in Hudson's Bay would not only eat all kinds of animal food, but also drink freely of the wash or pot-liquor intended for the hogs. We are assured by the best authorities, that in Iceland not only black cattle, but also the sheep, are almost entirely fed on fish and fish-bones during the winter season. Even in the isles of Orkney, and that in the summer, the sheep attend the ebbing of the tide as regularly as the Eskimaux curlew, and go down to the shore which the tide has left to feed on the sea-weed. This, however, is through *necessity*, for even the famous island of Pomona will not afford them an ex-

source of error in this and other departments of natural history, but there is one circumstance peculiar to the history of the beaver, which has thrown over it more delusion than in the case of almost any other animal. The fur-traders, Indian interpreters, and Indians themselves, have furnished the greater part of the information which we possess of the habits and manners of this animal. To these persons the beaver is a most important object, and regarded with a degree of admiration and superstition exactly proportioned to their ignorance. Hence they have in numerous instances been led to magnify facts actually observed, and to state their own notions of the sagacity of the animal as realities, not intending to deceive, although they have deceived themselves. To become acquainted with the peculiarities of a species both nocturnal and exceedingly timid and vigilant, requires years of patient and assiduous attention. It is not surprising, therefore, that persons seeking information should resort to those who are devoted to the pursuit of the animal, and receive their statements given with seriousness and minute detail as worthy of credit. In addition to the errors which spring from the ignorance of these observers, there is a worse evil to which inquirers are exposed. The traders, hunters, and interpre-

---

istence above high-water-mark."—*Hearne*, 8vo. *p.* 245. It must always be borne in mind that observations made on the diet of captive animals, will not at all apply to them when they are free to follow the dictates of nature. It is, however, highly interesting to know how far they can accommodate themselves to necessity.

ters have, for various reasons, considerable jealousy of all those who are too inquisitive about their peculiar concerns, and it is an occurrence of almost daily repetition, that when they are questioned on these subjects, they take a malicious pleasure in palming, with truly Indian gravity and patience, the most false and marvellous relations upon their auditor.

This is frequently done with so much art as by no means to outrage probability, and the whole is made to appear so consistent, and is to the eager inquirer so highly interesting, as to prevent him for a moment from supposing that the whole is an extempore fable. We have been informed by an ear witness on one such occasion, that he was astonished to hear a trader giving a long account, full of the most extraordinary and interesting particulars, of the habits of the beaver, to an ardent inquirer, who was writing it down with great delight. As soon as the collector of notes on natural history had retired, after listening to the whole story with the most unsuspecting confidence, the other inquired of the trader how it happened that he never had before given this information, which he must have known would have been so very acceptable. The answer to this question was a roar of laughter, and an assurance that there was not a word of truth in the whole statement; but that, having been exceedingly annoyed by the inquisitiveness of the individual, he had chosen to get rid of him at once by appearing to tell him all he knew.

As the reader is already in possession of all the well attested facts to be procured in illustration of its habits and character, we may safely present a few of the marvellous relations which have been heretofore given of the beaver, leaving him to separate the great mass of fiction from the few truths with which they may be mingled. We therefore begin with the most ancient of these fictions, and come down to the latest writers who have contributed to the perpetuation of such erroneous views.

"The castor, or beaver, when in the rivers, feeds upon shell fish and such other prey as it can catch. This variety of food is the reason why its hinder parts, to the ribs, have the taste of fish, and that they are eaten upon fast days, and all the rest has the taste of flesh, so that it is not used at other times.

"It has pretty large teeth, the under standing out beyond their lips about three fingers breadth; the upper about half a finger, being very broad, crooked, strong and sharp, growing double, very deep in their mouths, bending circular, like the edge of an axe, and are of a yellowish red. They take fishes upon them as if they were hooks, being able to break in pieces the hardest bones. When he bites he never loses his hold until his teeth meet together. The bristles about their mouths are as hard as horns; their bones are solid and without marrow; their forefeet are like a dog's, and their hinder like a swan's. Their tail is covered over with scales, being, like a soal, about six inches broad and ten inches long, which he uses as a rudder to steer with when he swims to catch fish; and though his teeth are so terrible, yet when men have seized his tail they can govern the animal as they please.

"The beavers make themselves houses of square timber, which they gnaw down with their teeth almost as even as if they were sawed, and almost as equal as if it were measured. They lay these pieces across, and each is let down by large notches into the other, so that, having dug a hole for their foundation, they build several stories, that they may rise higher or lower, according to the fall of water."*

"Amongst the beavers some are accounted masters, some servants. They are cleanly in their houses; for the making of which, they draw the timber on the belly of their ancients, they lying on their backs."†

"While some are engaged in cutting down large trees for the dam, others traverse the vicinity of the river and cut smaller trees, some as thick as one's leg, and others as large as the thigh. They trim these and gnaw them in two at a certain height to make stakes: they bring these pieces first by land to the edge of the stream, and there float them to the dam; they then form a sort of close piling, which is still farther strengthened by interlacing the branches between the stakes. This operation supposes many difficulties vanquished; for to prepare these stakes, and place them in a nearly perpendicular situation, they must raise the large end of the stake upon the bank of the river, or against a tree thrown across it, while others at the same time plunge into the water and dig a hole with their forefeet for the purpose of receiving the point of the stake or pile, in order to sustain it erect. In pro-

---

* Pomet, Hist. of Drugs.
† Lemery.

portion as some thus plant the piles, others bring earth, which they temper with their fore-feet and beat with their tails; they carry it in their mouths and with their fore-feet, and convey so large a quantity that they fill all the intervals of the piling. This pile work is composed of several ranges of stakes of equal height, all planted against each other, extending from one side of the river to the other: it is piled and plastered throughout. The piles are planted vertically on the side next the water-fall; the whole work is sloping on the side sustaining the pressure, so that the dam, which is ten or twelve feet wide at base, is only two or three feet thick at the summit. It has therefore not only all the solidity necessary, but the most convenient form for raising the water, preventing it from escaping, sustaining its weight, and breaking its violence. At the top of the dam, that is at the thinnest part, they make two or three sloped openings for the discharge of the superfluous water, and these are enlarged or closed up as the river swells or diminishes, &c.

"It would be superfluous after such an exposition of their public works, to give a detail of their private edifices, if in a history it were not necessary to relate *all* the *facts*, and if this first great work were not done with a view to render their little dwellings more commodious. These dwellings are cabins or rather little houses, built in the water on close piles, near the edge of the pond, having two doors or issues, one on the land and the other on the water side. Sometimes they are found to have two or three stories, the walls being as much as two feet thick, elevated perpendicularly upon the piles which serve at the same time for the foundation and floor of the house, &c. The walls are covered with a sort of stucco, so well tempered and so properly applied, that it appears as if it had been done by human hands. Their tail serves them as a trowel for applying this mortar, which they temper with their feet, &c.

" These retreats are not only very secure, but also very neat and commodious; the floor is strewed with verdure; boughs of box and fir serve for a carpet, upon which they never leave the least dirt. The window which looks out upon the water serves them for a balcony for the enjoyment of the air, or to bathe during the greater part of the day. They sit with the head and anterior parts of the body elevated and the posterior plunged in water; the opening is sufficiently elevated never to be closed by the ice, which, in the climates where the beavers reside, is sometimes three feet thick; they then lower the shelf by cutting the piles upon which it rested aslope, and make an opening into the water below the ice!!

" The habit which they have of continually retaining the tail and hinder parts in the water, appears to have changed the nature of their flesh. Thus the fore parts, as far as to the loins, has the quality, taste and consistence of land animals; that of the thighs and tail has the odour, savour, and all the qualities of fish; this tail, a foot long, an inch thick, and five or six broad, is really an extremity, a true portion of a fish attached to the body of a quadruped.

" However admirable, or marvellous the statements we have made on the labours and society of the beaver may appear, we dare to say that no one will doubt their reality.* All the relations made by different witnesses, at various times, agree together as to the facts we have related; and if our statement differ from some among them, it is only at points where they have swelled the marvellous, surpassed the truth, and even transcended probability!†

"Beavers are most industrious animals; nothing equals the art with which they construct their dwellings. They choose a small piece of ground with a rivulet running through it. This they form into a pond by making a dam across, first by driving into the ground stakes five or six feet long, placed in rows, walling each row with pliant twigs, and filling the interstices with clay, ramming it down close."‡

"They have a chief or superintendant in their works, who directs the whole. The utmost attention is paid to him by the whole community. Every individual has his task allotted, which they undertake with the utmost alacrity. The overseer gives a signal, by a certain number of smart slaps with his tail, expressive of his orders. The moment the artificers hear it they hasten to the place thus pointed out, and perform the allotted labour, whether it is to carry wood, or draw the clay, or repair any acci-

---

* O! magnus posthac inimicis risus!—Uterne
Ad casus dubios fidet sibi certius? Hor. *Serm. lib. ii.*
† Buffon ed. Sonnini, vol. xxvi. p. 102.
‡ Pennant's History of Quadrupeds. The whole of the observations in that work on the habits of the beaver are transcribed from Buffon.

dental breach. They have also their centinels, who, by the same kind of signal, give notice of any apprehended danger. They are said to have a sort of slavish beaver among them (analogous to the drone) which they employ in servile works and domestic drudgery."*

" In 1792, Capt. G. Cartwright published a journal of transactions, &c. on the Labrador coast, where he had resided nearly sixteen years. In this he apprises the reader that his account will appear *very different* from what Buffon and others have written on the subject, and begs it may be remembered that they wrote chiefly from hearsay, but what he advances is the result of his own actual observation."†

Yet, with a very trifling exception, this actual observer repeats all the trash of preceding *hearsay*-writers, nearly in their own words, only expressing doubts about the tail being used as a trowel, or a sledge upon which they haul stones and clay. The following is his version of Buffon's account of the solitary or hermit beaver: "Sometimes a single beaver lives by itself, and is then called a *hermit* or *terrier*. Whatever may have been the cause which has separated these individuals from society, it is certain that they always have a black mark on the inside of the skin upon their backs, which is called a saddle, and distinguishes them from the others. Cartwright supposes this separation from society may arise from their fidelity and constancy to each other,

---

* Pennant's Arctic Zoology, p. 117. vol. i.
† Church's Cabinet of Quadrupeds.

and that, having by some accident lost their mate, they will not readily pair again. He thinks, likewise, that the mark on the back may proceed from the want of a companion to keep that part warm."*

"Three beavers were seen cutting down a large cotten-wood tree: when they had made considerable progress one of them retired to a short distance and took his station in the water, looking steadfastly at the top of the tree. As soon as he perceived the top of the tree begin to move towards its fall, he gave notice of the danger to his companions, who were still at work, gnawing at its base, by slapping his tail upon the surface of the water, and they immediately ran from the tree out of harm's way."†

"It is difficult for a traveller to publish his travels without speaking of the beaver, although he should have travelled only in Africa, where there are none. I should wish to avoid repetition, but I have no recollection of what those gentlemen individually, even Buffon from his closet, have written. I will communicate what I have seen and learned on the spot, respecting this surprising animal. If I say the same that others have said, it will serve to con-

---

* Church, Cab. Quad. Bachelors of the human species have good cause to rejoice that their backs are clothed, if Capt. Cartwright's doctrine holds good throughout, otherwise their forlorn condition would be at once indicated by something like the aforesaid *saddle.*

† Long's Exped. to the Rocky Mountains, vol. i. p. 464. It is but just to state that this is given in that work as a "*hunter's story,*" which is too often synonymous with an English word of three letters.

firm you the more in what you already know, if there should be any thing new, you will be obliged to me for adding to your stock of information.

"On the west side a small stream enters the lake. The beavers have barricaded the mouth of it, by means of a causeway, which a regiment of engineers could not have made better; the water is thrown back and forms a pond, where they have erected their town. It must be observed that they know that this river is never dry; for otherwise, they would not have chosen it.

"The stakes planted in the earth, and the trunks of trees which cross them, are of a considerable thickness and length. It is incredible how this little animal could transport such enormous pieces; but what is most astonishing, they never use trees thrown down by the wind or felled by men, but they make their own selection, and cut those which seem to them best adapted for their buildings.

"Whilst five or six of them cut or gnaw with their teeth at the foot of the tree, another remains in the middle of the river, and informs them, either by a whistle, or by a blow with his tail on the water, when he observes the top inclining, in order that, continuing not the less their labour, they may be cautious and remain on their guard. Observe, they never gnaw the tree on the land-side, but always on that next the water, in order that it may certainly fall in that direction.

"All the *tribes* then unite their efforts and float it to the place proposed. Then, with their teeth, they sharpen the stake, with their claws they make deep holes in the earth. and with their paws they plant and drive it in. They place branches of trees crossways against these stakes, they then fill up the interstices with mortar, which some prepare, while others are cutting the trees, or are occupied with other labours,—for the tasks are so distributed that none remain idle. This mortar becomes harder and more solid than the celebrated cement known among the Romans.

" When the causeway is completed and they have tried it, in order to know if it answers their purposes, they work out at the lower part of it an opening, in the nature of a sluice, which they open and shut at need, in order to let the river flow again; they then begin to build their house in the midst of the ground destined to form the pond. They never build the house before the causeway, lest this last should not succeed according to their wishes, and they should thus lose their time and trouble.

" Their house, built likewise of wood, and plastered, is of two stories, and double. It is long in proportion to the number of the tribe which are to inhabit it.

" The first story serves them in common, as a magazine for provisions, and is under the water; the second is above, and serves them for lodging-rooms, where each family has its apartment.

" Under the foundation of the house they work out a number of passages, by means of which they enter and go out under ground, without being perceived even by the most vigilant Indian; these open at some distance from the house, and at that part which forms the pond, or at the lakes or rivers, near which they commonly establish themselves, in order to have the choice of taking that direction which may be most convenient to them, or least dangerous in the different incidents of their life.

"The beavers are divided into tribes, and sometimes into small bands only, of which each has its chief, and order and discipline reign there, much more, perhaps, than among the Indians, or even among civilized nations.

"Their magazines are invariably provisioned in summer, and no one touches them before the scarcity of winter is felt. unless extraordinary circumstances render it absolutely necessary, but never in any case does any one enter except by the authority and in the presence of the chief. Their food consists in general of the bark of trees, principally that of willow, and of all the trees which belong to the poplar family. Sometimes when bark is not found in sufficient quantity, they collect the wood, and in this case they cut it into bits with their teeth.

"Each tribe has its territory. If any stranger is caught trespassing, he is brought before the chief, who for the first offence punishes him *ad correctionem*, and for the second deprives him of his tail, which is the greatest misfortune that can happen to a beaver, for their tail is their cart, upon which they transport, wherever it is desired, mortar, stones, provisions, &c. and it is also the trowel, which it exactly resembles in shape, used by them in building. This infraction of the laws of nations is considered among them as so great an outrage, that the whole tribe of the mutilated beaver side with him, and set off immediately to take vengeance for it.

"In this contest the victorious party, using the rights of war, drives the vanquished from their *quarters*, takes possession of them, and places a provisional garrison, and finally establishes there a colony of young beavers. With respect to this point, another particularity of these admirable animals will not appear less astonishing.

"The female of the beaver produces her young usually in the month of April, and has as many as four. She nourishes them, and carefully instructs them during a year, that is to say until the family is about to have another increase, and then these young beavers, obliged to give place, build a new dwelling by the side of the paternal mansion, if they are not very numerous, otherwise they are obliged to go with others in order to form elsewhere a new tribe and a new establishment. If then, at this time, the enemy is driven from his quarters, the victors, if their young of that year are arrived at the period of emancipation, (that is to say of governing themselves) instal them there. The Indians have related to me in a positive manner another trait of these animals, but it is so extraordinary that I leave you at liberty to believe or reject it. They assert, and there are some who profess to have been ocular witnesses, that the two chiefs of two belligerent tribes sometimes terminate the quarrel by single combat, in the presence of the two hostile armies, like the people of *Mediéve*, or three against three, like the Horatii and Curiatii of antiquity. Beavers marry, and death alone separates them. They punish infidelity in the females severely, even with death.

"When they are sick, they are carefully nursed.

The sick have also their plaintive cries, like human beings. The Indians hunt them in the same manner in which, as you have seen in our sixth promenade, they hunt the musk-rat. The musk-rat is a beaver of the second degree. He has the same form in miniature, and many of his qualities, although his fur is inferior in beauty and fineness. The Indians, moreover, in winter make holes in the ice which cover the ponds surrounding the houses of the beaver, watch for the moment when they put out their heads to take the air, and shoot them.

" The Great Hare at Red Lake wished to make me believe that, having come to the spot where two tribes of beaver had just been engaged in battle, he found about fifteen dead or dying on the field, and other Indians, Sioux and Chippeways, have also assured me that they have obtained valuable booty in similar circumstances. It is a fact that they sometimes take them without tails. I have seen such myself. In fine, these animals are so extraordinary, even in the eyes of the Indians themselves, that they suppose them men, become beavers by transmigration, and they think in killing them to do them a great service, for they say they restore them to their original state."*

We may advantageously conclude the fabulous history of the beaver by introducing the judicious observations made on the subject by HEARNE, whose excellent remarks on this animal have been, hitherto, altogether overlooked.

---

* Beltrami; La Decouverte des Sources du Mississippi, &c. 1825.

OF THE BEAVER.

"I cannot refrain from smiling when I read the accounts of different authors who have written on the economy of these animals, as there seems to be a contest between them who shall most exceed in fiction. But the compiler of the Wonders of Nature and Art, seems in my opinion to have succeeded best in this respect, as he has not only collected all the fictions into which other writers on this subject have run, but has so greatly improved on them, that little remains to be added to his account of the beaver, besides a vocabulary of their language, a code of their laws, and a sketch of their religion, to make it the most complete natural history of that animal which can possibly be offered to the public.

"There cannot be a greater imposition, or indeed a grosser insult on common understanding, than the wish to make us believe the stories of some of the works ascribed to the beaver; and though it is not to be supposed that the compiler of a general work can be intimately acquainted with every subject of which it may be necessary to treat, yet a very moderate share of understanding would be sufficient to guard him against giving credit to such marvellous tales, however smoothly they may be told, or however boldly they may be asserted by the romancing traveller."*

Most of the wonders related of the beaver are to be found in Gesner's work, *De Quadrupedibus*, which contains a collection of all the statements made anterior to this time. These extravagances will be found, with slight variations, repeated down to the

---

* Octavo ed. 1796, p. 231.

present day, by Buffon and his successors. We subjoin a few of these, which it is unnecessary to translate, as specimens of the close repetition indulged in by various writers, who should have drawn more largely upon nature instead of aiding in the diffusion of fictions and error.

" Morsu potentissimum adeo ut cum hominem invadit, conventum dentium non prius laxet quam concrepuise persenserit ossa fracta: *Plin.* et *Solin.* Apud Gesnerum.

" Gaudent enim, ripis magnorum fluvium cum animal sit amphibium, non solum ut reliqua quibus hoc nomen tribuitur quæ victus tantum gratia aquas petunt, sed etiam quadam natura affinitate, ut jam in caudæ et pedum posteriorum mentione diximus quæ ad piscium naturam accedunt.

" Castores gregatim ad sylvas lignatum pergunt imponunt autem ligna super ventrem resupinati unius qui pro vehiculo sit et inter crura ejus artificiose componunt: qui ne delabantur compressis ea cruribus ante et retro stringit; hunc sic onustum cæteri cauda ad casas usque pertrahunt. Hanc injuriam fieri negant nisi peregrino castori qui aliunde ad eos confugerit aut fortuito pervenerit ad castores loci alicujus incolas: illum enim hoc pacto in servitutem ab eis redigi. Alii non peregrino sed natu grandi et laboribus confecto qui propter dentes obtusos lignis secandis ineptus jam sit, hoc fieri aiunt. Ita tractati castores in dorso glabrescunt, quo signo a venatoribus agniti illæsi interdum dimittuntur.—

" Falsum est quod agitatus a venatore castret seipsum dentibus ac testes projici at et postea si ab alio venatore urgeatur erecto corpore, castratum se ostendat, ut sæpe in regionibus nostris compertum est."—*Alb. Mag.**

Description of the Beaver.

The beaver is about two feet in length, having a thick and heavy body, especially at its hinder part. The head is compressed and somewhat arched at the front, the upper part being rather narrow, and the snout, at the extremity, quite so; the neck is very short and thick. The eyes are situated rather high up on the head, and have rounded pupils; the ears are short, elliptical, and almost entirely concealed by the fur. The whole skin is covered by two sorts of hair; one which is long, rather stiff, elastic, and of a gray colour for two-thirds of its length next the base, and terminated by shining, reddish, brown points, giving the general colour to the pelage; the other is short, very fine, thick, tufted and soft, being of different shades of silver gray or light lead colour. On the head and feet the hair is shorter than elsewhere. The tail, which is ten or eleven inches long, is covered with hair similar to that of the back, for about one-third of its length nearest the base, the rest of it is covered by hexagonal scales, which are not imbricated.†

---

* "Imitatus castora, qui se Eunuchum ipse facit, cupiens evadere damno testiculorum. Juvenalis xii. liv. xxxiv.

† When the beaver sits erect upon its hinder limbs, as in the act of conveying his food to the mouth with his fore

The only species of beaver known is the one we have described; all the others which have been noticed are varieties of this species.

During the first year of their lives, the beavers are termed pappooses by the hunters; when two years old, small meddlers; at three years of age, large meddlers. In their fourth year they are called beavers, and after that old or great beavers.*

---

paws, like the squirrel. the tail is doubled under, or thrown forwards, lying between the legs.

Castoris penis modo profecto singulari ab ano, copulandi gratia protensus est; dehinc inter ista, et *monotremata*, sive animalia unico communi que foramine prædita, similitudo.

* It was our intention to have concluded the account of the beaver, by presenting a sketch of the history of the *American* fur trade, so intimately connected with this animal. But the difficulty of collecting the necessary data is so great, and our inquirers thus far have been so unproductive of satisfaction, that we are reluctantly obliged to defer our observations on this interesting subject until a future period.

1. Beaver. 2. Muskrat.

CHAPTER III.

Genus XXI. Musk-Rat; *Fiber;* Ill.

Germ. Zibethratze: Bisambiber; u. s. f.   *Fr.* Rat Musqué.
Swed. Desmansrotta.   *Eng.* Musk-Beaver.

GENERIC CHARACTERS.

The head is rather long and blunt at the snout, with eyes of a moderate size and short ears. The limbs are short, the anterior having four toes, not united, but bordered by a membranous edging somewhat fringed. All the toes are furnished with incurvated nails of moderate size. The tail is long, compressed, or flattened vertically, covered with a naked granulous integument, with a few hairs interspersed. The teats, which are six in number, are placed on the belly. A peculiar matter, having a strong musky odour, is secreted by glands situated in the pubic regions.

Dental System.

16 Teeth: { 8 Upper { 2 Incisive / 6 Molar.   8 Lower { 2 Incisive / 6 Molar. }

The teeth of this genus do not differ from those of the Campagnole Arvicola; *Lacep.* hereafter to be described, except in having distinct roots, and in the lower jaw the first molar having two triangles, one on each side, more than the Arvicolæ.

Species I.—*Musk-Rat.*

Fiber Zibethicus.

Rat Musqué du Canada: Briss. Reg. An. p. 136.
Castor Zibethicus: L. Erxl. Bod.
Mus Zibethicus, Gmel.
Ondatra: Buff. x. pl. i.
Ondatra Zibethicus: Say, Long's Exped. to the Rocky Mountains.
Fiber Zibethicus: Sab. App. p. 659.
Musquash of the Traders and Indians.

The musk-rat, which is so closely allied in form and habits to the beaver, does not, like that timid animal, retire from the vicinities inhabited by man, but, relying on its peculiar instinct for concealment, remains secure, notwithstanding the changes induced by cultivation, and multiplies its species in the very midst of its enemies. Thus, while the beaver has long since entirely disappeared and become forgotten in the Atlantic states, the musk-rat is found within a very short distance of our largest and oldest cities,* and bid fair to maintain its place in such situations during an indefinite future period.

The musk-rat owes this security to its nocturnal and aquatic mode of life, as well as to the peculiar mode in which its domicile is constructed. Along small streams, mill-races and ponds, where the banks

---

* Within a mile of Philadelphia, on every side, this animal may be found, along the banks of all the streams emptying into the Delaware and Schuylkill.

are of some elevation and strength, the musk-rats form large and extensive burrows. These have the entrance always in the deep water, so as to be entered or left without betraying the presence of the animal. The mouth of the burrow ascends from its commencement near the bottom, and slopes upwards until it is above the level of the high water. The burrow then extends to great distances, according to the numbers or necessities of the occupants. Like most other animals residing in such burrows, they frequently excavate them beneath the roots of large trees, where they are perfectly secure from being disturbed by having their burrow broken into from above.

The injuries done by the musk-rat to the banks thrown up to exclude the tide from meadows and other grounds, are frequently very extensive. The tide encroaches more and more on the burrow as the soil softens and is washed away; the animals extend their excavations in various directions, in order to free themselves from the inconvenience of the water, and at length, from the co-operation of both causes, the bank caves in and the water is allowed free access, often laying waste the most valuable parts of the farm. To understand the extent to which such mischief may be carried, it is sufficient to take a walk along the banks thrown up to protect the meadows on the Delaware, on both sides of the river. Similar, though not as extensive injury, is produced along the borders of ponds, races and small streams, by the caving in of the burrows formerly tenanted by the musk-rat.

Where musk-rats frequent low and marshy situations, they build houses, which, in form and general appearance, resemble those made by the beaver. These edifices are round, and covered at top in form of a dome, and are built of reeds, flags, &c. mingled with mud. Instead of one place of entrance and exit there are several subterraneous passages, leading in different directions, and as these are extensive, the musk-rats when disturbed take refuge in them. Numerous individuals, composing several families, live together during the winter season; but, in the warm weather the house is entirely deserted, and the musk-rats live in pairs and rear their young, of which they have from three to six at a litter.

The musk-rat builds in a comparatively dry situation, at least not in a stream or pond of water, but in the marsh or swamp. He requires no dam, and does not, like the beaver, lay up a stock of winter provision, neither does he erect so strong and durable a dwelling, as it is not to be repaired, but deserted for a new one the following season.

Speaking of the musk-rat, as observed by him in the Hudson's Bay country, HEARNE remarks, that "instead of making their houses on the banks of ponds or swamps, like the beaver, they build on the ice, as soon as it is skinned over, and at a considerable distance from the shore, always taking care to keep a hole open in the ice to admit them to dive for their food, which consists chiefly of the roots of grass. The materials made use of in building their houses are mud and grass, which they bring up from the bottom. It sometimes happens in very cold winters that the holes in their houses freeze over, in spite of all their efforts to keep them open. When

THE MUSK-RAT.

that is the case, and they have no provision left in the house, the strongest prey upon the weakest, till by degrees only one is left in a whole lodge. I have seen several instances sufficient to confirm the truth of this assertion: for when their houses were broke open, the skeletons of seven or eight have been found and only one entire animal. Though I have before said that they generally build their houses on the ice, it is not always the case: for in the southern parts of the country, particularly about Cumberland House, I have seen, in some of the deep swamps that were over-run with rushes and long grass, many small islands that have been raised by the industry of those animals, on the tops of which they had built their houses like the beaver, some of which were very large. The tops of these houses are favourite breeding places for the geese, which bring forth their young brood there without the fear of being molested by foxes, or any other destructive animal, except the eagle."

The musk-rat feeds upon the roots, &c. of aquatic plants, and is especially fond of the acorus verus, or calamus aromaticus, which grows abundantly in most of the marshy vicinities inhabited by the musk-rat. It has been imagined that this animal feeds also upon fish, merely from its habit of living much in the water. There is the same reason for believing that the beaver is piscivorous, an opinion which the structure of the teeth, stomach and intestines of both animals, sufficiently contradict.

The musk-rat is an excellent swimmer, dives well, and remains for a considerable time under water. It is rare to have an opportunity of seeing the animal during the day time, as it lies concealed in its burrow, but by watching during moonlight nights, in situations not much frequented by human visitors, the musk-rat may be seen swimming in various directions, and coming on shore for the sake of seeking food, or for recreation.

The musk-rat has its nose thick and blunt at the end, and short ears, nearly concealed in fur. Its body and head very much resemble those of the beaver, but differ from it in colour, being a reddish brown. The belly and breast are ash colour, mingled slightly with ferruginous. The feet and tail of the musk-rat are also remarkably different from those of the beaver; all the toes are free and unconnected. On the hinder, instead of a web uniting the toes, there is a stiff fringe of bristly hair, closely set and projecting from the sides of the toes. The tail is thin at the edges, compressed so as to be vertically flattened, covered with small scales, having a slight intermixture of hair, and is about nine inches long, being nearly of the length of the body, which measures about twelve inches from the end of the nose to the root of the tail. The powerful odour of musk renders the flesh of the musk-rat of little value, and few can eat it. The skin is highly valued on account of the fineness of its fur.*

---

* " The musk-rat is never seen in Carolina, Georgia, or Florida, within one hundred miles of the sea coast, and very few in the most northern parts of these regions; which must be considered as a most favourable circumstance by the people in countries where there is so much banking and draining of the land, they being the most destructive creatures to dykes."—*Bartram's Travels*, p. 281.

CHAPTER IV.

Genus XXII. Field Mouse; *Arvicola;* Lacep.

Germ. Feldmaus: Heerdenmaus; u. s. f.
Fr. Campagnol: Rat des champs.
Ital. Topo Terrajuolo: Campagnuolo.
Swed. Molle.

GENERIC CHARACTERS.

The animals belonging to this genus are in general appearance very similar to the common rat, have a rudiment of a thumb on the anterior feet, and the toes of these feet armed with slender nails. The posterior feet have five toes, provided with nails, and destitute of connecting membrane, or fringing of hairs on their edges. The teats, which vary in number from eight to twelve, are situated on the chest and belly. The tail is nearly of the length of the body, round, and covered by a velvet-like tegument.

Dental System.

16 Teeth:
- 8 Upper { 2 Incisive / 6 Molars.
- 8 Lower { 2 Incisive / 6 Molars.

This system of dentition is composed of small triangles, surrounded by enamel, and disposed alternately on each side of a common axis, so that there is a triangular vacancy between each of them, that forms a deep groove on the outside of the tooth.

In the upper jaw the incisors are even, and slightly rounded on their anterior surface. The first molar is composed of five triangles,—one anterior, two external, and two internal, and these correspond to the interval left between the others, so that they are closer than those of the anterior triangle. The second is composed of four triangles, one anterior, two external and two on the inner side, corresponding to the vacant spaces which separate the two others. The third is composed also of four triangles, one anterior, one external, one internal, and one posterior; the latter is irregular, being narrow, elongated, and the lines forming it sinuous. These three teeth diminish gradually in size, from the first to the last.

In the lower jaw the forms of the teeth are the same as in the upper: the incisors are even, and slightly rounded on the anterior surface. The first molar has five angles, or rather five divisions; the first is in form of a trefoil, then come two small internal triangles, an external and a posterior larger than the middle ones. The second is also composed of five triangles: one small anterior, two internal, one exterior and one posterior. The third appears to have only three or four triangles, placed nearly one behind the other, and joined by their angles.

Species I.—*The Meadow-Mouse.*

Arvicola Xanthognatus; Leach.

Arvicola Xanthognatus: Sabine, App. p. 660. Say, Long's Exped. to the Rocky Mountains, i. 369.
Campagnol aux joues fauves: Desm. Mammal. p. 282.

Were we to confine our attention to an individual of this species, its diminutive size, delicacy of limbs and evident feebleness, might lead us to consider it as altogether insignificant, and equally incapable of benefitting or injuring mankind. In this, as in various analogous instances, nature has compensated for individual feebleness by numerical force, and endowed this species with a fecundity which not only preserves it amidst numerous vigilant and destructive enemies, but enables it to multiply so extensively as to become a severe tax, and occasionally a scourge to the farmers.

The meadow-mouse is found in various degrees of abundance throughout this country, and, as implied by its name, prefers the meadow and grass fields to other situations. The banks of drains, and those thrown up to keep off the tide, or overflow of streams, are the favourite places for their burrows, which are both numerous and extensive, being continued in various directions and to considerable depths. These burrows are frequently causes of injury similar to that resulting from those of the musk rat, the tide gradually enlarging the cavities, and the bank finally falling in, until a fair breach is made, through which the grounds are injuriously inundated.

During the temperate and warm seasons of the year, the meadow-mice spend the greater part of their time above ground, travelling about through little lanes and alleys among the grass. These small roads are so frequently travelled, that after the hay-harvest, when they are left exposed, they have something of the appearance of little burrows among the grass-roots. At the season of the first hay-harvest their nests are found in great numbers on the surface of the ground. These are made very similar to a small bird's nest, of soft grass, and generally contain six or eight young ones. Recollecting that this species breeds more than once a year, we shall find no difficulty in understanding how the meadow-mice may become very injurious by excessive multiplication, notwithstanding their defenceless condition and numerous enemies. Besides being preyed upon by owls, hawks, cats, &c. the country people are very vigilant in putting them to death, and the hay-makers consider mouse hunting as one of the most enlivening circumstances connected with their labours.

Thus far the mischief of which the species is notoriously guilty, appears not to be compensated by any peculiar good quality; but although we are unable to state the precise degree of service rendered, the fact of its existence is sufficient evidence of importance in the great scale of creation, whatever difficulty there may be in discovering or acknowledging it. No doubt this, among other species, was destined to limit the undue increase of the vegetable kingdom; various other creatures in an analogous manner subsist by the destruction of meadow-mice, while the great destroyer man seems to be the last in the chain of destructiveness, since he is not only in the habit of extinguishing vegetable and brute animal life, but of extending his ravages to his own kind.

The general colour of this species is a reddish yellow, mingled with black on the upper part of the body, and a clear cinereous gray beneath. The sides of the head are fulvous; the tail is black above and white beneath; the paws are brownish on their superior surface, and white beneath. Its length, including the tail, is about five inches.

Species II.—*The Marsh Campagnol.*

Arvicola Riparius; Ord.

Arvicola Riparius: Ord. Journal of the Academy of Natural Sciences, iv. part ii. p. 305.

This species, like the preceding, makes its burrows in the meadow banks, and resembles it in various other respects. All that we know of it is derived from the account given by Mr. Ord, in the work above quoted.

"This species (says he,) is fond of the seeds of the wild-oats (*zizania aquatica*,) and is found in the autumn in those fresh water marshes which are frequented by the common rail, (*Gallinula Carolina Lath.*) When the tide is high the animal may be observed sitting upon the fallen reeds, patiently waiting for the recession of the water. From its position when at rest it has much the appearance of a lump of mud, and is commonly mistaken for such by those who are unacquainted with its habits. It swims and dives well."

The head of the marsh campagnol is large, with a thick obtuse snout,—having small eyes and short roundish ears, nearly concealed by the hair on the cheeks. The fore legs are very short; the posterior parts of the body are more slender and weaker than the anterior. The tail is thinly covered with hair, and tufted or penciled at tip, and is longest in the male. The upper parts of the body are of a tawny brown colour mixed with black, the lower parts of an ash or gray colour. The female has four pectoral and four abdominal teats, and brings forth eight young at a litter.

Species III.—*The Cotton-Rat, or Hairy Campagnol.*

Arvicola Hispidus; Ord.

Sigmodon Hispidum: Say & Ord, Journal of the Academy of Natural Sciences, iv. part ii. p. 354.

This animal was discovered in East Florida in the year 1818, by Mr. Ord, whose description was not published until 1825. He found its burrows in the deserted plantations lying on the river St. Johns, East Florida, especially in the gardens, where they are seen in every direction. It is highly probable, he thinks, that this animal will be found a source of much injury and vexation to the future settlers of that country.

The head of the hairy campagnol is thick, and the snout elongated, having eyes of considerable size, and large round ears; the tail is nearly as long as the body. The ears are slightly covered with hair; the fore legs are short; the hind feet are large and strong, with short lateral toes and stout claws. The upper parts of the body and head are of a pale, dirty, yellow ochre colour, mixed with black; the lower parts are cinereous. On the upper parts of the body and sides the hair is long, plentiful and coarse. The animal is six inches long from the tip of the snout to the insertion of the tail, which is four inches long. In the adult animal yellow is the predominant colour; the young are generally black.*

The cotton-rat obtains its name from the circumstance of making its nest with cotton, which it collects for the purpose in large quantities; the nest is generally placed within a hollow log, or else in a chamber at the extremity of a burrow.

Species IV.—*The Wood-Rat.*

Arvicola Floridanus; Ord.

The Wood-Rat: Bartram, Travels in E. Florida, p. 124.
Mus Floridanus: Say, Long's Exped. to the Rocky Mountains, i. p. 54.
Neotoma Floridana: Say and Ord, Journ. of the Acad. of Nat. Sciences, iv. part ii. p. 352.

This beautiful animal was once thought to be peculiar to Florida, and received its scientific name from that circumstance. But it is now highly pro-

---

See note at the end of the next species.

bable that it is to be found throughout this country in certain situations; by Say it was obtained on the Missouri.

From all that we can learn relative to this animal, it is of a gentle, timid disposition; harmless in its manners and inoffensive in its mode of living. Far from having any of that peculiar cunning and distrustful air that is so remarkable in the common rat, it shows few signs of fear when approached, and allows itself to be made prisoner or killed without difficulty. They burrow under stones and among the ruins of buildings, and feed on vegetable substances. They construct their nests with large quantities of brush and rubbish.

The wood campagnol is about sixteen inches long, including the tail, which measures seven inches. The head gradually diminishes in size from the ears to the snout, and is of a lead colour intermingled with gray. Its ears are nine-tenths of an inch long, rounded, prominent and open, having but few hairs on their back part and on the margin within. The eyes are of a moderate size and prominent; the whiskers are arranged in six longitudinal series, the longest of them surpassing the tips of the ears. The tail is hairy and brown above; the legs are stout and of nearly equal length, with white feet, having the toes annulated beneath, and the nails concealed by the hair. The thumb is minute, and the palms of the fore feet have five tuberculous prominences; in the soles of the hind feet there are six tubercles, of which the three posterior are distant from each other.

" The wood-rat (says Bartram,) is a very curious

THE WOOD-RAT. 71

animal; they are not half the size of the domestic rat, and of a dark brown or black colour; their tail slender and shorter in proportion, and covered thinly with short hair. They are singular with respect to their ingenuity and great labour in the construction of their habitations, which are conical pyramids about three or four feet high, constructed with dry branches, which they collect with great labour and perseverance, and pile up without any apparent order, yet they are so interwoven with one another, that it would take a bear or wild cat some time to pull one of these castles to pieces, and allow the animals sufficient time to secure a retreat with their young."*

The wood-rat has, beyond doubt, been as common throughout this country at a former period, as it is at present in Florida and on the Missouri.† It has very universally given place to the black-rat, and both have disappeared before the Norway rat, as we

---

* Page 125.

† "In turning over some of the baggage we caught a rat somewhat larger than the common European rat, and of a lighter colour; the body and outer part of the legs as well as the belly, feet and ears, are white; the ears are not covered with hair, and are much larger than those of the common rat; the toes are also longer; their eyes black and prominent; the whiskers very long and full; the tail rather longer than the body, and covered with fine fur and hair, of the same size with that on the back, which is very close, short and silky in its texture. This was the first we had met, although its nests are frequent among the clifts of rocks and hollow trees, where we also found large quantities of the shells and seeds of the prickly pear, on which we conclude they chiefly subsist."—*Lewis and Clarke, i. p.* 289.

shall soon have occasion to state. The wood-rat soon learns to infest the houses of the settlers, and to do nearly if not quite as much mischief as the common rat. It is highly probable that some of these rats still remain in the remote and barren parts of the Atlantic states, or in situations analogous to those occupied by this species in the southern and western country.*

---

* In the Journal of the Academy of Natural Sciences. (vol. iv. part ii. p. 345, 352,) Messrs. Say and Ord propose to establish two new genera for the reception of this and the preceding species, under the names of *Sigmodon* and *Neotoma*, in consequence of the differences they have observed in the dentition of these animals. These differences are the following: In the hairy campagnol (*arvicola hispidus**) " the different arrangement of the folds of the enamel, and the circumstance of the molars being divided into radicles, certainly exclude it from the genus arvicola." In relation to the wood-rat (*arvicola Floridanus*†) they remark, " that the grinding surface of the molars differs somewhat from that of the molars of the genus arvicola, but the large roots of the grinders constitute a character essentially different." With due deference to the opinion of our respected friends, we are decidedly of an opposite belief. This variation of dental arrangement may be sufficient to indicate modifications or differences in the regimen or feeding of these animals, but cannot of themselves suffice to establish generic distinctions, when the external characters and habits of the animals are so strikingly similar to the genus arvicola. The wood-rat certainly is closely related to the genus *mus*, and

* Sigmodon Hispidum of Say and Ord.
† Neotoma Floridana of Say and Ord. We have, as in all similar instances, referred the species to the original proprietors or describers, notwithstanding the changes produced by arranging them under other genera.

CHAPTER V.

Genus XXIII.—Lemming; *Lemmus;* Link, Cuv.

This genus is closely allied to the preceding, and differs from it principally in the conformation of the fore feet and the shortness of the tail. The fore feet are five toed in some instances, and four toed in others, being provided with nails fit for burrowing; the hind feet are five toed. The tail is not very acute at its extremity, is shorter than the body, and covered by a velvety integument.

The *dental system* is the same as that of the genus Arvicola.

Species I.—*The Hudson's Bay Lemming.*

Lemmus Hudsonius.

Lemmus Hudsonius: Sabine, App. p. 661.
Mus Hudsonius; Pallas, Glires, p. 208. pl. 26.
Rat de Labrador: Encycl. pl. 69, fig. 6.
Lemming de la Baie d'Hudson: Desm. Mammal. p. 289.
The Hair Tailed Mouse: Hearne, 8vo. ed. p. 385.

The Hudson's Bay Lemming is covered by a very fine, soft and long hair, which is of an ash colour,

---

might with great propriety be considered as a distinct subgenus of arvicola, as Say and Ord suggest that some naturalists may consider it. The arrangement proposed by them

with a tinge of tawny on the back, having along its middle a dusky stripe, and on each side a pale tawny line. The limbs are quite short, and the fore feet being formed for burrowing, are very strong. The two middle claws of the male, which are compressed, thick and strong, appear to be bifid or double, because the skin of these toes is callous, and projects from beneath the nail.

The mode of life peculiar to this species is but little known; the Lapland lemming is very notorious for its extensive migrations, but nothing of the same kind has been observed of the Hudson's Bay species.

"The hair-tailed mouse, (says HEARNE) is the largest in the northern parts of the bay, being little inferior in size to a common rat. They always burrow under stones on dry ridges, are very inoffensive, and so easily tamed, that, if taken when full grown, some of them will in a day or two be perfectly reconciled, and are so fond of being handled that they will creep about your neck or into your bosom. In summer they are gray, and in winter change to white, but are by no means so beautiful as a white ermine. At that season they are infested with multitudes o small lice, not a sixth part so large as the mites in a cheese; in fact, they are so small that at first sight they only appear like reddish brown dust, but on closer examination are all perceived in motion. In

---

we esteem to be in the highest degree artificial, unnatural, and by consequence unnecessary, and therefore not to be adopted; at least in a work in which nature and useful ness are the supreme objects of regard.

one large and beautiful animal of this kind, caught in the depth of winter, I found those little vermin so numerous about it, that almost every hair was covered with them as thick as ropes with onions, and when they approached near the ends of the hair they may be said to change the mouse from white to a faint brown. At that time I had an excellent microscope, and endeavoured to examine them, and to ascertain their form, but the weather was so exceedingly cold that the glasses became damp with the moisture of my breath before I could get a single sight. The hind feet of these mice are exactly like those of a bear, and the fore feet are armed with a horny substance,* (that I never saw in any other species of the mouse,) which is wonderfully adapted for scraping away the ground where they wish to take up their abode. They are plentiful on some of the stony ridges near Churchill factory, but never approach the house or any of the out-offices. From appearances they are very local, and seldom stray far from their habitations, even in summer, and in winter they are seldom seen on the surface of the snow—a great proof of their being provident in summer to lay up a stock for that season.†"

---

* The description given of this "horny substance," which is a mere induration of the cuticle covering the palms, and caused by the act of scraping among the stones, &c. sufficiently indicates the species.

† "I observed with astonishment long ridges of mouse-dung, several inches deep, extending for above two miles. By what means this could have arrived here I was at a loss to conceive, as I did not see any mouse-holes or other traces of these animals: besides which they live in stony dry

CHAPTER VI.

Genus XXIV.—Rat; *Mus;* L.

GENERIC CHARACTERS.

The head is conical, more or less short, having a pointed snout, rather large eyes, and almost naked ears. There are no cheek pouches; the neck is short and the body thick, having from ten to twelve teats, part situated upon the chest and part upon the belly. The toes are free, or unconnected by membrane, and provided with hooked nails. The anterior feet have four digits and a rudimental thumb, covered by a blunt nail. The tail is naked, scaly and tapering; the body is covered by long, stiff hairs, intermingled with a close fine fur.

Dental System.

16 Teeth:
- 8 Upper. — 2 Incisive, 6 Molar.
- 8 Lower. — 2 Incisive, 6 Molar.

In the upper jaw the incisors are smooth and flat, and rise from the sides of the anterior part of

---

places, and this was a swamp. It is possible, however, that this accumulation of the excrements of mice may be from the mus *(Lemmus,)* Hudsonius, occasionally migrating in the same wonderful manner as the lemmer of Lapland." *Lyon's Private Journal. p.* 432.

the maxillary bone. The three molars diminish in size from the first to the last; they are very remarkable for being inclined from before backwards. The first molar is composed of six tubercles, which, considered in a transverse order, present themselves thus:—two in front, one larger corresponding to the middle of the tooth, and the other at the inside; then three, two small ones on the edges, the largest in the middle, and finally one at the posterior part of the tooth, and of the size of the middle tubercle of the three preceding. This arrangement of the great tubercles in the middle, and of the small ones on the edge, gives the form of a trefoil to the undulating line they produce. The second molar is formed of four tubercles, one in front on the inside, two in the middle, arranged obliquely from without inwards, and from before backwards, and the fourth at the posterior part on the outside. The last has also four tubercles arranged like those of the second molar.

In the lower jaw the incisors are similar to those of the upper jaw; they arise far behind and above the molars, from the middle of the ascending branch of the jaw-bone, where its bulb produces a little projection. The molars diminish in size from the first to the third, and are inclined in a direction opposite to those of the upper jaw; that is, they lean forwards and are equally formed of tubercles. The first has five, one small anterior, two middle, and two posterior; the second has four, also arranged in pairs, two before and two behind. The last has but three, a single one in front, followed by a pair

Species I.—*The Common, Brown or Norway Rat.*

Mus Decumanus; Pall.

Mus Sylvestris: Briss. Reg. An. 170, No. 3; *Mus Norvegicus:* ibid. p. 173, No. 8.
Mus Decumanus: L. Gmel. Schreb. pl. 178, Encycl. pl. 67, fig. 9.
Le Surmulot: Buff. 8, pl. 27.
Brown-Rat: Penn. Quad. No. 298, Arct. Zool. i. 151, No. 57.

It must be confessed that this rat is one of the veriest scoundrels in the brute creation, though it is a misfortune in him rather than a fault, since he acts solely in obedience to the impulses of nature, is guided by no other law than his own will, and submits to no restraints except such as are imposed by force. He is, therefore, by no means as bad as the scoundrels of a higher order of beings, who, endowed with superior powers of intelligence, and enjoying the advantages of education, do still act as if they possessed all the villainous qualities of the rat, without being able to offer a similar apology for their conduct.

Among quadrupeds this rat may be considered as occupying the same rank as the crow does among birds. He is one of the most impudent, troublesome, mischievous, wicked wretches that ever infested the habitations of man. To the most wily cunning he adds a fierceness and malignancy of disposition that frequently renders him a dangerous enemy, and a destroyer of every living creature he can master. He is a pure thief, stealing not merely articles of food, for which his hunger would be a sufficient justification, but substances which can be of no possible utility to him. When he gains access to the library he does not hesitate to *translate* and appropriate to his own use the works of the most learned authors, and is not so readily detected as some of his brother pirates of the human kind, since he does not carry off his prize entire, but cuts it into pieces before he conveys it to his den. He is, in short, possessed of no one quality to save him from being universally despised, and his character inspires no stronger feeling than contempt, even in those who are under the necessity of putting him to death.

The common, brown, or Norway rat, now so extensively diffused over this country, is not indigenous to our soil, but was introduced from Europe, which received it from Asia in the eighteenth century, as late as the year 1750. There are few parts of the world now visited by navigators where this animal has not been introduced, and the immediate consequence of its introduction has been, that all the native rats have been destroyed, or obliged to withdraw beyond the reach of this subtle and implacable enemy.

Prior to the year above mentioned this rat, now so notorious for its ravages, was almost, if not wholly, unknown in Europe.* It was conveyed to England,

---

* Specie ob hoc (*ratto*) diversus, *mus decumanus*, Persia ut videtur et vicini orientis indigena, vix ante alterum tertiumve sæculi præteriti decennium Europæ invasisse fertur* et ubi agmina eorum consedere, domesticum contra *rattum* sensim defecisse constat. *Norvegicum* plures nuperorum zoologicorum vocant quam vero appellationem Zimmermanus id improbat quod faunæ scandinavicæ nullam ejus men-

* R. Smith's Rat Catcher, p. 5, 1768.

about the period above mentioned, in the timber-ships from Norway, and hence it has received one of its common names. Many years subsequently it was brought to this country in European ships, and has been gradually propagated from the sea-ports over the greater part of the continent.

The brown rat takes up its residence about wharves, store-houses, cellars, granaries; &c. and destroys the common black rat and mouse, or entirely expels them from the vicinities it frequents. To chickens, rabbits, young pigeons, ducks, and various other domestic animals, it is equally destructive when urged by hunger and opportunity. Eggs are also a very favourite article of food with this species, and are sought with great avidity; in fact, every thing that is edible falls a prey to their voracity, and can scarcely be secured from their persevering and audacious inroads. In the country they take up their abodes according to convenience and the abundance of provision, infesting especially mills,

---

tionem faciant. At enim vero hoc sane idoneis testibus evictum est ipsissimum hunc Rattum decumanum ante annum MDCCXXX, Anglis plane ignotum, tum temporis primum et quidem quod expresse asserunt, *ex Norvegia* navibus onerariis quæ lignorum materiam inde advehebant illatum esse.* Cumque tum temporis in universa Germania boreali nullibi adhuc visus fuerat inficetum, corruit asseclarum petitoris quondam regni Anglici figmentum quo illum murem ex Hanoverianis terris in Brittanniam translatum esse fabulabantur.—*Jo. Frid. Blumenbach. Com. Soc. Goett.* 1285.

* Espriella's (Southey's) Letters, i. p. 285, ed. 3.

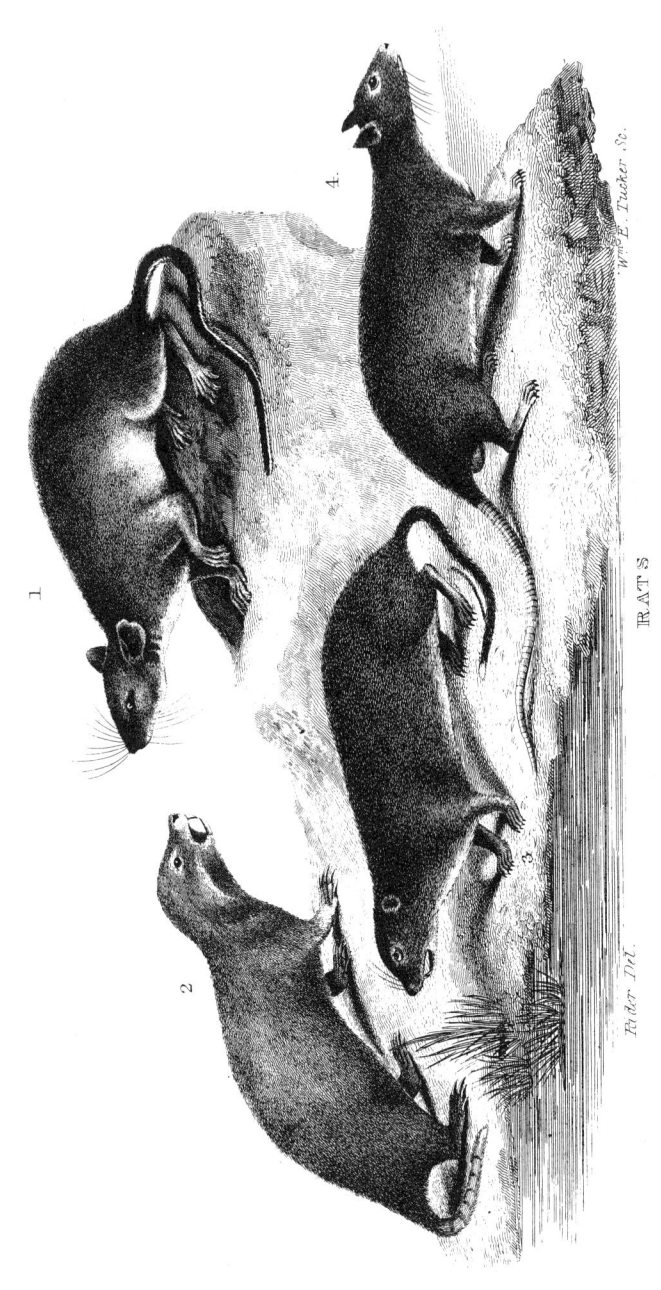

RATS.
1. Florida. 2. Peached. 3. Water. 4. Common.

barns and out-houses, or residing in holes along the banks of races or other water-courses.

The brown rat swims with great facility, and dives with vigor, remaining under water for a considerable time, and swimming thus to some distance. When attacked and not allowed an opportunity of escaping, he becomes a dangerous antagonist, leaping at his enemy and inflicting severe and dangerous wounds with his teeth. The most eager cat becomes immediately intimidated in the presence of one of these rats thus penned up, and is very willing to escape the dangers of an encounter.

The brown rat is amazingly prolific, and but for its numerous enemies, and its own rapacious disposition, would become an intolerable pest. Happily for the world, in addition to man, to the weazel, cat, some species of dog, &c. rats frequently find destructive enemies in each other, both in the adult and young state, their numbers thus being prevented from becoming such an intolerable grievance as they otherwise necessarily would. The strongest of the species prey upon the weaker, and are the most merciless destroyers of their own kind.* The weazel and the terrier are the most efficient rat-killers, as the first

---

* "It is a singular fact in the history of these animals that the skins of such of them as have been devoured in their holes have frequently been found curiously turned inside out, every part being completely inverted to the ends of the toes. How the operation is performed it would be difficult to ascertain; but it appears to be effected in some peculiar mode of eating out the contents."—*Bewick, Hist of Quadrupeds.*

can pursue the enemy to his most secret retreat, and the second derives from his superior strength and activity a very decided advantage in the contest. The cat, though in general a very useful auxiliary in lessening the number of this species, is very liable both to be foiled and worsted in her attempts. Bringing forth from twelve to eighteen at a litter, we have good reason to rejoice that so many animals have an instinctive animosity against so noxious a marauder.*

The cunning of these rats is not less than their impudence; it is almost impossible to take them in traps after one or two have been thus caught, as the rest appear perfectly to understand the object of the machine, and afterwards avoid it with scrupulous care, however tempting may be the bait it contains. The surest way to remove them is by poison, which, however, they frequently detect and avoid. The powder of nux vomica, mixed with some Indian corn or oat-meal, and scented with oil of rhodium, is found very effectual in destroying them. Arsenic is very commonly used in the same way for this purpose, but the fatal accidents which frequently occur when this poison is kept about the house, in consequence of the label being removed or changed, and the arsenic administered to members of a family instead of some other medicine, render it a very objectionable resource.

---

*The name of this species, *decumanus*, was given on account of its great size, and is equally applicable to its great mischievousness. The word originally was decimanus, and eventually by custom became synonymous with magnus or great. See Callipœnus. Litleton. &c.

The brown rat measures about nine inches, and is of a light brown colour, intermingled with ash and tawny. The colour of the throat and belly is of a dirty white, inclining to gray. It has pale, flesh-coloured, naked feet, with a tail of the same length as the body, and covered with small dusky scales, with short hairs thinly scattered between.

Species II.—*The Black Rat.*

Mus Rattus; L.

Mus Rattus: Pall. Schreb. &c.
Mus Domesticus Major: Ray, Quad. Sp. 217.
Le Rat: Buff. 7, pl. 36.
Black Rat: Penn. Quad. ii. p. 176. Arct. Zool. i. p. 150.

This rat was much more common previous to the introduction of the brown rat than at present. It is now found only in situations to which the brown rat has not extended its emigrations, and is almost as injurious and destructive, resembling it closely in manners and habits. It is of a deep iron gray, and indeed nearly of a black colour above, and of an ash colour on the lower parts of its body. Its legs are nearly naked, and on its fore feet instead of the rudimental thumb it has a claw. The length from the nose to the root of the tail is seven inches; the tail itself is almost eight inches long.

It has been a matter of dispute, whether this animal was received here from Europe, or was originally taken hence to that quarter of the world. Blumenbach, who has devoted much attention to the subject, states it as his opinion that the black rat was carried from Europe to America.* Garcilasso de la Vega states, that it was first introduced into South America by the Europeans, about the year 1544, and Geraldus Cambrensis speaks of them in Europe previous to the discovery of America.

Species III.—*The Common Mouse.*

Mus Musculus; L.

Mus Musculus: Erxl. Bod. Schreb. Ac.
Mus Domesticus Vulgaris. Ray, Quad.
Mus Sorex: Briss. p. 169, Sp. 2.
La Souris: Buff. viii. pl. 39, id. suppl. viii. pl. 20.

Like the two preceding species the Common Mouse is not an original inhabitant of this country.

---

* De primigenio et patrio *Ratti* vulgariter *domestici* habitaculo, diversimodi disputatum est. Mirum videtur paradoxam Linnei opinionem, qui eum ex Indis occidentalibus in Europam advectum fuisse putarat, vel ipsi Pallasio ideo non improbabilem visam esse quod apud antiquos licet Musculi frequens, mentio nulla occurrat Ratti. Etsi enim nullus veterum, sive Græcorum sive Romanorum Rattum memoret, medii tamen ævi scriptores, iique de historia naturali perbene meriti, diu ante orbem novum detectum de *Ratto* nostrate agunt, in quibus egregius Silvester Geraldus anno **MCLXXXVIII** expresse *mures majores* nominat qui vulgariter Ratti dicuntur.* Probabile vero videtur huic Rattum primitus Europam mediam incoluisse donec occasione commerciorum et præsertim navigationem per universum quæ Europæis patet orbem adeo propagatus est ut quondam inter prodigia relatum sit nonnullos Germanicæ urbes eo plane caruisse.—*Blumenbach, Act. Soc. Gœtting. v.* 1823.

* Itinerar. Cambriæ.

but was brought here from Europe, and has long since become perfectly naturalized throughout the continent, having been conveyed in every direction by persons moving their household goods, even to the most remote frontier settlements.

The common mouse, from its size and feebleness, is to be regarded rather as a troublesome than a very injurious inmate of our dwellings, but always likely to effect much mischief on account of its fecundity, which is full as remarkable as that of any of its kindred species.* It is a timid and vigilant creature, yet confides to a considerable extent in its swiftness and watchfulness, coming out after various trials, and stealing about a room even when there are several persons present, provided they are silent and do not move.

The mouse makes a nest very similar to that of a bird, having the inside lined with some soft material, such as wool, cotton, &c. and brings forth her young several times during a year, generally from six to ten at each litter. At birth her offspring are naked and helpless, but in about fifteen days they are able to shift for themselves, and the mother is soon at liberty to prepare for another family.

The mouse is a very beautiful little animal, when seen not alarmed and at perfect liberty. Its long and

---

* " The propagation of mice, ($\mu\nu\varepsilon\sigma$) in comparison with that of other animals, is very remarkable both for quickness and profuseness. A pregnant female was shut up in a chest of grain; in a short time a hundred and twenty individuals were counted."—*Aristotle, Hist. of Animals, Book* vi. *chap.* 37

slender whiskers, which extend in numerous and graceful lines from around the fore part of the head. its bright prominent eyes, delicate ears, and slight limbs, with its peculiar movements in search of food, or while sporting with its companions, are all such as to render it a pleasing and interesting animal. It is generally, however, viewed with great disgust on account of prejudices connected with its mischievousness, and the peculiar smell which is more or less prevalent in the places where the species is most numerous.

The common mouse has frequently been tamed, and exhibits a considerable degree of attachment to its feeder. Instances are on record of prisoners who have amused themselves by feeding one of these little animals, until it has become quite tame, and appeared immediately, whenever called by its master. Among other circumstances connected with the history of the mouse, it is generally rumoured that this animal is peculiarly susceptible to impressions produced by music, and some very wonderful accounts have been published on the subject. The following story may serve as a specimen of the manner in which *facts* may be stated with perfect accuracy, and yet *conclusions* entirely unfounded be thence deduced. It is related by a gentleman who heard it from another " of undoubted veracity."

" One evening, in the month of December, as a few officers on board of a British man of war, in the harbour of Portsmouth, were seated around the fire, one of them began to play a *plaintive* air on the violin. He had scarcely performed ten minutes when a mouse, apparently frantic, made its appearance in the centre of the floor, near the large table which usually stands in the ward-room, the residence of the lieutenants in ships of the line. The strange gestures of the little animal strongly excited the attention of the officers, who, with one consent, resolved to suffer it to continue its singular actions unmolested. Its exertions now appeared to be greater every moment. It shook its head, leaped about the table, and exhibited signs of the most extatic delight. It was observed that in proportion to the gradation of tones to the soft point, the extacy of the animal appeared to be increased, and *vice versa*. After performing actions which an animal so diminutive would at first sight seem incapable of, the little creature, to the astonishment of the delighted spectators, suddenly ceased to move, fell down and expired, without evincing any pain."*

Of the truth of this narration we are thoroughly satisfied, but we should explain it differently. The mouse, under the influence of disease, and almost in the agonies of death, came out of its hole when the musician was performing, and after struggling for a time, until exhausted by convulsions, died. The inferences of its " extacy," &c. are, for any thing to the contrary contained in the above account, entirely gratuitous, and we are much mistaken if the filing of a saw, the scraping of a gridiron, or the whetting of a scythe, would not in this instance have been accompanied by a similar degree of " extacy" in proportion as the

---

Barton's Medical and Physical Journal, i. p. 58.

"gradation of the tones" approached the "soft point."

The common mouse is about three inches and a-half long, and has a long, nearly naked tail. Its colour varies considerably, but is generally of an ashy brown. It has four digits on its anterior feet, and a rudimental thumb, destitute of a claw: the hind feet are five toed. The mouse is preyed on by cats, weazels, owls, rats, &c. &c.

Species IV.—*The Rustic Mouse.*

Mus Agrarius; Gmel.

Mus Agrarius: Gmel. Pall. p. 341. Say, Long's Exped. to the Rocky Mountains, p. 369.

This little mouse is very common throughout this country, and is found in great abundance in places favourable to their multiplication. They are occasionally very injurious to the farmers by the destruction of the small grain, the heads of which they cut off and convey to their subterranean hoards, which differ very little, if at all, from those made by the meadow-mouse or campagnol.

The rustic mouse is about three inches long, and has a streak of a mixed dusky and ferruginous colour along the back: the spaces between the ears (which are large, open and naked) and sides are of an orange colour, while the whole of the under parts of the body, legs and feet, are of a pure white; the tail is dusky above and whitish beneath. The whiskers are long, and some of the hairs are white, some black.

CHAPTER VII.

Genus XXV.—Pouched-Rat; *Pseudostoma;* Say.

GENERIC CHARACTERS.

The head and body are large, giving to the animal a clumsy appearance; the cheek-pouches are very extensive, situated outside of the mouth, separated therefrom by the common integuments, and are profoundly concave, opening downwards, towards the mouth. The legs are short, the fore feet large and armed with very long claws; the hind feet are small.

Dental System.

20 Teeth: { 10 Upper { 2 Incisive / 8 Molar. ; 10 Lower { 2 Incisive / 8 Molar.

In the upper jaw the incisors, always exposed to view, are strong and truncated in their entire width at tip, marked by a deep longitudinal groove near the middle, and by a smaller one at the inner margin. The molars, eight in number, penetrate to the base of their alveolæ without separating into roots, as in the genera *Arvicola Lepus, &c.* having simply discoidal, transversely oblong, oval crowns, margined by enamel, resembling in general form the molars of the genus Lepus, but without the appearance of either a groove at their ends or of a dividing crest of enamel. The posterior tooth is rather more rounded than the others, and that of the upper jaw has a small prominent angle on its posterior face; the anterior tooth is double, in consequence of a profound duplicature in its side, so that its crown presents two oval disks, of which the anterior one is smaller, and the lower one somewhat angulated. All these teeth incline obliquely backward, thus resembling those of the preceding genus.

In the lower jaw the teeth are similar to those in the upper, except that the molars are inclined forwards.*

Species I.—*The Pouched-Rat.*

Pseudostoma Bursarium; Say.

Pseudostoma Bursaria: Say, Long's Exped. to the Rocky Mountains, i. 406.
Canada Rat: Shaw, Gen. Zool. ii. part i. p. 100.
Mus Bursarius: Linn. Trans. v. p. 227, pl. 8.
Mus Saccatus: Mitchill, New York Med. Rep. Jan. 1821. Lewis and Clark, ii. 180.
Cricetus Bursareus: Desm. Mammal. 312.

[*Vulgarly called Salamander; Pouched-Rat; Sand-Rat; &c.*]

The pouched-rat, though long since noted by various observers, is still but little known, even in the vicinities where it is most common. Its peculiar mode of life, its nocturnal habits and vigilance,

---

* This dental system, &c. is from Say. See Long's Expedition to the Rocky Mountains. i. p. 407.

unite to secure it from the view of incidental observers, and those who are desirous of becoming acquainted with this rat in a state of nature, must be prepared to exercise the most untiring patience as well as the most assiduous attention.

In Florida, Georgia, &c., and the plains adjacent to the Missouri, the pouched-rat is to be found in great numbers; their burrows are exceedingly numerous in various places, and give an appearance to the plains similar to that produced by ploughing. Over their burrows, hillocks of loose earth are raised, resembling in some respects those thrown up by the shrew-mole. These hillocks consist of about ten or twelve pounds of loose earth, which appears as if it had been emptied out of a flower-pot on the spot; no hole is to be discovered under this mass of loose soil, but if it be carefully removed, it is seen that the earth has been broken in a circle of an inch and a-half in diameter, within which space the ground is loose, but still without any distinct opening.

This species is rendered peculiar in its appearance by the cheek-pouches exterior to the mouth, its short fore legs and long claws. By the aid of the latter the animal is enabled in a light soil to burrow with great rapidity, and is seldom or never dug out, since it can escape through the ground as fast as one person can dig in pursuit.

As the pouched-rat is so entirely subterranean in its mode of life, it is not surprising that very little should have been learned of its history; neither can we hope that our knowledge will be much increased on the subject until some one, who is sufficiently acquainted with natural history, will devote himself assiduously to the investigation of the manners of this animal in its native haunts. Except the slight notices given by Bartram, Lewis & Clark, and Say, nothing satisfactory on the habits of the pouched-rat has yet been published.*

The pouched-rat is covered by a reddish brown hair, which is lead-coloured at base; on the under parts of the body the colour is somewhat paler; the feet are white. The eyes are black; the ears scarcely prominent, and the cheek-pouches, which are hairy internally and externally, are very capacious. The whiskers are numerous, slender and whitish. The feet are five-toed, the anterior pair being robust, with large, elongated, somewhat compressed nails, exposing the bone on the inner side; the middle nail is much the longest, then the fourth, then the second, then the fifth, the first being very short. The hind feet are very slender, with nails concave beneath and rounded at tip, the exterior one being very small; the tail is short, hairy at base, and nearly naked at its tip.†

---

* The figures which are given of this animal most commonly represent it with the cheek-pouches *inverted* in a most unnatural manner.

† Say, as above quoted.

CHAPTER VIII.

GENUS XXVI.—JUMPING-MOUSE; *Gerbillus;* DESM.

GENERIC CHARACTERS.

The head is elongated, and the ears are rounded and of moderate size. The anterior extremities are short, have four digits, furnished with small nails and a rudimental thumb; the posterior limbs are either long or very long, have five digits, each of which is supported by a distinct metatarsal bone, and provided with a nail. The tail is long and covered by hair.

Dental System.

16 Teeth:
- 8 Upper { 2 Incisive, 6 Molar.
- 8 Lower { 2 Incisive, 6 Molar.

In the upper jaw the incisor, which arises from the middle part of the maxillary bone, is divided into equal parts by a longitudinal furrow. The molars diminish in size from the first to the last. The first is composed of three transverse prominences, formed by two intervening furrows, not so deep in the middle as at their extremities; these prominences are slightly depressed in the middle, but the anterior and posterior are narrower than the middle. The second molar is composed of two prominences, formed by an intervening depression, the posterior being the narrowest. The third molar is similar to the second, but smaller, especially its posterior prominence. Hence these teeth differ from those of the Hamster (*Cricetus*) principally in the breadth of their prominences. When worn down these teeth are remarkably like those of the Hamster in the same condition. They present an even surface, with depressions on the internal and external edges, which are traces of the extremities of the furrows; the difference in the breadth of the prominences may still be recognized.

In the lower jaw the incisor is even; the molars diminish in breadth from the first to the last, the reverse of what we find in the hamster. The first molar has three prominences and two furrows, but the first is very narrow and almost circular. The second has two prominences and a furrow, and the third is so small as to be scarcely more than rudimental.

Species I.—*The Jumping-Mouse.*

Gerbillus Canadensis; Desm.

Dipus Canadensis: Davies, Linn. Trans. iv. 155.
Dipus Americanus: Barton, An. Philos. Trans. iv. 114.
Canadian Jerboa: Shaw, Gen. Zool. pl. 2d. i. 192.
Gerbille du Canada: Desm. Mammal. p. 132.

This little animal is very remarkable for the great length of its hind legs and its mode of progression, in both of which it bears some resemblance to the kanguroo of Australasia, and the jerboa of the old continent. When not in motion the jumping-mouse might be mistaken for the common field-mouse, as its general aspect is very similar. To rectify such an erroneous view, it is sufficient that an attempt be made to capture it, when the force and celerity of its leaps soon remove it from danger, and the pursuer is astonished at seeing so small a creature, with very slight apparent effort, eluding his most eager speed, by clearing five or six feet of ground at every spring. When the jumping-mouse is pursued by one or two persons, and permitted to advance in one direction, its movements resemble those of a bird rather than a quadruped, so high does it leap into the air, so great is the distance it measures at every bound, and so light and quick is its ascent and descent. The jumping-mouse, however, does not exclusively move in this manner, but is capable of running on all its feet with considerable speed; hence it frequently excites the wonder of the country people, or gives them much labour in vain when they attempt to run it down.

The jumping-mouse is found in this country from Canada to Pennsylvania, and no doubt still farther south. It is in size nearly the same as the common mouse. The head, back, and upper parts of the body, generally, are of a reddish brown colour, somewhat approaching to yellow. On the back the brown is darker than elsewhere. The under parts of the body throughout are cream-colour, as well as the inner parts of all the limbs. Near the lower part of the nostrils there is a band or yellow streak, which runs on each side along the whole length of the head and the superior and inferior side of the fore limbs, whence, passing along the body, it terminates at the joint of the thighs. The upper jaw projects considerably beyond the lower, and the nostrils are open. The ears are small, rather oval and hairy; the whiskers are long. The fore limbs are short, and have four digits, provided with long and very sharp nails; there is also a minute tubercle instead of thumb, which is entirely destitute of nail. The posterior extremities are very long, especially from the heel to the ends of the toes, which are five in number, long, slender, and the three middle ones nearly of equal length. The external and internal toes are much shorter; the inner one is shortest of all. The tail considerably exceeds the body in length, and gradually decreases in size from its origin to its extremity, being finely ciliated or clothed with hair throughout, and terminating with a fine pencil of hairs. On the upper side it is of a slate-brown colour, beneath it is of a yellowish cream-colour, and composed of very numerous joints.

The jumping-mouse is found in the grain and grass fields, like the other little plunderers heretofore described, and feeds on the same substances. It breeds very fast, and may occasionally become injurious to the farmer. It is not usual, however, to find them in great numbers in Pennsylvania, though in some vicinities they are quite common.

At the commencement of cool weather, or about the time the frost sets in, the jumping-mice go into their winter-quarters, where they remain in a torpid state until the last of May or first of June. They are dug up sometimes during winter from a depth of twenty inches, being curiously disposed in a ball of clay about an inch thick, and so completely coiled into a globular form as to conceal the figure of the animal entirely.

Species II.—*The Labrador Jumping-Mouse.*

Gerbillus Labradorius; Sab.

Mus Labradorius. Sab. App. to Franklin's Exped. p. 661.
Labrador Rat: Penn. Quad. ii. 173, Arct. Zool.
Gerbillus Hudsonius: Rafin. Prodr. de Somiol.

This species, which closely resembles the preceding in its mode of living, is found in the Labrador and Hudson's Bay country. It is about four inches in length, exclusive of the tail, which is two inches and a-half long. The general colour of the superior parts of the body is brown; of the inferior parts white. The front is very much arched or projecting, so that the nostrils present towards the earth. The mouth, which is far below, is small, with the upper lip bifid, and long black whiskers projecting in two tufts. The ears are rounded and situated far back on the head. The hind legs are an inch and a-half long, covered with short hair, and five-toed, the inner one being the shortest, the others nearly equal. The tail is covered with black hair above and white below.

CHAPTER IX.

GENUS XXVII.—MARMOT; *Arctomys;* GMEL.

Germ. Murmelthier.   *Fr.* Marmotte.
Ital. Marmotto.   *Swed.* Mormoldjuret.

GENERIC CHARACTERS.

THE head, which is thick and flattened, has a blunt and somewhat compressed snout, with eyes of a moderate size, and short ears. The trunk of the body is thick, the limbs short, and the feet robust. The fore feet have four digits, not united by membrane, and a rudimental thumb. The posterior extremities have five digits, which are also free; all the toes are furnished with strong, hooked, compressed nails. The tail is short, or of moderate length, and covered with hair.

Dental System.

$$22\text{ Teeth:}\begin{cases}12\text{ Upper}\begin{cases}2\text{ Incisive}\\10\text{ Molar.}\end{cases}\\10\text{ Lower}\begin{cases}2\text{ Incisive}\\8\text{ Molar.}\end{cases}\end{cases}$$

IN THE UPPER JAW the incisive is rounded and smooth in front, and rises from the anterior and inferior part of the maxillary bone above the first molar. The first molar is a simple tubercle with one root; the three following, which are of the same size, are divided transversely by two depressions, which produce three prominences; the first of these depressions traverses the tooth entirely, but the second is obstructed by a spine or internal spur, which unites the two posterior prominences. These teeth have three roots, two external and one internal. The last, or fifth, resembles the others, except in its posterior prominence, which is extended posteriorly in a sort of spur, which corresponds to the root analogous to the second external root of the preceding molars.

In the lower jaw the incisive is similar to that of the upper jaw, and rises below the last molar. The four molars are of equal size and entirely similar in form. They present a groove on their outside, on the inside a depression which comprises the whole width of the tooth, and at their antero-inferior edge a narrow and very salient tubercle, which diminishes in size from the first to the last. The first of these teeth has, besides, at its neck and on its anterior face, a hollow bordered by a small spine.

When these teeth are worn to a certain degree, all their projections disappear, and their crowns become entirely smooth; but both subsist during the whole life of the animal.

Species of this genus are found in various parts of the old continent, and in the Bahama islands. In this country the greater number of species are found far to the north. The habits of the genus are detailed at length in describing the following species:

Species I.—*The Maryland Marmot.*

Arctomys Monax; Gmel.

Bahama Coney: Catesby, Carolina, ii. 79. *Marmota Americana:* ibid. App. 28.
Monax, or Marmotte of America: Edwards, Nat. Hist. ii. 104.
Glis Fuscus; Marmota Bahamensis: Briss, Reg. An. 4to. 163. *Marmota Americana:* ibid. 164.
Maryland Marmot: Penn. Synops. 270. Quad. ii. 398. Arc. Zool. i. 111. Shaw's Zool. iii. 117.
Le Monax ou Marmotte du Canada: Buff. Hist. Nat. xiii. 136. Supp. iii. 175. pl. 28. Ed. Sonnini. xxxii. 22.

As the *Maryland* Marmot is no where more common than in Virginia, New Jersey, Pennsylvania, and indeed all the temperate parts of this country, we state the fact in commencing the history of this interesting species, to prevent readers from drawing the erroneous conclusion that the *name* is a correct indication of the *place* to which the animal exclusively belongs. In doing this we cannot refrain from once more expressing our unavailing regret that the importance of bestowing right names is still so little felt or understood, and that in heedless haste an original observer should be allowed permanently to establish designations, which uniformly betray the ignorant into error, and prove sources of vexation to all who feel their inappropriateness. The abuse of terms, however, has long been justly esteemed as one of the most abundant sources of human mistake and suffering, and if with all advantages of knowledge men persist in occasionally calling brutal rudeness by the name of *candour* and *bluntness*,—swaggering, *courage*,—and a destitution of good breeding and honesty, *imprudence*, the student of natural history can scarcely expect that much attention will be paid to the evils he endures from the impediments thrown in his way by the same prolific source of mischief.

The scientific name of the genus to which the Maryland marmot belongs, is excellent, if the species now under consideration be taken in illustration, for a first glance at the animal is sure to bring to mind the idea of a bear and a rat, of both of which this creature is a curious miniature resemblance. The thickness of its body, entirely plantigrade walk, posture when engaged in listening, and heavy gait, are such as vividly to excite a recollection of the bear, while the form of the head, teeth, position and appearance of the eye, and general aspect, equally remind the observer of the rat. In some of its actions it more nearly resembles the squirrel, especially when in feeding it employs the fore paws, yet in this it also exhibits a marked similitude to the bear, as it frequently uses one paw at a time with the same awkward facility that appears so singular in bruin. Among the country people it bears the name of *woodchuck* and ground-hog, the latter being expressive of its habit of burrowing and peculiar voracity.

This marmot is the cause of great injury, especially to the farmers engaged in the cultivation of clover, as their numbers become very considerable, and the quantity of herbage they consume is really surprising. They are the more capable of doing mischief from the circumstance of their extreme vigilance and acute sense of hearing, as well as from the security afforded them by their extensive subterranean dwellings.

When about to make an inroad upon a clover-field, all the marmots resident in the vicinity quietly and cautiously steal towards the spot, being favoured in their march by their gray colour, which is not easily distinguished. While the main body are actively engaged in cropping the clover-heads and gorging their ample cheek-pouches, one or more individuals remain at some distance in the rear as sentinels. These watchmen sit erect, with their fore-paws held close to their breast, and their heads slightly inclined to catch every sound which may move the air. Their extreme sensibility of ear enables them to distinguish the approach of an enemy long before he is sufficiently near to be dangerous, and the instant the sentinel takes alarm he gives a clear shrill whistle, which immediately disperses the troop in every direction, and they speedily take refuge in their deepest caves. The time at which such incursions are made is generally about mid-day, when they are less liable to be interrupted than at any other period, either by human or brute enemies.

The habitations of this marmot are formed by burrowing into banks, the sides of hills, or other similar situations, and are generally inclined slightly upwards from the mouth, by which the access of water is prevented. In forming the burrow, where the ground is soft, the fore paws are the principal agents; the strength of the animal's fore limbs is very great. Where the soil is hard and compact the long cutting teeth are very freely and efficiently employed, and we have been surprised to see large stones and lumps of hardened clay dug out in this way. As the burrow is deepened the earth is brought out in the following manner:—The marmot first throws the earth, with his fore paws, under his belly, and when it has accumulated to a certain degree, he rests on his fore paws and kicks the dirt forcibly onwards with the hind ones, and thus going backwards to the mouth of his den he finally throws it to a considerable distance from the entrance. It is very easy to determine when one of these animals has been engaged in forming a new burrow, as his whiskers are worn close to the head, in proportion to the hardness of the soil in which he has worked, and his teeth and the edges of his upper lip show evident marks of the hard service they have performed. The paws are admirably adapted for burrowing, both on account of the length of the toes and nails, and the peculiar arrangement of the skin of the palms and soles of the feet, which is extended between the toes so as to make them distinctly semi-palmated or webbed, especially in the hind feet. This circumstance is not commonly noted by the writers on natural history, but we have repeatedly examined the living animal, and find the character uniformly present. That this structure has reference to burrowing is evident, as the animal shows a great repugnance to water, very seldom drinking, and then in but small quantity; he suffers exceedingly from exposure to rain.

The burrows extend to great distances under ground, and terminate in various chambers, according to the number of inhabitants. In these, very comfortable beds are made by the marmot, of dry leaves, grass, or any soft dry rubbish to be collected. It is really surprising to see the vast quantity of such material an individual will cram into his mouth to carry off for this purpose. He first grasps with the teeth as much as he possibly can; then sitting erect, with both fore paws he stuffs the mass projecting on each side deeper into the mouth, and having arranged it satisfactorily, takes up successive portions, which are treated in like manner; during the whole time, the head is moved up and down to aid in filling the mouth to the very utmost. This is repeated until every fragment at hand is collected, and the whole transferred to the sleeping apartment, into which the marmot retires towards the decline of the day, and remains there until the morning is far advanced. At some seasons of the year this marmot is seen out on moonlight nights at a considerable distance from the burrow, either in search of better pasture or looking for a mate; on such occasions, when attacked by a dog, the marmot makes battle, and when the individual is full grown, his bite is very severe. The teeth of the dog give him vast superiority in the combat, as when once he seizes, he is sure of the hold until the parts bitten are torn through, while the marmot can merely pinch his fore teeth together, and must renew his attempts very frequently. The fight is also soon ended by the dog seizing the marmot by the small of the back, and crushing the spine so as to disable his antagonist effectually.

There is no animal so perfectly cleanly in its habits as this marmot; not only the fragments of its food and the litter of its bed are carefully removed, but the loose earth about the mouth of the burrow is carefully scraped away. However numerous they may be in any vicinity, their excrement is not seen, nor any offensive odour perceived. Whenever the calls of nature are felt, this animal seeks a spot at some distance from his dwelling, and having dug a hole of two or three inches in depth, and performed his evacuations, he covers it up with extreme care, and not content with placing a thick layer of earth over it, he presses it, or rather rams it down with the end of his nose, striking with a force which seems very extraordinary when thus applied.

The Maryland marmot, as we have already mentioned, eats with great greediness and large quantities. To the wild animal red clover is a very favourite food, and, when it can be obtained, lettuce, cabbage, and various other garden vegetables. In captivity it eats of almost every vegetable offered, is exceedingly fond of bread and milk, and will display the most violent anger, by erecting its hair, growling and yelping, if it see a cat or other animal fed with this substance. One which we kept for a long time in a state of domestication, would, on such occasions, become almost furious, and never desist from his efforts until he had broken his chain, when he would rush to the spot, drive off the cat by a severe bite, or bite the person who attempted to withhold him from the dish. Yet on other occasions he did not interfere with the cats, even when feeding within his reach, though he would at any time bite them if they came immediately in his way. This marmot would eat the parts about the joints of the legs of fowls, when thrown to him, and occasionally a small piece of salt-fish,—but, as a general rule, refused animal food of every description.

This individual was very tame, playful and cunning, having the freedom of the yard, and the privilege of performing all his operations unmolested. He was very fond of being handled and petted, and would play with great good humour, though in a clumsy and awkward manner. Every thing fit to make a bed of, that he could get at, was sure to be carried under ground, and when clothes were missed, which had been hung out to dry, it was only necessary to fasten a hook to a long stick and draw them out of his burrow. When this was to be effected, it was necessary to tie the marmot up short, as he appeared to understand perfectly what was to be done, and was by no means willing that his bed should be rendered less comfortable. Although he would not attempt to bite the person engaged in removing his plunder, he would rush to the entrance and endeavour to make his way in, as if to secure his prize, or remove it to a still greater distance. On one occasion he carried off and stowed at a distance of six feet from the entrance, eight pairs of stockings, a towel, and a girl's frock, and had he not been discovered in the act, would have made a still larger transfer of materials to form a more luxurious bed.

In whatever action engaged, the vigilance of this animal was unceasing, and his ear appeared the sense almost exclusively relied on. By observing him closely it was evident that every variation of sound, however slight, or from whatever different sources, was immediately perceived. While earnestly engaged in eating, and making no inconsiderable noise in munching lettuce, or other crisp vegetables, the least noise would be sufficient to suspend his hunger and excite all his vigilance, and if it were one to which he was unaccustomed, or loud enough to alarm him, he would run with great precipitation until he arrived at the edge of his hole, where he would sit up for an instant in an attitude of the profoundest attention, and either return to his food, or take refuge in his hole, as he might feel satisfied that there was or was not danger to be apprehended.

To look at the ear of this marmot without close examination, placed on the side of the head, high up and far back, with very little external cartilaginous projection, and a wide orifice leading to the internal ear, it would seem very inappropriate to the subterranean mode of life, since it appears to be so placed as to allow the dirt ready access. But no such inconvenience takes place, as the ear is provided with a muscular apparatus, by which the upper portion is brought down, and the sides of the lower portion are so accurately pressed against each other, as effectually to exclude the smallest particles of dirt or dust.

At the commencement of the cold weather the marmot goes into winter quarters; having blocked up the door from within, he there remains until the return of the warm season revives him again to renew his accustomed mode of life. The female produces five or six young at a litter.

The body of the Maryland marmot is about the size of that of a rabbit, and covered by long rusty brown hair, generally gray at the tips; the face is of a pale bluish ash-colour. The ears are short, but broad, and as if they had been cropped at their superior edges; the tail is about half the length of the body, and covered with dark brown hairs, somewhat bushy at its extremity. The feet and claws are black; the claws are long and sharp.

All the figures which have been heretofore published of this animal, (with the exception of one given in the English translation of Cuvier, borrowed from a drawing by LESUEUR,) have been copied from Edwards', which is altogether unlike the animal.

SPECIES II.—*The Quebec Marmot.*

Arctomys Empetra; GM. SCHREB.

Marmotte du Canada: Encycl. pl. 67, fig. 4.
Quebec Marmot: PENN. Synopsis, 270, pl. 24, fig. 2, Quad. ii. 397, pl. 412, ed. 3, ii. 129, pl. 741. Arct. Zool. i. 111, Shaw's Zool. iii. 119.
Arctomys Empetra: SCHREB. Quad. 743, pl. 210. GMEL. Syst. Nat. i 143. SABINE, App. to Franklin, p. 662. IBID. Trans. Lin. Society xiii. 584.
Marmotte de Quebec: DESM. Nouv. Dict. d'Hist. Nat. xix. 314.

[*Called Siffleur by the Canadians.*]

The *Quebec* Marmot is found throughout the northern parts of this country, and in its habits closely resembles the preceding species. Its entire length, from the tip of the nose to the extremity of the tail, is about twenty-six inches, of which the tail forms six inches.

The general colour of the upper part of the body is grayish, the hairs being thus coloured; at the base they are dark, in the middle yellowish, near their tops black, and white at their tips; near the tail the white is not so remarkable. On the cheeks and chin the hair is short, and inclines to gray, on the nose dark or blackish, the top of the head is dark brown:

the whiskers and long hairs growing over the eyes
are black. The throat, legs, and all the under parts
of the body are of a dark chestnut-colour. The hair
on the tail is dusky throughout, longer than on the
back and darker at the end. The toes are covered
with short hairs, which are black. The inner toes
on the hind feet and the outer on the fore feet,
are shorter than the others; there is a rudimental
fifth toe on the inside of the fore feet. All the toes
are provided with long and sharp claws, those on
the fore feet being longest and most arched.

Species III.—*Franklin's Marmot.*

Arctomys Franklinii; Sab.

Arctomys Franklinii; Gray *American Marmot:* Sabine, Trans. Linn.
Society, xiii. 587, App. to Franklin, p. 662.

This interesting animal was found near Fort Enterprize, by the expedition under command of the
intrepid and adventurous Capt. Franklin, in honour
of whom the scientific designation was bestowed by
Sabine, whose own name is invariably associated by
scientific readers with profound and philosophical
research, illumined and adorned by a mind richly
imbued with the most valuable learning. As the
trivial name, *Gray American Marmot,* is equally applicable to other species, we have preferred to translate the scientific appellation, which more definitely
refers to the species in question.

In size this animal equals a large rat, measuring
eleven inches from the nose to the insertion of the

tail; the latter, to the end of the hair at its extremity, is five inches long. Its face is broad and nearly covered with rigid black and white hairs, which give it a gray colour; the nose is very blunt, and the ears are broad and covered with short hairs. The whiskers on the cheeks are short and black, and similar hairs thinly distributed grow above and below the eyes.* The upper part of the body is covered with short hairs, dark at the base, dingy white in the middle, then first black, next yellowish white, and tipped with black, the whole forming a variegated dark yellowish gray. On the sides the hair is longer, not so black, and destitute of the yellow tinge; on the belly it is dark at the base, and dingy white at tip. The tail is covered with long hairs, banded with black and white, and tipped with white, the whole appearing indistinctly striped with black and white. The feet are rather broad, the toes being thin and covered with gray hairs. On the fore feet the second toe from the inside is longest; the outer shortest and placed far back; the three centre hind toes nearly of an equal length, the extremes shorter and far back. The claws, which are of a horn-colour, are long and sharp on the fore feet, and on the hind feet shorter.†

---

* The upper fore teeth are short and reddish yellow; the lower fore teeth are twice the length of the upper, and paler.
† See Sabine's paper as above quoted.

Species IV.—*Tawny American Marmot.*

Arctomys Richardsonii; Sab.

Arctomys Richardsonii: Sabine, Linn. Soc. Trans. xiii. 589.

This marmot was found by Franklin's expedition, near Carlton-house, in the Hudson's Bay country, and was named in honour of Dr. John Richardson, who, on that perilous journey, was so highly distinguished for his scientific zeal, and his intrepid and philanthropic spirit.

The tawny American marmot, or Richardson's marmot, is nearly of the size of the foregoing species, but more slender. The top of the head is covered with short hairs, dark at the base and light at their tips. The nose is tapering, sharp, bare at the end, and covered above with short light brown hairs, joining and mixing with those on the top of the head. The ears are short and oval; the cheeks swollen and clothed with light brown hairs; the whiskers are short, growing from the cheeks, and there are a few rigid hairs above the eyes. The throat is of a dirty white colour; the upper part of the body is covered with short soft hairs, dark at the base and fulvous at their extremities; in the middle of the back the hairs are like those on the top of the head, but lighter. The hair on the sides is longer, when raised appearing dark at the base, the ends being of a smoky white; the under parts are similar but dashed with a little rust-colour. The tail is three inches and a-half long to the end of the hair, slender and thinly covered with long hairs, which, at the base, are of the same colour as the body, but above of three distinct hues,—first black, next dark, and lastly light at the upper extremity. The legs are rather long and slender, with narrow feet, furnished with sharp, arched, horn-coloured claws. The fore feet have on the inside a small toe placed far back, with a blunt claw, which gives it a character different from the general character of the genus. The outer toe and claw of the fore feet much shorter than the remaining three, the middle one of which is the longest. Of the hind toes the two extremes shorter and placed back, the other three of nearly the same length.*

Species V.—*Hood's Marmot.*

Arctomys Tridecemlineatus; Mitchill.

Sciurus Tridecemlineatus; Federation Squirrel: Mitchill, Med. Rep. 1821.
Ecureuil de la Federation: Desm. Mammal. p. 339.
Arctomys Hoodii: Sab. Trans. Linn. Society, xiii. 590.

This beautiful marmot is an inhabitant of the northern and western parts of this country, and when first discovered was thought to be a squirrel, and classed near the *sciurus striatus*, or ground-squirrel, to which it exhibits considerable analogy in the arrangement of its stripes. Though now properly removed to the genus Arctomys, we retain the *specific* name first proposed by our distinguished countryman Professor Mitchill, derived from the number of stripes on the back of the animal, being the same as

---

* Sabine as above quoted.

that displayed in the "star spangled banner" of our federation.* For the trivial name we have adopted a translation of the scientific appellation proposed by Sabine, in honour of Lieut. Hood, so truly meritorious for his exertions on the expedition commanded by Franklin, and remarkable for having been so cruelly murdered by one of his fellow travellers.

From the nose to the root of the tail Hood's marmot is about seven inches and a-half long, and the tail itself two inches. The top of the head is broad, flat, and obscurely marked with alternate stripes of dark brown and dingy white. The nose is tapering and very sharp, being covered with light brown hairs. The ears are small and very short, the cheeks tumid and clothed with dingy light, coloured hairs, the throat being of the same colour; the whiskers are rather long and grow from between the nose and the eyes; some small rigid hairs, similar to the whiskers, also grow over the eyes. The whole of the upper part of the body is marked on each side longitudinally with three alternate dark brown and dingy white stripes, the dark being twice as broad as the light, and dotted in the centre, at equal distances throughout their whole length, with small spots of dingy white. In the centre of the back there is a dark stripe, rather broader than the others. The lowest stripe on each side is not so well marked or distinctly spotted. All the under parts are of a dingy white or slightly tawny colour. The tail is

---

* To the kindness of Dr. J. E. Dekay of New-York, we are indebted for an opportunity of examining a fine specimen of this marmot.

indistinctly banded with dark brown and dingy white, being of the latter colour at tip.

The fore legs, which are short and small, are covered with light hairs; the outer toe and claw are small and placed back, the centre toe is the longest of the other three. On the inside there is also a rudimental toe with a small obtuse claw, but this is not so remarkable as in the tawny marmot. The hind legs are longer than the fore legs, and clothed with light hairs; the extreme toes and claws are nearly of equal length and placed far back; the three others are also of equal length with each other. The claws are dark horn-colour, small and light at their ends, the fore ones being the longest.*

SPECIES VI.—*The Prairie Marmot.*

Arctomys Ludovicianus; ORD.

Petit chien: LEWIS & CLARKE, i. 67.
Wistonwish: PIKE, Exped. &c. 156.
Arctomys Ludovicianus: ORD. in Guthrie, ii. 302. [1815.]
Arctomys Missouriensis: WARDEN, Descr. des Etats Unis. v. 567.
Arctomys Ludovicianus: SAY, Long's Exped. to the Rocky Mountains, i. 451.

[*Le Petit chien des Voyageurs; commonly called Prairie-dog.*]

The vast solitudes of our remote territories, where man has not yet established his abode, are generally overshadowed by dense forests, which, during an unknown lapse of ages, have there successively flour-

---

* See Sabine, as above quoted.

1. Hood's Marmot. 2. Louisiana Marmot.

ished and decayed; imparting to the landscape a character of grand though sombre uniformity, broken only by the courses of rivers, the ruggedness and sterility of some portions of soil, or where the furious hurricane has swept along, prostrating the giant sons of earth with a destructiveness proportioned to their resistance. The traveller who, impelled by curiosity, advances beyond the " father of western rivers," with delighted admiration finds himself gradually emerging from these apparently interminable shades, and entering upon a new world. Before him, spreading as far as vision can extend, he beholds fields of richest verdure, interspersed with clumps of slight and graceful trees, as if with an exclusive view to ornament, and discovers the far distant windings of the river as it steals through the plain, by the cottonwood and willows fringing its banks. After traversing such scenes, enlivened by numerous herds of browsing animals, that here find a luxurious subsistence, and arriving at the higher and more barren parts of the tract, he is startled by a sudden shrill whistle, which he may fear to be the signal of some ambushed savage; but on advancing into a clearer space, the innocent cause of alarm is found to be a little quadruped whose dwelling is indicated by a small mound of earth, near which the animal sits erect in an attitude of profound attention. Similar mounds are now seen to be scattered at intervals over many acres of ground, and the whole forms one village or community, containing thousands of inhabitants, whose various actions and gambols awaken the most pleasing associations.

In some instances these villages are limited, or at most occupy but a few acres, but still nearer to the Rocky Mountains, where they are entirely undisturbed, they are found to extend even for miles. We may form some idea of the number of these animals when we learn that each burrow contains several occupants, and that frequently as many as seven or eight are seen reposing upon one mound. Here in pleasant weather they delight to sport, and enjoy the warmth of the sun. On the approach of danger, while it is yet too distant to be feared, they bark defiance, and flourish their little tails with great intrepidity. But as soon as it appears to be drawing rather nigh, the whole troop precipitately retire into their subterranean cells, where they securely remain until the peril be past. One by one they then peep forth, and vigilantly scrutinize every sound and object before they resume their wonted actions. While thus near to their retreats they almost uniformly escape the hunter, and if killed they mostly fall into their burrows, which are too deep to allow their bodies to be obtained.

The villages found nearest the mountains, have an appearance of greater antiquity than those observed elsewhere. Some of the mounds in such situations are several yards in diameter, though of slight elevation. These, except about the entrance, are overgrown by a scanty herbage, which is characteristic of the vicinity of these villages. Say has observed on this subject, that it is not easy to assign a reason for the preference shown by the prairie marmot, which lives on grassy and herbaceous plants,

THE PRAIRIE MARMOT. 117

in selecting the most barren places for its dwelling, "unless it be that he may enjoy an unobstructed view of the surrounding country, in order to be seasonably warned of the approach of wolves or other enemies." This reason may be sufficiently valid of itself, but we would suggest another in the difference of soil, rendering such barren places fitter for the burrows. It is by no means necessary to suppose that this marmot obtains its food exclusively near its own dwelling. We know that this is not the case with the Maryland marmot, which so closely resembles this species in every respect, and goes to considerable distances in search of food, even in the immediate vicinity of man.*

The mound thrown up by the prairie marmot consists of the earth excavated in forming the burrow, and rarely rises higher than eighteen inches, though measuring two or three feet in width at the base. The form of the mound is that of a truncated cone, and the entrance, which is a comparatively large hole, is at the summit or in the side, the whole surface, but especially the top of the mound, being well beaten down like a much used foot-path. From the entrance the hole descends perpendicularly for a foot or two, and then is continued obliquely or somewhat spirally downwards, to a depth which has not been determined.

This marmot, like his kindred species, passes the winter in a state of torpidity, and to secure himself comfortably against the effects of the cold, he closes

---

* Pike says of the prairie marmot, that "they never extend their excursions more than half a mile from their burrows."

accurately the mouth of the burrow, and constructs at the bottom of it a neat globular cell, of fine dry grass, having an aperture at top sufficiently large to admit a finger, and so compactly put together that it might almost be rolled along the ground uninjured.

This active and industrious community of quadrupeds (like every other society,) is infested by various depredators who subsist by plunder, or are too ignorant or indolent to labour for themselves. Hence a strange association is frequently observed in their villages, for burrowing-owls (*Stryx Hypugea* of Bonaparte,*) rattle-snakes, lizards and land-tortoises, are seen to take refuge in their habitations. The burrowing-owl, however, appears to appropriate an excavation to his own use, as is evinced by its decayed and dilapidated condition, while those frequented by the marmot are always neat and in good repair. The young of the marmot most probably become the prey of this singular bird. The rattle-snakes also exact their tribute with great certainty, and without exciting alarm, as they can penetrate the inmost recesses of the burrow, and a slight wound inflicted by their fangs is followed by the immediate extinction of life.†

---

* See his splendid work on American Ornithology, vol. i.

† "It is extremely dangerous to pass through their towns, as they abound with rattle-snakes, both of the yellow and black species; and strange as it may appear, I have seen the *wiston-wish* (prairie marmot,) the horn-frog (orbicular lizard,) and a land-tortoise all take refuge in the same hole. I do not pretend to assert that it was their common place of resort, but I have witnessed the above facts in more than one instance."—*Pike*, p. 156.

The prairie marmot is about sixteen inches long from the tip of the nose to the root of the tail, which is two inches and three-quarters in length. The head is broad and depressed above with large eyes, having dark brown irides. The ears are short and truncated; the whiskers black and of moderate length; there are a few bristles above the eye, and a few also on the side of the cheek; the nose is rather short and compressed. The general colour is a light dingy reddish brown, intermingled with some gray, and a few black hairs, which are dark or dusky at base, then bluish white, then light reddish, and finally gray at tip. The under parts of the body are of a dirty white colour; the hair on the anterior legs, that on the throat and on the neck, is not dusky at base.*

All the feet are five-toed, clothed with very short hair, and armed with rather long black nails; the outer one on the fore feet reaches nearly to the base of the next, and the middle one is nearly half an inch long. The thumb has a conical nail, three-tenths of an inch in length; the tail is banded with brown near the tip, and the hair, except that next the body, is not plumbeous at base.

---

* The description of this species is from Say, who has given the best account of the habits of this animal hitherto published

Species VII.—*Parry's Marmot.*

Arctomys Parryii; Richardson.

Arctomys Parryii, Gray Arctic Marmot: Richardson, App. to Franklin.
Arctomys Alpina: Parry's 2d Voyage, p. 61.

This species was brought in by the expedition under Capt. Franklin, and was named by Dr. Richardson in honour of Capt. Parry. It is rather larger than the Arctomys Franklinii, and measures to the root of the tail twelve or fourteen inches; the tail itself is four inches long, and the hair at its extremity five inches and a-half in length.

Parry's marmot has a broad and flattened body, with thick legs; flattish head and blunt nose, covered with a close coat of short brown hairs. The margin of the mouth is hoary; the eyes are large and black coloured. The ear is very short, consisting of a flat semi-oval cartilage, projecting about the sixth of an inch over a large auditory passage. The cheek-pouches, which are very large, open into the mouth anterior to the grinders.

The body is covered with a soft fur, consisting of a soft down of a dark smoky gray at the roots, pale clear gray in the middle, and yellowish gray at the tip. This arrangement causes a crowded assemblage of ill defined, irregular, and confluent whitish spots, margined and separated by black. The throat and all the under parts of the body are brownish red and brownish yellow, or rather an intermediate colour, blending with the colours of the back. The tail is flattish and subdistichous, and is at the will of the animal expanded like a feather; it is then brown along the middle, tipped and margined for two-thirds of its length with black. The feet are furnished with five toes, having short flattened claws, which are large, blackish, slightly arched and grooved underneath. On the inside of the fore feet, and high up, there is a small toe or thumb, armed with a small nail; the palms are naked and have callous protuberances, three of them at the base of the toes, from the largest of which the thumb rises.*

---

* Richardson, as cited above.

CHAPTER X.

Genus XXVIII.—Squirrel; *Sciurus;* L.

Gr. Σκιυρος. *Germ.* Eichorn.
Fr. Ecureuil. *Ital.* Scojattolo.

GENERIC CHARACTERS.

The head is somewhat elongated, with a sharp muzzle, moderately long ears, and large eyes. The upper lip is divided, and the cheeks destitute of pouches, the neck of a middling length, and the body rather slender. The teats are eight in number, two being situated on the chest and six on the belly. The posterior are much longer than the anterior extremities, which have four digits and a rudimental thumb; the external digit is short, the others are long. The posterior feet are five-toed, with a short internal and external digit: the three intermediate toes are long and slender. The next to the external toe is the longest of all, both in the anterior and posterior feet, the digits of which are furnished with curved acute nails, with the exception of the rudimental thumb, which is blunt and naked. The tail is long and clothed with long and thickly set hairs.

Dental System.*

22 Teeth:
- 12 Upper { 2 Incisive / 10 Molar.
- 10 Lower { 2 Incisive / 8 Molar.

IN THE UPPER JAW the incisive is smooth and rounded in front, and rises from the sides of the anterior part of the maxillary bone. The first molar is a rudimental and cylindrical tooth, which falls out very early, and which is placed against the antero-internal surface of the second; this latter, which is sometimes a little smaller than the following, has, like them, a central depression, and another smaller one at both extremities: from these three depressions results a small spine on the anterior edge of the tooth, then two prominences, separated from each other by the central depression, and, finally, another small crest or spine on its posterior edge.

On the outside of these depressions prominences and spines remain distinct, but on the inside they are reunited by a large and circular crest. This crest embraces the second molar rather less than the others. which thus differs in being narrower internally than externally: the same is the case with the last, which differs by the prolongation of its postero-external part.

---

* Frederic Cuvier introduces this dental system by remarking that it is evidently similar to that of the marmot and spermophili, all forming one and the same family; they differ, however, in some circumstances, which are uniform in their recurrence, and by consequence are characteristic.

IN THE LOWER JAW the incisor is like that of the upper jaw, but is narrower, and rises from below and behind the last molar. The third molar is a third smaller than the others, which gradually increase in size to the last, but all formed alike, presenting in their middle a circular depression, and on their periphery a crest divided by a groove at the internal and at the external edge: from the centre of each of these grooves a small tubercle arises. Age, however, soon effaces these fugitive characters, and then these teeth exhibit a nearly smooth surface.

The species comprised in this genus are in different degrees remarkable for their sprightly agility and graceful movements, as well as for their personal beauty and neatness. The forest is their appropriate residence, and nature has provided them not only with the means of rapidly ascending the loftiest trees, but with teeth capable of opening the way to food, which is effectually secured from almost every other creature. The hardest nuts found in the woods afford ample provision to the squirrels, and the number of nuts destroyed by these animals, though small when compared with the whole quantity produced, must have some effect in preventing the superabundant increase of forest trees.

The muscular strength displayed by these animals is very great, when compared with their size. They make astonishing leaps from branch to branch, and from tree to tree, when engaged in sporting with each other, or endeavouring to escape from pursuit. At such times, when no tree is sufficiently near to be reached by a single spring, the squirrel unhesitatingly drops from the greatest height to the ground, and falls with a force apparently sufficient to crush him; but no injury is experienced, and a few seconds are sufficient for his escape into the top of the nearest tree.

The actions of most of these animals are marked by a peculiar vivacity and playfulness. When moving on the ground, squirrels advance by a succession of short leaps, while the long bushy tail, waving in graceful undulations, renders their whole appearance very interesting. When engaged in listening, they sit erect on their hinder limbs, having the tail beautifully raised against the back, and falling in an easy curve at its extremity towards the ground. In eating, the position is much the same; the food is held in the fore paws, principally between the rudimental thumbs and the adjoining part of the palms. The facility with which they cut through the covering of the hardest nuts is very remarkable; they first turn the nut about until they get it into the most favourable position, and then examine it by gnawing slightly in different places. If the nut be withered or rotten it is speedily thrown aside and another sought. When a good one is obtained, and the proper place for opening it is selected, (which is the thinnest part, immediately over the kernel,) a small linear opening is first made, which at length admits the points of the lower front teeth. These are now inserted, and the hole enlarged by breaking off successive pieces of the shell in the direction of the kernel. A hickory-nut is thus frequently cut down on four sides from end to end, leaving the intermediate thick portions untouched. After satisfying his hunger the squirrel generally buries the superfluous food; previous to the approach of winter large hoards of nuts and grain are collected and secured in the ground for future use. Their nests are at no great distance from these store-houses, and are built of small sticks and leaves in the top branches of forest trees, or in hollows of their trunks, except in the case of a few species which inhabit burrows at all times. All the squirrels are peculiarly cleanly, and are frequently seen to rub their heads and faces with their fore paws as if for the purpose of washing. When they accidentally step into water they make use of their bushy tail for the purpose of drying themselves, passing it several times through their hands.

Like most of the animals belonging to this order, they are very prolific, and multiply until from their numbers large districts of country are injuriously overrun by them. They then invade and literally lay waste the cornfields, consuming vast quantities of grain, and destroy nearly as much as they eat by breaking it down and scattering it on the ground. On such occasions the farmers in thinly settled districts severely suffer, and are deprived of a large share of the fruits of their industry. The efforts of a whole family are occasionally insufficient to drive off or destroy these busy plunderers, as new crowds appear to be continually arriving to renew the depredation.

While travelling through the state of Ohio, in the autumn of 1822, we had an opportunity of witnessing something of this sort. Parts of the country appeared to swarm with squirrels, which were so numerous that, in travelling along the high road, they might be seen scampering in every direction; the woods and fields might be truly said, in the country phrase, to be "alive with them." A farmer, who had a large field of Indian corn near the road, informed us, that notwithstanding the continued exertions of himself and his two sons, he feared he should lose the greater part of his crop, in addition to his time and the expense of ammunition used in killing and scaring off the little robbers. This man and his sons frequently took stations in different parts of the field, and killed squirrels until their guns became too dirty longer to be used with safety; yet they always found, on returning, that the squirrels had mustered as strongly as before. During this journey we frequently met squirrel-shooters heavily laden with this game, which in many instances they had only desisted from slaying from want of ammunition or through mere fatigue.

Fortunately for the farmers these animals are not at the same time equally numerous in different parts of the country. We found the squirrels in 1822, most numerous throughout the country lying between the Great and Little Miami rivers; they became evidently fewer as we advanced towards Chillicothe, and beyond that place were so rare as to be seldom seen. During some seasons they appear to move in mass, deserting certain districts entirely, and concentrating upon others. In such migrations vast numbers are drowned in crossing the rivers, and numbers are also destroyed by beasts and birds of prey, and various other causes.

Species I.—*The Fox Squirrel.*

Sciurus Vulpinus: Gmel.

Sciurus Vulpinus: Gmel. Turton's L. i. 91.
The Fox Squirrel: Lawson's Carolina, 124.

This fine squirrel is found throughout the southern states, where it frequents the pine forests in considerable numbers, and derives its principle subsistence from the seeds of the pine. In the tops of these lofty trees it is almost out of the reach of danger, except from the pine-marten or other climbing beasts of prey, and possibly some large predacious birds. The fox squirrel displays a consciousness of his security by the fearless manner in which he usually looks down upon those who pass under the tree on which he is placed. When alarmed, like many of his kindred species, he immediately resorts to the artifice of spreading himself out, or lying flat on the upper surface of a branch on the side opposite to the apprehended danger, where he patiently clings until he has no longer cause to fear. Under such circumstances it is very difficult to discover his position, or to distinguish him from the branch on which he lies.

The nest of this species is placed in the top of the high pine trees, and is made of twigs and small sticks, lined with leaves, or the long soft moss which is found so commonly streaming from the branches. The season of their sexual intercourse is the month of January. The young, which are from five to seven in number, are seen abroad as early as the month of March.

The fox squirrel measures about fourteen inches, and the tail is sixteen inches in length. The colour varies from white to pale gray and black; various shades of red, mottled, (like the cats called "tortoise shell,") and in short, of all the intermediate hues. This is fully shown in the Philadelphia Museum, where nearly all the varieties just mentioned may be seen. It is therefore not surprising that those who deem colour a sufficient indication of specific difference, should make a number of species of this one. Perhaps many, at present considered well established, will be found to rest on no better foundation, and require to be stricken out of the catalogue.*

Species II.—*The Cat-Squirrel.*

Sciurus Cinereus; L. Gmel.

Cat-Squirrel, B Penn. Arct. Zool. i. 137.
Sciurus Capistratus: Bosc. An. du Mus. i. p. 205.
Ecureuil à Masque: C. Régne Anim. i. 205.
Ecureuil Capistrate: Desm. Mammal. p. 332, Sp. 529.

The cat-squirrel is one of our largest species, and is found in great abundance throughout the oak and chestnut forests of this country. It is generally

---

* We have much pleasure in acknowledging the receipt of a letter on the subject of the squirrels of this country, from Capt. J. Le Conte, U. S. A. The time, we hope, will soon arrive, when this accomplished naturalist will find leisure to give the scientific world the full benefit of his valuable researches relative to American Natural History.

about eleven inches long, having a tail fourteen inches in length.

This squirrel is comparatively heavy and slow in its movements, running up the trunks of trees and among the branches with more apparent effort than any of the other species; its appearance also is by no means as pleasing as that of any of its kindred. It is rarely seen to leap from tree to tree, or even from branch to branch, except when closely pursued or much alarmed. In building its nest, and in general habits, it is very similar to the other species. The size is the only circumstance which distinguishes it positively from the fox-squirrel.

As to colour, it is impossible to state all the shades and variegations exhibited by this species. In the Philadelphia Museum a great variety may be seen, of almost every colour, from a light gray to black and spotted, pale reddish brown and nearly white. Three individuals, taken from the same nest, are so differently coloured as to be entirely unlike, one having all the marks attributed to the *capistratus*, and the others strongly resembling the common black squirrel.

Species III.—*The Common Gray Squirrel.*

Sciurus Carolinensis; Gmel.

Sciurus Carolinensis et Cinereus: Gm. Schreb. tab. 213.
Petit Gris: Buff. 10, pl. 25, Encycl. pl. 74, fig. 3.
Ecureuil gris de la Carolina; Bosc. ii. p. 96, pl. 29: F. Cuv. Mam. Lithog. livr. 11e.
Gray Squirrel: Penn. Arct. Zool. i. 135, Hist. Quad. No. 272.

This species, still exceedingly common throughout the United States, was once so excessively multiplied as to be a scourge to the inhabitants, not only consuming their grain, but exhausting the public treasury by the amount of premiums given for their destruction. " Pennsylvania (says Pennant,) paid from January, 1749, to January, 1750, *eight thousand pounds* currency; but on complaint being made by the deputies that their treasuries were exhausted by these rewards, they were reduced to one half;— [from three pence to a penny and a half.] How improved must the state of the Americans then be; in thirty-five years to wage an expensive and successful war against its parent country, which before could not bear the charges of clearing the provinces from the ravages of these insignificant animals!"

The gray squirrel prefers the oak, hickory and chestnut woods, where it finds a copious supply of nuts and mast, of which it provides large hoards for the winter. Their nests are placed chiefly in tall oak-trees at the forks of the branches; these nests are very comfortable, being thickly covered and lined with dried leaves. During cold weather the squirrels seldom leave these snug retreats, except for the purpose of visiting their store-houses, and obtaining a supply of provisions. It has been observed that the approach of uncommonly cold weather is foretold when these squirrels are seen out in unusual numbers, gathering a larger stock of provisions, lest their magazines should fail. This, however, is not an infallible sign, at least in vicinities where many hogs are allowed to roam at large, as these keen nosed brutes are very expert at discovering the winter hoards of the squirrel, which they immediately appropriate to their own use.

If the gray squirrels confined themselves to the diet afforded by the forest trees, the farmers would profit considerably thereby. But, having once tasted the sweetness of Indian corn and other cultivated grains, they leave acorns and such coarse fare to the hogs, while they invade the corn-fields, and carry off and destroy a very large quantity.

This species is remarkable among all our squirrels for its beauty and activity. It is in captivity remarkably playful and mischievous, and is more frequently kept as a pet than any other. It becomes very tame, and may be allowed to spend a great deal of the time entirely at liberty, where there is nothing exposed that can be injured by its teeth, which it is sure to try upon every article of furniture, &c. in its vicinity. This squirrel, when domesticated, drinks frequently, and a considerable quantity of water at each draught.

The gray squirrel varies considerably in colour, but is most commonly of a fine bluish gray, mingled with a slight golden hue. This golden colour is especially obvious on the head, along the sides, where the white hair of the belly approaches the gray of

1 Chickaree S. 2 Grey S. 3 Black S.

the sides, and on the anterior part of the fore and superior part of the hind feet, where it is very rich and deep. This mark on the hind feet is very permanent, and evident even in those varieties which differ most from the common colour. There is one specimen in the Philadelphia Museum of a light brownish red on all the superior parts of the body.

Species IV.—*The Black Squirrel.*

Sciurus Niger; L.

Black Squirrel: Penn. Arct. Zool. i. 138. Hist. Quad. No. 273. Brown's Zool. tab. xvii.
Sciurus Mexicanus: Hernan. 582.
Black Squirrel: Catesby's Carolina, ii. p. 73.

This species is very common, but is liable to be confounded with the black varieties of the squirrels heretofore described. From the black varieties of the cat-squirrel, *S. Cinereus*, it may be easily distinguished by its smaller size and the softness of its fur. The proportional length of the tail, together with the difference in number of the jaw teeth, will distinguish it from the fox-squirrel, *S. Vulpinus*, which has five above and four below, while the black squirrel has four above and four below.

The black squirrel very seldom varies; in the summer the pelage is rather gray on the back and sides, though the whole colour of the body is a black, intermingled with a small quantity of gray, and of a dark reddish brown on the under parts. In the winter the colour is a pure black, varying slightly in intensity on any part of the body.

SPECIES V.—*The Great Tailed Squirrel.*

Sciurus Macroureus; SAY.

*Sciurus Macrourus:** SAY. Long's Exp. to the Rocky Mountains, i. 115.

This species, which is a fine one, is the most common on the Missouri, where it was first observed by SAY, who describes it as displaying all the graceful activity so much admired in the common gray squirrel.

The total length of this species from the tip of the nose to the end of the tail (exclusive of the hair) is nineteen inches and three-quarters, of which the tail makes nine inches and one-tenth. The following description of its colouring, &c. is drawn up from that given by SAY, in the work above quoted:

The body above and on each side is of a mixed gray and black; the fur is plumbeous, black at base, then pale cinnamon colour, then black, and finally cinereous, with a long black tip. The ears, three-

---

* As the term Macrourus was previously given to the Ceylon squirrel, (see Pennant's History of Quadrupeds, ii. p. 140, No. 330,) we have taken the liberty to change the name given to the present species, by the addition of a single letter, which is sufficient to render further change unnecessary.

THE GREAT TAILED SQUIRREL. 135

fourths of an inch long, are behind of a bright ferruginous colour, extending to the base of the fur, which, in the winter dress, is prominent beyond the edge; on the inside of the ear the fur is of a dull ferruginous hue, slightly tipped with black. The sides of the head and orbits of the eyes are pale ferruginous; beneath the ears and eyes the cheeks are dusky. The whiskers are composed of about five series of rather flattened hairs, the inferior ones are more distinct. The mouth is margined with black; the teeth are of a reddish yellow colour. The under part of the head and neck, and the upper part of the feet, are ferruginous; the belly is paler, the fur being plumbeous at base. The tail is of a bright ferruginous colour below, and this colour extends to the base of the fur with a submarginal black line. On its upper part it is ferruginous and black. The fur within is of a pale cinnamon colour, with the base and three bands black; the tip is ferruginous. The palms of the fore feet are black, and the rudimental thumb, which is very short, is covered by a broad flat nail.

" The fur of the back in the summer dress is from three-fifths to seven-tenths of an inch long; but in the winter dress the longest hairs of the middle of the back are from one inch to one and three-fourths in length. This difference in the length of the hairs, combined with a greater portion of fat, gives to the animal a thicker and shorter appearance; but the colours continue the same, and it is only in this latter season that the ears are fringed, which is the necessary consequence of the elongation of the hair. The species was not an unfrequent article of food at our frugal yet social meals, at Engineer cantonment, and we could always immediately distinguish the bones from those of other animals by their remarkably red colour. The tail is even more voluminous than that of the *S. Cinereus;*" (cat-squirrel.)

Species VI.—*The Line-Tail Squirrel.*

Sciurus Grammurus; Say.

This species is most remarkable for the peculiar coarseness and flattened form of its fur, and by three black lines on each side of the tail, which are united over the surface of it, as in the Barbary squirrel, *S. Getulus.*

The Line-Tail squirrel inhabits the Missouri country, about the naked parts of the sand-stone cliffs, where there are but few bushes. Its nest is found in holes and crevices of rocks, and it appears not to be in the habit of ascending trees, unless driven. It feeds on the buds, leaves, and fruits of the plants growing in the situations we have mentioned.

The line-tail squirrel measures eleven inches and a-half, and its tail is nine inches long. The general colour of the body is cinereous, variously tinged with rust-colour. The fur is very coarse, much flattened, canaliculate above; it is lead coloured or blackish at base, then whitish or ferruginous, with a brownish tip. The whitish colour prevails above the neck and shoulders, while the ferruginous is in greatest quantity from the middle of the back, sides, and exterior surface of the legs: above and below the orbits of the eyes the fur is whitish, the tail is whitish, being marked by three black lines, the base and tip of each hair being whitish, beneath, the colour is whitish, tinged with ferruginous.

Species VII.—*The Four-Lined Squirrel.*

Sciurus Quadrivittatus; Say.

Sciurus Quadrivittatus: Say. Long's Ex. to the Rocky Mountains, ii. 45.

This handsome little squirrel is found on the Rocky Mountains adjacent to the sources of the rivers Arkansa and Platte. Of its habits we know nothing but what is given in the following sentences, by Say, in the work above quoted:

"It does not seem to ascend trees by choice, but nestles in holes and on the edge of the rocks. We did not observe it to have cheek-pouches. Its nest is composed of a most extraordinary quantity of the burrs of the xanthium, branches and other portions of the large upright cactus, small branches of pine trees, and other vegetable productions, sufficient in some instances to fill an ordinary cart. What the object of so great, and apparently so superfluous, an assemblage of rubbish may be, we are at a loss to conjecture, we do not know what peculiarly dangerous enemy it may be intended to exclude by so much labour. Their principal food, at least at this season, is the seeds of the pine, which they readily extract from the cones."

The four-lined squirrel is four inches and a quarter long, from the tip of the nose to the root of the tail; the tail is three inches in length. The head is of a brown colour, mixed with tawny, having four white lines; the upper one on each side passes from the tip of the nose immediately over the eye to the superior base of the ear, and the lower one passes immediately beneath the eye to the inferior base of the ear. The ears are of a moderate size and half oval. On the back there are four broad white lines, and alternate, mixed black and ferruginous ones. The sides are tawny; the under part of the body whitish. The hair of the tail is black at base, then tawny, then black in the middle, and paler tawny at tip. Beneath it is fulvous, having a submarginal black line. On the anterior feet there is a prominent tubercle in place of a thumb. The striped head, less rounded ears, and bushy tail, which is neither banded nor striated, together with its smaller size and the presence of the thumb warts, in SAY's opinion, sufficiently distinguish this species from the *S. Getulus,* or Barbary squirrel of Linné.

SPECIES VIII.—*The Hudson's Bay Squirrel.*

Sciurus Hudsonius: FOSTER. Royal Soc. Trans. lxii. 378.
Hudson Squirrel: PENN. Arct. Zool. i. 134. No. 48. Hist. Quad. No. 274
Sciurus Hudsonius: GMEL. SCHREB. tab. 214.
The Common Squirrel: HEARNE, 8vo. ed. 378.

[*Commonly called Chickaree.*]

This beautiful species is very common in the northern and western parts of this country, and, where seldom disturbed, are so fearless as to allow themselves to be approached almost within reach,

They resemble the European more closely than any of our squirrels, and are remarkable for having tufts on the ears like that species, *S. Vulgaris.* This arrangement of the hair on the ears has been hitherto regarded as peculiar to European squirrels, and Pennant, in his Arctic Zoology, has prefixed to his description of the Hudson's Bay squirrel the following: " N. B. The ears of the *American* squirrels have no tufts," which is rather unfortunately placed before an *American species,* possessing these appendages in a very conspicuous degree.*

The Hudson's Bay squirrel is, perhaps, more remarkable for its neatness and beauty than any of its kindred species, which, in habits and manners, it closely resembles. It is between seven and eight inches long, having a tail five inches in length. Its whiskers are very long and black; the superior parts of the body are of a reddish brown colour, varying in intensity, and shaded with black. On the inferior parts the general colour is a tarnished or yellowish white.— The under part of the head and front of the fore limbs are reddish brown, like the back; the insides of the thighs are coloured like the belly; on each flank there is a distinctly marked black line, separating the colours of the back and belly. The tail is of a reddish brown colour, and is very beautiful.

---

* Other American species of squirrels have tufts on their ears, when in full pelage; none, however, so remarkably as the Hudson's Bay squirrel. Next to this species, Say's great tailed squirrel (*S. Macroureus,*) has them longest.

"The common squirrels are plentiful in the woody parts of this (the Hudson's Bay) country, and are caught by the natives in considerable numbers with snares, while the boys kill many of them with blunt-headed arrows. The method of snaring them is rather curious, though very simple, as it consists of nothing more than setting a number of snares all around the body of the tree in which they are seen, and arranging them in such a manner that it is scarcely possible for the squirrels to descend without being entangled in one of them. This is generally the amusement of the boys. Though small, and seldom fat, yet they are good eating.

"The beauty and delicacy of this animal induced me to attempt taming and domesticating some of them, but without success; for though several of them were so familiar as to take any thing out of my hand, and sit on the table where I was writing, and play with the pens, &c. yet they never would bear to be handled, and were very mischievous, gnawing the chair bottoms, window-curtains and sashes to pieces. They are an article of trade in the company's standard, but the greatest part of their skins, being killed in summer, are of very little value."*

Hearne, as above cited.

Species IX.—*The Red-Belly Squirrel.*

Sciurus Rufiventer; Geoff.

Sciurus Rufiventer; Geoff. Coll du Mus. Desm. Nouv. Dict. d'Hist. Nat. ton. x. 103.

An individual of this species, brought from the vicinity of New Orleans, belongs to the valuable collection of the Philadelphia Museum. It is about seven or eight inches long, having the tail shorter than the body. Its general colour is dark grayish brown above, with a bright yellowish red beneath. The tail at its base is of the colour of the back, about its middle it is of nearly the same colour as the belly, and at the extremity it is yellowish.*

---

* The following is the description of this species, given by Desmarest, p. 333:—The pelage is of a reddish brown, pricked with black on the head, neck, flanks and paws; all the hairs covering these parts being of a gray slate-colour at their bases, then clear brown or yellowish, and deep brown at their tips; the lower jaw, under part of the neck, throat, belly, and inner surface of the paws, of a nearly pure red. The neck is as if marked with transverse brownish lines; whiskers black and as long as the head; ears reddish and covered with short hairs; extremities of the paws of a deep brown, without mixture of yellow; tail bushy, brown at its base and yellow at its extremity.

Species X.—*The Ground Squirrel.**

Sciurus Striatus; Klein.

Sciurus Striatus: Klein, Pall. Glires, 378. Gmel. Schreb. tab. 221.
Sciurus Lysteri: Ray, Lyn. Quad. 216.
Sciurus Carolinensis: Briss. Reg. An. 155, No. 9.
Écureuil Suisse: Desm. 339, Sp. 547.

[*Commonly called Hacky, or Hackee, Ground, or Striped Squirrel.*]

Few persons have travelled through our delightful country without becoming acquainted with the pretty animal we are now to describe,—which, though very different in its general appearance from its kindred tenanting the lofty forest-trees, still approaches to them so closely in personal beauty and activity, as always to command the attention of the most incidental observer.

This squirrel is most generally seen scudding along the lower rails of the common zigzag or "Virginia" fences, which afford him at once a pleasant and secure path, as in a few turns he finds a safe hiding place behind the projecting angles, or enters his burrow undiscovered. When no fence is near, or his retreat is cut off, after having been out in search of food, he becomes exceedingly alarmed, and runs up the nearest tree, uttering a very shrill cry or whistle, indicative of his distress, and it is in this situation that

---

* This and the following species belong to the subgenus *Tamias,* of Illiger: having cheek-pouches.

1. Flying Squirrel. 2. Great-tailed S. 3. Ground S.

he is most frequently made captive by his persecuting enemies, the mischievous school-boys.

The ground-squirrel makes his burrow generally near the roots of trees, along the course of fences and old walls, or in banks adjacent to forests, whence he obtains his principal supplies of food. The burrows frequently extend to very considerable distances, having several galleries or lateral excavations, in which provisions are stored for winter use. The burrow has always two openings, which are usually far distant from each other; it very rarely happens that the animal is dug out, unless it be accidentally during the winter season.

The ground-squirrel appears to suffer more when made captive than any other squirrel with which we are acquainted. We have several times endeavoured to tame individuals of this species, but without success. In losing its liberty, the ground-squirrel appears to lose all vivacity, becomes a dull and melancholy animal, and can yield very little amusement or satisfaction to its keeper, whom it always flies, or bites severely, if not permitted to get out of his reach.

The ground-squirrel is rather more than five inches in length, from the nose to the root of the tail; the last is about two inches and a-half long.— The general colour of the head and upper parts of the body is reddish brown, all the hairs on these parts being gray at base. The eye-lids are whitish, and from the external angle of each eye a black line runs towards the ear, while on each cheek there is a reddish brown line. The short rounded ears are covered with fine hairs, which are on the outside of a reddish brown colour, and within of a whitish gray. The upper part of the neck, shoulders, and base of the hair on the back, are of a gray brown, mingled with whitish.

On the back there are five longitudinal black bands, which are at their posterior parts bordered slightly with red. The middle one begins at the back of the head, the two lateral ones on the shoulders; they all terminate at the rump, whose colour is reddish. On each side two white separate the lateral black bands. The lower part of the flanks and sides of the neck are of a paler red; the exterior of the fore feet is of a grayish yellow; the thighs and hind feet are red above. The upper lip, the chin, throat, belly, and internal face of the limbs, are of a dirty brown. The tail is reddish at its base, blackish below, and has an edging of black.

Species XI.—*The Rocky Mountain Ground-Squirrel.*

Sciurus Lateralis; Say.

Sciurus Lateralis; Say, Long's Exped. to the Rocky Mountains, ii. 46.

The Rocky Mountain ground-squirrel was first seen by Lewis and Clark, while on their celebrated expedition to the Pacific Ocean; they, however, merely mention it in their journal, without appending a particular description. Say has given a description of the species, but no account of its habits, which we may infer to be generally similar to those of the common ground-squirrel, to which this species is nearly allied.

The Rocky Mountain ground-squirrel may be distinguished from the common species by being of rather larger size, entirely destitute of the line along the middle of the back, by the lateral lines commencing anterior to the humerus, where they are broadest, by the longer nails on the fore feet, and the broad nails on the thumb tubercles. It is, however, most closely allied to the *S. Bilineatus* of Geoffroy.

The body is of a brownish ash colour, intermixed with blackish above. On each side of the back there is a dull yellowish-white dilated line, which is broader before, and margined above and beneath with black; these lines commence on the neck, anterior to the fore limbs, and terminate before they reach the tail. There is no vertebral line. The top of the head, neck, anterior to the tip of the white line, and the thigh, are tinged with rust-colour; the orbit is whitish. The sides are of a dull yellowish-white; the colour beneath is pale mixed with blackish. The tail is short and thin, having a submarginal black line beneath. The nails on the fore feet are elongated, and the thumb tubercles are furnished with broad nails.*

Say: loco citato.

CHAPTER XI.

Genus XXIX.—Flying-Squirrel; *Pteromys;* Ill.

GENERIC CHARACTERS.

The head is short and thick, having small or moderate sized ears, large prominent eyes, a somewhat blunted snout, and the upper lip divided. The trunk of the body is proportionally shorter and thicker than in other squirrels, and the skin of the sides is extended from the fore to the hind limbs, so as to form a sort of sail, which, in most of the species, is spread out by an additional bone on the anterior extremities, articulated with the wrist. The tail is either long, or of moderate length, flattened and distichous.

The *dental system* of this genus is the same as the preceding.

Species I.—*The Common Flying-Squirrel.*

Pteromys Volucella.

Sciurus Volucella: L. Gmel. Pall. Schreb. pl. 222.
Pteromys Volucella: Desm. Mam. 343, Encycl. pl. 77. fig. 4.
Le Polatouche: Buff. x. pl. 21. Shaw, Gen. Zool. ii. pt. 1, p. 155.
Assapannick: Smith's Virginia, p. 27.
Asapan: Fred. Cuv. Mammal. Lithog. livr. 8.

Nature has endowed this beautiful animal with an instrument to facilitate its passage from place to

THE COMMON FLYING-SQUIRREL. 147

place in the easiest and most pleasant manner. Capable of moving on the bodies and limbs of trees, like other squirrels, it does not require an equal degree of muscular strength to leap from tree to tree, or from great elevations to the ground, but launching itself from a lofty bough into the air, and extending its limbs and the intervening membranes, its body is buoyed up as by a parachute, and sails swiftly and obliquely downwards, passing over a very considerable space. To aid in this sailing movement, we find the whole body covered with a short and silky fur, which lies close to the skin, and the hairs on the tail, which partake of the same quality, lie close, and form a flattened and feather-shaped rudder.

During the day-light the flying-squirrel is rarely to be met with abroad, unless it has been disturbed. Occasionally large troops are seen together, and their sailing leaps have been said to present to the inexperienced the appearance of a large number of leaves blown off the trees. Their peculiar construction and habit render them very unfit for living on the ground, and they speedily regain the nearest tree, when at any time they fall short of the object towards which they may have leaped. They always take advantage of the wind, when about to leap to any distance, and then they appear to deserve the name of flying-squirrels, from the ease and velocity of their movements.

This species is very common throughout the United States, and individuals are frequently tamed as pets, but are more admired on account of the softness of their fur and the gentleness of their dispositions, than for any of the frolicsome and amusing actions that characterize other squirrels. Their nocturnal habits, more than their fondness for warmth, or the persons of their keepers, make them always desirous to hide themselves in the pockets, &c. When confined in a cage with a reel appended, they continue running almost uninterruptedly throughout the night.

The flying-squirrel makes its nest in hollow trees, where it brings forth three or four young at a litter. It is very easy to ascertain whether this squirrel has a nest in any hollow tree, by knocking against the trunk with a stone or stick; as soon as the jarring is felt, the animal comes to the opening and endeavours to escape. In this way the young are very commonly discovered and taken.

The flying-squirrel is quite small, being little more than four inches and a-half long, the tail being three inches and a-half in length.

The general colour is a brownish ash, with rounded, nearly naked ears, and large prominent black eyes. The under parts of the body are white, with a yellowish margin where the colour of the back and belly approach each other.

1 Maryland Marmot. 2 American Porcupine.

CHAPTER XII.

SECTION II.—INCLAVICULATA.

The Clavicles incomplete, or entirely wanting.

GENUS XXX.—PORCUPINE; *Hystrix;* L.

Germ. Stachelthier; Stachelschwein.
Fr. Porc-epic.   *Port.* Ouriço-cacheiro.

GENERIC CHARACTERS.

THE head is rather short, with an obtuse and somewhat compressed snout, long whiskers, short rounded ears, and small eyes; the upper lip is cleft, and the tongue set with scaly spines. The covering of the body is partly of bristles and partly of prickles or spines. The neck is thick, the belly large, and the limbs of equal length; the anterior have four, and the posterior five digits, armed with long, stout, curved nails. The tail is either short or of moderate length, and not prehensile.

Dental System.

$$20\ \text{Teeth:} \begin{cases} 10\ \text{Upper} & \begin{cases} 2\ \text{Incisive} \\ 8\ \text{Molar.} \end{cases} \\ 10\ \text{Lower} & \begin{cases} 2\ \text{Incisive} \\ 8\ \text{Molar.} \end{cases} \end{cases}$$

IN THE UPPER JAW the incisors are rounded and even in front, and they arise from the anterior and inferior part of the maxillary bone. The molars are of nearly the same size from the first to the last, and they are especially remarkable for the elevation of the crown above the neck of the root. The outline they present is very irregular. In the young animal they are traversed with various degrees of irregularity, by grooves, which, after being worn to a certain extent, begin to be interrupted, and then they exhibit a depression in front on the inside, and another at the back part on the outside; in front, as at the back part, one or two ellipses are seen, the remains of primitive grooves or tubercles. In old animals we find teeth with only one depression, and in the middle, three or four insulated figures, more or less irregular.

In the lower jaw the incisors resemble those in the upper, and take root some lines below the condyles. The molars have a great general resemblance to those of the upper jaw, and a precise idea can only be obtained by actual inspection, as description cannot convey a knowledge of such irregular and variable forms as are presented at different stages in the course of attrition.

Species I.—*The Canada Porcupine.*

Hystrix Dorsata; L. Gmel.

Hystrix Dorsata: Erxl. Schreb. pl. 169, Sab. App. 664.
Hystrix Hudsonis: Briss. 128.
Cavia Hudsonis: Klein, Quad. 51.
Hystrix Pilosus Americanus: Catesby, Car. App. 30.
Urson: Buff. xii. pl. 52.

The American porcupine exhibits none of the long and large quills which are so conspicuous and formidable in the European species, and the short spines or prickles which are thickly set over all the superior parts of its body are covered by a long coarse hair, which almost entirely conceals them.— These spines are not more than two inches and a-half in length, yet form a very efficient protection to our animal against every other enemy but man. Too slow in its movements to escape by flight, on the approach of danger, the porcupine places his head between his legs, and folds his body into a globular mass, erecting his pointed and barbed spines. The cunning caution of the fox, the furious violence of the wolf, and the persevering attacks of the domestic dog, are alike fruitless. At every attempt to bite the porcupine, the nose and mouth of the aggressor is severely wounded, and the pain increased by every renewed effort, as the quills of the porcupine are left sticking in the wounds, and the death of the assailant is frequently the consequence of the violent irritation and inflammation thus produced.

In the remote and unsettled parts of Pennsylvania the porcupine is still occasionally found, but south of this state it is almost unknown. According to Catesby it never was found in that direction beyond Virginia, where it was quite rare. In the Hudson's Bay country, Canada, and New England, as well as in some parts of the western states, and throughout the country lying between the Rocky Mountains and the great western rivers, they are found in great abundance, and are highly prized by the aboriginals, both for the sake of their flesh and their quills, which are extensively employed as ornaments to their dresses, pipes, weapons, &c.

The porcupine passes a great part of its time in sleep, and appears to be a solitary and sluggish animal, very seldom leaving its haunts, except in search of food, and then going but to a short distance. The bark and buds of trees, such as the willow, pine, ash, &c. constitute its food during the winter season; in summer, various wild fruits are also eaten by this animal.

Dr. Best, of Lexington, Ky. in a letter to the author of this work, observes that " the porcupine is seldom found in the state of Ohio, south of Dayton; but they are numerous on the river St. Mary. During winter they take up their residence in hollow trees, whence it appeared to me in several instances, from their tracks in the snow, they only travel to the nearest ash-tree, whose branches serve them for food. In every instance which came under my observation, there was no single track, but a plain beaten path, from the tree in which they lodged to the ash from which they obtained their food. I cut down two trees for porcupine, and found but one in each; one of the trees also contained four raccoons, but in a separate hollow, they occupied the trunk, the porcupine the limbs."

The following are Hearne's observations on this species:—" Porcupines are so scarce to the north of Churchill river, that I do not recollect to have seen more than six during almost three years residence among the northern Indians. Mr. Pennant observes, in his Arctic Zoology, that they always have two at a time, one brought forth alive, and the other still-born, but I never saw an instance of this kind, though in different parts of the country I have seen them killed in all stages of pregnancy. The flesh of the porcupine is very delicious, and so much esteemed by the Indians, that they think it the greatest luxury their country affords. The quills are in great request among the women, who make them into a variety of ornaments, such as shot-bags, belts, garters, bracelets, &c.* They are the most forlorn animals I know; for in those parts of Hudson's Bay where they are most numerous, it is not common to see more than one in a place. They are so remarkably slow and stupid, that our Indians, going with packets from fort to fort, often see them in the trees, but not having occasion for them at that time, leave them till their return, and should their absence be for a week or ten days, they are sure to find them within a mile of the place where they had seen them before."

The patience and ingenuity displayed by the Indian women in ornamenting dresses, buffaloe robes, moccasins, &c. can scarcely be appreciated by those who have never seen any of the articles thus adorned. We have already mentioned that these quills rarely exceed two inches and a-half, or at most three inches in length, and are not larger in circumference than a moderate sized wheat straw. Yet we find large surfaces worked or embroidered in the neatest

---

* Modus illis copulandi (testante Hearne,) profecto singularis est. Femina super marem dorso recubantem, a capite usque ad caudam ambulat, donec genitalia mutuo tangunt, sic, spinis acutis evitatis, veneris suaviis, fruuntur: aliquando ambobus lateribus resupinatis, actum est.

and most beautiful manner with these quills, which are dyed of various rich and permanent colours. In making this embroidery they have not the advantage of a needle, but use a straight awl. Some of their work is done by passing the sinew of a deer or other animal through a hole made with the awl, and at every stitch wrapping this thread with one or more turns of a porcupine-quill. When they wind the quill near to its end, the extremity is turned into the skin, or is concealed by the succeeding turn so as to appear, when the whole is completed, as if but a single strip had been used. In other instances the ornament is wrought of the porcupine-quills exclusively, and is frequently extremely beautiful, from its neatness and the good taste of the figures into which it is arranged. In general, however, the strong contrast of colours is the most remarkable effect aimed at. On some of the articles of dress figures of animals, exhibiting much ingenuity, are formed by embroidering with these quills. The Philadelphia Museum, so rich in objects of natural history, also boasts a most splendid and valuable collection of articles of dress, and implements of peace and war, peculiar to the various aboriginals of our country. Whoever wishes to see to what extent the quills of the porcupine are employed by these interesting people, and also to form a better idea of the number of porcupines that must be found in the trans-Mississipian regions, may be fully gratified by visiting this great institution.

CHAPTER XIII.

Genus XXXI.—Hare; *Lepus;* L.

Fr. Lièvre.   *Germ.* Hase.

GENERIC CHARACTERS.

The head is narrow and compressed, having a rather acute snout, large, prominent, laterally-placed eyes, and long ears, situated close together. The upper lip is cleft, and the inside of the cheeks covered with hair: in each groin there is a fold of the skin that forms a sort of pouch. The fore limbs are slender and short, and have five digits, which are below, covered with a soft, velvety hair; the posterior limbs are very long, and have four digits, the soles being covered with hair similar to that in the palms. The teats are from six to ten in number; the tail is very short and turned upwards.

Dental System.

28 Teeth:
- 16 Upper { 4 Incisive / 12 Molar.
- 12 Lower { 2 Incisive / 10 Molar.

It is known how anomalous the hares are in the order of *gnawers*, by the number and singular arrangement of their upper incisor teeth. They are equally so in the structure of the head, and in many other organic peculiarities, which do not allow them to be naturally approximated to any other group of this order.*

In the upper jaw the anterior incisor is flat on its anterior surface, and unequally divided by a longitudinal depression, nearer to its internal than its external edge. Behind this tooth another small one is found, divided at its extremity by a transverse groove, and in very young individuals we find a third tooth behind the second, but it soon falls out, and the alveole disappears; these two last teeth are placed in the intermaxillary bones. The molars have nearly the same structure, but differ in size. They are twice as long as their breadth; the first, smaller than the succeeding one, exhibits two folds of enamel on its anterior surface, but all the parts of these rejoin, and are solidified together. The four following are of the same size, and divided longitudinally in their middle by two folds of enamel, which arise at their extremities and approach each other, so that the laminæ composing them, though entirely reunited, leave no intervening vacancy proper to be filled by the cortical matter. The internal fold is the most profound. The last molar, which is extremely small, appears to have no fold, and to be of a simple structure,—that is, it presents the form of a very elongated ellipsis, surrounded by enamel.

---

* These animals have an exceedingly large cœcum, which has a spiral valve running though its whole length. Beneath the orbit of the eye there is, in the skull, a space at the inner angle, which is cribriform, or pierced by a great number of small holes.

In the lower jaw the incisor is smooth and flat. The molars are formed after the same system as those of the upper jaw, but differ slightly from each other. The first, which is the largest, has three sides on its external face, and a slight depression on its anterior face, although it is only divided into two parts by a deep fold of enamel, the plates of which reunite. The three following are similar: they are of the same size and divided by a deep fold of enamel, the plates joining each other only on the outside, which leaves a deep depression on their inner face. The fifth is a third smaller than the preceding, and divided into two unequal parts by two lateral grooves, the anterior of which is the largest.

Species I.—*The American Hare.*

Lepus Americanus; L. Gmel.

Lepus Americanus: Schœpf. Natur. fig. 20, p. 20.
Lepus Hudsonius: Pall. Glir. pt. 1, p. 30.
American Hare: Penn. Arct. Zool. i. 109, No. 38, Hist. Quad. No. 243; Hearne, Journey, &c. 8vo. ed. 385; Sabine, App. to Franklin's Exped. 665.

[*Commonly, but improperly called Rabbit.*]

The American Hare is found throughout this country to as far north as the vicinity of Carlton House, in the Hudson's Bay country. According to the statement of Hearne " they are not plentiful in the eastern parts of the northern Indian country, not even in those parts that are situated among the woods; but to the westward, bordering on the south ern Indian country, they are in some places pretty numerous, though by no means equal to what has been reported of them at York Fort, and some other settlements in the Bay." In various parts of the Union this hare is exceedingly common, and large numbers are annually destroyed for the sake of their flesh and fur.

The timidity and defencelessness characteristic of the genus, are well illustrated in this species, which has no protection against its numerous enemies, and can escape by flight alone. Its peculiar colour must, however, minister to its safety, as it is so similar to the general colour of the soil as to require a close attention to distinguish the animal, which is usually passed without being observed by such as are not especially in search of it. Yet the swiftness and other natural advantages of the hare, insufficient to secure it from the artifices of man, or from being preyed upon by various beasts and birds, would not prevent the species from soon being extinguished, were it not for its remarkable fecundity.

During the day time the hare remains crouched within its form, which is a mere space, of the size of the animal, upon the surface of the ground, cleared of grass, and sheltered by some overarching plant: or else its habitation is in the hollowed trunk of a tree, or under a collection of stones, &c.

It is commonly at the earliest dawn, while the dew-drops still glitter on the herbage, or when the fresh verdure is concealed beneath a mantle of glistening frost, that the timorous hare ventures forth in quest of food, or courses undisturbed over the plains. Occasionally during the day, in retired and little frequented parts of the country, an individual is seen to scud from the path, where it has been basking in the sun; but the best time for studying the habits of the animal is during moon-light nights, when the hare is to be seen sporting with its companions in unrestrained gambols, frisking with delighted eagerness around its mate, or busily engaged in cropping its food. On such occasions the turnip and cabbage fields suffer severely, where these animals are numerous, though in general they are not productive of serious injury. However, when food is scarce they do much mischief to the farmers, by destroying the bark on the young trees in the nurseries, and by cutting valuable plants.

The flesh of the American hare, though of a dark colour, is much esteemed as an article of food. During the summer season they are lean and tough, and in many situations they are infested by a species of œstrus, which lays its eggs in their skins, producing worms of considerable size. But in the autumnal season, and especially after the commencement of the frost, when the wild berries, &c. are ripe, they become very fat, and are a delicious article of food. In the north, during winter, they feed on the twigs and buds of the pine and fir, and are fit for the table throughout the season. The Indians eat the contents of their stomachs, notwithstanding the food is such as we have just mentioned.

The American hare never burrows in the ground like the common European rabbit; (*L. Cuniculus.*) When confined in a yard, our animal has been known to attempt an escape by scratching a hole in the earth near the fence or wall, but there are few wild animals, whatever may be their characters, that will not do the same, under similar circumstances, though in their natural condition they may never attempt to burrow. Such is the fact in relation to the American hare, which never burrows while it is a free tenant of the fields and woods. It has been said that this animal also occasionally ascends trees, which must be understood solely of its going up within the trunks of hollow trees, which it effects by pressing with its back and feet against opposite sides of the hollow, ascending somewhat in the same manner that a sweep climbs a chimney.

The hare is not hunted in this country as in Europe, but is generally roused by a dog, and shot, or is caught in various snares and traps. In its movements our hare closely resembles the common hare of Europe, bounding along with great celerity, and would no doubt, when pursued, resort to the artifices of doubling, &c. so well known to be used by the European animal. The American hare breeds several times during the year, and in the southern states even during the winter months, having from two to four or six at a litter.

In summer pelage the American hare is dark brown on the upper part of its head, a lighter brown on the sides, and of an ash colour below. The ears are wide and edged with white, tipped with brown, and very dark on their back parts; their sides approach to an ash colour. The inside of the neck is slightly ferruginous; the belly and the tail is small, dark above, and white below, having the inferior surface turned up. The hind legs are covered with more white than dark hairs, and both fore and hind feet have sharp pointed, narrow, and nearly straight nails.

THE AMERICAN HARE.

In winter the pelage is nearly twice the length of what it is in summer, and is altogether, or very nearly, white. The weight of the animal is about seven pounds.

This species is about fourteen inches in length. The hind legs are ten inches long, by which circumstance it is most strongly distinguished from the common rabbit of Europe.*

---

* "The hare and rabbit so nearly resemble each other in form and structure, that it has puzzled the most experienced zoologists to assign definite distinguishing marks. Yet there are many circumstances in which they differ (besides the colour of their flesh when boiled, and their manner of escaping from their foes) in reference to their reproductive system. The nest of the hare is open, constructed without care, and destitute of a lining of fur. The nest of the rabbit is concealed in a hole of the earth, constructed of dried plants, and lined with fur, which is pulled from its own body. The young of the hare, at birth, have their eyes and ears perfect, their legs in a condition for running, and their bodies covered with fur. The young of the rabbit, at birth, have their eyes and ears closed, are unable to travel, and are naked. The maternal duties of the hare are few in number, and consist in licking the young dry at first, and supplying them regularly with food. Those of the rabbit are more numerous, and consist of the additional duties of keeping the young in a state of suitable cleanliness and warmth. The circumstances attending the birth of a hare are analogous to those of a horse, while those of a rabbit more nearly resemble the fox."—FLEM. *Philosophy of Zoology*, ii. p. 140.

The *rabbit* is not a native of this country, but has frequently been introduced in a domesticated state, from England, &c. The species above described we have already stated to be improperly called "the rabbit."

Species II.—*The Polar Hare.*

Lepus Glacialis; Sab.

Lepus Glacialis: Leach. Miscel. Sabine, App. to Franklin, p. 664; in App. to Parry's Voyage of 1819, 1820.

The Polar Hare is found in greatest abundance at the extreme northern part of this continent, along the southern coast of Barrow's strait, and in the North Georgian islands. Capt. Sabine, who found the animal in considerable numbers on Melville island, has pointed out, in the Appendix above quoted, (whence the following description is taken) the differences existing between this species and the *L. Variabilis,* with which it had been previously confounded.

The polar hare is larger than the alpine or varying hare, next to be described, and weighs about eight pounds. Its colour, in winter dress, is white, having the ears black at their tips and longer than the head. The nails are strong, broad and depressed.

" The ears are longer, in proportion to the head, than those of the common hare, (*L. Timidus*) and much longer than those of the alpine hare (*L. Variabilis.*) The ears of the common hare are usually considered one-tenth longer than the head, those of the present species are from one-fifth to one-seventh. The fore teeth are curves of a much larger circle, and the orbits of the eye project much more than those of either of the other species; the claws are broad, depressed and strong: those of the *L. Timidus* and *Variabilis* being, on the contrary, compressed and weak; the hind leg is shorter, in proportion to the size of the animal, than in the alpine, (*Variabilis;*) the fur is exceedingly thick and woolly, of the purest white in the spring and autumn, excepting a tuft of long black hair at the tip of the ears, which is reddish brown at base; the whiskers are also black at the base for half their length. In some of the full grown specimens, killed in the height of summer, the hair of the back and sides was a grayish brown towards the points, but the mass of fur beneath still remained white. The face and the front of the ears were a deeper gray; the fur is interspersed with long, solitary hairs, which in many individuals were, in the middle of summer, banded with brown and white. The hares which Mr. Hearne describes, in his northern voyage, as inhabiting the continent of America, as high as the seventy-second degree of latitude, are stated to weigh fourteen or fifteen pounds when full grown and in good condition. The largest hare killed at Melville island did not weigh nine pounds; were it not for this difference in size, they might be supposed, from other parts of their description, to be the same species."*

Through the kindness of that zealous friend of science, Charles L. Bonaparte, we have had an opportunity of examining and preparing a description

---

* In the Appendix to Franklin's Journey, p. 665, we have the following observations on this species:—" The polar hare appears to vary much in size, and consequently in weight:

of a hare, from specimens in winter and summer pelage, belonging to his valuable collection. This species, which appears to be the same with that indicated by Lewis and Clarke, and after them by Warden, has also been *proposed* as a new species, under the name of *Lepus Virginianus.* That it is a species distinct from the *L. Glacialis* and *Variabilis*, remains yet to be established, since differential characters have not been adduced to prove the fact.— We shall first give a description of the animal in summer and winter dress, and then examine whether any differential characters have been given, or, under existing circumstances, can be offered, to entitle it to rank as a new species.

The general colour of this hare, in summer dress, is a light reddish brown, which is lighter on the breast and head, becoming darker from the superior parts of the shoulders to the posterior parts of the body. The hairs are coloured in the following manner:—They are plumbeous at base, then light yellowish, then dusky, then reddish brown, and finally black at tip. The under jaw is white, and this colour extends backwards until opposite the bases of the ears. The belly and legs are white, faintly

---

this, perhaps, may be caused by the quantity and quality of the food it can command. Dr. RICHARDSON observed that the polar hare is never seen in woods; it frequents the barren grounds, living chiefly on the berries of the *arbutus alpina* and the bark of a dwarf birch. It sits, like the common hare, on the whole length of the metatarsal bones, but in running its hind feet make a round print in the snow, similar to that made by the fore ones."

tinged with light reddish brown; the tail is whitish, which colour is superiorly mingled with bluish or lead colour. The ears are externally bluish white, and darker at tip; internally they are of a faint reddish white.

The following measurements of a recent specimen of this animal, were carefully made by the distinguished individual before mentioned:

Total length,	2 ft. 7 in.
Height to the top of the fore shoulder,	" 10
———— to the top of the thigh,	1 2
Length of the head,	" 4
———— of the ears,	" 4
Distance from the eyes to the end of the nose,	" $1\frac{3}{4}$
Length of the fore arm,	" 4
———— of the fore paw,	" $2\frac{3}{4}$
———— of the thigh,	" 6
———— of the hind foot,	" 6
———— of the tail,	" $1\frac{1}{2}$

In winter dress the general colour is pure white, the fur being long, soft, fine, and in greatest quantity upon the breast. The hairs in the summer, as in winter pelage, are plumbeous at base, but are then reddish, and at tip of a snowy whiteness. The ears are slightly tipped with dark lead colour, and edged within by brown and white hairs intermixed. The whiskers are entirely white, or black at base and white at tip. The feet are thickly clothed with hair, which conceals the slightly curved nails, which are long and narrow at base.

When we compare this animal with the polar hare, *L. Glacialis* of Sabine, and with the *L. Variabilis*, or alpine hare, we shall be convinced that distinctive characters have not yet been given to establish the supposed new species, as well as that such distinctive characters are very few and difficult of discovery.

The essential or distinctive characters ascribed by Sabine to the polar hare, are as follows:—Colour white, ears black at tip, longer than the head; nails robust, broad and depressed.

The essential characters of the *L. Variabilis*, as given by Desmarest, are,—pelage grayish yellow in summer, white in winter; ears shorter than the head, and black at all times; tail white in winter and gray in summer.

The "characters essential" given of the animal under consideration as a *new species*, entitled *Lepus Virginianus*, are as follows:—" Grayish brown in summer; the orbits of the eyes surrounded by a reddish fawn colour at all times; ears and head of nearly equal length; tail very short."

As the colour of the pelage is common to several species, both in summer and winter, it is peculiarly insufficient as a differential character in the establishment of the proposed new species. The second character laid down in the last definition, concerning the permanent fawn colour surrounding the orbit, is incorrect. One of the specimens above described has the orbits of the eyes surrounded by a very different colour; neither is the statement, that the ears are *nearly* equal in length to the head, of any avail in establishing the specific difference, since the ears of the *Variabilis* are also *nearly* equal in length to the head, being somewhat shorter. If it be meant that the ears of the supposed new species are, in the same sense, nearly of the length of the head, it is incorrect, since the head of the animal in its recent state measured four inches, and the ears were of the same length. The shortness of the tail is as characteristic of the *Variabilis*, in which it is but one inch and three-quarters, while the proposed new species has a tail one inch and a-half long.

In the present state of our knowledge, the only truly differential character that can be given is the equality existing between the length of the ears and head. The toe-nails differ from those of the polar hare described by Sabine, but they are very similar to those of the common hare, and may also be similar to those of the *Variabilis*, which are not minutely described, even by DESMAREST; hence no positive conclusion can be deduced. Neither can the relative height of the hind and fore parts aid in distinguishing this hare from the alpine, (*L. Variabilis*) in which the hind are to the fore parts as fourteen to twelve, while in the proposed new species the proportion is the same, being as twelve to ten; the polar hare (*L. Glacialis*) has the hind limbs proportionally shorter than the *Variabilis*, though their actual length is not given: this being equally true of the supposed new species, we cannot infer any specific difference therefrom. The weight of these hares is a circumstance equally inefficient in deciding this doubtful matter; the polar hare weighs from seven to nine pounds, (*Sab.*)—the alpine seven to seven and a-half, (*Penn.*)—the hare described by Lewis and Clarke, seven to eleven pounds. The weight given by the latter observers inclines us to believe that this animal is the same as that described by HEARNE, as the varying hare, which SABINE says differs from the polar hare only in weight.*

---

* The following is HEARNE's account of this animal:— "The *varying hares* are numerous, and extend as far as latitude 72° N., probably farther. They delight most in rocky and stony places, near the borders of woods, though many of them brave the coldest winters on entirely barren ground. In summer they are nearly of the colour of our English wild rabbit, but in winter assume a most delicate white all over, except the tips of the ears, which are black. They are, when full grown and in good condition, very large, many of them weighing fourteen or fifteen pounds; and, if not too old, are good eating. In winter they feed on long rye grass and the tops of dwarf willows, but in summer eat berries and different sorts of small herbage. They are frequently killed on the south side of Churchill river, and several have been known to breed near the settlement at that place. They must breed very fast, for, when we evacuated Prince of Wales' fort, in 1783, it was common for one man to kill two or three in a day, within three miles of the new settlement. But partly, perhaps, from so many being killed, and partly from the survivors being so frequently disturbed, they have shifted their situation, and at present are as scarce near the settlement as ever. The northern Indians pursue a singular method of shooting those hares; finding, by long experience, that these animals will not bear a direct approach, when the Indians see a hare sitting, they walk round it in circles, always drawing nearer at every revolution, till by degrees they get within gun-shot. The middle of the day, if it be clear weather, is the best time to kill them in this manner; for before and after noon the sun's altitude being so small makes a man's shadow so long on the snow as to frighten the hare before he can approach near enough to kill it."—8vo. ed. p. 385.

In the specimen in summer dress (which we have described in beginning this article) the tail is nearly white, and in the hares observed by Lewis and Clarke, presently to be quoted, the tail was likewise white during the summer. Should this colour of the tail prove to be uniformly permanent, it may be added to the only other differential character, drawn from the ears. But until more decisive evidence can be adduced, it will be safest to consider this hare as at most a variety of the alpine hare, the *Lepus Variabilis* of authors.

It is found throughout the mountainous regions of the Union, and on the plains and in the woods of the western territories. To the north it is known as far as observation has yet extended. Lewis and Clarke, in the second volume, p. 178, of their extremely interesting journal, give the following account of this animal:—" The hare on the western side of the Rocky Mountains inhabits the great plains of the Columbia. On the eastward of those mountains they inhabit the plains of the Missouri. They weigh from seven to eleven pounds; the eye is large and prominent, the pupil of a deep sea-green, occupying one-third of the diameter of the eye; the iris is of a bright yellow and silver-colour; the ears are placed far back and near each other, which the animal can, with surprising ease and quickness, dilate and throw forward, or contract and hold upon his back at pleasure; the head, neck, back, shoulders, thighs, and outer parts of the legs and thighs, are of a lead colour; the sides, as they approach he belly, become gradually more white: the belly, breast, and inner parts of the legs and thighs are white, with a light shade of lead-colour; the tail is round and bluntly pointed, covered with white soft fur, not quite so long as on the other parts of the body; the body is covered with a deep, fine, soft, close fur. The colours here described are those which the animal assumes from the middle of April to the middle of November; the rest of the year he is of a pure white, except the black and reddish brown of the ears, which never changes. A few reddish brown spots are sometimes mixed with the white, at this season, (February 26) on their heads and the upper parts of their necks and shoulders; the body of the animal is smaller and longer, in proportion to its height, than the rabbit; when he runs he conveys his tail straight behind, in the direction of his body. He appears to run and bound with surprising agility and ease: he is extremely fleet, and never burrows nor takes shelter in the ground when pursued. His teeth are like those of a rabbit, (*L. Americanus*) as is also his upper lip, which is divided as high as the nose. His food is grass, herbs, and in the winter he feeds much on the bark of several aromatic herbs growing on the plains. Capt. Lewis measured the leaps of this animal, and found them commonly from eighteen to twenty-one feet: they are generally found separate, and never seen to associate in greater numbers than two or three."

Warden, in a note to his " Description des Etats Unis," p. 632, says, " the varying hare of the southern parts of the United States is distinguished from the American rabbit (*Lepus Americanus*) by changing from a gray brown, which is its colour in spring and summer, to a full white in winter. Its ears are also shorter and marked with black, and its legs more slender. The largest varying hares are about eighteen inches long, and weigh from seven to eight pounds. They are very prolific, as the female litters several times a year, having three or four young each time. The flesh of this animal is represented to be agreeable and nutritious. It frequents the marshes and prairies, but never burrows; its colour is similar to that of the European rabbit, and the female equally conceals her young from the male. When pursued, they mount as high as possible within a hollow tree."

CHAPTER XIV

Order IV.—Bruta;* L. *Animals destitute of Cutting-Teeth.*

In North America no living animal belonging to this order has yet been found, but gigantic fossil remains of extinct species have been occasionally disinterred in different parts of the Union. The circumstance first stated may appear the more singular when the fact is recollected, that the greater num-

---

* Brisson first established an order, under the title of *Edentata*, which comprised the animals having no teeth; he made a second order, of *Dentata*, embracing those possessed of molars: which division was adopted by Lacepede. Storr disapproving this arrangement, formed a single order of all these animals, which he called *Mutici*, and Boddaert subsequently changed the name to that of *Edentes*, which was afterwards changed to *Edentata* by Cuvier. Various changes have been proposed by other writers, founded on their peculiar views, (of the structure, &c. of these animals) which it is needless to detail. We have adopted the Linnean name for the order, as it conveys no incorrect idea, which all the others do, by calling the order Toothless, when only one genus is in that predicament. The place in the system of classification is that given to the order by Cuvier, because these beings have some analogy to the digitigrade animals, in the circumstance of their toes being terminated by large and long claws, &c.*

* Vedi Ranzani; *Elementi di Zoologia*, tomo. IIdo. parte IIda. p. 473.

EXTINCT GIANT SLOTH. 173

ber of the living genera and species, comprised by this order, are, at present, inhabitants of the southern division of this continent.

The animals of this order are characterized by the exceeding slowness of their movements, dependant on the singular structure and proportions of their limbs. They have the orbits of the eyes and the temporal cavities opening into each other, so as to form one cavity in the skeleton; and their limbs are terminated by digits, (varying in number in different genera and species) armed with large and hoof-like claws. Such of the genera as have molar or jaw-teeth, feed on bark of trees, &c. others, entirely destitute of teeth, feed exclusively on insects. Some of them use their claws for climbing and clinging to the branches of trees; others for the purpose of burrowing.

Family I.—Tardigrada; *Sloths*.

Genus I.—Megatherium; C. *Extinct Giant Sloth.*

GENERIC CHARACTERS.

Unlike the living members of this family, the present genus has complete zygomatic arches, yet it again closely resembles the existing genera in having at the anterior basis of the zygoma, a large descending process. The bones of the upper jaw are much prolonged; the nasal bones are very short: the lower jaw has very large ascending branches, and at its anterior extremity, or chin, it is salient, and hollowed within. The spine, composed of twenty-six vertebræ, has seven belonging to the neck, sixteen to the back, and three to the loins. It cannot be positively stated that these animals had no tail, though it is probable;—if it did exist, it is presumed to have been very short. The posterior limbs exceed the anterior in size considerably; all the feet have five toes, yet three only on the fore feet are provided with large claws, the other two being rudimental. On the hind feet but one toe is furnished with an enormous claw; the other four are nearly rudimental.

Dental System.

16 Teeth: { 8 Upper / 8 Lower } Molars.

"The twelve posterior teeth are larger than the others, each of them being nearly two inches square: they present rounded angles, and between each of these angles there is a small canal. Each tooth has four angles, two internal and two external. The lower part, which is imbedded in the alveolar process, diminishes gradually, becoming only two inches broad, of a square form, having beneath a pyramidal cavity separated by four points, which buries itself sufficiently forward in the tooth. The four first teeth weighed exactly twenty ounces; the others as much as twenty-six."*

---

* Don JUAN B. BRU; description of the skeleton from Paraguay, in the Madrid Museum; translated by Bonpland in Cuv. oss. foss. tom. iv.

"Their remarkable structure, so much unlike any before observed, is still more deserving of particular description. The tooth is covered externally with a coating of enamel, extremely thin, and uniformly so on all sides, and which does not extend over any part of the crown. Within is a coating of bone or ivory, which, at the sides of the tooth, is as thin as the enamel; but where it is parallel to the cutting edges, is nearly a-quarter the thickness of the whole tooth. Enclosed within this is a second coating of enamel, which, like the first two, has two sides very thin. The other two sides are more than a line thick, and terminate in the cutting process, which by this means are kept constantly sharp and prominent, by the wearing away of the softer ivory on each side of them. Where these laminæ of enamel terminate on the anterior side of the higher process, may be observed a semilunate truncation, which is not seen on the lower process, although terminated in a similar manner. The whole solid part of the tooth thus represents a prism of bone, enveloped within three *cases*, two of enamel, and the third of a substance similar to itself."*

[That the reader may be better prepared to understand the peculiar character of these fossil remains, we subjoin Cuvier's observations on the construction of the existing animals, to which these extinct species were closely allied, and which they must have resembled in all their general habits.

---

* Annals of the Lyceum of Natural History of New York, vol. i. p. 114. A highly interesting paper by W. Cooper on the Megatherium found in Georgia.

as well as in conformation. This comparison may also prove of advantage to the inquirer, (independent of satisfying him of the correctness of the opinions advanced, relative to the similarity of these animals) should it awaken his curiosity to become better acquainted with the works of the great naturalist quoted. His writings, though principally occupied with the relics of former worlds—with animals that ceased to be before the foundations of human society were laid, nevertheless overflow with the energies of an immortal intellect, and expand the mind of the student with those sublime ideas of the God of Nature, which are not to be equalled by any mere effort of imagination, since they are inspired by the most extraordinary facts, beheld under the powerful illumination of disciplined genius.

"In considering these beings, we find so few relations with ordinary animals—the general laws of existing organizations apply so little to them—the different parts of their bodies are so much in contradiction to the rules of co-existence established throughout the animal kingdom, that we might really believe them to be the remains of another order of beings, the living fragments of that antecedent nature, whose other ruins we are obliged to seek in the bosom of the earth, which by some miracle have escaped the catastrophies that destroyed their cotemporary species.

"With the solitary exception of the elephant, there is not, perhaps. among all the quadrupeds, an animal which so widely departs from the general plan of nature, in the formation of that class, as the sloths; still, the deviations from that plan correspond with each other so reciprocally as to correct their bad effects, and produce a concordant whole; but in the sloths, each singularity of organization appears to have no other result but weakness and imperfection, and the inconveniences they cause the animal are not compensated by any advantage.

"The mere aspect of the skeleton of the *ai*, (three-toed sloth) in some sort indicates deficiencies of proportion. The arms and forearms taken together are almost twice as long as the thigh and leg, so that when the animal moves on all four limbs it is obliged to crawl upon its elbows, and when it raises itself upon its claws, the entire hand may still be placed against the ground. There are some apes alone which approach this disproportion; but they often keep themselves erect, or walk with the aid of a staff, which cannot be done by the *ai*, since its hind feet are so peculiarly articulated that they cannot sustain the body. The pelvis, moreover, is so large, and the cotyloid cavities (or sockets for the heads of the thigh bones) are turned so far backwards, that the knees cannot be brought together, and the thighs are kept forcibly separate.

"Animals, when they run, receive their principal impulsion from their hind feet; hence, the best runners have the longest hind legs, as the hares, jerboas, &c. The length of the fore legs serve merely to embarrass, and hence crabs are forced to move backward. Sloths can scarcely employ their fore limbs, except for the purpose of clinging to objects and then dragging forwards their hinder parts.

"In the other quadrupeds, the *os sacrum* is only attached to the ossa ilia, or haunch bones, by a small portion of its sides in front; all the rest is free, and the interval between the posterior part of the sacrum and ossa innominata is vacant, for the reception of the muscles and other soft parts, bearing the name of the great ischiatic notch. In the sloth there is a second posterior union between the sacrum and tuberosity of the ischium, and instead of the ischiatic notch there is nothing but an opening like a second obturator foramen.

The joint which attaches the hind foot and leg, " appears to be expressly arranged to deprive the animal of the use of the foot." In other animals the articulation is such as to allow the foot to be flexed upon the leg, but the foot of the sloth turns upon the bones of the leg " like a weathercock upon its pin, but cannot be flexed. Hence it results that the body of the foot is nearly vertical when the leg is so, and that the animal cannot place the sole of the foot on the ground unless by separating the leg so far as to render it almost horizontal. From these two peculiarities the absolute weakness of the foot is derived, and the total impossibility of its affording a solid point of support to the body." On the fore and hind feet " the skin envelopes all the parts except the nails, which are separate, and the whole of the remainder of the digits is united, being without interval or mobility between them; they, therefore, can only be flexed or extended together.

" The nails of the sloth are of an enormous length, and the dreadful weapon they furnish is doubtless the mean by which these animals defend themselves with sufficient success to compensate for all the disadvantage of the rest of their organization. Nearly as sharp as those of the cat, it is necessary for their preservation in that condition that they should be protected from friction against the ground. It is by withdrawing them between their toes, having the points turned upwards, that those of the cat are preserved. The sloths cannot do the same, because their digits, being united by the integument, leave no interval; besides, these long reverted points would be very inconvenient, and might wound the throat and belly. When not in use they are kept recurved, and placed with their convexity on the ground; this, as in the cat, is effected without fatigue to their muscles, and by the simple elastic action of the ligaments; the muscles have only to act to extend them.

"From this difference, another results in the form of the articulation. The last phalanges of the cat, like those of the sloth, are at the back part hollowed into an arc of a circle, since they must move as pullies upon the next to the last bone. But in those of the cat the most salient part of the arc is below; in the sloth it is above, always on the side towards which the nail is not carried. By this circumstance we may distinguish, at the first glance, even a single phalanx of either of these genera. We may also distinguish them by the osseous sheath which retains and overlaps the base of the nail. Both genera equally have them, because both require solidity in so long a weapon; but in the sloth it is the lower part of this sheath which is the most prolonged, while in the cat it is rather the superior part."*

---

* Recherches sur les Oss: Foss. tom. iv.

Species I.—*Cuvier's Giant Sloth.*

Megatherium Cuvieri.

Mégathére: C. Ann. du Mus. v. 176, pl. 24, 25. Recherches sur les Ossem. Foss. tom. iv. Bru, Descr. &c. trad. par Bonpland, Ejusdem, tom. iv. Descr. d'un squelette conservé dans le Mus. de Madrid; trad. de Garriga. Mitchill; Ann. of the Lyceum of Nat. History of New York, vol. i. Cooper on the Megatherium of Georgia, Ann Lyceum, vol. i.
Megatherium Cuvieri: Desm. Mammal. 365.

The first discovered skeleton of this extraordinary animal was obtained from some excavations made on the banks of the river Luxan, near a town of the same name, situated about three leagues W. S. W. of Buenos Ayres. It was found at the depth of a hundred feet from the surface, in a sandy soil, and is the most perfect specimen of this animal yet procured. It was sent to Spain by the viceroy of Buenos Ayres, the Marquis of Loretto, where it was mounted in the museum of Madrid by Don. J. B. Bru, who first published a description of it. Another specimen was sent to the same cabinet in 1795. from Lima, and a third was discovered in Paraguay.[*] The only skeleton yet found in North America was first indicated by our celebrated countryman, Dr. Mitchill, and subsequently more fully detailed by that ardent votary of natural science, W. Cooper, of New York, in the work above quoted. Having but a few mutilated fragments of this skeleton in the cabinets of this country, it is impossible, by describing them alone, to give the reader any proper idea

---

[*] Garriga, as quoted by Cuvier.

of the animal. We shall therefore introduce Cuvier's account of the species, drawn up principally from the work of Garriga, and add thereto the observations made on the American specimen recently discovered in Georgia.

"A first glance at the head of the megatherium gives us the most marked relations with that of the sloth, especially the *ai* (three-toed sloth.) The most striking feature of resemblance is the long descending apophysis placed at the anterior base of the zygomatic arch. It is proportionally as long in the ai as in the megatherium; but the latter has the zygomatic arch entire, while in both species of sloth, even when adult, it is not continuous.

"The ascending branch of the lower jaw sufficiently resembles that of the sloth, but its inferior part forms a convexity, to which we find but a slight resemblance even in that of the elephant. The osseous snout is more salient in the megatherium than in the *ai;* this arises from an advance of the symphysis of the lower jaw, (chin) which is also found in the two-toed sloth, (*unau*) and from a corresponding advance of the intermaxillary bones.— The bones of the nose are very short, which, after the example of the elephant and tapir, might lead us to suspect that this animal had a trunk.

"This might also be inferred from the multitude of holes and small canals with which the anterior part of the snout is pierced, which must have served to give passage to vessels and nerves destined to nourish some organ of considerable size. However, if such a trunk existed, it was doubtless very short, judging by the length of the neck, which appears very natural, and not owing to the introduction of vertebræ, belonging to larger individuals in forming the skeleton. The head not being disproportionately large, and especially being without tusks, a long neck would not be as prejudicial as it would have been in the elephant.

"The molar teeth are four in number, on each side, both above and below, as in the *ai*, and, like the teeth of that species, of a prismatic form, and the crown traversed by a groove. They are only closer together, and have no pointed canine in front, as the *ai* has one at least in the upper jaw, and the unau in both upper and lower. Yet that is scarcely sufficient to distinguish a genus, for in the *unau* itself the canines differ little from the molars, which are as pointed as in that species.

"If the number of seven cervical vertebræ, seen in this skeleton, be correct, as analogy with other animals induces us readily to believe, the megatherium differs much in this respect from the three-toed sloth, which itself is separated from all known quadrupeds by the length of its neck. The megatherium has sixteen dorsal vertebræ, and by consequence sixteen ribs on each side, and three lumbar vertebræ. The number is exactly the same in the *ai*.

"The relative proportion of the extremities is not the same as in the sloth, where the anterior have nearly double the length of the posterior limbs: in this animal the inequality is much less. But in return, the disproportionate thickness of the thigh and leg bones (indications of which are found in the sloths, tatous, and especially the pangolins) is carried here to an excessive degree, the thigh-bone being in height only double its greatest thickness, which renders it larger than that of any other animal known, not excepting the mastodon.

"This general disposition of the extremities leads to the conclusion that this animal had a slow and equal gait, and advanced neither by running nor leaping, like animals having the fore limbs shorter, nor in crawling, like those which have them longer, and especially the sloths, to which they otherwise are so closely similar. The shoulder-blade has generally the same proportions as those of the sloths.— It has a clavicle, as in one of them, (the two-fingered or *unau*) which, together with the length of the phalanges supporting the nails, proves that this animal also employed its fore feet to seize and even to climb with. The presence of clavicles separates our giant sloth from all the animals which might be confounded with it on account of their size, as the elephant, rhinoceros, and all the large ruminants, none of which have these bones.

"The arm of the megatherium is very remarkable for the breadth of its inferior part, which is owing to the great surface of the spines placed above its condyles. Hence, the muscles which originate there, and serve, as is known, to move the hand and fingers, must have been very considerable; this is another proof of the great use made by our animal of its inferior extremities. This great breadth of the lower part of the humerus is peculiarly found in the ant-eater, which is known to employ its powerful claws to suspend itself from trees, or to tear open the solid nests of the termites. It is in the ant-eater three-fifths of its length—while in our animal the breadth is one-half; which is also the proportion in the long-tailed scaly ant-eater, or *phatagin*. In the rhinoceros this breadth is only a third, and in the elephant a fourth, of the length. Ruminant animals, which scarcely make any use of their toes, have hardly any thing of these spines.

" The length of the olecranon (point of the elbow) must have given to the extensor muscles of the forearm, an advantage which they have not in the sloths, whose olecranon is extremely short, which contributes not a little to the imperfection of their movements. The radius turns freely upon the ulna; but it should be remarked that this bone has been inverted in the skeleton, and the figures published represent it in this erroneous manner. The shortness of the metacarpus shows that the palm was entirely placed on the ground in walking. The digits, which were apparent and armed with nails, were three in number, and the two others concealed under the skin, as there are two in the ai, three in the *unau* and two fingered ant-eater.

" The last phalanges were composed of an axis which carried the claw, and of a sheath which enclosed its base absolutely, as in the great clawed animals compared with this. But the bones of the metacarpus were not solidified together, as they are in the *ai*. The proportion of these bones, as well as those of the *megalonyx*, (*Jefferson's giant sloth*) are very different from those of the sloths, being the same as in the ant-eaters.

" The pelvic bones are very different in our animal from those of the kindred species. The haunch bones are the only ones preserved in the Madrid skeleton; they form a half pelvis, broad and hollowed out, the mid-plane of which is perpendicular to the spine, resembling somewhat that of the elephant, and especially of the rhinoceros. The broad part of these bones have a peculiarly striking analogy with that of the latter animal, by the proportion of its three lines; but their narrow part, and near the cotyloid cavity, is much shorter. This form of pelvis indicates that the megatherium had a large belly, and accords, with the form of the teeth, to indicate that its subsistence was vegetable matter.

" The pubis and ischium are wanting in the Madrid skeleton, but, in my opinion, these were lost at the time of the exhumation. However, if this defect be natural to the species, it is still in an *edentous* animal (the two-toed ant-eater) that we find the first, though a slight indication of it. The ossa pubis and ischium of this ant-eater do not unite in front, and remain always separate.

" The tibia and fibula are united by bony matter at their two extremities, a circumstance absolutely peculiar to this animal: they present also by their union a disproportionately broad surface. In this respect the leg of the megatherium resembles considerably that of the *ai*, which is very broad, because its two bones each form a convexity on their sides, thus separating from each other. The figures lead to the belief that the articulation of the leg and foot is not so singular as it is in the *ai*, and that it is much more solid.

" The megatherium having a broad astragalus, articulated with a tibia equally so, and strengthened farther by the lateral position of the fibula, stood more solidly than the sloths, and in this respect must have resembled most other quadrupeds.

"We find but a single toe on the hind feet of the Madrid skeleton, which was armed with claws; but in this respect I think there is less certainty than relative to the fore feet; especially as the figures represent but two other toes, which have no claws; and my researches have uniformly established as a rule without exception, that all unguiculated animals have five digits, whether externally visible, concealed beneath the skin, or reduced to simple osseous rudiments.

"The tail is wanting in the Madrid skeleton, and the smallness of the posterior face of the body of the sacrum, leads to the conclusion, that it was very short in this animal.

"The comparison of the bones of the megatherium and megalonyx, (Jefferson's giant sloth) results in establishing almost the absolute identity of forms, at least in the parts yet discovered of the latter; but the size is different. The bones of the megatherium are a third larger than those of the megalonyx, and as the latter bear all the characters of the adult age, we can only attribute the difference of size to difference of species: we may add that the claw-sheaths are longer and more complete in the last phalanges of the megatherium. These two animals then should form two species of the same genus, belonging to the *Edentous* family, being intermediate to the sloths and ant-eaters, though nearer the former than the latter."*

* Oss. Foss. tom. iv.

After this long extract from Cuvier, we deem it most advantageous to the reader to present the account of the fragments of the North American specimen described in the Annals of the Lyceum of Natural History of New York, in a paper entititled " On the Remains of the Megatherium recently discovered in Georgia, by William Cooper." In giving this paper nearly entire,* we feel satisfied that its zealous and scientific author will lose nothing by having his researches on this subject immediately contrasted with those of the illustrious zoologist above quoted.

" It has been already announced that remains of the great fossil animal of Paraguay exist within the limits of the United States, and under a latitude nearly as far north, as they have hitherto been found south of the Equator. We are indebted for the first intelligence of this discovery, which possesses so much interest for the lovers of natural science, to our learned associate, professor Mitchill, distinguished by his previous contributions to the knowledge of the fossil productions of this country. In a paper contained in the present volume of these Annals, that gentleman has given an account of two fragments of teeth brought to him from an island on the sea-coast of Georgia, which, at the same time that they differed totally from those of any quadruped now known to exist, presented the most striking resemblance to those of the *Megatherium*. To an animal

---

* We have already quoted, in the dental system of this species, a part of this paper. The comparison with Bru's description, &c. not being necessary at present, is also omitted.

of this very extraordinary, and now extinct, species, he accordingly does not hesitate to refer them.

"The information thus given, however, was calculated rather to stimulate than to satisfy the curiosity of naturalists. Although the fact of these remains existing in North America might perhaps be considered as thereby established, yet its connexion with the most difficult problems in zoology and geology rendered it highly desirable to obtain other and more entire parts of the skeleton, and with them to institute a more extensive comparison. By means of this we might expect to discover any difference possibly existing between them, or else to determine, in the most unquestionable manner, the specific identity of the animal of Georgia with that of Paraguay.

"These considerations induced me to address a letter to my friend, *Dr. Wm. R. Waring*, of Savannah, begging him to make inquiry whether any more of these relics had been found, and, if possible, to procure me some of them. His answer informed me that his friend, *Dr. Joseph C. Habersham*, of the same place, had, with much trouble, and at some expense, assembled a collection of the bones found in the marshes of Skidaway Island, and at his request consented to allow them to be sent to this city, under the condition that they should be placed where they might be publicly viewed. They were transmitted to me in the month of March last, and in compliance with the wishes of the owner, are now deposited in the cabinet of the Lyceum.

"The collection was found to consist of parts of several members of the skeleton, which, as nearly as their very mutilated and disconnected state would enable me to determine, were as follows:

"A portion of the posterior part of the *right* side of the lower jaw.

"Another portion which had been continuous with the preceding.

"A considerable portion of the anterior part of the same jaw.

"A fragment of the *left* side of the same jaw, about three inches square.

"Five fragments belonging to three different teeth.

"The vertebra dentata, with nearly one half broken off.

"Three other vertebræ, two of which appear to be dorsal, and the third either the last dorsal or the first lumbar. None of these are entire.

"A fragment undetermined, but supposed to be of the ilium.

"Eight pieces belonging to three or four different ribs. Three of these pieces have the heads attached to them, and two seem to have belonged to the left side, and the remainder to the right.

"The head of the lower extremity of the humerus, with both condyles nearly entire.

"Two pieces with a concavity at one end, perhaps the superior parts of a radius and ulna.

"A bone supposed to be tarsal, much broken.

"Two carpal bones adapted to each other.

"The heads of both femora; and a fragment, apparently the lower condyle of a femur.

"Part of a bone about seven inches long, supposed to be part of a fibula.

"Besides these were four or five other small pieces of bone, but so imperfect as not to be easily referred to their proper places in the skeleton.

"In addition to the foregoing should be enumerated the two fragments of teeth from which professor Mitchill drew up his description. On being compared with Dr. Habersham's collection, one of them was found to correspond with a fragment supposed to be of a *fourth* molar, of which it formed the posterior process. The other, as it fitted with great exactness into what remained of the socket of the *third* molar, appeared to have occupied that place in the jaw. Thus it is rendered extremely probable that all the relics of the *Megatherium* yet discovered, as far as we know, in North America, have belonged to a single individual.

"I shall first endeavour to bring together some of these fragments so as to show what has been their original state: after which they may be compared with the figure and description of the animal of Paraguay, as given by M. Cuvier in the *Annales du Museum*, vol. v., and in the *Recherches sur les Ossemens Fossiles*, vol. iv. first edition.

"*Restoration of part of the lower jaw.*—A and B (see plate) formed one continuous piece. Of this there can be no doubt, as the edges of the fracture, though very irregular, correspond perfectly with each other. These two portions compose the greater part of the *right* side of the lower jaw, and contain parts of the sockets of all the four molar teeth.

"The plate represents two views of the jaw as partly restored, reduced to one-fifth their natural size. Fig. 1, is an oblique view of the inside of the jaw. Fig. 2, a profile of the outside. The dotted line represents the part supposed to be broken off.

"C also belongs to the lower jaw. It consists of the anterior part, comprising the symphysis, with part of the elongation, and parts of the sockets of the two first molars. It has been continuous with B.

"D (not in the plate) is a fragment of the *left* side of the same jaw. This is evident from its containing parts of the sockets of the two last molars, part of the opening for the passage of the maxillary vessels, and the origin of the ascending branch of the jaw.

"The teeth had fallen out of all the sockets except one, which contained the body of the second molar with the crown and fangs broken off, apparently by recent violence. I attempted, therefore, to find the places of the four remaning teeth. Two of them I perceived to be alike in all respects, and therefore concluded that they had occupied corresponding situations in opposite sides of the jaw. Both are broken in two across, and consist of the crown and part of the body, as far down as below the commencement of the internal pyramidal cavity. The longer of the two is about four inches, the other somewhat less. On trying the first of these, it was found to fit with great exactness into the socket, of which part remains in B, and part in C, that is, the socket of the *first* molar. This, it may be observed, corroborates the approximation of these two fragments. Its form also showed this to have been its place; its diameter in the direction of its cutting edges being less than the contrary diameter, and its being narrowed anteriorly, proved its situation to have been in the thinner and more tapering part of the maxillary bone.

"The *second* molar of the same (that is, the *right*) side, remained in its socket as already mentioned. It is remarkable for its rhomboidal form, the diagonal through its left anterior internal, and right posterior external angle, being the greatest.

"The remaining two teeth appeared to belong to the *left* side of the same jaw. One of them I conjectured to be the *third;* 1st, from its fitting into a part of this socket remaining in D; and 2dly, from its form, which shows the passage between the rhomboidal figure of the second molar, and the flattened shape of that which I suppose to be the *fourth.* This last is more flattened, that is, broader in the direction of its cutting edges than any of the others; and from this, as well as from its agreeing with the form of the fourth socket, partly remaining in D, I have referred it hither. This tooth may, however, have belonged to the upper jaw.

"The fragments of teeth in Dr. Habersham's collection, for there is not one entire, agree with Bru's description of those in the skeleton of Madrid, so far, at least, as it is given in the French abridgment. There are the sockets of four in the right lower side, and consequently eight teeth in all, in the lower jaw, the six posterior being the greatest. They are square, with rounded angles, and a groove between on the inner and outer sides, and are longitudinally striated. The inferior pyramidal cavity may be observed with advantage in the right second molar, which remains in the socket: but the terminating points are broken off from this as well as from all the others. Consequently, we are not enabled to ascertain their precise length, but it appears to have been at least seven inches, and probably more.

"The heaviest of our teeth, which is the first of the right side, weighs nine and a-quarter ounces. The fourth of the left side weighs nine ounces. To make them agree with the weights of the corresponding teeth, as stated by Bru, we must suppose that more than half has been broken off the former, and from the latter nearly two-thirds. This, from a comparison with the sockets, I should hardly suppose to be the case, at least with the latter.

"The peculiar form of the crown of these teeth is not well represented in any figure I have seen, excepting that given by Professor Mitchill, to which the reader is referred. Their posterior crest is higher than the anterior. The posterior crest is known by the curvature of the tooth corresponding with that of the socket. This peculiarity does not appear in the figure in the 'Ossemens Fossiles,' but rather the contrary.

"Fig. 3 represents a transverse section, natural size, of one of the first molars, showing the arrangement and relative thickness of the coats.

"Fig. 4, a longitudinal section of part of a larger tooth, showing the manner in which the interior enamel terminates the cutting process.

"Of the four vertebræ, three have little remaining besides the body, the processes being almost all entirely broken off. The other, which appears to be one of the dorsal, perhaps the third, is tolerably entire. It agrees with Bru's description of those of the *megatherium*, excepting that I am not able to find the two holes which he describes in the atlas, and which, he says, are common to all the other vertebræ. As, however, this bone is much incrusted with various shells, they may possibly be covered or filled up.

"The ribs, also, are too much injured to afford any very distinctive characters. Neither can I observe any thing peculiar in the condyles of the humerus, as we have supposed them to be, for nothing more of this bone remains besides the inferior articulating extremity. The remarkable enlargement described in the Madrid skeleton is entirely wanting. The two fragments, conjectured to be the superior extremities of the radius and ulna, are in the same state, and present nothing but smooth and even concavities, with their edges partly broken. That supposed to be the radius exhibits on one side a smooth facet, where it may have played upon the ulna.— One of these pieces is six, the other four inches long; the diameter of their cavities about four inches.

"Of the two supposed to be carpal bones, the first, which is of a triangular figure, is the smaller. One side is convex and the other concave, with a slight elevation crossing it about the middle, which adapts itself to a corresponding depression in the other bone. It measures nearly five inches in length, and nearly three and a-half in breadth, and is about an inch thick. The second is of a singular figure: one side is convex, as in the first; the other side has one half concave, while the other half swells out into a hemispherical knob. Its outline is quadrangular, and it is a little longer and broader than the first, with its concave end about as thick, and the other nearly three times that thickness, measuring through the knob.

"The heads of the two femora are both nearly entire, and would perhaps be sufficient of themselves to prove the identity of our animal with the South American species. They are, as observed by M. Bru, 'perfectly spherical, and with a superficies very smooth,' and measure full twenty-three inches in circumference. The dimensions of the skeleton of Madrid are not given in detail in the French abridged description. Even if we had not the evidence afforded by the teeth, these huge condyles would indicate an animal of much superior bulk to the *megalonyx;* for we can hardly imagine that a creature not larger than an ox, which is conjectured to have been the size of this quadruped of Virginia, could be furnished with thigh-bones of such disproportionate bigness. Indeed, they would seem calculated to encumber rather than support even the *megatherium*, whose size is supposed by M. Cuvier to have equalled that of the *rhinoceros*.

"The other fragments being small and much broken, nothing satisfactory could be determined with respect to them.

"My inquiries have not, as yet, enabled me to give any very precise information respecting the locality of these bones, or the character of the formation in which they were found. Their appearance, however, indicates that they have been overflowed by the sea; and they seem to have had one side imbedded in the earth or mud, while the other was washed by the salt water. They are thinly incrusted in some places with *Flustræ* and other zoophytes, and have recent shells of the genera *Balanus, Ostrea* and others, adhering to them. All are remarkably hard and heavy, and of a deep black colour. They do not retain any part of their animal matter.

" Drs. Waring and Habersham state that these bones are still to be procured in *great quantity*, by some labour and expense, at the same place. They add, that bones of the same kind may be obtained at two other places, one called Whitebluff, said to be also on the seacoast; the other is at some distance up the Savannah river. We may hope, through the zeal and exertions of the same gentleman, to whom the scientific public generally is so much indebted for the preservation of the remains which have formed the subject of these remarks, to have these interesting deposites further explored; and in a manner worthy of the great questions, which a proper examination of their contents would contribute so much to elucidate."

Species II.—*Jefferson's Giant Sloth.*

Megatherium Jeffersonii.

Megalonyx: Jefferson, Transact. of the Am. Philos. Society, iv. 246.
Megalonyx: C. Annals du useum, v. 358, pl. 23; Recherches sur les Ossemens Fossiles, iv.
Megatherium Jeffersonii: Desm. Mammal. 366, Sp. 580.

To the author of the Declaration of American Independence the scientific world is indebted for the first account of the extraordinary and interesting re-

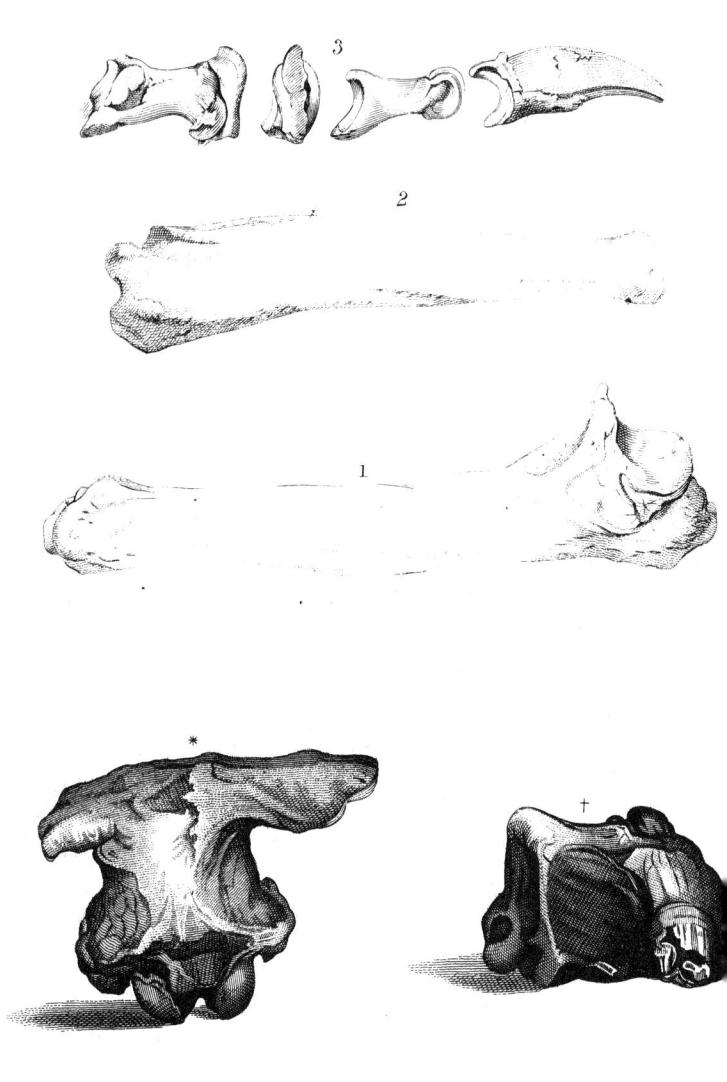

Fig. 1.2.3 Bones of Jeffersons Giant Sloth
Fossil Deer. †Fossil Deer. (See Supplement D. Vol. 3.)

lies, which indubitably establish the fact, that at some very early period this country contained a second species of quadruped of gigantic size, resembling the sloths in structure and manners. The only fragments yet obtained of the skeleton of this extinct species were discovered in a saltpetre-cave, belonging to Mr. Frederic Cromer, in Green Briar county, Va. where they were found about three feet below the surface of the cave's floor. "The importance of the discovery (says the distinguished author first above cited) was not known to those who made it, yet it excited conversation in the neighbourhood, and led persons of vague curiosity to seek and take away the bones. It was fortunate for science that one of its zealous and well informed friends, Col. JOHN STEWART, of that neighbourhood, heard of the discovery, and, sensible from the description that they were not of any animal known, took measures without delay for saving those which still remained. He was kind enough to inform me of the incident, and to forward me the bones from time to time as they were recovered. To these I was enabled accidentally to add some others, by the kindness of Mr. Hopkins, of New York, who had visited the cave."

The bones thus obtained consisted of a fragment of an arm or thigh-bone, a complete radius, and an ulna, which was broken in two, but not otherwise injured; three of the phalanges on which the claws were sustained, and several bones belonging to the fore or hind feet.

In the absence of every opportunity for making a proper comparison of these bones, we are not surprised that JEFFERSON should, in the first instance, have compared them with the skeleton of the lion, as described by DAUBENTON; or that he should come to the conclusion that this unknown species was "*more* than three times as large as the lion; that he stood pre-eminently at the head of the column of clawed animals, as the mammoth stood at that of the elephant, rhinoceros and hippopotamus; and that he may have been as formidable an antagonist to the mammoth as the lion to the elephant." In a postscript to the same memoir, the author makes some observations on a very imperfect account of the *megatherium*, which prove that nothing but the want of proper materials for comparison prevented him from referring his *megalonyx*, or great claw, to its proper place.*

The late professor WISTAR, so justly distinguished for his zeal in the cause of science, drew correct, though not altogether positive conclusions in relation to these bones. After giving a detailed description

---

* "P. S. March 10, 1797. After the preceding communication was ready to be delivered in to the society, in a periodical publication from London, (Monthly Magazine, Sept. 1796) I met with an account and drawing of the skeleton of an animal dug up near the river La Plata, in Paraguay, and now mounted in the cabinet of Natural History of Madrid. The figure is not so done as to be relied on, and the account is only an abstract from that of Cuvier and Roumé. This skeleton is also of the clawed kind, and having only four teeth on each side, above and below, all grinders is, on this account, classed in the family of the unguiculated quadrupeds destitute of cutting teeth, and receives the new denomination of *megatherium;* having nothing

of them, he makes the following observations:— "from the shortness of the metacarpal bone, and the form and arrangement of the other bones of the paw, and also from the form of the solitary metatarsal bone, it seems probable that the animal did not walk on the toes; *it is also evident that the last phalanx was not* retracted. The particular form of the second bone, and its connexion with the first and third, must have produced a peculiar species of flexion in the toes, which, combined with the greater flexion of the last phalanx upon the second, must

---

of our animal but the leg and foot-bones, we have few points of comparison between them. They resemble in their stature, that being twelve feet nine inches long, and six feet four and a-half inches high, and ours by computation, five feet 1.75 inches high: they are alike in the colossal thickness of the thigh and leg-bones also. They resemble, too, in having claws: but those of the figure appear very small, and the verbal description does not satisfy us, whether the claw-bone, or only its horny cover, be large. They agree too in the circumstance of the two bones of the forearm being distinct and moveable on each other; which, however, is believed to be so usual as to form no mark of distinction. They differ in the following circumstances, if our relations are to be trusted:—The megatherium is not of the cat-form, as are the lion, tiger and panther, but is said to have striking relations in all parts of its body with the bradypus, darypus, pangolin, &c. According to analogy, then, it had not the phosphoric eye nor leonine roar. But to solve satisfactorily the question of identity, the discovery of fore teeth, or a jawbone, showing it [the megalonyx, or Jefferson's animal, both jaws of the megatherium having been figured] had, or had not such teeth, must be waited for and hoped with patience. It may be better in the mean time to keep up the difference of name."—*Phil. Soc. Trans.* p. 259.

have enabled the animal to turn the claws under the soles of the feet; from this view of the subject there seems to be some analogy between the foot of this animal and that of the Bradypus [Sloth]—having no specimens of that animal, I derive this conclusion from the description of its feet given by M. Daubenton."*

Cuvier was the first to establish, from sufficient data. the true place and character of this animal; from all his comparisons and investigations he lays down the following positions:

" 1st, That the animals which furnish these fossil bones were not carnivorous;—2d, that they had, in large, all the forms and all the details of organization that the sloths exhibit in small, and that the details of these organizations must have been similar;—3d, that if they are separated from them in some unimportant particulars, it is only in approaching the nearest allied genus, that of the ant-eaters;—4th, that the approximation of these fossil animals to the sloths, and their classification in the *Edentous* family, in general, are not arbitrary, nor founded on artificial characters, but that they are the necessary result of the intimate identity in the nature of both."

The great size of this animal precludes the idea of its living upon trees, exactly in the manner of living sloths, but every thing discovered of its structure forbids us from thinking that its mode of life was widely different. A sloth of the size of an ox, would find few trees whose branches would be capable of

---

* Am. Phil. Trans. vol. iv. 530.

sustaining so great a weight; but in not climbing it would not differ more from the sloth than species of other genera do from each other.

We subjoin the measurements of these bones, and deem it unnecessary to describe them individually with minuteness. The figures given in the plate will convey a better idea of them than we possibly could by words.

	Inches.
Length of the ulna,	20.1
Breadth to tip of its coronoid process,	9.55
———— in the middle of the bone,	3.8
Length of the radius,	17.75
———— at its head,	2.65
Breadth near the carpal extremity,	4.5*
Length of the metacarpal bone, a	3.5
———— of the first phalanx, b	1.25
———— of the second, c	2.25
———— of the third, d	7.

---

* Jefferson, in Philos. Trans. ut supra.

CHAPTER XV.

Order V.—Belluæ; L. *Dense Skinned Animals;* C.

The animals pertaining to this order cannot flex their digits, nor lay hold of objects, their feet being exclusively destined to support their weight: they are, therefore, not provided with clavicles; (collar-bones.) The fore arm always remains in a state of pronation; (with the palm against the earth.) They feed on vegetable matters, and do not ruminate; the stomach is membranous and simple, or merely divided by membranous bands.*

Family I.—Proboscidia; *Having a Trunk and Tusks.*

In the skeleton all the feet are distinctly five-toed, but in the living animal these are entirely concealed by thick and callous integument, which shows no external mark of their existence, except by the nails, which border this sort of hoof. There are no true incisive nor canine teeth in the upper jaw, but

---

* The term used by Linné as the name of this order, is applicable to every wild beast of great size, strength, &c. Cuvier calls the order *Pachydermata*, the translation of which we use as a trivial name for the order, instead of a

two great tusks, growing from the intermaxillary bones, project externally and increase to a vast size. The magnitude of the sockets required for these tusks renders the upper jaw so high, and shortens the bones of the nose so much, that in the skeleton the nostrils are found near the upper part of the face; but in the living animal they are prolonged into a cylindrical trunk, composed of thousands of small muscles, variously interlaced, moveable in every direction, endowed with an exquisite sensibility, and terminating by an appendix somewhat in form of a finger. The skull has large vacant spaces between its plates, by which a greater extent is given for the origin of muscles, without unnecessary increase of weight to the head. The lower jaw has no incisive teeth; the intestines are very large; the cœcum enormous; the teats, two in number, are placed upon the chest. The young of the elephant, the only living animal of this family, sucks with the mouth and not with the trunk.*

---

better, which cannot well be obtained, however much it is to be desired. Other animals, meriting either of the names used for the order, may be found, yet by no means corresponding with the definition above given. The *name*, therefore, in this case, as in numerous others, must be associated with the characters of the order, as laid down, without reference to its own etymological signification.

* Cuvier Règne Animal, i. 228.

Genus II.—Mastodon; *Mastodon;* C.

GENERIC CHARACTERS.

The form of the superior part of the head still remains unknown, the whole mass above the level of the zygomatic process being destroyed. The intermaxillary bones are long, and have at their extremities the openings of very large sockets for the tusks, which are very large and long. The lower jaw, ending in a point at the symphisis, is hollowed into a sort of canal; the neck is very short, the limbs long and five toed, and the ribs nineteen in number. The tail was moderately long.

Dental System.

10 Teeth:
- 6 Upper { 2 Incisive, (in form of tusks) 4 Molar.
- 4 Lower { 4 Molar.

The incisive teeth, very analogous to those of the elephant, are formed of ivory, which, when transversely cut, exhibits curvilinear lozenges, produced by the intersection of lines of a harder bony substance. The molars have rectangular crowns, somewhat straighter in the hind than in the fore teeth. They are composed of but two substances, the external being a thick enamel, and the internal bony matter, without cement or cortical substance, each tooth weighing about twelve pounds.

The crowns of these teeth are divided by very open trenches into transverse eminences, and each eminence is itself divided into two great, obtuse, ir-

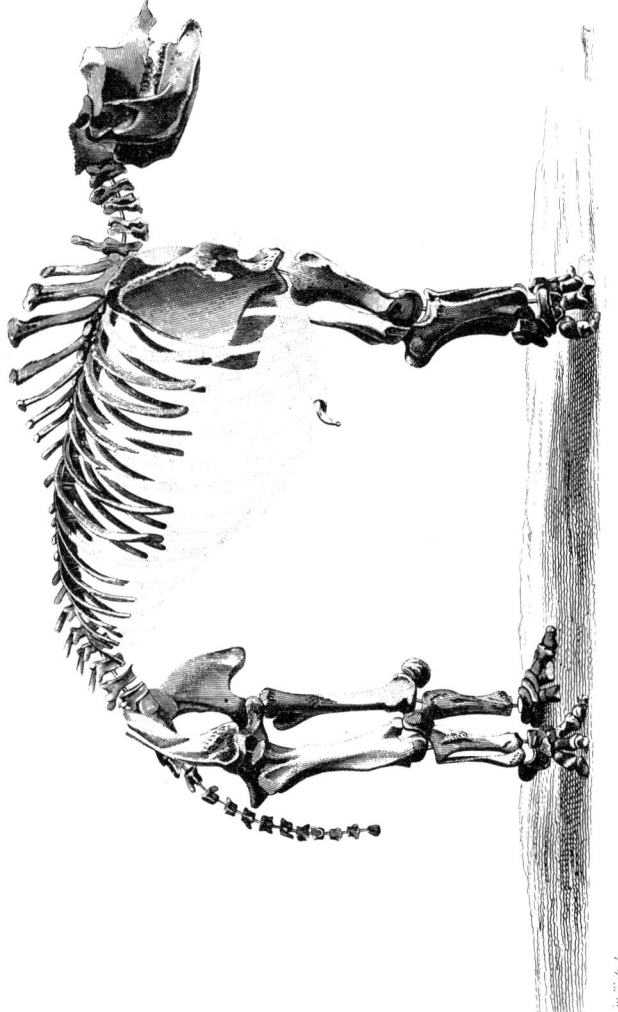

Great Mastodon

regularly formed points, constituting slightly rounded quadrangular pyramids. When the crown is not worn it is studded with knobs, or points disposed in pairs, from six to ten in number.

Species I.—*The Gigantic Mastodon.*

Mastodon Giganteum; C.

Animal Incognitum: Remb. Peale, Account of the Skeleton of the Mammoth, 4to. 2d. ed. Lond. 1806.
Mastodon Giganteum: C. An. du Mus. Recherches Sur less Oss. Foss. ed 2, i. p. 206.
Mastodon Giganteum: Account of the Discovery of the Skeleton of the Mastodon Giganteum Extracted from the Report made to the Lyceum of Natural History, by Messrs. Dekay, Van Rensselaer, and Cooper. Annals Lyceum of Nat. Hist. of N. York, v. i. p. 143.

[*Improperly called Mammoth.*]

In various parts of North America single bones of extraordinary size had been occasionally disinterred, without exciting more than temporary curiosity, or leading to any thing better than wild and unsatisfactory speculation. Some persons regarded them as the relics of a gigantic race of men, of whose existence no other traces remained; others, who appeared willing to surpass all absurdity, suggested that they might have belonged to the *angels* who were expelled their celestial habitations; while a third, and more rational party, concluded that they were the bones of an animal still in existence, or belonged to a larger variety of the well known elephant species.— The inquiry generally ceased when the novelty of their discovery passed away; those by whom they were found were in pursuit of other objects, and very frequently neglected to preserve the fragments already obtained. But when situations were explored where they were procured in greater abundance, and the curiosity of European naturalists was awakened, these relics were eagerly sought for, until nearly a whole skeleton was obtained, the fact satisfactorily established, that these bones belonged to a peculiar race never before known, and, what was still more surprising, that the whole race was utterly extinct.

We find, as early as the year 1712, a letter from Dr. Mather to Dr. Woodward, published in the Philosophical Transactions, announcing that some bones and teeth of a monstrous size had been discovered at Albany, in New York.

In the year 1739, some savages belonging to the company of a French officer named LONGUEIL, who was descending the Ohio to the Mississipi, found, at a short distance from the river, at the edge of a marsh, some bones, grinders, and tusks, belonging to this unknown animal. The year after LONGUEIL took to Paris a thigh-bone, the extremity of a tusk, and three grinders, which are still preserved there. Since that time these bones have been discovered in many places; though, in consequence of the notice first attracted by the specimens found on the Ohio river, the name of Animal of the Ohio had been bestowed on this creature, yet this name, and that of *Mammoth*, have at length been entirely superseded by that proposed by CUVIER.

About the year 1740, vast numbers of these bones, which had been washed up by the current of the Ohio, or were purposely digged for, were found in Kentucky. The eagerness to procure them, and the

THE GIGANTIC MASTODON.

haste with which they were sent to Europe, retarded the knowledge of the true character of the animal—as it became impossible to procure or recognize the bones belonging to different skeletons, or to determine their exact numbers and proportions. Over France, England and Germany, they were in this manner scattered in confusion; and we need not be surprised that naturalists were long in forming just ideas of the character of the animal, or indulged so much the disposition to maintain theories established on such slight foundations.

The force of prejudice may be clearly seen in the perseverance with which BUFFON, and some other scientific men, maintained that these bones belonged to a variety of the elephant race; for if he admitted that they did not belong to that kind, he must have acknowledged that they were the bones of an extinct genus, which was an idea not then proposed, but has since most amply been proved true, and a vast number of extinct species discovered.

In consequence of some large bones having been previously found in Siberia, that were really *elephantine*, the idea readily became prevalent that the great bones of the Ohio and other parts of North America were similar. Hence the name *mammoth* (said to be a corruption of the Hebrew word Behemoth) was applied to the American animal, and continued to be generally used, until the extreme difference of its structure induced naturalists to consider it properly, raise it to the rank of a distinct genus, and bestow on it a name expressive of its most striking characteristic, the form of the teeth.

It was not until the year 1801, a period of eighty-nine years from the first discovery of the bones at Albany, that any hopes were entertained of finding an entire skeleton of this wonderful and interesting animal.

In the year 1824 a considerable part of a skeleton was raised in New Jersey by some scientific gentlemen of New York; but they have not discovered any thing more than was previously made known by the exertions of Messrs. PEALE; the head, which is the only important part wanting, was too much decomposed to enable them to form any idea of its figure.

The emotions experienced, when for the first time we behold the giant relics of this great animal, are those of unmingled awe. We cannot avoid reflecting on the time when this huge frame was clothed with its peculiar integuments, and moved by appropriate muscles; when the mighty heart dashed forth its torrents of blood through vessels of enormous caliber, and the mastodon strode along in supreme dominion over every other tenant of the wilderness. However we examine what is left to us, we cannot help feeling that this animal must have been endowed with a strength exceeding that of other quadrupeds, as much as it exceeded them in size; and, looking at its ponderous jaws, armed with teeth peculiarly formed for the most effectual crushing of the firmest substances, we are assured that its life could only be supported by the destruction of vast quantities of food.

Enormous as were these creatures during life, and endowed with faculties proportioned to the bulk of their frames, the whole race has been extinct for ages. No tradition nor human record of their existence has been saved, and but for the accidental preservation of a comparatively few bones, we should never have dreamed that a creature of such vast size and strength once existed,—nor could we have believed that such a race had been extinguished forever. Such, however, is the fact—ages after ages have rolled away—empires and nations have arisen, flourished, and sunk into irretrievable oblivion, while the bones of the mastodon, which perished long before the periods of their origin, have been discovered, scarcely changed in colour, and exhibiting all the marks of perfection and durability.

That a race of animals so large, and consisting of so many species, should become entirely and universally extinct, is a circumstance of high interest;—for it is not with the mastodon as with the elephant, which still continues to be a living genus, although many of its species have become extinct:—the entire race of the mastodon has been utterly destroyed, leaving nothing but the " mighty wreck" of their skeletons, to testify that they once were among the living occupants of this land. Into the probable causes of this extinction we shall hereafter make a fuller inquiry.

The situations whence these bones have been most commonly obtained, appear to have greatly contributed to their preservation. They have generally been dug from beneath a considerable mass of mud, or marle, where they have long soaked in fluids charged with saline and other impregnations. Thus they have been equally protected against the effects of detrition and vicissitudes of weather, and most of the bones found are in every respect perfect, with the exception of an unimportant change in colour. This circumstance is almost universally observed of the bones contained in the different cabinets of this country; when scraped or cut they exhale an odour remarkably similar to that produced by the same treatment of a recent bone.

There are several circumstances leading us almost to despair of ever procuring the upper part of the skull, which, on account of its comparative thinness and weakness, as well as the fact of its being always found much nearer the surface, must be among the first parts to decay, and be irrecoverably lost. No specimen has yet been obtained more perfect than the one in the Philadelphia Museum, and this has no part of the skull above the level of the zygomatic arches. In this, as in all the individuals discovered, the top of the head was so far decayed and destroyed as to prevent the least idea being formed as to its figure or elevation.

Enough of the head has fortunately been preserved to make us fully acquainted with the dentition of this great animal, and enable us to decide on the general nature of its food and habits of living. Without the aid derived from this source we should still be in doubt, and have nothing to guide us to a satisfactory conclusion, although the analogy in size and general configuration might have served to produce the inference, that the animal was, in other respects, most nearly allied to the elephant, rhinoceros, or hippopotamus.

The circumstances attending the exhumation of the most perfect skeleton ever obtained of this great animal, are deeply interesting to every votary of natural science; and the author believes that he cannot more effectually minister to the gratification of his readers, than by introducing in this place the account written by his father-in-law, an eye-witness and enthusiatic co-operator in that enterprise, which has secured to the scientific world one of its most interesting and instructive possessions. In addition to the authenticity of this record, (prepared almost on the spot, by so competent a hand) it is drawn up with a raciness and vigour which imparts to the reader's mind an excitement, not to be awakened by any cause, short of truth, breathed forth with the vivid energy of enthusiasm.

Narrative of the discovery and exhumation of the skeleton of the Mastodon; by REMBRANDT PEALE.

In the spring of 1801, receiving information from a scientific correspondent in the state of New York, that in the autumn of 1799 many bones of the MAMMOTH had been found in digging a marle-pit in the vicinity of Newburgh, which is situated on the river Hudson, sixty-seven miles from the city of New York, my father, Charles Wilson Peale, immediately proceeded to the spot, and through the politeness of Dr. Graham, whose residence on the banks of the Wall-kill enabled him to be present when most of the bones were dug up, received every information with respect to what had been done, and the most probable means of future success. The bones that had been found were then in the possession of the farmer who discovered them, heaped on the floor of his garret or granary, where they were occasionally visited by the curious. These my father was fortunate to make a purchase of,* together with the right of digging for the remainder, and, immediately packing them up, sent them on to Philadelphia. But as the farmer's fields were then in grain, the enterprise of further investigation was postponed for a short time.

The whole of this part of the country abounding with morasses, solid enough for cattle to walk over, containing peat, or turf and shell-marle, it is the custom of the farmers to assist each other, in order to obtain a quantity of the marle for manure. Pits are dug generally twelve feet long and five feet wide at the top, lessening to three feet at the bottom.— The peat or turf is thrown on lands not immediately in use; and the marle, after mellowing through the winter, is in the spring scattered over the cultivated fields—the most luxuriant crops are the consequence. It was in digging one of these, on the farm of John Masten, that one of the men, thrusting his spade deeper than usual, struck what he supposed to be a

---

* They consisted of all the neck, most of the vertebræ of the back, and some of the tail; most of the ribs, in greater part broken; both scapulæ; both humeri, with the radii and ulnæ; one femur; a tibia of one leg, and a fibula of the other; some large fragments of the head; many of the fore and hind feet bones; the pelvis, somewhat broken; and a large fragment, five feet long, of one tusk, about mid-way. He therefore was in want of some of the back and tail bones, some of the ribs, the under jaw, one whole tusk and part of the other, the breast bone, one thigh, and a tibia and fibula, and many of the feet bones.

THE GIGANTIC MASTODON. 213

log of wood, but on cutting it to ascertain the kind, to his astonishment, he found it was a bone: it was quickly cleared from the surrounding earth, and proved to be that of the thigh, three feet nine inches in length, and eighteen inches in circumference, in the smallest part. The search was continued, and the same evening several other bones were discovered. The fame of it soon spread through the neighbourhood, and excited a general interest in the pursuit: all were eager, at the expense of some exertions, to gratify their curiosity in seeing the ruins of an animal so gigantic, of whose bones very few among them had ever heard, and over which they had so often unconsciously trod. For the two succeeding days upwards of an hundred men were actively engaged, encouraged by several gentleman, chiefly physicians, of the neighbourhood, and success the most sanguine attended their labours: but, unfortunately, the habits of the men requiring the use of spirits, it was afforded them in too great profusion, and they quickly became so impatient and unruly, that they had nearly destroyed the skeleton; and, in one or two instances, using oxen and chains to drag them from the clay and marle, the head, hips, and tusks were much broken; some parts being drawn out, and others left behind. So great a quantity of water, from copious springs, bursting from the bottom, rose upon the men, that it required several score of hands to lade it out, with all the milk-pails, buckets and bowls, they could collect in the neighbourhood. All their ingenuity was exerted to conquer difficulties that every hour increased upon their hands; they even made and sunk a large coffer-dam, and within it found many valuable small bones. The fourth day so much water had risen in the pit, that they had not courage to attack it again. In this state we found it in 1801.

It was a curious circumstance attending the purchase of these bones, that the sum which was paid for them was little more than one-third of what had been offered to the farmer for them by another, and refused, not long before. This anecdote may not be uninteresting to the moralist, and I shall explain it. The farmer, of German extraction—and like many others in America, speaking the language of his fathers better than that of his country—was born on his farm; he was brought up to it as a business, and it continued to be his pleasure in old age; not because it was likely to free him from labour, but because profit, and the prospect of profit, cheered him in it, until the end was forgotten in the means.— Intent upon manuring his lands to increase its production, (always laudable) he felt no interest in the fossil-shells contained in his morass; and had it not been for the men who dug with him, and those whose casual attention was arrested, or who were drawn by report to the spot, for him the bones might have rotted in the hole in which he discovered them; this he confessed to me would have been his conduct, certain that after the surprise of the moment they were good for nothing but to rot as manure. But the learned physician, the reverend divine, to whom he had been accustomed to look upwards, gave importance to the objects which excited the vulgar stare of his more inquisitive neighbours: he therefore joined his exertions to theirs, to recover as many of the bones as possible. With him, hope was every thing:

with the men, curiosity did much, but rum did more, and some little was owing to certain prospects which they had of sharing in the future possible profit. It is possible he might have encouraged this idea; his fear of it, however, seems to have given him some uneasiness; for when he was offered a small sum for the bones, it appeared too little to divide; and when a larger sum, he fain would have engrossed the whole of it, or persuade himself that the real value might be something greater. Ignorant of what had been offered him, my father's application was in a critical moment, and the farmer accepted his price, on condition that he should receive a new gun for his son, and new gowns for his wife and daughters, with some other articles of the same class. The farmer was glad they were out of his granary, and that they were in a few days to be two hundred miles distant; and my father was no less pleased with the consciousness, and on which every one complimented him, that they were in the hands of one who would spare no exertions to make the best use of them. The neighbours, who had assisted the farmer in this discovery, envious of his good fortune, sued him for a share in the profit; but they gained nothing more than a dividend of the costs; it appearing that they had been satisfied with the gratification of their curiosity, and the quality and quantity of the rum; no one could prove that he had given them reason to hope for a share in the price of any thing his land might happen to produce.

Not willing to lose the advantage of an uncommonly dry season, when the springs in the morass were low, we proceeded on the arduous enterprize.

In New York every article was provided which might be necessary in surmounting expected difficulties; such as a pump, ropes, pullies, augers, &c.; boards and plank were provided in the neighbourhood, and timber was in sufficient plenty on the spot.

Confident that nothing could be done without having a perfect command of the water, the first idea was to drain it by a ditch; but the necessary distance of perhaps half a mile, presented a length of labour that appeared immense. It was therefore resolved to throw the water into a natural bason, about sixty feet distant, the upper edge of which was about ten feet above the level of the water. An ingenious millwright constructed the machinery, and, after a week of close labour, completed a large scaffolding and a wheel twenty feet diameter, wide enough for three or four men to walk a-breast in: a rope round this turned a small spindle, which worked a chain of buckets regulated by a floating cylinder; the water, thus raised, was emptied into a trough, which conveyed it to the bason; a ship's pump assisted, and, towards the latter part of the operation a pair of half barrels, in removing the mud. This machine worked so powerfully, that in the second day the water was lowered so much as to enable them to dig, and in a few hours they were rewarded with several small bones.

The road which passed through this farm was a highway, and the attention of every traveller was arrested by the coaches, wagons, chaises, and horses, which animated the road, or were collected at the entrance of the field: rich and poor, men, women and children, all flocked to see the operation; and a swamp always noted as the solitary abode of snakes and frogs, became the active scene of curiosity and bustle: most of the spectators were astonished at the purpose which could prompt such vigorous and expensive exertions, in a manner so unprecedented, and so foreign to the pursuits for which they were noted.—But the amusement was not wholly on their side; and the variety of company not only amused us, but tended to encourage the workmen, each of whom, before so many spectators, was ambitious of signalizing himself by the number of his discoveries.

For several weeks no exertions were spared, and the most unremitting were required to insure success; bank after bank fell in; the increase of water was a constant impediment, the extreme coldness of which benumbed the workmen. Each day required some new expedient, and the carpenter was always making additions to the machinery; every day bones and pieces of bones were found between six and seven feet deep, but none of the most important ones. But the greatest obstacle to the search was occasioned by the shell marle which formed the lower stratum; this, rendered thin by the springs at the bottom, was, by the weight of the whole morass, always pressed upwards on the workmen to a certain height, which, without an incalculable expense, it was impossible to prevent. Twenty-five hands at high wages were almost constantly employed at work which was so uncomfortable and severe, that nothing but their anxiety to see the head, and particularly the under jaw, could have kept up their resolution. The patience of employer and workmen was at length exhausted, and the work relinquished without obtaining those interesting parts, the want of which rendered it impossible to form a complete skeleton.

It would not have been a very difficult matter to put these bones together, and they would have presented the general appearance of the skeleton; but the under jaw was broken to pieces in the first attempt to get out the bones, and nothing but the teeth and a few fragments of it were now found; the tail was mostly wanting, and some toe-bones. It was, therefore, a desirable object not only to procure some knowledge of these deficient parts, but if possible to find some other skeleton in such order as to see the position, and correctly to ascertain the number of the bones. In the course of eighteen years there had been found within twelve miles of this spot, a bone or two in several different places; concerning these we made particular inquiries, but found that most of the morasses had been since drained, and consequently either the bones had been exposed to a certain decay; or else so deep, that a fortune might have been spent in the fruitless pursuit. But through the polite attention of *Dr. Galatan*, we were induced to examine a small morass, eleven miles distant from the former, belonging to Capt. J. Barber, where, eight years before, four ribs had been found in digging a pit. From the description which was given of their position, and the appearance of the morass, we began our operations with all the vigour a certainty of success could inspire. Nearly a week was consumed in making a ditch, by which all the water was carried off, except what a hand pump could occasionally empty: the digging, therefore, was less difficult than that at Masten's, though still tedious and unpleasant; particularly as the sun, unclouded as it had been for seven weeks, poured its scorching rays on the morass, so circumscribed by trees, that the western breeze afforded no refreshment; yet nothing could exceed the ardour of the men, particularly of one, a gigantic and athletic negro, who exulted in choosing the most laborious tasks, although he seemed melting with the heat. Almost an entire set of ribs were found, lying nearly together, and very entire; but as none of the back bones were found near them (a sufficient proof of their having been scattered) our latitude for search was extended to very uncertain limits; therefore, after working about two weeks, and finding nothing belonging to the head but two rotten tusks, (part of one of them is with the skeleton here) three or four small grinders, a few vertebræ of the back and tail, a broken scapula, some toe-bones, and the ribs, found between four and seven feet deep—a reluctant terminating pause ensued.

These bones were kept distinct from those found at Masten's, as it would not be proper to incorporate into one skeleton any other than the bones belonging to it; and nothing more was intended than to collate the corresponding parts. These bones were chiefly valuable as specimens of the individual parts; but no bones were found among them which were deficient in the former collection, and therefore our chief object was defeated. To have failed in so small a morass was rather discouraging to the idea of making another attempt; and yet the smallness of the morass was, perhaps, the cause of our failure, as it was extremely probable the bones we could not find were long since decayed, from being situated on the rising slope at no considerable depth, unprotected by the shell-marle, which lay only in the lower part of the bason forming the morass. When every exertion was given over, we could not but look at the surrounding unexplored parts with some concern, uncertain how near we might have been to the discovery of all that we wanted, and regretting the probability that, in consequence of the drain we had made, a few years would wholly destroy the venerable objects of our research.

Almost in despair at our failure in the last place, where so much was expected, it was with very little spirit we mounted our horses, on another inquiry. Crossing the Walkill at the falls, we ascended over a double swelling hill into a rudely cultivated country, about twenty miles west from the Hudson, where, in a thinly settled neighbourhood, lived the honest farmer Peter Millspaw, who, three years before, had discovered several bones: from his log-hut he accompanied us to the morass.—It was impossible to resist the solemnity of the approach to this venerable spot, which was surrounded by a fence of safety to the cattle without. Here we fastened our horses, and followed our guide into the centre of the morass, or rather marshy forest, where every step was taken on rotten timber and the spreading roots of tall trees, the luxuriant growth of a few years, half of which were tottering over our heads. Breathless silence had here taken her reign amid unhealthy fogs, and nothing was heard but the fearful crash of some mouldering branch or towering beach. It was almost a dead level, and the holes dug for the purpose of obtaining manure, out of which a few bones had been taken six or seven years before, were full of water, and connected with others containing a vast quantity; so that to empty one was to empty them all; yet a last effort might be crowned with success; and, since so many difficulties *had been* conquered, it was resolved to embrace the only opportunity that now offered for any farther discovery. Machinery was accordingly erected, pumps and buckets were employed, and a long course of troughs conducted the water among the distant roots to a fall of a few inches, by which the men were enabled, unmolested, unless by the caving in of the banks, to dig on every side from the spot where the first discovery of the bones had been made.

Here alternate success and disappointment amused and fatigued us for a long while; until, with empty pockets, low spirits, and languid workmen, we were about to quit the morass with but a small collection, though in good preservation, of ribs, toe and leg-bones, &c. In the meanwhile, to leave no means untried, the ground was searched in various directions with long-pointed rods and cross-handles: after some practice we were able to distinguish by feeling, whatever substances we touched harder than the soil; and by this means, in a very unexpected direction, though not more than twenty feet from the first bones that were discovered, struck upon a large collection of bones, which were dug to and taken up, with every possible care. They proved to be a humerus, or large bone of the right leg, with the radius and ulna of the left, the right scapula, the atlas, several toe-bones, and the great object of our pursuit, a complete UNDER JAW!

After such a variety of labour and length of fruitless expectation, this success was extremely grateful to all parties, and the unconscious woods echoed with repeated huzzas, which could not have been more animated if every tree had participated in the joy. "Gracious God, what a jaw! how many animals have been crushed by it!" was the exclamation of all; a fresh supply of grog went round, and the hearty fellows, covered with mud, continued the search with increasing vigour. The upper part of the head was found twelve feet distant, but so extremely rotten that we could only preserve the teeth and a few fragments. In its form it exactly resembled the head found at Masten's; but, as that was much injured by rough usage, this, from its small depth beneath the surface, had the cranium so rotted away as only to show the form around the teeth, and thence extending to the condyles of the neck; the rotten bone formed a black and greasy mould above that part which was still entire, yet so tender as to break to pieces on lifting it from its bed.

This collection was rendered still more complete by the addition of those formerly taken up, and presented to us by Drs. Graham and Post. They were a rib, the sternum, a femur, tibia and fibula, and a patella or knee-pan. One of the ribs had found its way into an obscure farm-house, ten miles distant, to which we fortunately traced it.

Thus terminated this strange and laborious campaign of three months, during which we were wonderfully favoured, although vegetation suffered, by the driest season which had occurred within eight years. Our venerable relics were carefully packed up in distinct cases; and, loading two wagons with them, we bade adieu to the vallies and stupendous mountains of Shawangunk: so called by their former inhabitants, the Indians of the Lenape tribe. The three sets of bones were kept distinct: with the two collections which were most numerous it was intended to form two skeletons, by still keeping them separate, and filling up the deficiencies in each by artificial imitations from the other, and from counterparts in themselves. For instance, in order to complete the first skeleton, which was found at Masten's, the under jaw was to be modelled from this, which is the only intire one that has yet been discovered, although we have seen considerable fragments of at least ten different jaws: while, on the other hand, in the skeleton just discovered at Barber's, the upper jaw, which was found in the extreme of decay, was to be completed, so far as it goes, from the more solid fragment of the head belonging to the skeleton found at Masten's. Several feet-bones in this skeleton were to be made from that; and a few in that were to be made from this. In this the right humerus being real, the imitation for the left one could be made with the utmost certainty; and the radius and ulna of the left leg being real, those on the right side would follow, of course, &c. The collection of ribs in both cases was almost entire; therefore, having discovered from a correspondence between the number of vertebræ and ribs in both animals, that there were nineteen pair of the latter, it was necessary in only four or five instances to supply the counterparts, by correct models from the real bones. In this manner the two skeletons were formed, and are in both instances composed of the appropriate bones of the animal, or exact imitations from the real bones in the same skeleton, or from those of the same proportion in the other. Nothing in either skeleton is imaginary; and what we have not unquestionable authority for, we leave deficient; which happens in only two instances, the *summit* of the head, and the *end* of the tail.

We now proceed to describe the parts composing the skeleton of the mastodon, and give in detail the measurements we have very carefully made on the excellent specimen in the Philadelphia Museum. To naturalists this will be the more acceptable as it has not heretofore been done throughout; and it will enable the general reader to form more definite conclusions relative to the animal, by furnishing positive data for the basis of an accurate comparison between the bones of this skeleton and those belonging to other large quadrupeds.

The Skull.—The upper parts of the skull are entirely lost, as already stated, down to the level of the anterior part of the zygomatic arch, except at the back of the skull, where the occipital bone rises above the level stated, and is eleven inches and a quarter high. The lower halves (or rather more) of the intermaxillary bones, and nearly the whole of the superior maxillary and cheek bones, are also preserved. The zygomatic arches are complete, and

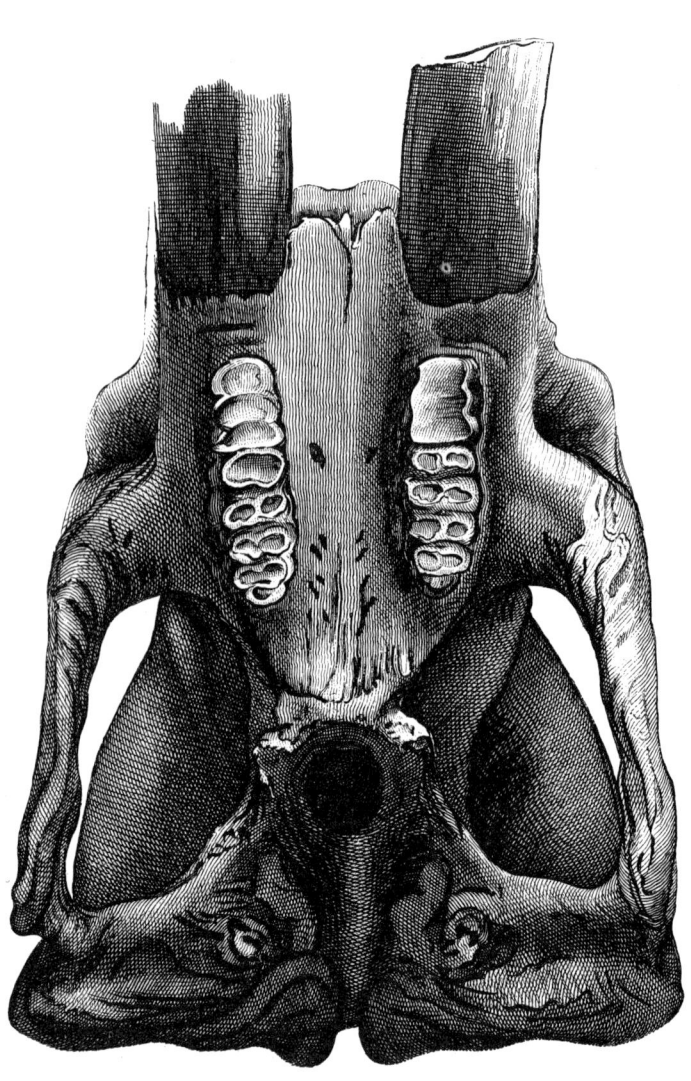

the junction between the jugal process of the temporal and that of the malar bone is very strong, the process of the temporal bone forming little more than one-third of the arch.

The posterior part of the skull is the broadest portion, being thirty-two inches across. When the skull is placed on the ground inverted, and we look upon its inferior surface, (as exhibited in Mr. Lesueur's very accurate drawing) from the extreme points of this widest part of the head, the outlines of the sides of the skull speedily converge so as to run within the zygomatic arches, and continue to become narrower until traced to the posterior surface of the facial bases of the zygomatic arch, where the skull is narrowest. The outline formed by the external surface of the zygomatic arches, from the origin to the angles of the occipital bone, give the whole inferior and posterior surface of the skull a peculiarly square form. All the parts of the skull are exceedingly massive and hard, appearing to have undergone very little change.

The intermaxillary, or incisive bones, are nearly entire in their inferior portions, and on the left side, were most perfect, the part forming the sockets for the tusk measures thirteen inches and a-half in circumference, beginning opposite to the ante-orbitar foramen, and extending to a line continuous with the centre of the palate.

The maxillary, or upper jaw-bones, are entire, and the palate plate remarkably strong and compact in texture. The alveolar processes are situated very near the outer edge, and rise very slightly above the plane of the roof of the mouth, and diverge considerably from the posterior to the anterior part of the range. Hence, at the back of the mouth the distance from the inside of the last molar to the same place on the opposite side, is six inches and one-eighth, while from the inside of the first molar to the correspondent tooth on the other side the space is eight inches. Immediately in front of the first molar, measuring from the external edge of its alveolar process to the same on the opposite side, the width of the palate plate is fifteen inches and four-eighths. Behind the last molar, and midway to the pterygoid processes of the sphenoid, the palate-bone is seven and a-half inches wide. The length of the pterygoid process, to the base of the skull, is seven inches.

The malar or cheek-bones, forming the prominences at the superior and external part of the face, are nearly entire. From the edge of the infra, or ante-orbitar foramen, to the zygomatic or temporal fossa, its width is five inches and an eighth; its height, measured within the fossa, is eight; its greatest breadth externally is six inches, and the narrowest portion of its zygomatic process three-quarters of an inch. Its length, from the foramen to the extremity of the zygomatic process, is seventeen inches and seven eighths.

The temporal bone is entire, except in its thin superior portion. The length of its zygomatic process is seven and a-half inches. The distance of the auditory foramen, from the cavity for the articulation of the lower jaw, is one inch. The cavity for the reception of the condyles of the lower jaw is one inch and seven-eighths, measured through the centre transversely.

The occipital bone is remarkably square on its posterior surface, which is thirty-two inches broad, and eleven inches and a-quarter high. How much higher the bone ascended cannot now be determined. The distance between the posterior extremities of the occipital condyles is two inches and one-sixteenth; the breadth of the condyle is three inches and one-eighth. The foramen magnum (for the exit of the spinal marrow) is two and a-half inches in diameter.

The first vertebræ of the neck, or atlas, receiving the condyles of the occipital bone, is eleven inches broad. Its length from the tip of one transverse process to the other is eighteen inches.

The most remarkable peculiarities of the mastodon skull are summed up by Cuvier in the following manner:*

1st. The molars of the mastodon *diverge* in front, while those of the living elephant converge, more or less, and those of the fossil elephant (the true mammoth of Siberia) are nearly parallel. The hog and hippopotamus are the only animals which, in this respect, resemble our animal.

2d. The bony palate extends far beyond the last tooth; among herbivorous animals the Ethiopian boar (*Phacochærus*) alone possesses this character.

3d. The pterygoid apophysis of the palate bones have a size unexampled among quadrupeds.

4th. The depression anterior to this apophysis has some relation with that of the hippopotamus, which, however, is straighter.

See R. Peale's Disquisition, heretofore cited.

5th. There is no visible trace of the orbit at the anterior part of the zygomatic arch, whence the eye must have been much higher than in the elephant.

6th. The maxillary bones have much less vertical elevation than in the elephant, and resemble ordinary animals more strongly.

7th. The zygomatic arch, for the same reason, is much less elevated in front, which corresponds with the form of the lower jaw. The position of the ear depends on that of this arch.

8th. This position has much influence on the position of the occipital condyles, which are, in the elephant, considerably elevated above the level of the palate; in the mastodon they are nearly on the same level.

The Tusks.—As the bones of various other animals were discovered in the same place where the first tusks of this animal were found, some doubt was entertained of their belonging to the same skeleton, which contained the tuberculated molar teeth. Dr. W. HUNTER stated in the Philosophical Transactions, his belief that they pertained to the same animal. But all doubt was dispersed by the discovery of the great skeleton obtained in New York by Messrs. PEALE, which was entirely alone, or separated from the bones of all other quadrupeds.

The tusks of the mastodon bear a considerable resemblance to those of the elephant, but present some appearances different from those observed in the generality of tusks of that animal; though these are by no means greater than may be found in different individuals of that genus.

These tusks are rooted in the intermaxillary bones.

the sockets being eight inches in depth. The tusk belonging to the skeleton we are describing is ten feet seven inches long, measuring from the base to the tip, following the outside of the curvature; the point is not exactly in the same plane with the base, owing to the peculiar spiral twist of the anterior portion of the tusk. The direction of the tusk in leaving the socket is rather more oblique in front than in the elephant. The diameter of the tusks at base is seven inches and three-quarters; in the middle their substance is very similar to that of the elephant tusk, composed of an ivory. the grain of which is arranged in curvilinear lozenges. The external part of the tusk is hard, and differs considerably in appearance from common ivory; the internal is of the texture of ivory, but is of much softer consistence.

R. Peale dwells with much force on these circumstances, as well as on the roundness and peculiar curve of the mastodon tusk, in forming his conclusions relative to their position in the head, (which he believed to be with the convexity forwards, and the point turning downwards and backward) as well as in deciding on the mode of living of the animal. Cuvier has, however, satisfactorily shown that the differences are neither so uniform nor so remarkable as was believed, and that the difference in consistence of the ivory is accidental, or attributable to the circumstances under which these remains, during so great a lapse of time, were situated. As an immediate consequence of the great similarity existing between the skeletons and tusks of the elephant and mastodon, we form the inference that they were as *analogous* in their modes of living as in their conformation.

The tusks of our animal were placed with their convexity in front, and their points curving downwards and backwards, in the specimen mounted in the Philadelphia Museum. This position is certainly unnatural, as Cuvier has clearly shown, by reference to the length of limb of this animal, the impossibility of its using the tusks, thus arranged; and from the fact that the Siberian mammoth (elephant) has tusks equally curved, and their points unequivocally turned upwards.

The morse, which has tusks pointing downwards, (see vol. i. p. 351) is an animal possessing very short limbs, and destined to an aquatic life. A conclusion drawn from the tusks of this animal is inapplicable, since we must believe the mastodon (like the elephant) to have been a terrestrial animal. Nothing therefore can justify us in placing these tusks otherwise than in the elephant, unless we find a skull which has them actually implanted in a different manner.*

The Under Jaw of the mastodon is remarkable for its massiveness and solidity, and the form of it is peculiar to this animal. It is two feet ten inches long, and weighs sixty three and a-half pounds. The anterior part or chin is inclined so as to terminate conically, being marked by numerous rough prominences; where the two sides of the jaw unite in front, there is an intervening furrow or depression. The outline of the lower jaw is formed by three lines touching each other so as to form three different

---

* See Cuvier; Oss. Foss. tom. ii.

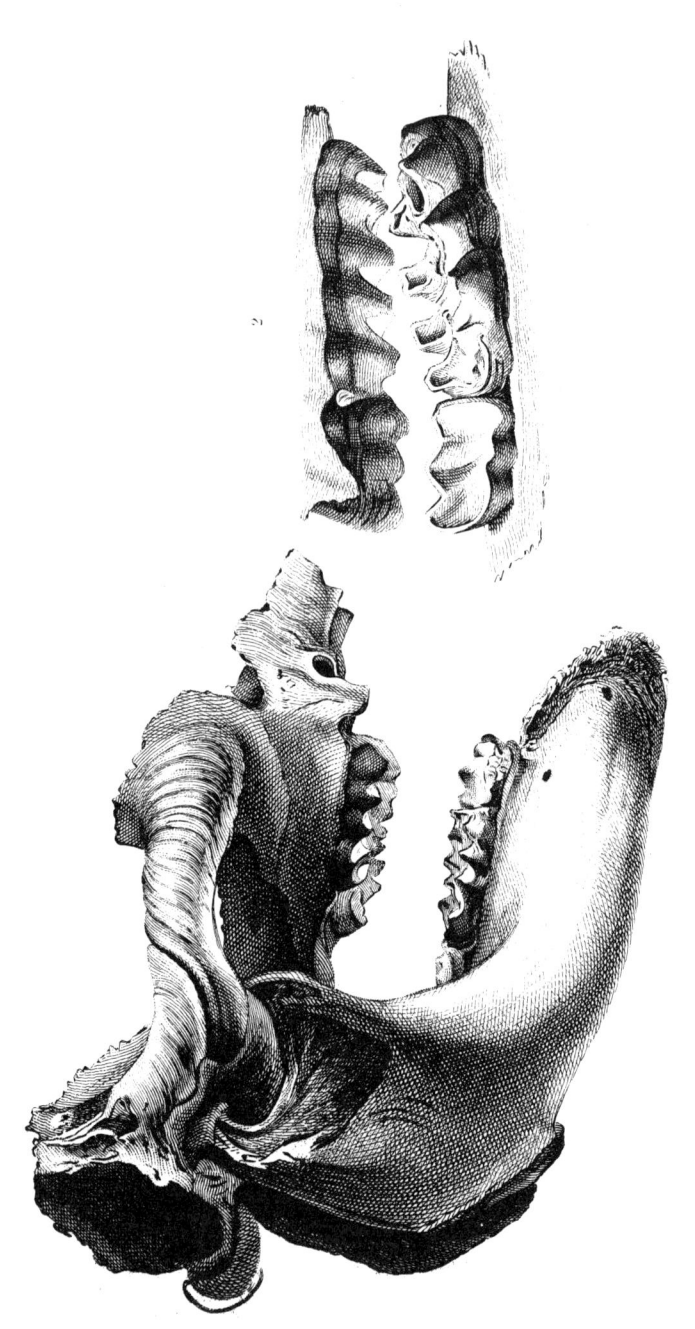

angles; the first extends from the top of the condyloid process for twelve inches towards the angle of the jaw:—the second, commencing at this point and terminating in a protuberance, which is at the inferior and anterior part of the angle, and the third passing thence almost horizontally, terminates with the anterior extremity of the jaw.

The condyloid or articulating surfaces are five and a-half inches wide, and stand on very strong processes; the coronoid processes for the insertion of the temporal muscles are nearly on a level with the condyles, and are separated from them by a semilunar notch, six and a-quarter inches in width. The general figure of the lower jaw, at the posterior part extending forwards to the base of the coronoid process, bears considerable resemblance to the same parts in the human jaw.

The teeth in the lower jaw are arranged so as to be very nearly parallel to those in the upper jaw, and the two ranges are most widely separate at the fore part. They are not disposed parallel to the direction of the sides of the jaw, but diagonally, from the inner to the outer part. Thus these teeth do not meet the superior teeth fully crown to crown, but obliquely crossing each other, the lower teeth being worn most at the anterior part and on the outside, while the superior teeth are most worn on the inside and fore part of the mouth, as shown in the plate, figure 2.

From the size of the head, the thickness and solidity of the teeth, and the enormous magnitude of the tusks, we can at once perceive that the neck of the animal must of necessity have been short, in order to sustain so great a weight. These circumstances, considered in connexion with the length of the limbs, presently to be described, clearly indicate that the mastodon, like the elephant, had a long and flexible trunk for the purpose of conveying its aliment to the mouth; the shortness of the neck, and the projection and curvature of the tusks, would equally have prevented the approach of the mouth to the ground.

Bones of the Trunk.—The bones of the neck are similar in character to those of the elephant, and thus far support the opinion drawn from the preceding circumstances. According to the observation of R. PEALE the spinous processes of the three last vertebræ of the neck are not so long in the elephant.

The spinous processes of the second, third and fourth dorsal vertebræ, are exceedingly long. The longest of them measures eighteen or twenty inches, the whole length of the vertebræ being twenty-seven. The spinous processes of the back then rapidly diminish to the twelfth, and become so small as scarcely to be remarked, thence to the sacrum. This conformation, as Mr. PEALE has well pointed out, differs remarkably from that of the elephant, in which the processes are more uniform in their length:—those over the shoulders being shorter, and those of the back and loins much longer; hence the form of the back in the elephant is more arched. There are seventeen cervical, nineteen dorsal, and three lumbar vertebræ. CUVIER remarks that the elephant has one more dorsal vertebra, and one more pair of ribs; but suggests that the corresponding parts in the mastodon have been destroyed.

The ribs are not similar to those of the elephant.

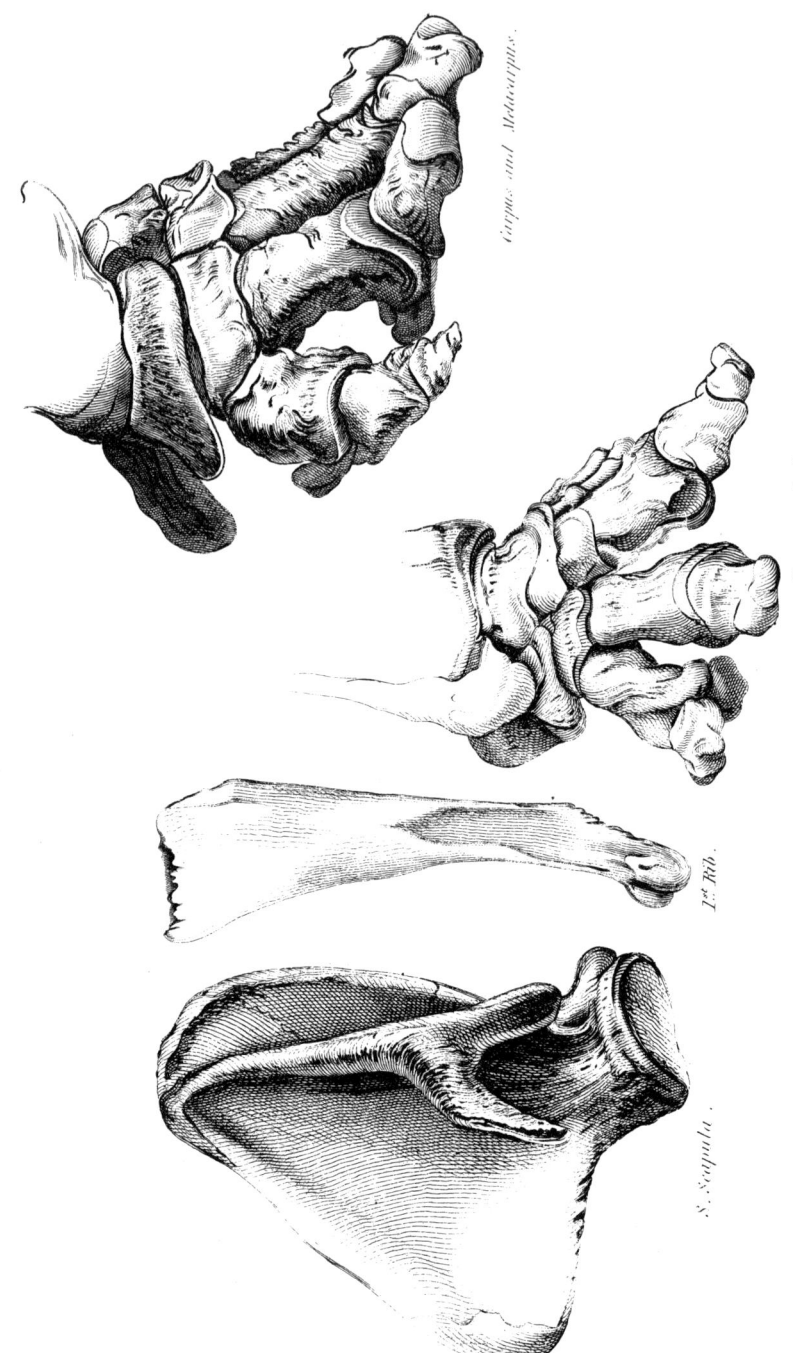

small near the head, broad as they approach the cartilage, and bent sidewise in an undulating manner; but they are slender near the cartilage, and thick and strong towards the back. The difference is peculiarly observable in the first rib. The six first pairs are remarkably strong, when compared with the remainder, which are proportionally short. This, joined to the flatness of the pelvis, shows the belly to have been less in the mastodon than in the elephant.

Scapula, or Shoulder-blade.—This bone has the characters peculiar to that belonging to the elephant; particularly the recurrent process, which is only found in the elephant and some of the gnawing animals. The length of the whole bone is thirty-seven inches. The acromion process is very long and pointed.

Arm, Forearm, &c.—The arm-bone, or *humerus*, is very thick, and, in proportion, much thicker than the thigh-bone; this difference, in proportion, is much more remarkable than in the elephant.* In length the humerus is two feet ten inches; its greatest circumference is three feet two inches and a-half, and its smallest part measures one foot five inches around.

The *ulna* is proportionally as massive as the humerus, and the olecranon (process forming the point of the elbow) is strong and knobbed at the end, being eight inches and a-quarter in circumference at base. The ulna is two feet five inches and a-half long, while its circumference around the elbow is three feet eight inches.

---

* Peale, Hist. Disquis. 8vo. 56.

The *radius* is a comparatively small bone, two feet four inches long, and is placed in such a manner as to cross obliquely from the outside above to the inside below, forming thus a greater angle than if the bones were slender, in which case the crossing would be scarcely observable; perhaps it is more remarkable in the mammoth than in any other animal.* Its carpal articulating surface is four inches and five-eighths broad.

The bones of the *carpus*, in the skeleton belonging to the Philadelphia Museum, are seven in number; the forms of those in the first row generally agree with those of the elephant, as figured by Cuvier.† The external faces of those belonging to the second row appear to differ by being proportionally larger and squarer than in the elephant. The metacarpal bones are strong and massive; their surfaces for articulation with the digital phalanges are extensive, and indicate that the toes were capable of very considerable flexion. The metacarpal of the first digit, or thumb, is two inches and a-quarter in length, of the second digit three inches; of the third and fourth four inches and a-half; and of the fifth external, or smallest, three inches.

The Pelvis.—This part of the skeleton has sustained a considerable degree of injury. The iliac or haunch bones at their superior parts being in a great degree lost. Still the quantity of sound bone remaining is quite sufficient to show the general form and dimensions of this part of the animal. On

---

* Peale, Hist. Disquis. 4to. 56.
† Ossemens Fossiles, vol. ii. ed. 1.

the left side the bone is uninjured, except along the border, from above the anterior superior spine. The width of the pelvis, measuring from this spine to the edge of the pubis at the symphysis, is two feet eleven inches, which gives a total breadth of the pelvis of five feet ten inches, without allowing for the cartilage, which must, in the living condition, have intervened at the pubic and sacro iliac symphyses. The pubis, from the anterior to the posterior edge, is six inches in extent. The longest diameter of the foramen thyroideum is eight inches; the transverse diameter five inches.

We were led to make this measurement of the pelvis with the greater care, because Cuvier makes the following remarks on the subject of its width:— "Mr. Peale states that the width of the pelvis of his skeleton is five feet eight inches, (Engl.) but I fear that this is a typographical error, or that he meant it for the measure of the circumference."*

The difference between the measurement of the pelvis stated by R. Peale, and that given by us, is owing to the circumstance of our having measured different skeletons. His measurements were made on the skeleton now in Baltimore; ours were carefully taken from that in the Philadelphia Museum.

It was first stated by R. Peale, and subsequently confirmed by Cuvier, that these bones are more depressed than in the elephant. This indicates, says this Zoologist, that the belly must have been smaller, and consequently the intestines less volu-

---

* Oss. Foss. tom. 2.

minous, than in the elephant; this, together with the structure of its teeth, concurs in causing the mastodon to be regarded as less exclusively herbivorous than from other circumstances is commonly inferred.

The Femur, or Thigh-Bone.—This bone is perfectly preserved, and is a fit column for the support of so large a superstructure. It is three feet seven inches long, and eight inches in diameter at the middle of its shaft; the whole of the middle part of the bone is peculiarly flattened. The neck of this thighbone, which is six inches and three-fourths in diameter, on a level with the top of the trochanter, is a very strong process, and is surmounted by a head seven inches in diameter. The great or external trochanter, projecting below and opposite the neck of the bone, is a strong and massive process, having a large depression at its basis on the posterior surface of the bone. The lesser or internal trochanter does not exist, except as a slightly extended roughness on the inner edge of the bone. The transverse diameter of the articulating surfaces or condyles of the femur is nine inches and five-eighths; of each condyle, four inches and a-half.

The Tibia and Fibula, or Leg-Bones.—The tibia is two feet long, and in strength and solidity is well proportioned to the femur; it is ten inches broad at its superior portion, and seven inches and seven-eighths at the inferior part. Its diameter in the middle is four inches and five-eighths.

The fibula is comparatively slight and slender, and occupies the same relative position in the mastodon as in the human subject. Its superior extremity is

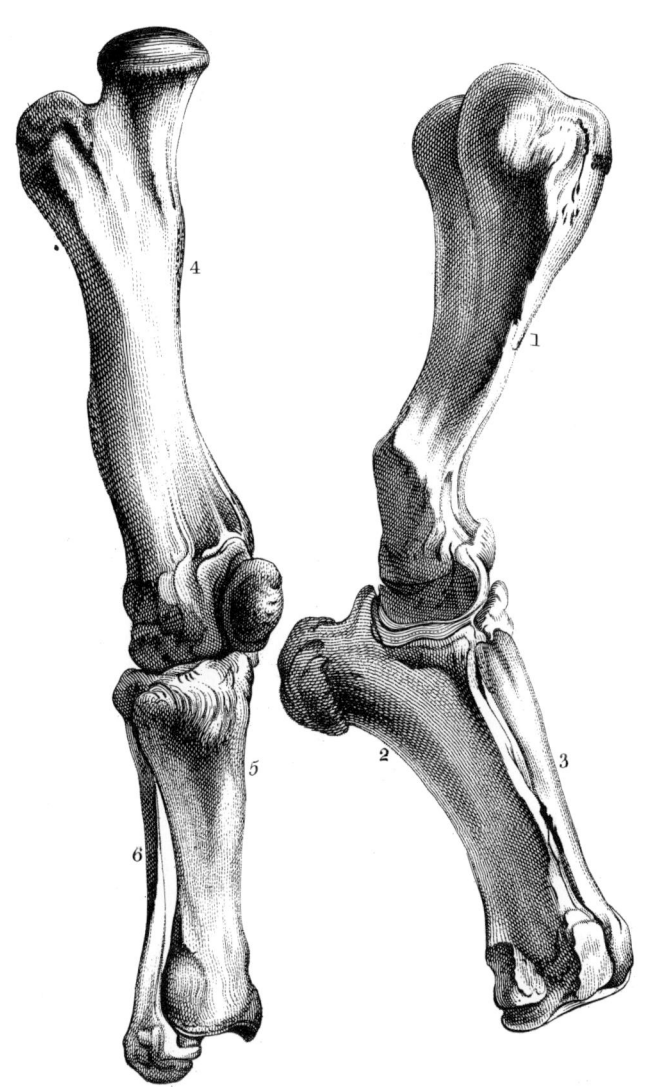

MASTODON

1. Humerus
2. Ulna
3. Radius
4. Femur
5. Tibia
6. Fibula

closely united to the superior and posterior part of the tibia; its inferior extremity passes below that of the tibia for three inches and a-half, constituting the support of the outer part of the ankle joint. The whole length of the bone is twenty-six inches.

The bones of the *tarsus* are very analogous to the same bones in the elephant, but appear flatter and thinner in proportion. The articulating surface of the astragalus is remarkably flat, and is five inches broad from the anterior to the posterior edge; the thickness of the bone, measured on the anterior surface, is two inches and three-quarters. The os calcis, measured on its inferior surface, is six inches long, and is a very large and strong bone. Its surface for articulation with the extremity of the fibula, is four inches and a-half in length, from its anterior to the posterior edge. The os naviculare is five inches long; its breadth in the middle is one inch and three-quarters. The internal cuneiform bone sustains the metatarsal of the internal or first toe; the middle and external cuneiform sustain a part of the second and medial metatarsal bones, while the cuboid receives both the external, or the fourth, and little toe. The length of the first metatarsal is three inches; of the second three and three-fourths; of the third five; of the fourth four and a-half; of the fifth four.

Localities whence Mastodon bones have been obtained in greatest abundance.

Among the earliest localities discovered was Bigbone Lick, in Kentucky, which derived its name from the great number of fossil bones there found. This celebrated lick is a morass, or marsh-like valley, surrounded by considerable hills, and is about four miles south-east of the Ohio, and nearly opposite to the mouth of the Great Miami river. The basis of this morass is a black and fetid mud, and the water that oozes through it is impregnated with saline matter. At present this spot is much frequented by inhabitants of the western country, on account of its excellent mineral springs, which are found useful in relieving the system from various states of disease, but more especially from the peculiar affections caused by the common autumnal fevers of the country. The principal spring yields large quantities of very limpid water, which is highly charged with sulphuretted hydrogen gas, immediately blackening silver vessels, or implements plated with that metal.

From the vast number of bones of various extinct and recent species obtained from this locality, it is evident that, previous to the peopling of the surrounding country, it was resorted to by animals for the purpose of gratifying their appetite for salt, so abundantly contained in the waters oozing from such places. From the position in which many of the bones, especially of the larger quadrupeds, have been found, it is highly probable that they may have perished in consequence of the peculiar softness of the soil:—their great weight causing them to sink deeply, while their strength was rendered entirely unavailing for want of firm materials, against which exertions could be effectually made.

In the year 1807, THOMAS JEFFERSON, then pre

sident of the United States, requested governor Clarke, previously so justly distinguished by his travels in company with Capt. Lewis to the Pacific Ocean, carefully to explore the ground at Bigbone Lick. His researches were rewarded by a large collection of bones belonging to various species, which were sent to the city of Washington, and by the philanthropist and sage then at the head of our government, they were afterwards divided between the American Philosophical Society, the National Institute of France, &c.

Mitchill informs us that bones of the mastodon "were found in July, 1817, in the east branch of the White River, a stream emptying into the Wabash, at a point distant forty-four miles in a right line from the mouth of the Wabash. This east branch unites with the west branch at a point twenty-nine miles in a direct line from the mouth of the White River. The intelligence was communicated by Josiah Meigs, Esq. commissioner of the general land office, in the treasury department of the United States, who received it from Mr. Spotts, living near the falls of the east branch. These consisted, it is stated, among others, of the upper jaw, whose width from outside to outside was twenty and a-half inches; length twenty-five inches; length of the posterior grinder (composed of five divisions in three rows) seven and three-fourths inches; breadth of the same across, five and a-half."*

---

* See the interesting observations on the Geology of North America, by this zealous and distinguished votary of science, appended to his edition of Cuvier's Theory of the Earth, p. 368.

According to Jefferson, these bones are also found on the North-Holston, (36° N. Lat.) a branch of the Tennessee river, west of the Alleghanies in Carolina, and were also obtained from a morass similar to those containing these bones in other places.*

Subsequently, bones of the mastodon were found still farther south in different parts of Louisiana west of the Mississippi, but always in the river Alluvions.†

In Wythe county, Virginia, near Green Briar county, whence the bones of the megatherium were obtained, a large number of bones, probably almost an entire skeleton, was obtained. This interesting discovery was first made known by Bishop Madison, in a letter to the late professor B. S. Barton, who published an account of it in his Medical and Physical Journal. "But what renders this discovery unique among others, (says Cuvier, vol. ii. ed. 1, Oss. Foss.) is that in the midst of the bones was found a half triturated mass of small branches, of gramina, and of leaves among which it was believed that a species of reed still common in Virginia could be recognized, and that the whole seemed to be enveloped in a sort of sac, which was considered as the stomach of the animal; so that there was no doubt but that these were the very substances upon which the animal had fed."

This information was communicated to the Zoologist by Dr. B. S. Barton in a letter which runs thus:—"Without further delay, I hasten to inform you of a recent discovery relative to the mammoth,

---

* Notes on Virginia.
† Am. Phil. Trans. vi. 40.

or American elephant. If the facts be as I state them, I think you will not hesitate to consider the discovery one of the most interesting that has been made for a long time. I may add that such a discovery was hardly to be expected by the most sanguine or enthusiastic zoologist. Very lately, in digging a well near a salt-lick in the county of Wythe, in Virginia, after penetrating about five feet below the surface of the soil, the workmen struck upon the *stomach* of one of those huge animals, best known in the United States by the name of the mammoth. The contents of the viscus were carefully examined, and were found to be ' in a state of perfect preservation.' They consisted of half-masticated reeds, (a species of Arundo or Arundinaria, still common in Virginia and other parts of the United States,) of twigs of trees, and of grass or leaves."*

The best comment that can be offered on this discovery is the original letter of the learned and excellent Bishop MADISON, from which it will appear that he never saw the *place* nor the *thing* which was supposed to be the *stomach*, neither is the evidence given with sufficient conclusiveness to establish *any opinion* on the subject. We take the liberty of proving this by italicising some parts of the bishop's letter:

"One of those facts has lately occurred, which the naturalist knows best how to appreciate, and which I, therefore, take a pleasure in communicating to you. It is now no longer a question whether the

---

* Barton's Med. & Phys. Journal, vol. iii. p. 23 of first supplement.

mammoth was a herbivorous or carnivorous animal. Human industry has revealed a secret, which the bosom of the earth had, in vain, attempted to conceal. In digging a well near a Salt-Lick, in Wythe county, Virginia, after penetrating about five feet and a-half from the surface, the labourers struck upon *the stomach of a mammoth*. The contents were in a state of *perfect* preservation, consisting of half-masticated reeds, twigs, and grass, or leaves. There *could be no deception;* the *substances* were designated by *obvious characters*, which *could not be mistaken*, and of which *every one* could judge; besides, the *bones of the animal lay around*, and added a silent, but sure confirmation. The whole rested upon a lime-stone rock. I *have not seen*, as yet, *any part of those contents; for, though I was within two days' journey of the place where they were found, I was so well satisfied with the narration of gentlemen who had seen them, and upon whose veracity, as well as accuracy, I could rely, that I thought the journey* UNNECESSARY; especially as I took measures to ensure the transmission of a sufficient quantity of the contents, together with all the bones, to Williamsburgh. When the contents arrive, a part shall be forwarded to you. I hope to form a complete skeleton of this vast animal, having given directions to spare no labour in digging up every bone.

[Then follows a sentence cited from Blumenbach, showing how the soft parts of animals may be preserved, after which the letter concludes in the following manner:]

"Whether this first kind of petrifaction, of which Blumenbach speaks, and which he calls *simplement calcinés*, has been the cause of the preservation of these substances, or whether it be the effect of the marine salt, with which the earth, where they were buried, has been constantly charged, must be left to future investigation. I pretend not to decide. *Had they been buried deep in the earth*, that circumstance alone might have prevented a decomposition; *but the depth of five or six feet* seems insufficient to arrest that chemical action, which changes the appearances of organized bodies. *The fact*, however, is *decisive, as to the principal question. It* has summoned the discordant opinions of philosophers *before a tribunal from which there is no appeal.*"*

Such is the letter of Bishop Madison; and thus the discovery remains at the present day. Dr. Barton's letter, first quoted, consists of little more than the same matter in other words. It is much to be regretted that the worthy prelate did not inspect this locality, and ascertain for himself, by scrutinizing all the collateral circumstances, whether this *stomach*, so confidently pronounced to be that of the mastodon, might not have belonged to some domestic animal, which had perished from disease, and been interred "at a depth of five or six feet," and by accident, in the immediate vicinity of mastodon bones. Neither the dimensions, figure, nor peculiar nature

---

* Bishop MADISON subsequently corrected the impression made by this letter, acknowledging that his information was inaccurate, and his conclusions too hastily adopted.

of this stomach is described, nor do we know that it was such, except upon the hearsay evidence contained in the bishop's letter. Dr. Barton tells us of the specific nature of part of the matter, but does not say whether he had examined it himself or not, though it is probable he had received a specimen. We think but one opinion can be formed on the subject—whatever this *stomach* may have been, or whatever was the nature of its contents, its connexion with the bones of the mastodon was altogether *accidental*. It would be something very singular indeed, when the strongest animal fibres, the ligaments, tendons, muscles, &c. had all disappeared, that the stomach alone " at a depth of five or six feet," should escape almost uninjured!

In the year 1817 Professor MITCHILL, in company with Dr. PETER S. TOWNSEND and several other gentlemen, explored a small meadow in the vicinity of Chester, near Goshen, in Orange County, New York. Ten years previously some bones had been disinterred at that place, and some of them still remained at the bottom of a ditch.

This ground had been successfully drained and converted into a meadow. The surface was covered by a fine grassy sward, beneath which was a soil composed of a bed of black peat turf, six feet in thickness. " The soil and sward were about four feet thick over the bones; beneath them, and immediately around them, was a stratum of coarse vegetable stems and films, resembling chopped straw, or rather drift stuff of the sea, for it seemed to be mixed with broken films of conferva, like those of the Atlantic shore."

As these bones were found in a peat-bog containing no marl, the bones were far more rotten than those obtained from the marl pits by Messrs. Peale. It was impossible to extract them entire, and it was equally so to reunite the fragments after their exhumation. The bones discovered consisted of parts of the feet, legs, shoulder-blade, back-bone, rump, lower jaw, upper jaw, teeth and tusks.

The teeth were uninjured, and more than half of the lower jaw was preserved; but the condyles and angles of the other side broke in pieces when handled. The upper jaw, with its teeth and tusks, were found retaining their natural connexions. When the mud was carefully removed from them, the palate bones and teeth were found to be uppermost, as if the animal had perished on its back. The tusks differed in size, length and curvature; the right one measured seven feet in length, and was thicker and blunter than the left, which was nearly nine feet long, more regular tapered and pointed.

Dr. Mitchill concludes his account of this interesting research, by stating that " the flatness of the cranium, the connexion of the tusks with the head by exsertion and not by gomphosis, and the insertion of the grinders into them at their origin, will not fail to attract the attention of zoologists."*

These circumstances do not, however, appear to us very extraordinary, when taken in connexion with the facts previously stated. The bog was one containing no marl, or other antiseptic agent, and

---

* Mitchill's Geological Appendix to Cuvier's Theory of the Earth, p. 379.

the soil covering these bones was a bed of black peat-turf, soaked in a large quantity of fluid. These bones were very rotten, and doubtless much decayed before the draining was begun. As the whole superincumbent soil settled, in consequence of the removal of the water, it appears highly probable that at the same time the water was withdrawn from the mud within the decayed skull, and the whole mass of decayed bone was flattened by the general pressure caused by the subsiding soil. Thus we would account for the appearances so faithfully noted in the above mentioned instance. In a recent conversation with Dr. Townsend, who aided professor Mitchill on that occasion, and made accurate drawings of the bones as they were found, he expressed his entire belief in the probability of the explanation here given.

The *apparent* exsertion of the tusks we would attribute to the entire destruction and removal of the inferior portions of their sockets, formed by the intermaxillary bones. To the causes above stated, we attribute the *apparent* insertion of the grinders into the roots of the tusks, producing the softening and subsequent compression of the alveolar processes, together with the approximation of all the inferior and superior walls of the skull. We are fully convinced that these bones were in the relative positions so well described by Mitchill. But that such was the relation of parts in the living animal, or in the sound skeleton, is forcibly denied by the two nearly entire skeletons of Philadelphia and Baltimore, as well as by fragments preserved in various cabinets of natural history.

In Rockland county, N. Y. eleven miles west of

a spot where fragments of bones belonging to land-animals were found imbedded in sand-stone, by Dr. Mitchill, at a distance of thirty-two miles from the city of New York, the remains of a mastodon were found in July 1817. They consisted of a set of grinders, which were accidentally discovered by a ditcher, in mud only three feet below the surface. They were large, having remarkably white and glossy surfaces; the roots were much decayed. Mr. Edward Suffern, Jr. who presented these teeth to Professor Mitchill, informed him that the cavities of these teeth contained a fatty substance when they were first discovered. This, however, had entirely disappeared before they were received by Dr. Mitchill.

In the year 1811, the remains of a mastodon of the species we have been describing was found on the banks of York river, about six miles east of Williamsburgh, in Virginia. They lay upon marsh-mud, or buried a few feet within it, and were surrounded by the roots of cypress trees. The trees which these roots once supported had long been removed, and the difference between the level where the bones and roots are found and the top of the adjacent bank, is more than twenty feet. This locality was carefully examined by the Rev. Bishop Madison, then president of William and Mary College, Va. who gave the details of the discovery to Dr. Mitchill.* The parts of the skeleton obtained were the pelvis bones, a thigh-bone, two vertebræ, two ribs, nearly perfect,

---

* See the Medical Repository, (N. Y.) xv. 388.

two tusks, not greatly injured, and seven molar teeth, four of which were in their sockets, apparently part of the lower jaw. The largest tooth weighed seven and a-quarter, the smallest three to four pounds.

Various other localities have been mentioned, where bones of the mastodon and other large quadrupeds have been disinterred. At the Salines of Great Osage river they are said to be very abundant, as much so as at Bigbone Lick, or in the vicinity of the Waalkill.* Darby, the geographer, states in a letter to Dr. Mitchill, that while in Louisiana, in 1804, he visited Opelousas, within a few days after the exhumation of part of an under jaw and teeth of a mastodon.† We have been informed by our friend Dr. Griffin, of Virginia, that the greater part of a skeleton of this animal was disinterred a few years since. in Bottetourt county, Va. These bones were in very good preservation.

There is every reason to believe that the bones of this gigantic animal, as well as the relics of various other extinct species, will be procured in great abundance as the internal improvement of our happy country advances. The magnificent works already completed have given no inconsiderable earnest of what may be expected from numerous others now in progress, and the means which are intended to facilitate the intercourse of our citizens, and bind them more firmly together by mutual interests, may also contribute in a powerful degree to shed light on some of the most obscure and interesting topics connected

---

* Vide Breckenridge's View of Louisiana.
† Mitchill's Geological Appendix.

with the history of the globe. It should, therefore, be deeply impressed upon the minds of those who superintend the construction of canals, tunnels, roads, &c. that the fragments of organic remains which they might feel inclined to neglect as insignificant, may prove of the highest importance to science, when viewed in connexion with all the circumstances of their characters and positions, the peculiar nature of the superincumbent soil, and the general aspect of the surrounding country. When opportunities are presented, no pains should be spared, in order to procure bones, or other relics of animal bodies, with the least possible delay; and every attendant circumstance should be noted with the most scrupulous accuracy.

The last mastodon skeleton disinterred, was found in Monmouth county, New Jersey, three miles southwest of Long Branch. A grinder was presented by Mr. William Moore to the Lyceum of Natural History of New York, accompanied by information that the whole skeleton lay buried near the spot whence this tooth was obtained.*

This skeleton was accidentally discovered on Poplar farm, which is about two miles distant from the sea-beach, in 1823, fourteen months previous to the researches of Messrs. Cooper, Dekay and Van Rensselaer. The proprietor of the farm walking over a reclaimed marsh, observed something projecting

---

* See the Report of Messrs. Dekay, Van Rensselaer and Cooper, vol. i. p. 143 of the Annals of the Lyceum, whence this account is derived.

through the turf, which he struck loose with his foot and found to be a grinder tooth. Two other teeth, some pieces of the skull, the spine, humeri, and other bones, were afterwards exhumed.

The country adjacent to this farm is generally level, but a low and broad ridge, running parallel with the sea-coast, lies between it and the sea. At Poplar this high ground very gradually slopes on its western side, now disencumbered of its forests, and intersected by ditches, so that at some seasons it is nearly dry. It is stated to have been watery at a former period, and abounded in a species of poplar, whence the place obtained its name. Near the border of this marsh there was a shallow cavity containing a little water; the left foreleg had been removed therefrom, and several broken vertebræ and fragments of other bones were scattered on the surrounding turf. Having obtained permission to make farther explorations, these zealous inquirers commenced operations, and during two days, with the aid of some gentlemen who accompanied them, succeeded in obtaining all the bones of this skeleton which remain undecomposed. This valuable collection was added to the cabinet of the Lyceum, previously enriched by the specimens obtained at Chester county in New York.

The following is the account given by these gentlemen of the position in which the animal appeared to have been placed at the time of its extinction, whatever may have been the catastrophe which destroyed the whole of its race:

The surrounding soil " was a stratum of what is, by the German geologists, called *geest*, that is, a black, soft, shining earth, abounding near the surface in vegetable roots and fibres. Before the time of our visit, the skull, broken into many fragments, as well as the greater part of the spine, most of the ribs, both scapula, the left radius, ulna, and the humerus of the right arm, had been removed. Of the situation of these, therefore, we cannot speak from our own observation; but Mr. Croxson informed us, that, as before mentioned, part of the head had protruded itself through the surface of the ground. In consequence probably of this, it was so much decayed that he could find but three of the teeth, and no trace whatever of the tusks, nor was the search we afterwards made for these latter more effectual. The vertebral column, with all its joints, and the ribs attached to them in their natural position, lay about eight or ten inches below the surface. The scapulæ rested upon the heads of the humeri, and these in a vertical position upon the bones of the fore arm, as in life. We found the right fore arm still buried. It inclined a little backwards, and the foot, which was immediately below it, was placed a little in advance of the other, as it would be if the animal had been walking.

"At the depth of about ten inches, and immediately below the matting of turf, which forms the surface of the meadow, we came to the sacrum, with the pelvis remaining united to it, though in a very decayed state. The femora lay adjoining, but, unlike the bones of the fore legs, in a position nearly horizontal, the right less so than the left, and both at right angles with the spine. These were also, from their proximity to the surface, much decayed, so that the left fell to pieces on being lifted from its bed. Both tibiæ, each with its fibula, stood nearly erect under the extremities of the femora, and under them the bones of both hind feet in their proper relative position. We found no caudal vertebræ. The marsh had been cleared and drained about three years, and during that period, as the proprietor informed us, the level of its surface had lowered about two feet. To this may have, perhaps, been owing the horizontal position of the thigh bones, which would naturally be forced out of their originally erect position by the pressure of the heavy superincumbent bed of turf. The bones of the fore leg, however, do not appear to have been thus acted upon.

"The deeper we penetrated the sounder we found the bones, so that those of the foot, lying lowest, were obtained in a state of perfect preservation. The greater part of the bones had, adhering to their surface and in their cavities, the phosphats of iron and lime, and the sulphat of lime in very small quantities, the last in minute crystals. There were also considerable masses of oxyd of iron or bog-ore, which, however, abounded in various places in the marsh. Immediately underlying the stratum of black earth, we came to another of sand, having a ferruginous tinge, and containing numerous rolled quartz pebbles. Upon this sandy stratum the skeleton seemed to stand, so that the upper side of the foot was covered by the black earth; the sole rested immediately upon the sand. We found all the feet placed thus, the surface of the sandy stratum being apparently quite level."*

---

* Annals of the Lyceum, &c. vol. i. p. 145.

CHAPTER XVI.

Genus III.—Elephant; *Elephas;* L.

GENERIC CHARACTERS.

The head is of great size, supported upon a singularly short neck; the eyes are small, the ears of great extent, and, like the eyes, placed laterally. The snout is drawn out or extended to form the flexible trunk, through which is continued two canals leading from the nostrils: the extremity of the trunk is furnished with a small process, which both from its figure and mobility appears to perform the duties of a finger. The openings to the true nasal cavities are situated very high upon the head, and the bones of the nose are very thick, small, and triangular. The sinuses or cavities between the plates of the frontal and maxillary bones are enormously large, and increase to a great degree the volume of the skull. The lower jaw is massive and rounded at its angle: its branches terminate at the chin in a pointed extremity, between the sides of which there is a deep gutter or furrow.

The body is very large, and sustained at a considerable height from the ground by long and strong bones, whose articular surfaces are arranged upon a vertical line. The head of the thigh bone is in the axis of its shaft, and the cotyloid cavity for its reception is situated far forward, or rather on the inferior surface of the pelvis. The limbs are five-toed, but the digits are entirely concealed by the integuments, though their situations are designated externally by an equal number of horny plates, or small hoofs, on the inferior surfaces of the feet. On the hind feet one or two of these plates are sometimes deficient. The tail is of moderate length, and terminated by a brush or tuft of coarse hairs. The stomach is simple, the intestines of great volume, and the cœcum of vast magnitude: the liver has two lobes, but no gall-bladder. The teats, two in number, are situated upon the chest between the fore limbs.

Dental System.

10 Teeth:
- 6 Upper { 2 Incisive, (in the form of great tusks) 4 Molar.
- 4 Lower { 4 Molar.

The superior incisive teeth are, in this genus, represented by tusks of ivory, which are frequently of great length and thickness. They are cylindrical, arcuated downwards, and turned up at the point. Their texture consists of a peculiar osseous matter of a fine grain, which is intermingled with a harder and more compact substance, arranged in convergent curved lines, which cross each other so as to form very regular curvilinear lozenges. There is, moreover, a slight covering of true enamel to these tusks.

The molar teeth are made up of vertical and transverse layers, each of which is formed of osseous matter, surrounded by a plate of enamel, and the whole connected together by a solid inorganic substance or cement. These teeth grow obliquely from the posterior to the anterior part of the jaw.*

Species I.—*The Fossil Elephant.*

Elephas Primogenius; Blumenb.

Elephant Fossile, ou du Mammouth des Russes: C. Ossem. Foss. Nouv. ed. i. 75. Mitchill, Geological Appendix to Cuvier's Theory of the Earth.

The discovery of elephant bones in North America is a curious fact, which forcibly arrests attention and invokes a train of far-extending reflections relative to the mutations produced in the animal world, by the irresistible causes which, at various periods, have entirely changed the condition of the earth's surface. In the early ages of the world the fossil elephant, now utterly extinct, must have been extensively and abundantly distributed over the earth, as fragments of its skeleton have been disinterred not only in Asia, and throughout Europe, but in va-

---

* See Desmarest, *Mammalogie,* p. 381. F. Cuvier, *Dents,* etc. p. 221, and for a most luminous and ample account of the dentition of this genus, the reader may with great profit refer to the paper on Living and Fossil Elephants, contained in the first volume (new edition) of Cuvier's Recherches sur les Ossemens Fossiles. Corse, in the Philosophical Transactions, 1799, has given a great number of interesting details. Blake's work on the structure and formation of the teeth in man and various animals, is also of great value, in relation to these teeth.

rious parts of North America. From the greater numbers of bones which have been discovered, and the fortunate preservation of the entire animal, in the almost eternal ice of Siberia, less doubt is felt concerning the peculiar characters of this than any other extinct species.

Two living species of elephant are well known as inhabitants of Asia and Africa, whence they are named; the varieties of these species are neither numerous nor remarkable. The Asiatic is distinguished from the African by superior size and other peculiarities, the most striking of which is the arrangement of the perpendicular plates in the huge grinders; these in the first named species exhibit transverse undulating ribbons of enamel, while those of the African display on their crowns a succession of lozenge-shaped lines. The teeth of the fossil elephant resemble the Asiatic, but have straighter and narrower ribbons of enamel.

The localities whence the fossil elephant bones have been generally procured in this country, have in numerous instances been the same as those indicated in speaking of the mastodon. Scarcely any remains, except the teeth, have been discovered in these situations; the other bones having altogether decayed, would indicate that this elephant must have perished anterior to the remote period in which the mastodon bones were deposited in the same places. Kentucky, so remarkable for containing great numbers of the mastodon, has furnished the largest number of the teeth of the fossil elephant, but the state of South Carolina has thus far been found to contain the greatest quantity of other parts of the skeleton

Mitchill has given a figure of a fossil elephant-tooth, obtained in Monmouth Co. N. J.*

Drayton informs us, in his views of South Carolina, that Col. Senf, in 1794, discovered teeth of the elephant in Biggin Swamp, not far from the head of the west branch of Cooper river. They were found at a depth of eight or nine feet. A good figure of one of these grinders is given in Drayton's work.

According to Catesby, teeth of an elephant were found at Stono in Carolina, which were recognized by the Negroes (natives of Africa) as the grinders of that animal. This statement of Catesby is unnecessarily criticised by Cuvier, after Dr. B. S. Barton, since Catesby does not say that the Negroes recognized them as teeth of the African species of elephant, but merely that they were teeth of *an* elephant.

Dr. H. Hayden, of Baltimore, in his Geological Essays, gives an account of an elephant-tooth, which was found on the eastern shore of Maryland, in Queen Ann's county. This tooth differs considerably from the tooth either of the living or fossil species, resembling each in a certain degree. The distance from the crown to the roots of the tooth is nine inches; the grinding surface is also nine inches long, and the breadth four inches and a-half. Its present weight is ten pounds, and from the convexity of its outer surface, it is thought to be a grinder of the upper jaw.

The collection of the Philadelphia Museum is enriched with various specimens of fossil elephant-

---

* Mitchill, libro. citato.

teeth; and the cabinets of the American Philosophical Society, and of the Academy of Natural Sciences. contain numerous fragments of the skeleton of this animal.*

The characters by which the skeleton of this elephant is to be distinguished from the others, have been laid down by Cuvier, after a very extended and minute examination of vast numbers of perfect and mutilated specimens. The head is oblong, the forehead concave; the sockets for the tusks very large, and the molar teeth of great size. They are marked on the surface by parallel plates of enamel, very closely approaching each other. The lower jaw is obtuse in front. The tusks are exceedingly long, more or less arcuated spirally, and directed upwards.

We cannot offer any facts from which a sufficiently satisfactory conclusion can be drawn, relative to the time or manner in which this species became extinct; but the evidence afforded by the specimen obtained from the Siberian ice, renders highly probable the supposition that it was adapted to a much more northern climate than either of the elephants now known. The skin of this animal was covered with a long and coarse hair,† and by a finer and wooliy

---

* See Appendix, E.

† Cuvier, who received a piece of the skin of this animal, states that there are two, and even three, sorts of hair. The longest are from twelve to fifteen inches, of a brown colour, and about the thickness of horse-hair. Others are nine or ten inches long, rather more slender, and of a fawn colour. The wool, which seems to have been placed at the roots of

THE EXTINCT ELEPHANT. 259

hair, which is shorter and applied more closely to the surface.

The number of the relics of this animal found in Siberia is very great, and it is highly probable that the northern parts of this continent may hereafter furnish us with sufficient proofs of its abundant diffusion in the species. The explorations annually made in different parts of our southern and western country will doubtless enlarge our knowledge of this species, and afford data upon which opinions may hereafter be more advantageously based.

[We shall conclude this article by inserting a translation of great part of Mr. MICHA L ADAMS' account of his visit to the Siberian mammoth, or extinct elephant, which was through his zealous exertions preserved from final destruction, and at present belongs to the museum of St. Petersburg.

" I was informed at Yakoutsk, by M. Popoff, who is at the head of a company of merchants of that town, that they had discovered upon the shores of the Frozen Sea, near the mouth of the river Lena, an animal of extraordinary size, having the flesh, skin and hair in good preservation. It was believed that the fossil production known as mammoth-horns must have belonged to an animal similar to this. I commenced my journey on the 7th of June. 1806; on the 16th I arrived at the small town of Schigarsk, and near the end of the month reached Kumak surka, whence my excursion was made to search for the

---

the long hairs, is four or five inches long, somewhat fine and soft, and slightly curled, at its root especially: this is of a clear fawn colour.

mammoth. Accompanied by a Tonguse chief, *Ossip Schoumakoff*, and by Bellkoff, a merchant of Schigarsk, together with my huntsman, three Cossacs, and ten Tonguse, we set out upon our journey, mounted upon reindeer.

"On the third day of our journey we pitched our tents a few hundred paces from the mammoth, upon a hillock called *Kembisugashaeta*, signifying the stone with a broad side. Schoumakoff related the history of the discovery of the mammoth to me, in nearly the following words:

"The Tonguse, who are a wandering people, seldom remain long in one place. Those who live in the forests often spend ten years and more in traversing the vast regions among the mountains—during which period they never visit their homes. Each family lives separated from the rest; the chief takes care of them, and knows no other society. If, after several years of absence, two friends casually meet, they then mutually communicate their adventures, the various success of their hunting, and the quantity of peltry they have acquired. After spending some days together, and consuming their small stock of provisions, they separate cheerfully, charge each other with messages to their respective friends, and trust to chance for their future meetings. The Tonguse who inhabit the coast differ from the rest in having more regularly built houses, and in assembling at certain seasons for fishing and hunting. In winter they inhabit cabins built close to each other, so as to form small villages. It is to one of these annual excursions of the Tonguse that we are indebted for the discovery of the mammoth.

THE EXTINCT ELEPHANT.

"Towards the end of August, after the fishing in the Lena is over, Schoumakoff is in the habit of going, along with his brothers, to the peninsula of Turmut, where they employ themselves in hunting, and where the fresh fish of the sea furnish them with wholesome and agreeable nourishment.

"In 1799 he built for his women some cabins upon the shores of the lake Onroul; and he himself coasted along the sea-shore in order to seek for mammoth-horns. One day he observed, in the midst of a rock of ice, an unformed block, which by no means resembled the pieces of wood usually found there. He clambered up the ice and examined the new object on all sides. The ensuing year he found at the same spot the carcase of a walrus, and remarked that the mass he had formerly examined was freer from the ice, and by the side of it he perceived two similar pieces, which he afterwards found were the feet of the animal. About the close of the next summer, the entire flank of the animal, and one of the tusks, had distinctly come out from under the ice. Upon his return to the shores of the lake Onroul, he communicated this extraordinary discovery to his wife and some of his friends; but their manner of regarding the subject overwhelmed him with grief. The old men related, on this occasion, that they had heard their forefathers say that a similar monster had formerly shown itself in the same peninsula, and that the whole family of the person who had discovered it had become extinct in a very short time. In consequence of this, the mammoth was regarded as auguring a future calamity, and the Tonguse chief felt so much inquietude from it that he fell dangerously ill; but recovering again, his first suggestions were of the profit he might gain by selling the tusks of the animal, which were of extraordinary size and beauty. He therefore gave orders that the place where the animal was found should be carefully concealed, and all strangers removed from it under various pretexts, charging at the same time some trusty dependents not to suffer any part of this treasure to be carried away.

"The summer proved colder and more windy than usual, and kept the mammoth sunk in the ice, which scarcely melted all that season. At last, about the end of the fifth year afterwards, the ardent desires of Schoumakoff were happily accomplished: the ice which enclosed the animal having partly melted, the level became sloped, and this enormous mass, pushed forward by its own weight, fell over upon its side on a sand-bank. Of this, two Tonguse, who accompanied me in my journey, were witnesses. In the month of March, 1804, Schoumakoff came to his mammoth, and having cut off the tusks, exchanged them with the merchant Baltounoff for goods of the value of fifty rubles. On this occasion a drawing of the animal was made, but it was very incorrect; they described it with pointed ears, very small eyes, horse's hoofs, and a bristly mane along the whole of his back, so that the drawing represented something between a pig and an elephant.

"Two years afterwards, being the seventh from its first being discovered, a fortunate circumstance caused my visit to these distant and desert regions, and I congratulate myself upon having had it in my power to ascertain and verify a fact which would otherwise be thought so improbable. I found the animal still in the same place, but exceedingly mutilated. The prejudices against it having been dissipated by the Tonguse chief's recovery, the carcase might be approached without difficulty: the proprietor was content with the profit he had derived from it, and the Yakouts of the neighbourhood tore off the flesh, with which they fed their dogs. Ferocious animals, polar bears, gluttons, wolves, and foxes, preyed upon it also, and their burrows were seen in the neighbourhood. The skeleton, almost unfleshed, was entire, with the exception of one of the fore feet. The back-bone, from the head to the os coccygis, the pelvis, and the remains of the three extremities, were still firmly attached by the ligaments of the joints, and by strips of skin on the exterior side of the carcase. The head was covered with a dry skin; one of the ears, well preserved, was furnished with a tuft of bristles. All these parts must necessarily have suffered by a carriage of several thousand miles. The eyes, however, are preserved, and we can still distinguish the ball of the left eye. The tip of the under lip has been eaten away, and the upper part being destroyed, the teeth were laid bare. The brain was still within the cranium, but appeared dry.

"The parts least damaged are a fore foot and a hind one; they are covered with skin, and still have the sole attached. According to the assertion of the Tonguse chief, the animal had been so large and well fed that its belly hung down below the knee-joints. This animal was a male with a long mane at his neck, but it has no tail and no trunk.* Three-fourths of the skin were obtained; the whole is of a dark gray, and covered with reddish hair and black bristles. The humidity of the soil where the animal had lain so long has deprived the bristles of some part of their elasticity. The entire skeleton is about nine feet and a-half high,† and is fourteen feet in length from the tip of the nose to the coccyx.‡ The tusks are nine feet long, and weigh, each two hundred pounds.§ The head alone weighs four hundred and sixty pounds.

"The bones were separated and arranged with scrupulous care; and I had the satisfaction of finding the other shoulder-blade, which lay in a hole. I afterwards caused the skin to be stripped from the side upon which the animal had lain: it was in good preservation. This skin was of such extraordinary weight, that ten persons, who were employed to carry it to the sea side, to stretch it upon floating wood, moved it with great difficulty. After this was accomplished, I caused the ground to be dug in various places, in order to see if there were any bones around, but chiefly for the purpose of collecting all the bristles, which the white bears might have trodden into the wet ground on devouring the flesh. This operation was attended with difficulty, on account of the deficiency of proper tools for digging; however,

---

* These parts were, doubtless, removed by the animals which fed upon the carcase.
† Four archines. ‡ Seven archines. § Each five poods.

THE FOSSIL ELEPHANT. 265

we succeeded in procuring more than forty pounds* of bristles.

"The place where I found this animal is sixty paces distant from the sea shore, and about one hundred paces distant from the ice, whence it had fallen down. The fracture in the ice is exactly in the middle between the two points of the isthmus, and is three wersts long, and in the place where the body of the animal was situated, the rock of ice has a perpendicular elevation of one hundred and eighty or one hundred and ninety feet. Its substance is a clear ice, but of a nauseous taste; it slopes towards the sea. Its summit is covered by a bed of moss and friable earth, more than a foot in thickness. During the heat of the month of July a part of this crust melts, but the other remains frozen. Curiosity prompted me to ascend two other hillocks equally distant from the sea; they were of the same composition, and also slightly covered with moss. At intervals I saw pieces of wood of an enormous size and of all the species produced in Siberia; and also mammoth horns (elephant tusks) in great quantities frozen between the fissures of the rocks. They appeared to be of an astonishing freshness."]

---

* More than one pood.

CHAPTER XVII.

Order VI.—Pecora; *Ruminant Animals.*

These animals are peculiarly distinguished by having no incisive teeth in the upper jaw. the intermaxillary bone, covered by a hardened gum, being opposed to the incisors of the lower jaw, which are almost universally eight in number. Between these and the molar teeth there is a vacant space, except in certain genera having one or two canines. There are very uniformly six molars on each side of both jaws; these have their crowns marked by two double crescents, the convexity of which in the upper jaw is turned inwards, and in the lower jaw outwards.

The feet are all two toed, and these toes are covered by two hoofs, which approach each other by flat surfaces, whence they have the appearance of a single hoof cleft in the middle, a circumstance which has obtained for these animals, in various languages, the designation of cloven footed, &c. In some genera, there are behind these hoofs two small ones or rudimental hoofs, which are the only traces of lateral toes. The two bones of the metacarpus and metatarsus are consolidated to form one bone, which is called the *cannon bone.*

The most singular faculty possessed by these animals is that of rumination, or of returning the food to the mouth to subject it to a second mastication after it has been once swallowed. This process depends on the number and peculiar arrangement of their complicated stomachs.

The first stomach is called *rumen* or paunch,* which is divided externally at its extremity into two saccular appendices, and slightly separated into four parts on the inside, having a vast number of flattened papillæ over the internal surface.

The second is called *reticulum* or honeycomb,† and is distinguished from the first by its small and globular appearance, and by the beautiful arrangement of its internal membrane, which forms polygonal acute-angled cells.

The third stomach is the smallest of all, and is termed *omasum* or feck.‡ Its internal membrane is arranged in longitudinal folds, varying in breadth, in a regular alternate order.

The fourth stomach is called *abomasum* or reed,§ is next in size to the paunch, and is of an elongated pear-shape, having its internal membrane simply wrinkled longitudinally like the human stomach.

The three first named stomachs are connected with each other and a groove-like continuation of the œsophagus in the following manner. The groove-like continuation enters where the paunch, reticulum, and omasum, approach each other, and thence it is continued with the groove which ends in the third stomach. The groove is therefore open to the first

---

* Also ingluvies, magnus venter, penula.
† Ollula, bonnet, king's hood, &c.
‡ Echinus, conclave, centipellis, manyplies, book, feuillet.
§ Faliscus, ventriculus intestinalis.

stomachs which lie to its right and left. The thick and prominent margins of this groove allow them to be drawn together, so as to form a complete tube, and then the œsophagus is continued direct into the third stomach.*

The most generally received opinion on the act of rumination is, that the food is coarsely broken at the first mastication, and when swallowed passes into the paunch. It is thence gradually passed into the second stomach, where it undergoes a certain degree of maceration in the fluids of the organ, and is formed into little balls, which by a sudden contraction of this stomach are impelled through the œsophagus or gullet to the mouth. It is then subjected to the second more effectual mastication,† is again swallowed and passes directly into the third stomach, and after remaining in this for a certain time it finally enters the fourth, simple or true digestive stomach. This account of the stages of the act of rumination is adopted by BLUMENBACH, CUVIER, &c. TOGGIA‡ in part following the doctrine of BRUGNONE, sustains the opinion that the food, after the first mastication, enters the *paunch* only, and not the reticulum or second stomach. In the paunch, moreover, by the fluids which are poured out from its internal surface, and by the structure and regular movements of its parietes, the mass is softened, divided and formed into small pellets, which are brought by the contractions

---

* See Blumenbach's Comp. Anat. p. 137.

† Vide Cuvier, Regne Animal, 247.

‡ Della ruminazione e digestione de'Ruminanti; Turino 1819, 8vo. op. cit. per Ranzani.

of the organ to its cardia, and ascend the œsophagus to the mouth for the second mastication. Then the food is returned to the reticulum by means of the groove-like continuation; there it remains for a certain time, unless the matter be mixed or fluid, in which case it passes at once into the third or fourth stomach. Toggia is persuaded that it occurs in this and in no other way, because, 1st, when he had attentively examined the structure of the groove, he was convinced that nothing but finely comminuted food could pass through it, and not herbage but once and imperfectly masticated. 2d, When he examined the stomachs of ruminant animals killed either at the commencement of the rumination, during this process, or immediately after it, he found the food which had been only once masticated, in the paunch alone; the food reduced by the second mastication was contained in the reticulum or second stomach; that which was imbued with fluid in the omasum or third, and finally, abundantly mixed with fluids or in a semifluid state, in the abomasum or fourth stomach.*

The rumen or paunch is comparatively small in the young or suckling animal, and does not acquire its enormous size, until it has been for some time the receptacle of food. The intestinal canal is very long in ruminant quadrupeds, but not voluminous in the larger portion; the cœcum itself, is long and rather even. The teats are situated between the thighs.

The fat of these animals is remarkable for its

---

* Ranzani Elementi di Zoologia, tomo 2do. parte 3a.

hardness when cooled; it may then be broken into pieces. It is well known in commerce and the arts under the name of *tallow*.

To this order of animals man is more largely indebted than to all the rest of animated nature. The mass of his food, is obtained from their flesh, and there is no part of their bodies from which he does not derive additions to his comforts, and assistance to his arts. Their hides, horns, bones, hair, flesh, fat, milk, and even their blood are in hourly demand. Many of them during their lives yield him valuable services as beasts of draught and burthen, and contribute amply to his sustenance and luxury when they are finally slaughtered. Peaceful and patient in their dispositions, they feed exclusively on the verdure which is scattered over the earth, and prepare this vegetable matter most efficiently for the use of man and other creatures, by converting it into their own flesh, which is edible throughout all the members of the order, and in a large proportion is delicious food.

CHAPTER XVIII.

Genus IV.—Cervus; L. *Deer*.

Gr. Ελαφος. *Fr.* Cerf.
Lat. Cervus. *Germ.* Hirsch.

GENERIC CHARACTERS.

The head, which is elongated, is not very large, and most generally terminates by a smooth membranous surface which is called the muzzle; the nostrils are acutely oval and laterally situated; the eyes large and well proportioned, having the pupils transversely extended. At a short distance below the inner angle of the eye a peculiar pouch or cavity is found in most of the species, which secretes an unctuous humour in small quantities; these cavities are called *larmiers* by the French naturalist. The ears are large and pointed; the neck is of moderate length, the body plump, and the limbs slender, though strongly knit. The teats are inguinal, and four in number: the gall bladder does not exist in these animals. The tail is short.

The hair is very similar in colour throughout the species of this genus, and is dry and harsh; the young deer or fawns are mostly spotted with white upon a brownish yellow ground.

The males of this genus are all provided with horns, which are variously branched, or palmated, and are annually caducous. These horns are remarkable for being composed of *bone*, which is solid, throughout, and in its first or growing state is covered by a velvet-like membrane, through which blood circulates with great freedom. The horn commences its growth from a basis or peduncle which is attached to the frontal bone, having something of the form of a truncated cone; a short distance above this, on the level of the outer surface of the skin of the head, the horn is expanded in the form of an irregular tuberculous ring, which is called the *burr*,* above which the solid part of the horn rises to form the various branches or plantations, according to the species. The blood-vessels going to the horn are very large at the commencement and during its growth, and the extension of the velvet-like membrane is as rapid as the advance of the bone or horn. As soon as the horn attains its full growth the blood-vessels contract and diminish until they cease to convey blood to the velvet membrane, which then dries, loses its sensibility, and gradually flakes off. After the rutting season a slight tumescence occurs at the edge of the peduncle, and the whole horn is at length detached and falls off.

Dental System.

32 Teeth:
- 12 Upper { 0 Incisive, 0 Canine, 12 Molars. } { 6 False, 6 True Molars. }
- 20 Lower { 8 Incisive, 0 Canine, 12 Molars. } { 6 False, 6 True Molars. }

IN THE UPPER JAW the three first molars are bordered by a thick crest at their internal edges; the

---

* The part commonly used for cane-heads, &c.

two following are formed of two parts, each of which is composed of a single tubercle, having two crests in front, one on the outside, terminating abruptly, the other on the inside, which descends as far as the middle of the height of the tooth, and then rises upwards to rejoin the anterior border of the principal tubercle; between this crest and the tubercle there is a hollow. When the tubercle begins to wear, it exhibits a portion of a narrow circle, bordered by enamel. The last molar differs from the two preceding solely in being somewhat narrower, and in having thinner crescents.

In the lower jaw the first incisor is the largest, the second and third are somewhat less than the other, and the last is very small. They are all trenchant, inclined forwards, and separate themselves slightly from the median line. The two first false molars are simple, the third has a spur at its posterior part, and the three last differ very slightly from each other.

In their reciprocal position the inferior incisors correspond to the superior maxillary bone; the molars are alternate.

The writings of naturalists exhibit great confusion relative to the North American species of deer.—Much of this evil is attributable to the loose manner in which species have been proposed upon the authority of persons unqualified to distinguish between accidental varieties, dependent upon sex or age, and those permanent characteristics indicative of specific constitution.

Cuvier, with his usual acumen and amplitude of research, has turned his attention to this subject, with great advantage to students of natural history. Though he may not have been the first or only naturalist who knew and discriminated correctly the North American species, he is the first who has displayed his researches in such a manner as will enable every one to satisfy himself of the accuracy of his deductions.* He has admitted the following to be the species now inhabiting this country, all the others named as distinct in the books being mere varieties: *C. Alces*, the Moose; *C. Canadensis*, the American Elk; *C. Tarandus*, the Rein Deer; *C. Virginianus*, the Common Deer.

To this list must be added the *C. Macrotis*, Mule or Black-tail Deer, first indicated by Lewis and Clarke, and described by SAY, under the name just given, in Long's Expedition to the Rocky Mountains.

Species I.—*The Moose.*

Cervus Alces; L.

Alces, Achlis: Plin. Ald. Gesn. Jonst.
Original: Charlev. Nouv. France, iii. 126.
Elan: Buff. Hist. Nat. xii. Supp. vii.
Elk: Shaw, Gen. Zool. ii. pt. 2. 174.
Moose Deer: Dudley, Phil. Trans. No. 444. Warden, Descript. des Etats Unis. v. p. 636.
Elk: Penn. Hist. Quad. No. 42. *Moose:* Ib. Arct. Zool. i. No. 3. p. 18.

The Moose† is perhaps the only deer whose general appearance can be called ungraceful, or whose

---

* Ossemens Fossiles, nouv. ed. tome iv.
† This appellation is derived from *Musu*, the name given to the animal by the Algonquins.

proportions at first sight impress the beholder unfavorably. Its large head terminates in a square muzzle, having the nostrils curiously slouched over the sides of the mouth; the neck, from which rises a short thick mane, is not longer than the head, which in males is rendered still more cumbrous and unwieldy, by wide palmated horns: under the throat is found an excrescence from which grows a tuft of long hair; the body, which is short and thick, is mounted upon tall legs, and the whole aspect is so unusual that incidental observers are pardonable for considering it ugly. Yet as these singularities of structure have direct or indirect reference to peculiarities of use, an inquiry into the mode of life led by this species, may cause us to forget, in admiration of its adaptation to circumstances, prejudices excited by the comparative inelegance of its form.

The moose inhabits the northern parts of both continents;* on the American it has been found as far north as the country has been fully explored; its southern range, at former periods, extended to the shores of the great lakes and throughout the New England States. At present it is not heard of south of the state of Maine, where it is becoming rare. In Nova Scotia, the isle of Breton, the country adjacent to the bay of Fundy, and throughout the Hudson's Bay possessions, the moose is found in considerable numbers.

The dense forests and closely shaded swamps of

---

* It is, in Europe, called "Elk."

these regions are the favorite resorts of this animal, as there the most abundant supply of food is to be obtained with the least inconvenience. The length of limb and shortness of neck, which in an open pasture appear so disadvantageous, are here of essential importance, in enabling the moose to crop the buds and young twigs of the birch, maple, or poplar, or should he prefer the aquatic plants, which grow most luxuriantly where the soil is unfit to support other animals, the same length of limb enables him to feed with security and ease. We cannot avoid believing that the peculiar lateral and slouching position of the nostrils is immediately connected with the manner in which the moose browses. Their construction is very muscular, and seems well adapted for seizing and tearing off the twigs and foliage of trees, and conveying them to the mouth; it may also be designed to prevent the sense of smell from being at any time suspended by the prehension of food. The probability of this last suggestion is strengthened by the fact that the moose is endowed with an exquisite sensibility of smell, and can discover the approach of hunters at very great distances. When obliged to feed on level ground, the animal must either kneel or separate the legs very widely; in feeding on the sides of acclivities, the moose does so with less inconvenience by grazing from below upwards; the steeper the ground may be, so much the easier is it for this species to pasture. Yet, whenever food is to be procured from trees and shrubs, it is preferred to that which is only to be obtained by grazing.

The moose, like his kindred species, is a harmless

and peaceful animal, except in the season when the sexes seek each other. Then the males display a fierceness and pugnacity which forms a strong contrast to their ordinary actions; were they examined only during such seasons, the character of the species would be entirely misconceived. Under the influence of this powerful, though temporary excitement, the males battle furiously with each other, and resist the aggressions of man himself with vigour and effect.

In the summer the moose frequents swampy or low grounds near the margins of lakes and rivers, through which they delight to swim, as it frees them for the time from the annoyance of insects. They are also seen wading out from the shores, for the purpose of feeding on the aquatic plants which rise to the surface of the water. At this season they regularly frequent the same place in order to drink, of which circumstance the Indian hunter takes advantage to lie in ambush, and secure the destruction of the deer. At such drinking places as many as eight, or ten pairs of moose horns have been picked up.

During the winter the moose, in families of fifteen or twenty, seek the depths of the forest for shelter and food. Such a herd will range throughout an extent of about five hundred acres, subsisting upon the mosses attached to the trees, or browsing the tender branches of saplings, especially of the tree called moose-wood. The Indians name parts of the forest thus occupied moose-yards.

In Nova-Scotia, New-Brunswick, and the island of Grandmanan, the moose is generally hunted in the month of March, when the snow is deep, and sufficiently crusted with ice to bear the weight of a dog, not that of a moose, as has been stated. Five or six men, provided with knapsacks, containing food for as many days, and all necessary implements for building their "camp" at night, set out in search of a moose-yard. When they have discovered one, they collect their dogs and encamp for the night, in order to be ready to commence the chase at an early hour, before the sun softens the crust upon the snow, which would be the means of retarding the dogs, and facilitating the escape of the deer. At daybreak the dogs are laid on, and the hunters, wearing large snow shoes, follow as closely as possible. As soon as the dogs approach a moose, they assail him on all sides, and force him to attempt his escape by flight. The deer, however, does not run far, before the crust on the snow, through which he breaks at every step, cuts his legs so severely, that the poor animal stands at bay and endeavours to defend himself against the dogs by striking at them with his forefeet. The arrival of the hunter within a convenient distance soon terminates the combat, as a ball from his rifle rarely fails to bring the moose down.

Judging by the rapid diminution of this species within a comparatively few years, it is to be feared that it will, at no great distance of time, be exterminated. The moose is easily tamed, although of a wild and timid disposition; sometimes when taken very young they are domesticated to a remarkable degree. We are informed by our friend Mr. Vanbuskirk, of New-Brunswick, that he knew of one which was taken, when two days old, by an Indian, and presented to a gentleman in Nova-Scotia. The proprietor allowed it to suck a cow for three months, and afterwards fed it with different vegetables, until it was a year old. This moose displayed a singular animosity against one of the young ladies of the family, and would chase her with fierceness into the house. When the door was closed in time to exclude him, he would immediately turn round and kick violently against it.*

The horns of the moose spread out almost immediately from their base into a broad palmation: in old animals they increase to a great size, and have been known to weigh fifty-six pounds, each horn being thirty-two inches long. The horns are generally cast in the month of November; the Indians employ

---

* "In the year 1777, an Indian had two young moose so tame, that when on his passage to Prince of Wales' fort in a canoe, the moose always followed him along the bank of the river, and at night, or on any other occasion, when the Indians landed, the young moose generally came and fondled on them in the same manner as the most domestic animal would have done, and never offered to stray from the tents. Unfortunately, in crossing a deep bay in one of the lakes, on a fine day, all the Indians that were not interested in the safe landing of those engaging creatures, paddled from point to point; and the man that owned them not caring to go so far about by himself, accompanied the others in hopes they would follow him round as usual. But at night the young moose did not arrive, and as the howling of some wolves was heard in that quarter, it was supposed they had been devoured by them, as they were never afterwards seen." HEARNE, 8vo. Ed. p. 258.

them for various purposes, cutting them into spoons, scoops, &c.

When chased, the moose throws his horns towards his neck, elevates his nose, and dashes swiftly into the thickest of the forest; occasionally the horns prove the means of his destruction, by being entangled among vines, or caught between small trees. Where the moose runs over a plain, he moves with great celerity, although his gait is nothing better than a sort of long shambling trot: this, however, is rendered very efficient, by the great length of his limbs. While running in this manner the divisions of the hoofs, which are very long, separate as they press the ground, and close together as they are raised, with a clattering sound, which may be heard to some distance; this circumstance is also remarked in the rein-deer.

Notwithstanding the ease and swiftness of their movements, they would be easily captured, if pursued by horsemen and hounds, in a country adapted to such a chase, as they are both short breathed and tender footed.

The acuteness of their sense of hearing, thought to be that which is possessed by the moose in the greatest degree of perfection; together with the keenness of their smell, renders it very difficult to approach them. The Indians attempt it by creeping among the trees and bushes, always keeping to leeward of the deer. In summer, when they resort to the borders of lakes and rivers, the Indians often kill them while crossing the streams, or when swimming from the shore to the islands. "They are," says HEARNE, "when pursued in this manner, the most inoffensive of all animals, never making any resistance; and the young ones are so simple, that I remember to have seen an Indian paddle his canoe up to one of them and take it by the poll, without the least opposition: the poor harmless animal seeming, at the same time, as contented alongside the canoe as if swimming by the side of its dam, and looking up in our faces with the same fearless innocence that a house lamb would, making use of its forefoot almost every instant to clear its eyes of musquitoes, which at that time were remarkably numerous."

The flesh of the moose, though generally coarser and tougher than other venison, is esteemed excellent food, and the Indians, hunters and travellers, all declare they can withstand more fatigue while fed on this meat than when using any other. The large and gristly extremity of the nose is accounted an epicurean treat, and the tongue of the animal is also highly praised, notwithstanding it is not commonly so fat and delicate as the tongue of the common deer. As the moose feeds upon the twigs, buds and small branches of the willow, birch poplar, mosses, aquatic plants, &c. its flesh must be peculiarly flavoured. "The fat of the intestines is hard like suet, but all the external fat is soft like that of a breast of mutton, and when put into a bladder is as fine as marrow. In this they differ from all the other species of deer, of which the external fat is as hard as that of the kidneys."* The female moose never has any horns; they bring forth their

---

* Hearne.

young, "from one to three in number, in the latter end of April or beginning of May."*

The male moose often exceeds the largest horse in size and bulk; the females are much less than the males, and differently coloured. The hair of the male is long and soft like that of a common deer; it is black at tip, but within it is of an ash colour, and at the base pure white. The hair of the female is of a sandy brown colour, and in some places, especially under the throat, belly and flank, is nearly white at tip, and altogether so at base.

The skin of the moose is of great value to the Indians, as it is used for tent covers, clothing, &c. We shall defer the account of the methods of dressing these and other deer skins, until we treat of the common deer, when we shall describe the Indian modes of currying proper to each of these skins.

The moose, like other deer inhabiting the northern regions, is exceedingly annoyed by insects, which not only feast upon its blood, but deposit their eggs in different parts of its body, along the spine, within the cavities of the nose, mouth, &c. These eggs when hatched form large larvæ or maggots, that feed on the parts within which they are placed, until ready to assume their perfect or winged condition, when they perforate the skin and take flight. So great a number of such perforations are made at certain seasons, that the skins of the moose are rendered worthless to the hunter, unless it be for the purpose of cutting them into thongs for nets and other uses.

---

* Hearne.

Species II.—*The Rein-Deer.*

cervus tarandus; L.

Tarandus: Plin. Hist. Nat. viii. c. 34.
    Ælian, Anim. ii. c. 16.
Caribou: Charlev. Nouv. France, iii. p. 129; Dobbs, Hudson's Bay, 20.
Greenland Deer: Catesby's Carolina, App. p. 28.
Renne: Buff. Hist. Nat. xii.
Reinthier, Tharandthier: Gesn. Thierb. p. 206, 209.
Greenland Buck: Edw. Av. 1, tab. 51.
Rein-Deer: Penn. Arct. Zool. Ibid. Quad. p. 46.

This valuable animal is found in great abundance in the northern parts of both continents, and constitutes a very considerable part of the subsistence of the tribes inhabiting the regions it frequents. In the northern parts of Asia and Europe, the rein deer has been domesticated for a long time, and with the exception of the dog, is the only beast of draught or burthen possessed by the natives. The North American Indians, however, have never profited by the docility of this animal to aid them in transporting their families or property, though they annually destroy great numbers of them for the sake of their flesh, hides, horns, &c.

During the winter they take shelter in the forests, whence they are occasionally induced by the occurrence of a few fine days to pay a short visit to their favourite pastures on the barren grounds, which are covered with a profusion of mosses. Their great movement to the northward commonly begins towards the end of April, when the snow first melts from the sides of the hills; they are found on the banks of the Copper-Mine River early in May, at which time a considerable extent of ground is free from snow. In this spring migration the females take the lead, and bring forth their young on the sea coast about the end of May or beginning of June. They retire from the sea coast in July and August, but linger in the vicinity of the barren grounds as late as October, whence they seek their winter retreats in the woods.

In their migrations the whole herd frequently amounts to one or two thousand, and is separated into smaller herds, varying in number from ten to a hundred, as chance or their fears may determine them to unite or separate. The Indians have remarked that there are certain places which the rein-deer invariably visit in their migrations to and from the coast, and that they always travel against the wind. In the barren grounds the principal food of this species is the various lichens or mosses; the hay or dry grass found in the swamps during autumn is also eaten, and in the woods the mosses attached to the trees. " They are accustomed to gnaw their fallen antlers, and *are said* also to devour mice."

Some rein deer are never met with except in the woody country, and they are much larger than those which visit the coast. This variety is stated to weigh from 200 to 240lbs., while the weight of the common rein-deer, exclusive of the offal, varies from 90 to 130lbs. The large variety are found to have their skins as much perforated by the larvæ of the gadfly as the others, which is considered as a presumptive proof, by Capt. Franklin, that the smaller deer are not driven to the sea coast and islands of the Polar sea by the attacks of that insect. A few rein-deer killed in the spring are found to have their skins uninjured, and these are always fat, though all the other deer are lean at the same season.

As we have not had an opportunity either of becoming acquainted with this species in its native wilds, or of seeing any individuals in a state of captivity, we shall here introduce an account from the accurate observations of Hearne, confirmed by the recent and interesting remarks of Capt. Franklin, given in the narrative of his first and memorable journey to the shores of the Polar sea.

" When the Indians design to impound deer, they look out for one of the paths in which a number of them have trod, and which is observed to be still frequented by them. When these paths cross a lake, a wide river, or a barren plain, they are found to be much the best for the purpose; and if the path run through a cluster of woods, capable of affording materials for building the pound, it adds considerably to the commodiousness of the situation. The pound is built by making a strong fence of brushy trees, without observing any degree of regularity, and the work is continued to any extent, according to the pleasure of the builders. I have seen some that were not less than a mile round, and am informed that there are others still more extensive. The door or entrance of the pound is not larger than a common gate, and the inside is so crowded with small counter hedges as very much to resemble a maze; in every opening of which they set a snare made with thongs of parchment, deer skins, &c. twisted together, which are amazingly strong. One end of the snare is usually made fast to a growing pole; but if no one of sufficient size can be found near the place where the snare is set, a loose pole is substituted, which is always of such size and length, that a deer cannot drag it far before it gets entangled among the other woods which are all left standing, except what is found necessary for making the fence, hedges, &c.

" The pound being thus prepared, a row of small brushwood is stuck up in the snow on each side the door or entrance, and these hedgerows are continued along the open part of the lake, river or plain, where neither stick nor stump besides is to be seen. These poles or brushwood are generally placed at the distance of fifteen or twenty yards from each other, and ranged in such a manner as to form two sides of a long acute angle, growing gradually wider in proportion to the dimensions of the pound, which is sometimes not less than two or three miles, while the deer path is exactly along the middle, between the two rows of brushwood.

" Indians employed on this service always pitch their tent on or near to an eminence that affords a commanding prospect of the path leading to a pound; and when they see any deer going that way, men, women and children, walk along the lake or river side, under cover of the woods, until they get behind them, then step forth to open view, and proceed towards the pound in the form of a crescent. The poor timorous deer finding themselves pursued, and at the same time taking the two rows of bushy poles to be two ranks of people stationed to prevent their passing on either side, run straight forward in the path till they get into the pound. The Indians then close in, and block up the entrance with some brushy trees that have been cut down and lie at hand for that purpose. The deer being thus enclosed, the women and children walk round the pound to prevent them from jumping over the fence, while the men are employed in spearing such as are entangled in the snares, and shooting with bows and arrows those which remain loose in the pound. This method of hunting, if it deserves the name, is sometime so successful that many families subsist by it without having occasion to remove their tents above once or twice during the whole course of a winter; and when the spring advances both the deer and Indians draw out to the eastward, on the ground which is entirely barren, or at least what is so called in those parts, as it neither produces trees nor shrubs of any kind, so that moss, and some little grass, is all the herbage to be found on it.

" The great destruction of the deer in the month of August, for the sake of their skins, which are then fittest for use, is almost incredible; and as they are never known to have more than one fawn at a time, it is wonderful they do not become scarce. But so far is this from being the case, that the oldest northern Indian will affirm, that the deer are as plentiful now as they ever have been; and though they are remarkably scarce some years near Churchhill river, yet it is said, and with great probability of truth, that they are more numerous in other parts of the country than they were formerly. The scarcity or abundance of these animals in different places at the same season, is caused in a great measure, by the winds which prevail for some time before; for the deer are supposed by the natives to walk always in the direction from which the wind blows, except when they migrate from east to west, or from west to east, in search of the opposite sex.

"It requires the prime parts of from eight to ten deer skins to make a complete suit of warm clothing for a grown person during the winter; all of which should, if possible, be killed in the month of August or early in September, for after that time the hair is too long, and at the same time so loose in the pelt, that it will drop off with the slightest injury. Besides these skins, which must be in the hair, each person requires several others to be dressed into leather, for stockings and shoes and light summer clothing. Several more are also wanted in a parchment state to make *clewla*, as they call it, or thongs for the nettings of snow-shoes, snares for deer, sewing for their sledges, and in fact for every other use where strings or lines of any kind are required; so that each person on an average expends, in the course of a year, upwards of twenty deer skins in clothing and other domestic uses, exclusive of tent cloths, bags, and many other things. All skins for the above-mentioned purposes are, if possible, procured between the beginning of August and the middle of October; for when the rutting season is over, and the winter sets in, the deer skins are not only very thin, but in general full of worms and warbles, which render them of little use, except for thongs. Indeed, the chief use that is made of them in winter is for the purpose of food; and really, when the hair is properly taken off, and all the warbles are squeezed out, if they are well boiled they are far from being disagreeable. The Indians, however, never could persuade me to eat the warbles, (maggots of the gad-fly,) of which some of them are remarkably fond, particularly the children. They are always eaten raw and alive out of the skin, and are said by those who like them to be as fine as gooseberries. But the very idea of eating such things, exclusive of their appearance, (many of them being as large as the first joint of the little finger,) was quite sufficient to give me an unalterable disgust to such a repast.

"The month of October is the rutting season with these deer, and after the time of their courtship is over the bucks separate from the does; the former proceed to the westward, to take shelter in the woods during the winter, and the does keep out in the barren ground the whole year. This, though a general rule, is not without some exceptions, for I have frequently seen many does in the woods, though they bore no proportion to the bucks. This rule, therefore, only stands good respecting the deer to the north of Churchill river, for the deer to the southward live promiscuously among the woods as well as in the plains, and along the banks of the rivers, lakes, &c. the whole year.

"The old buck's horns are very large, with many branches, and always drop off in the month of November, which is the time they begin to approach the woods. This is undoubtedly wisely ordered by Providence, the better to enable them to escape from their enemies through the woods, otherwise they would become an easy prey to wolves and other beasts, and be liable to get entangled among the trees, even in ranging about in search of food. The young bucks in those parts do not shed their horns as soon as the old ones; I have frequently seen them killed at or near Christmas, and could discover no appearance of their horns being loose. The does do not shed their horns till the summer, so that when the buck's horns are ready to drop off, the horns of the doe are all hairy, and scarcely come to their full growth."*

"The haunches of the male rein-deer, in the beginning of the month of October, are covered to the depth of two inches or more with fat, which is beginning to get red and high flavoured, and is considered as a sure indication of the commencement of the rutting season. Their horns, which in the middle of August were yet tender, have now attained their proper size, and are beginning to lose their hairy covering, which hangs from them in ragged filaments. The horns of the rein-deer vary not only with its age and sex, but are otherwise so uncertain in their growth, that they are never alike in any two individuals. The old males shed theirs about the end of December; the females retain them until the disappearance of the snow enables them to frequent the barren grounds, which may be stated to be about the middle or end of May, soon after which period they proceed towards the sea coast and drop their young. The young males lose their horns about the same time with the females, or a little earlier, some of them as early as April. The hair of

---

* HEARNE, passim. 8vo. Ed

the rein-deer falls in July, and is succeeded by a short thick coat of mingled clove, deep reddish and yellowish browns—the belly and under parts of the neck, &c. remaining white. As the winter approaches, the hair becomes longer and lighter in its colours, and it begins to loosen in May, being then much worn on the sides from the animal rubbing itself against trees and stones. It becomes grayish and almost white before it is completely shed. The Indians form their robes of the skins procured in autumn, when the hair is short. Towards the spring the larvæ of the œstrus, attaining a large size, produce so many perforations in the skins that they are good for nothing. The cicatrices only of these holes are to be seen in August, but a fresh set of ova have in the mean time been deposited.*

"The herds of deer are attended in their migrations by bands of wolves, which destroy a great many of them. The Copper Indians kill the rein-deer in summer with the gun, or taking advantage of a favourable disposition of the ground, they enclose a herd upon a neck of land and drive them into a lake, where they fall an easy prey; but in the rutting season, and in the spring, when they are numerous on the skirts of the woods, they catch them in snares.

---

* "It is worthy of remark, that in the month of May a very great number of large larvæ exist under the mucous membrane at the root of the tongue and posterior part of the nares and pharynx. The Indians consider them to belong to the same species with the œstrus that deposits its ova under the skin; to us the larvæ of the former appeared more flattened than those of the latter."—*Dr. Richardson's Journal.*

The snares are simple nooses formed in a rope made of twisted sinew, which are placed in the aperture of a slight hedge, constructed of the branches of trees. This hedge is disposed so as to form several winding compartments,—and although it is by no means strong, yet the deer seldom attempt to break through it. The herd is led into the labyrinth by two converging rows of poles, and one is generally caught at each of the openings by the noose placed there. The hunter, too, lying in ambush, stabs some of them with his bayonet as they pass by, and the whole herd frequently becomes his prey. Where wood is scarce, a piece of turf turned up answers the purpose of a pole to conduct them towards the snares. The rein-deer has a quick eye, but the hunter, by keeping to leeward, and using a little caution, may approach very near, their apprehensions being much more easily aroused by the smell than the sight of any unusual object. Indeed, their curiosity often causes them to come close up and wheel round the hunter, thus affording him a good opportunity of singling out the fattest of the herd, and upon these occasions they often become so confused by the shouts and gestures of their enemy, that they run backwards and forwards with great rapidity, but without the power of making their escape. The Copper Indians find that a white dress attracts the most readily, and they often succeed in bringing them within gun shot, by kneeling and vibrating the gun from side to side, in imitation of the motions of a deer's horns when he is in the act of rubbing his head against a stone.

"The Dogrib Indians have a mode of killing these animals which, though simple, is very successful. It was thus described by Mr. Wentzel, who resided long amongst that people. The hunters go in pairs, the foremost man carrying in one hand the horns and part of the skin of the head of a deer, and in the other a small bundle of twigs, against which he, from time to time, rubs the horns, imitating the gestures peculiar to the animal. His comrade follows, treading exactly in his footsteps, and holding the guns of both in a horizontal position, so that the muzzles project under the arms of him who carries the head. Both hunters have a fillet of white skin round their foreheads, and the foremost has a strip of the same kind around his wrists. They approach the herd by degrees, raising their legs very slowly, but setting them down somewhat suddenly, after the manner of a deer, and always taking care to lift their right or left feet simultaneously. If any of the herd leave off feeding to gaze upon this extraordinary phenomenon, it instantly stops, and the head begins to play its part by licking its shoulders, and performing other necessary movements. In this way the hunters attain the very centre of the herd without exciting suspicion, and have leisure to single out the fattest. The hindmost man then pushes forward his comrade's gun, the head is dropt, and they both fire nearly at the same instant. The herd scampers off, the hunters trot after them; in a short time the poor animals halt to ascertain the cause of their terror, their foes stop at the same instant, and having loaded as they ran, greet the gazers with a second fatal discharge. The consternation of the deer increases, they run to and fro in the utmost confusion, and sometimes a great part of the herd is destroyed within the space of a few hundred yards."*

Species III.—*The Elk.*

Cervus Canadensis; Briss.

Cerf du Canada: Perrault, Mem. sur les Anim. ii. 45.
Cervus Major Americanus: Catesby, Carol. app. ii. 28.
Cervus Strongyloceros: Schreb. Saeugthiere.
Alces Americanus, cornibus teretibus: Jefferson, Virginia, 96.
The Elk: Lawson, New Voyage; Carver, Travels, 417.
The American Elk: Bewick, Quadrupeds, 112.
Cervus Wapiti: Barton, Med. and Physical Journ. iii. 36.
Wapiti: Warden, Descr. des Etats Unis, v. 368; Stag, Red Deer, Ibid. 367. *Wapiti*, Mitchill, Leach. Fred Cuvier, Mammif. Lithogr. liv. 21e.
Cerf Wapiti: Desm. Mamm. Sp. 664; Cerf Canadien, Ibid. Sp. 665.
Wewaskish [*Waskesse; Wawashkeesho*] Hearne, Journey, &c. 360.

[*Commonly called Stag, Red Deer, Gray Moose, Le Biche, Wapiti, American Elk, Round horn Elk, Elk, &c.*]

The stately and beautiful animal we are now to describe has been until very recently confounded with other species of deer, to which it bears but a slight resemblance, and from which it is distinguished by the most striking characters. The English name by which it is commonly known, and which we prefer to others, is the same as that given to the moose in Europe; hence this species was for a long time considered as a mere variety of the moose, if

---

* Franklin's Narrative of a Journey to the shores of the Polar Sea.

Elk.

1. Horns of the Animal in his 7th year. 2. Young horns in his 8th year.

not identically the same. A general resemblance to the European stag caused the application of the same name to our elk, and this circumstance led various writers into the error of considering our animal to be a variety of the *Cervus Elaphus,* or common stag of Europe. A reference to the synonymy we have prefixed to this article, will amply suffice to show how great a degree of confusion has hitherto existed upon this subject; a confusion rather increased than diminished by those who have attempted its removal by reconciling the discrepances of books instead of appealing to the proper and infallible authority, nature.

Hearne we believe to be justly entitled to the credit of having insisted upon the specific distinctness of this animal from the moose, by pointing out the error into which Pennant had fallen, in stating the Waskesse or Wewaskish to be of the same species. The description he gives of the Wewaskish sufficiently proves that it was our elk he described, and the characters he enumerates satisfactorily establish the specific differences between this animal and the moose.

Jefferson, in his valuable notes on Virginia, without being aware of Hearne's observations, proves very clearly that the elk of America ought to be regarded as identical neither with the moose nor stag of Europe, and proposed for our animal the name of *Alces Americanus.* Subsequently Dr. E. H. Smith published a very interesting paper in the New York Medical Repository, in which he described three individuals of this species, and gave a still more complete enumeration of their distinctive characters and history.

It would be as unprofitable as irksome to enter more extensively into the history of the different errors and changes respecting the classification of this deer. To us it appears sufficient to declare it to be now fully established that there is but *one* species of American Elk, upon which all the names prefixed, scientific and trivial, have been bestowed; that this species is second in size to the moose alone, and that in beauty of form, grace and agility of movement, and other attributes of its kind, it is not excelled by any deer of the old or new world.

The size and appearance of the elk are imposing; his air denotes confidence of great strength, while his towering horns exhibit weapons capable of doing much injury when offensively employed. The head is beautifully formed, tapering to a narrow point; the ears are large and rapidly moveable; the eyes are full and dark; the horns rise loftily from the front, with numerous sharp pointed branches, which are curved forwards, and the head is sustained upon a neck at once slender, vigorous and graceful. The beauty of the male elk is still farther heightened by the long forward-curling hair, which forms a sort of ruff or beard, extending from the head towards the breast, where it grows short and is but little different from the common covering. The body of the elk, though large, is finely proportioned; the limbs are small, and apparently delicate, but are strong, sinewy and agile. The hair is of a blueish gray colour in autumn; during winter it continues of a dark gray and at the approach of spring it assumes a reddish or bright brown colour, which is permanent throughout summer. The croupe is of a pale yellowish white or clay colour, and this colour extends about the tail for six or seven inches, and is almost uniformly found in both sexes. There is no very perceptible difference of colour between the male and female.

The female, however, does not participate in the "branching honours" of the male, which are found to attain, in numerous instances, a surprizing magnitude. It is not uncommon to see them of four and five feet in height, and it is said that they are sometimes still higher. Specimens of the largest size may be seen in the cabinets of the Philadelphia Museum, and of the Lyceum of Natural History of New York. These horns are said to consist of three principal divisions; 1st, The brow antlers, sometimes called "alters" by the hunters. 2d, The two middle prongs, named "fighting horns," and, 3d, The shaft or proper horns. The branches just mentioned are always placed on the front, outside, or posterior surface, never on the inner side of the horns, a circumstance which has been indicated as strikingly different from the arrangement of the branches of the horns of the common or Virginian deer, hereafter to be described.

The elk sheds his horns about the end of February or beginning of March, and such is the rapidity with which the new horns shoot forth, that in less than a month they are a foot in length. The whole surface of the horn is covered by a soft hairy membrane, which, from its resemblance to that substance, is called *velvet,* and the horns are said to be "in the velvet" until the month of August, by which time they have attained their full size.— After the horns are entirely formed, the membrane becomes gradually detached, and this separation is hastened by the animal, who appears to suffer some irritation, or itching, which causes him to rub the horns against trees, &c.

Almost all those who have written on this species, have dwelt upon the peculiar apparatus, situated beneath the eye, at the internal angle, which the French naturalists call *larmiers,* or sinus lacrymales. This apparatus is a slit, or depression, obliquely placed below the inner angle of each eye, and lined with a naked membrane, which secretes an unctuous matter, not unlike the cerumen or wax of the ear. Dr. Smith, in the paper we have above referred to, says that "the hunters assure us that the elk possesses the power, by strictly closing the nostrils, of forcing the air through these apertures in such a manner as to make a noise which may be heard at a great distance." This, however, is inaccurate; it is true that the elk, when alarmed, or his attention is strongly excited, makes a whistling noise at the moment that these lacrymal appendages are opened and vibrated in a peculiar manner. But having dissected these appendages in an elk, recently dead, we are perfectly assured that there is no communication between the nostril of the animal and these sacs. The bone behind these appendages is cribriform, or reticular, but we could discover no duct nor passage by which air or any fluid could find its way. The peculiar use and importance of this structure is still unknown; it exists in several species of the genus, as already indicated in the generic characters; and nothing but a close and careful examination of these animals in a state of nature will lead us to a correct understanding of their purpose. BARTON's notion, that "it seems in these animals to serve the purposes of an auxiliary breathing apparatus, and of an organ of smelling," is altogether speculation, founded upon a "*conjecture*," as to the structure of the sac and its connexion with the nostrils.

The elk has at one period ranged over the greater part, if not the whole, of this continent. JEFFERSON has stated that he " could never learn that the round-horned elk has been seen farther north than the Hudson river." But HEARNE has described the wewaskish in such a manner, as to leave no doubt of its existence as far north as the vicinity of Cumberland House, in lat. 53° 56'.*

---

* "The we-was-kish, or as some (though improperly) call it, the waskesse, is quite a different animal from the moose, being by no means so large in size. The horns of the wewaskish are *something similar* to those of the common deer [the *Rein Deer*; he distinguishes the common deer *C. Virginianus* (of the United States) as indian deer,] but are *not palmated* in any part. They stand more upright, have fewer branches, and want the brow antler. The head of this animal is so far from being like that of the moose, that the nose is sharp like that of a sheep. The hair is usually of a sandy red, and they are frequently called by the English who visit the interior parts of the country, red deer.—The person who informed Mr. PENNANT that the wewaskish and moose are the same animal, never saw one of them; and the only reason he had to suppose it, was the great resemblance of their skins." p. 360-1-2.

Elk are still occasionally found in the remote and thinly settled parts of Pennsylvania, but the number is small; it is only in the western wilds that they are seen in considerable herds. They are fond of the great forests, where a luxuriant vegetation affords them an abundant supply of buds and tender twigs; or of the great plains, where the solitude is seldom interrupted, and all bounteous nature spreads an immense field of verdure for their support.

The Elk is shy and retiring; having acute senses, he receives early warning of the approach of any human intruder. The moment the air is tainted by the odour of his enemy, his head is erected with spirit, his ears rapidly thrown in every direction to catch the sounds, and his large dark glistening eye expresses the most eager attention. Soon as the approaching hunter is fairly discovered, the elk bounds along for a few paces, as if trying his strength for flight, stops, turns half round, and scans his pursuer with a steady gaze, then, throwing back his lofty horns upon his neck, and projecting his taper nose forwards, he springs from the ground and advances with a velocity which soon leaves the object of his dread far out of sight.

But in the season when sexual passion reigns with its wonted influence over the animal creation, the elk, like various other creatures, assumes a more warlike and threatening character. He is neither so easily put to flight, nor can he be approached with impunity, although he may have been wounded. His horns and hoofs are then employed with great effect, and the lives of men and dogs are endangered by coming within his reach. This season is during August and September, when the horns are in perfect order, and the males appear filled with rage, and wage the fiercest war with each other for the possession of the females. During this season the males are said to make a loud and unpleasant noise, which is compared to a sound between the neighing of a stallion and the bellowing of a bull. Towards the end of May or the beginning of June, the female brings forth her young, commonly one, but very frequently two in number, which are generally male and female.

The flesh of the elk is highly esteemed by the Indians and hunters as food, and the horns, while in their soft state, are also considered a delicacy: of their hides a great variety of articles of dress and usefulness are prepared. The solid portion or shaft of the perfect horn is wrought by the Indians into a bow, which is highly serviceable from its elasticity, as well as susceptible of beauty of polish and form. Several of these bows may be seen in the extensive collection of Indian implements belonging to the Philadelphia Museum.*

---

* In a work devoted to the natural history of our country, a passing tribute to the memory of one who has done much for natural science, will not, we hope, be regarded as obtrusive.

But a few weeks have elapsed since the great debt of nature was paid by CHARLES WILSON PEALE, the founder of the Philadelphia Museum. If a long life, devoted with singular enthusiasm to the advancement of natural history, by the collection of objects in all the departments of natural science, be meritorious; if the establishment of an institution which has long been the pride, and promises hereafter to be an honour and ornament to our country, be valuable; if eigh-

The elk has occasionally been to a certain degree domesticated, and might possibly be rendered as serviceable as the rein-deer. A pair of these animals, represented in London under the name of Wapiti, were trained to draw in harness, or to bear the saddle, for the amusement of visitors. But these experiments are not sufficient to lead us to conclude, that the elk could be readily substituted for the rein-deer or horse.

With what little is known of this species from actual observation, several writers have mingled a great deal of fable, and have repeated the stories of "hunters" until they have at length passed for the truth. Thus we are told of "a small vesicle (on the outside of the elk's hind legs) that contains

---

ty-six years spent with unblemished integrity and consistency of character in the service of his friends and country, be worthy of respect, the memory of this good man will long continue to be dear to those who are capable of admiring unostentatious virtue, and appreciating the benefits which have already resulted, and will continue to flow from his labours. To the last moment of his existence he exemplified in the fullest degree the excellent effects of a temperate and industrious life; and in the benevolence of his disposition, the undisturbed serenity of his mind, and the unimpaired vigour of his intellect, showed how far the study of nature, in her curious and wonderful works, had refined and ennobled a mind which owed nothing to early education. To him death presented no terrors, for he had long considered it as the termination of his toils; he looked upon the grave but as the place in which he might yield his mortality to the beneficent source whence he sprung; and at peace with all mankind, he gently breathed his last, in cheering confidence of the mercy of the Most High. May he rest in peace!

a thin unctuous matter, which some of our hunters call the "oil." Various improbable uses are assigned to this unique and wonderful "oil spring," which it would be lost time to repeat or refute. We have inquired of those who have dissected several of these animals, and have been present at the dissection of one ourselves, but have never been able to discover any thing of this "vesicle." A friend who had one of these animals for several years living in his possession, states, that he never detected the presence of any such apparatus or oil. Until better proof be given than has yet been offered, we shall feel willing to rank this story among the "conjectures" which have been too often resorted to when there was a scarcity or difficulty of obtaining "facts."

We have already adverted to the warlike disposition of the elk during a particular season, but it may not be amiss to add, that at all times this animal appears to be more ready to attack with his horns than any other species of deer we have examined. When at bay, and especially if slightly wounded, he fights with great eagerness, as if resolved to be revenged. The following instance from Long's Expedition to the Rocky Mountains will, in some degree, illustrate this statement.

A herd of twenty or thirty elk were seen at no great distance from the party, standing in the water or lying upon the sand beach. One of the finest bucks was singled out by a hunter, who fired upon him, whereupon the whole herd plunged into the thicket and disappeared. Relying upon the skill of the hunter, and confident that his shot was fatal, several of the party dismounted and pursued the elk into the woods, where the wounded buck was soon overtaken. Finding his pursuers close upon him, the elk turned furiously upon the foremost, who only saved himself by springing into a thicket, which was impassable to the elk, whose enormous antlers becoming so entangled in the vines as to be covered to their tips, he was held fast and blindfolded, and was despatched by repeated bullets and stabs.

Species IV.—*The Black-tail Deer.*

Cervus Macrotis; Say.

The Black-tailed Fallow Deer: Lewis and Clarke, i. p. 30. *Mule Deer* Ibid. ii. 166.
Cervus Auritus: Warden, Descr. des Etats Unis, v. 640.
Cervus Macrotis: Say, Long's Exped. to the Rocky Mountains, ii. 88.

[*Commonly called Mule Deer.**]

The first indication of this fine deer was given by Lewis and Clarke, who found it upon the sea coast and the plains of the Missouri, as well as upon the borders of the Kooskoose river, in the vicinity of the rocky mountains. They inform us that the habits of this animal are similar to those of its kindred species, except that it does not run at full speed, but bounds along, raising every foot from the ground at the same time. It is found sometimes in the woodlands, but most frequently is met with in prairies and

---

* We avoid this name because it leads to an incorrect notion of the animal. The resemblance of its ears to those of the mule gave origin to the name.

in open grounds. Its size is rather greater than that of the common deer, (*C. Virginianus*) but its flesh is considered inferior to the flesh of that species.

According to Say's description, the horns are slightly grooved and tuberculated at base, having a small branch near thereto, resembling in situation and direction the first branch on the horn of the common deer. The front line of the antler is curved like that of the common deer, but not to so great a degree, and at about the middle of the entire length of the antlers they bifurcate equally, each of these processes again dividing near the extremity, the posterior being somewhat the shortest.

The ears are very long, being half the length of the whole antler, and extending to its principal bifurcation. The eye is larger than that of the common deer, and the subocular sinus is much larger. The hair is coarser, undulated and compressed, resembling that of the elk, (*C. Canadensis*) and is of a light reddish brown colour above. The sides of the hair on the front of the nose is of a dull ash colour; that on the back is intermixed with blackish tipped hairs, which form a distinct line on the neck, near the head. The tail is of a pale reddish ash colour, except at the extremity on its superior surface, where it is of a jetty black; beneath it is white, yet nearly destitute of hair. The hoofs are shorter and wider than those of the common deer, and more like those of the elk.*

---

* The following measurements are given by Say in the work above quoted.

Length from the base of the antlers to the origin of the na-

Species V.—*The Common Deer.*

Cervus Virginianus.

Fallow Deer: Catesby, App. ii. 28: Lawson, Carol. 123.
Caricon Femelle: Buff. 12, pl. 44.
Cerf de la Louisiane: C. Ossem. Foss. et Regne Anim. Fred. Cuv. Mammif. Lithogr. 4 fig.
Virginian Deer: Pennant, Quad.
Cerf de Virginie: Desm. Mammal. sp. 679, p. 442.

The Common Deer is the smallest American species at present known, and is found throughout the country between Canada in the north and the banks of the Orinoco in South America. In various parts of this extensive range, considerable varieties in size and colouring are presented by this species, though these being accidental and mutable, require no especial description.

The common deer is more remarkable for general slenderness and delicacy of form, than for size and vigour. The slightness and length of its limbs, small body, long and slim neck, sustaining a narrow and

---

sal process, two inches. Of the nasal process, two and a-half. From the nasal process to the principal bifurcation, four to five. Thence to the other two bifurcations, respectively, four and a-half to five and a-half. Terminal prongs of the anterior branch, four to four and a-half. Of the posterior branch, two and a-half to three. From the anterior base of the antlers to the tip of the upper jaw, nine and a quarter. From the anterior canthus of the eye to the tip of the jaw, six and a quarter. From the base of the antler to the anterior canthus, three. Of the ears, more than seven and a-half. Of the trunk of the tail, four. Of the hair at the tip of the tail, from three to four.

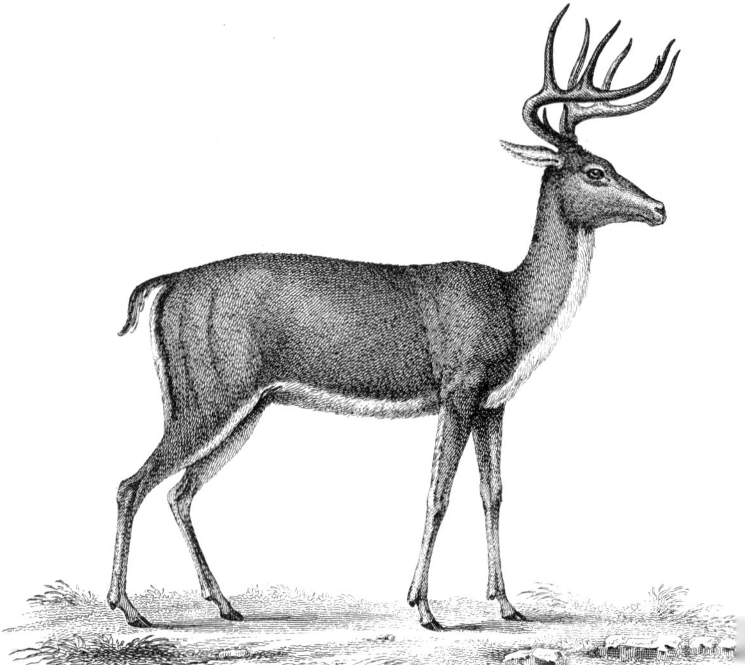

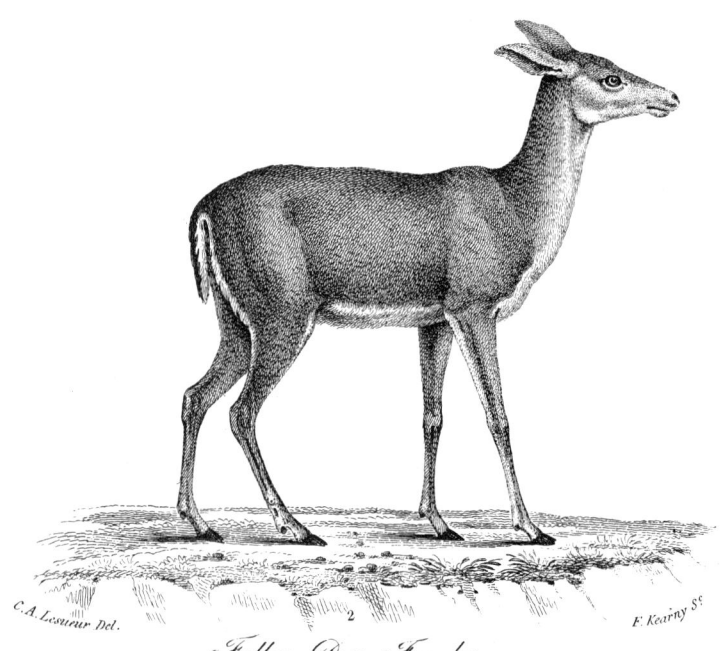

1

Virginia or Fallow Deer, Male

C. A. Lesueur Del. 2 F. Kearny Sc.

Fallow Deer, Female.

almost pointed head, give the animal an air of feebleness, the impression of which is only to be counteracted by observing the animated eye, the agile and playful movements, and admirable celerity of its course when its full speed is exerted. Then all that can be imagined of grace and swiftness of motion, joined with strength sufficient to continue a long career, may be realized.

The common deer has always been of great importance to the aborigines of America, as an abundant source of food and raiment, nor has its value been less to the pioneers of civilization in their advances into the untrodden solitudes of the west. The improvements in agriculture have long since rendered this supply of food of comparatively little value to the white man, yet vast numbers of this species are annually destroyed, equally for the sake of their flesh, hides and horns. Judging by the quantity of skins brought to our markets, and calculating the average number of common deer destroyed during the time which has elapsed since the settlement of the country, we may form an imperfect notion of the aggregate number and productiveness of this species; which, notwithstanding this extensive consumption, does not appear to be very rapidly diminishing, if we except the immediate vicinities of very thickly peopled districts. Even in these, where the destruction of deer during the breeding season is prevented by law, the increase seems quite equal to the demand, and such humane and judicious provisions will probably preserve this beautiful race to adorn the forest long after the species is exterminated in situations where it is not thus protected.

The common deer is possessed of keen senses, especially of hearing and smelling; the sight, though good, does not appear to equal in power the senses just named, upon which the safety of the animal most immediately depends.

It is therefore necessary for the hunter to approach the deer against the wind, otherwise he is discovered by the scent. at a great distance, and his objects are entirely frustrated. The slightest noise excites the attention of the deer, and his fears appear to be more readily awakened by this cause than any other; while, on the contrary, the sight of unaccustomed objects seems rather to arouse curiosity than to produce terror, as the animal will frequently approach, or stand gazing intently, until the hunter steals close enough to fire with fatal aim.

The deer, in herds of various numbers, frequent the forests and plains adjacent to the rivers, feeding principally upon the buds and twigs of trees and shrubs, though they are fond of grass when their favourite food is not more convenient. The herd is led by one of the largest and strongest bucks, who appears to watch over the general safety, and leads the way on all occasions. When any cause of alarm checks their progress, the leader stamps with his feet, threatens with his horns, and snorts so loudly as to be heard for a very considerable distance. So long as he stands fast, or prepares for combat, the rest of the herd appear to feel secure; but when he gives way they all follow with precipitation, and vie with each other in the race.

The salines, or licks, as they are commonly called, are eagerly sought for by these deer, as they have an equal fondness for salt with most other animals belonging to the same order. In licking the soil, through which the saline matter oozes to the surface, they take up very considerable quantities of the earthy matter, and this enables the hunter to discover when the deer have recently visited the spot, or that one of these places is not far distant, as the excrement of the animal then resembles small balls or pellets of hardened clay. The watchfulness of the leader of the herd, as above mentioned, has led the hunters to form an opinion, to which they pertinaciously adhere, that the deer, when they visit a salt lick, always post one of their number as a sentinel, who is to give the alarm in case of the approach of an enemy.

The common deer when startled from a resting place without being much alarmed, moves at first in a singular and amusing manner. With an apparent awkwardness, two or three springs are made, from which the animal alights on three feet, drawing up and extending the limbs in a stiff and peculiar manner. As the tail is erected this alternate resting upon the feet of opposite sides, causes the tail to describe a semicircle from side to side; a few high bounds are next made forwards, as if with a view to prepare for subsequent exertion, and then, if the cause of alarm be continued, the deer exerts his strength and dashes off in his swiftest career.

Although the common deer is generally a very shy and timid animal, the males are very much disposed to war with each other during the season of their sexual passion, and they are almost always inclined to fight when wounded or brought to bay.

At this time they fight with their forefeet as well as their horns, and inflict severe wounds by leaping forward and striking with the edges of their hoofs held together. If a hunter falls on the ground in attempting to close in and despatch a wounded deer with his knife, he is in great danger of being killed by such blows as we have described. This deer is also said by the hunters to evince a very strong degree of animosity towards serpents, and especially to the rattlesnake, of which it has an instinctive horror. In order to destroy one of these creatures, the deer makes a bound into the air, and alights upon the snake with all four feet brought together in a square, and these violent blows are rapidly repeated until the hated reptile is destroyed. The combats in which the males engage with each other are frequently destructive of the lives of both, in a way that would not readily be anticipated. In assaulting each other furiously, their horns come into contact, and being elastic, they yield mutually to the shock, so that the horns of one animal pass within those of the other and thus secure them, front to front, in such a manner that neither can escape, and they torment themselves in fruitless struggles until worn down by hunger, they perish, or become the prey of wolves or other animals. Heads of deer which have thus perished are frequently found, and there is scarcely a museum in this country which has not one or more specimens. The following instance is given by SAY in Long's Expedition to the Rocky Mountains. "As the party were descending a ridge, their attention was called to an unusual noise proceeding from a copse of low bushes, a few rods from the path. On arriving at the spot they found two buck deer, their horns fast interlocked with each other, and both much spent with fatigue, one in particular being so much exhausted as to be unable to stand. Perceiving that it would be impossible that they should extricate themselves, and must either linger in their present situations or die of hunger, or be destroyed by the wolves, they despatched them with their knives, after having made an unavailing attempt to disentangle them. Beyond doubt many of these animals must annually thus perish."

The common deer is fattest and in best condition in the months of October and November, when the rutting season commences, and continues about a month, terminating commonly about the middle of December. While this season continues, the neck of the male is enlarged or dilated.

The female commonly has one or two, and sometimes three* fawns, which are of a light cinnamon colour, spotted with white. While the fawns are still young, or from May until July, the doe very carefully conceals her offspring while she goes to feed; and this act of maternal fondness is not only done in a state of nature, but even when the common deer have been captive for some time and breed in parks. The hunters, however, turn this fondness to their own account, by imitating the cry of the fawn, either by the voice alone, or by a sort of pipe or

---

* "About the middle of March Mr. Peale shot a large doe, in the matrix of which were three perfectly formed young, of the size of rabbits." Long's Exped. to the Rocky Mountains. i.

reed which closely resembles the bleating of the animal. The parent soon relinquishes all fears for her own safety, in her desire to assist her offspring, and following the sound, approaches the ambush of the hunter, where a deadly shot insures her immediate destruction. When a doe is killed in company with her fawn, or the mother has been removed as above mentioned, the little animal is at once tamed, or exhibits no apprehension at the approach of man, but follows his captor with the most confiding simplicity, and soon becomes so attached to his feeder as to attend his steps at all times, and obey his voice.*

In the latter part of the summer the fawn loses the white spots, and in winter the hair grows longer and grayish, when the animal is said by the hunters to be *in the gray*. To this coat one of a reddish colour succeeds about the end of May and beginning of June; the deer is then said to be *in the red*. Towards the end of August, the old bucks begin to change to the dark bluish colour; the doe begins this change a week or two later, when they are said to be *in the blue*. This coat gradually lengthens until it finally returns to the gray. The skin is said to

---

* "From Capt. Parry I learned an interesting anecdote of a doe and her fawn, which he had pursued across a small inlet. The mother, finding her young one could not swim so fast as herself, was observed to stop repeatedly, so as to allow the fawn to come up with her, and having landed first, stood watching it with trembling anxiety as the boat chased it to the shore. She was repeatedly fired at, but remained immoveable, until her offspring landed in safety when they both cantered out of sight."—*Lyon's Narrative*, p. 80.

be toughest in the red, thickest in the blue, and thinnest in the gray; the blue skin is most valuable.*

In the month of January the males cast their horns; the new horns soon after commence their growth. They continue in the velvet until the end of September or beginning of October, so as to be in full condition for battle during their season of love and war. These horns are not very large, but are curved forwards in a peculiar manner. They have an antler placed high up on the inside of each shaft, which presents downwards, and two or three others on the posterior surface, turning backwards. In the fifth year, the horns consist of two cylindrical, whitish, and moderately smooth, shafts, separating at first slightly outwards and backwards, and then strongly curving forwards and downwards. From the second to the fifth year the variations of the horns consist in their gradual advance from single, slightly curved shoots, to three and four antlers.

From what has been already said of the changes occurring at different seasons, it will be perceived that no description of the pelage of any one can be generally applicable. It may be stated that the colour of the adults in summer is a fine fawn or yellowish brown above, with the under part of the lower jaw, throat, belly, lower part of the limbs, posterior edges of the fore-limbs, anterior part of the

---

* See Say, in Long's Expedition to the Rocky Mountains, i. p. 104.

thighs, and inferior surface of the tail, white. The front is rather gray, while the end of the muzzle is of a deep brown, with two white spots upon the upper lip; on the sides of the lower jaw, at the angles of the mouth, two triangular black spots are very generally found. Two-thirds of the upper surface of the tail is light brown, the outer third is black.

The total length of the common deer, exclusive of the hair at the tip of the tail, is five feet four or five inches. The tail, exclusive of the hair, is nine inches and a-half long. The hind foot, from the tip of the os calcis to the extremity of the toe, is sixteen inches and a-quarter. The fore arm eleven inches and seven-eighths. The weight, in the month of February, was 115 lbs.*

During the stay of Long's Expedition at Engineer Cantonment, three specimens of a variety of the common deer were brought in, having all the feet white near the hoofs, and extending to them on the hind part from a little above the spurious hoofs. This white extremity was divided upon the sides of the foot by the general colour of the leg, which extends down near to the hoof, leaving a white triangle in front, of which the point was elevated rather higher than the spurious hoofs. The black mark upon the lower lip, rather behind the middle of the sides, was strongly marked.

---

* SAY. Lewis and Clarke state that they saw common deer with tails seventeen inches in length

THE COMMON DEER.

The flesh of the common deer is justly esteemed as an excellent article of food, when killed in the proper season, which is the autumn. The Indians and hunters, whose necessities do not permit them to choose, feed upon these deer at all seasons. The markets of our large cities are supplied very abundantly with venison from this species every winter, and at so cheap a rate as to bring it within the means of almost every housekeeper.

The whole of the deer is used by the Indians, and, on pressing occasions, without the previous employment of fire. If a hunter kill a deer after a long and exhausting chase, he applies his mouth to the wound by which the animal was killed, in order to refresh himself by sucking some of the blood. When very hungry, they cut a hole in the side of the animal, thrust in their hands and tear out the kidneys, which are instantly devoured, though still quivering with life.*

---

* "After the hunters had been gone for about an hour, captain Lewis again mounted with one of the Indians behind him, and the whole party set out; but just as they passed through the narrows they saw one of the spies coming back at full speed across the plain, &c. The young Indian had scarcely breath to say a few words as he came up, and the whole troop dashed forward as fast as their horses could carry them; and captain Lewis, astonished at this movement, was borne along for nearly a mile, before he learned, with great satisfaction, that it was all caused by the spy's having come to announce that one of the white men had killed a deer. Relieved from his anxiety he now found the jolting very uncomfortable; for the Indian behind him, being afraid of not getting his share of the feast, had lashed the horse at every step since they set off; he therefore reined him in, and or-

The stomach of the deer, with its half digested contents, is a very favourite dish with almost all the savages, especially towards the north, where deer

dered the Indian to stop beating him. The fellow had no idea of losing time in disputing the point, and jumping off the horse, ran for a mile at full speed. Capt. Lewis slackened his pace, and followed at a sufficient distance to observe them. When they reached the place where Drewyer had thrown out the intestines, they all dismounted in confusion, and ran tumbling over each other like famished dogs: each tore away whatever part he could and instantly began to eat it; some had the liver, and some the kidneys; in short, no part on which we are accustomed to look with disgust escaped them. One of them who had seized about nine feet of the entrails, was chewing at one end, while with his hand he was diligently clearing his way by discharging the contents at the other. It was indeed impossible to see these wretches ravenously feeding on the filth of animals, and the blood streaming from their mouths, without deploring how nearly the condition of savages approaches that of the brute creation: yet, though suffering with hunger, they did not attempt, as they might have done, to take by force the whole deer, but contented themselves with what had been thrown away by the hunter. Capt. Lewis now had the deer skinned, and after reserving a quarter of it, gave the rest of the animal to the chief, to be divided among the Indians, who immediately devoured nearly the whole of it without cooking. They now went forwards towards the creek, where there was some brushwood to make a fire, and found Drewyer, who had killed a second deer: the same struggle for the entrails was renewed here, and on giving nearly the whole deer to the Indians, they devoured it, even to the soft parts of the hoofs. A fire being made, captain Lewis had his breakfast, during which Drewyer brought in a third deer; this, too, after reserving one-quarter, was given to the Indians, who now seemed completely satisfied and in good humour."—*Lewis and Clarke*, i. 375.

THE COMMON DEER.

feed in great degree on mosses and buds.* European travellers who have tasted of this substance have not found it disagreeable: the Indians eat it altogether, or with a very slight degree of preparation.† However shocking it may appear to us, the prejudice against raw meat is overcome with great ease when hunger pinches severely; and when once the prejudice is removed, a fondness for raw food is very readily acquired, even by those who have previously been fastidious in their tastes.‡

---

* "The stomach of no other large animal beside the deer is eaten by any of the Indians that border on Hudson's Bay. In winter, when the deer feed upon fine white moss, the contents of the stomachs are so much esteemed by the Indians, that I have often seen them sit round a deer where it was killed, and eat it warm out of the paunch. In summer the deer feed more coarsely, and therefore this dish, if it deserve that appellation, is then not so much in favour."—HEARNE, 318.

† "Of the *nerooka* [the contents of the deer's stomach] I also tasted a small portion, considering that no man who wishes to conciliate or inquire into the manners of savages should scruple to fare as they do while in their company. I found this substance acid and rather pungent, resembling, as near as I could judge, a mixture of sorrel and radish leaves. The smell reminded me of fresh brewer's grains; and the young grasses and delicate white lichens on which the deer feed were very apparent."—*Lyon's Narrative*, 242.

‡ "Dunn and myself, as an experiment, made our breakfast on a choice slice cut [*raw*] from the spine, and found it so good, that at dinner time we preferred the same food to our share of preserved meat, which we had saved from the preceding night. The windpipe is exceedingly good; and I

The skins of the common deer continue to form a very valuable article of commerce, and furnish a material better suited for the manufacture of gloves and various articles of dress, than the skin of any other animal with which we are acquainted. The Indian fashion of dressing these skins consists in depriving them of the hair and fleshy matter, and rubbing them sedulously with a lather made of the brains of the animal until they become uniformly soft, spongy and flexible. In this condition they impart to the touch a sensation of greater softness than that derived from the finest cloth.—Deer skins dressed in this way, however, are very liable to be spoiled by moisture, and rot with great rapidity if they continue for some time exposed to rain.

The buck-skin, as dressed for the use of our glovers, is remarkable for its thickness, softness and pliability,

---

am confident, that were it not from prejudice, raw venison might be considered a dainty."—Lyon, 242.

"The most remarkable dish among them, as well as all the other tribes of Indians in those parts, both northern and southern, is blood, mixed with the half digested food found in the deer's stomach or paunch, and boiled up with a sufficient quantity of water to make it of the consistence of pease pottage. Some fat and scraps of tender flesh are also shred small and boiled with it. To render this dish more palatable, they have a method of mixing the blood with the contents of the stomach in the paunch itself, and hanging it up in the heat and smoke of the fire for several days, which puts the whole mass into a state of fermentation, and gives it such an agreeable acid taste, that, were it not for prejudice, it might be eaten by those who have the nicest palates."—Hearne. 317.

and with these advantages it has the great superiority of not being liable to injury from moisture, as tannin is made use of in its preparation. In relation to its warmth, durability and agreeableness to the wearer, it appears to be much preferred to similar leather made from any other skins, whether of European or American deer. Within a few years past the use of buckskin shirts has very much increased among invalids, and often with great advantage. But it is generally believed that these shirts render the body extremely susceptible to changes of temperature, and, all things considered, do more injury than shirts made of flannel or other commonly used materials.

CHAPTER XIX.

GENUS V. ANTELOPE; *Antelope*. PALL. &c. &c.

Fr. Antèlope.   *Gër.* Antelope.

GENERIC CHARACTERS.

THE body, ears, eyes and lachrymal or sub-ocular sinuses, are very similar to those of the deer, and the limbs bear an equal resemblance thereto, except that some species of Antelope have tufts or brushes of long hair pending from the carpus. The outline of the front or face is nearly straight, and terminates in a muzzle, or half muzzle, though in some species this is absent. The teats are four or two in number, being sometimes two in one sex and four in the other. The gall bladder is uniformly present, a circumstance in which this genus differs remarkably from the deer.

The horns of both sexes (though in some species the horns are confined to the male) are placed upon a solid bony process of the os frontis. The horns are curved in various directions, being often marked with transverse bands, have a salient spiral line, or are bifurcated.

Dental System.

32 Teeth: { 12 Upper { 0 Incisive / 12 Molar. ; 20 Lower { 8 Incisive / 12 Molar.

Drawn by C.A. Lesueur. Eng.d by G.B.E.

1. The Mountain Goat. 2. Prong Horn Antilope.

Species I.—*The Prong-Horn Antelope.*

Antilope Americana; Ord.

Antelope: Lewis and Clarke, i. 75, 208, 369; ii. 169.
Antilope Americana: Ord. Guthrie's Geography, Philad. ed. 1815.
  Antilocapra Americana; ibid. Journal de Physique, 1818. Say.
  Long's Exped. to the Rocky Mountains, i. 363, 485.
Antilope Furcifer: Smith. Trans. of Linnæan Society, xiii. pl. 2.
Prong-Horned Antelope: Sab. App. p. 667.

Our adventurous countrymen who led the first expedition across the Rocky Mountains, were the first to call attention to this beautiful animal, and the first to call it by its true name. Notwithstanding the obviousness of all the other characters, the circumstance of its having an offset, or prong to its horns, kept *nomenclators* for years undecided as to what place it should occupy in their arrangements, and gave them an opportunity, by which they have not failed to profit, of multiplying *words* and *republishing* their own names, if they made no addition to our information on the subject. All that has been related concerning this animal which is worth repeating or remembering, was published in Lewis and Clarke's narrative, above quoted, and has since been confirmed by the observations of Dr. Richardson, appended to Franklin's Journey to the shores of the Polar Sea. Leaving to the nomenclators their disputations about what Dekay has happily called "the barren honours of a synonyme," we shall glean the few facts contained in the narrations of the above-mentioned accurate observers of nature.

The prong-horn antelope is an animal of wonderful fleetness, and so shy and timorous as but seldom to repose, except on ridges which command a view of the surrounding country. The acuteness of their sight, and the exquisite delicacy of their smell, render it exceedingly difficult to approach them; and when once danger is perceived, the celerity with which the ground is passed over appears to the spectator to resemble the flight of a bird rather than the motion of a quadruped.

In one instance Captain Lewis, after various fruitless attempts, by winding around the ridges, succeeded in approaching a party of seven that stood upon an eminence towards which the wind was unfortunately blowing. The only male of the party frequently encircled the summit of the hill, as if to announce any danger to the group of females which stood upon the top. Before they saw Capt. Lewis they became alarmed by the scent, and fled while he was at the distance of two hundred yards. He immediately ran to the spot where they had stood; a ravine concealed them from him, but at the next moment they appeared on a second ridge, at the distance of three miles. He could not but doubt whether these were the same he had alarmed, but their number and continued speed convinced him they were so, and he justly infers that they must have run with a rapidity equal to that of the most celebrated race horse.

Yet, notwithstanding the keenness of their senses, and surprising velocity of their course, the pronghorn antelope is often betrayed to his destruction by curiosity. When the hunter first comes in sight, his whole speed is exerted, but if his pursuer lies down, and lifts up his hat, arm or foot, the antelope trots back to gaze at the object, and sometimes goes and returns two or three times, until it comes within the reach of the rifle. This same curiosity occasionally enables the wolves to make them a prey; for sometimes one of them will leave his companions to go and look at the wolves, which, should the antelope be frightened at first, crouch down, repeating the manœuvre, sometimes relieving each other, until they succeed in decoying it within their power, when it is pulled down and devoured. But the wolves more frequently succeed in taking the antelopes when they are crossing the rivers, as they are not good swimmers.

"The chief game of the Shoshonees," say Lewis and Clarke, "is the antelope, which when pursued retreats to the open plains, where the horses have full room for the chase. But such is its extraordinary fleetness and wind, that a single horse has no possible chance of outrunning it, or tiring it down; and the hunters are therefore obliged to resort to stratagem. About twenty Indians, mounted on fine horses, armed with bows and arrows, left the camp; in a short time they descried a herd of ten antelopes: they immediately separated into squads of two or three, and formed a scattered circle round the herd for five or six miles, keeping at a wary distance, so as not to alarm them till they were perfectly inclosed, and usually selecting some commanding eminence as a stand. Having gained their positions, a small party rode towards the herd, and with wonderful dexterity the huntsman preserved his seat, and the horse his footing, as he ran at full speed over the hills and down the steep ravines, and along the borders of the precipices. They were soon outstripped by the antelopes, which, on gaining the other extremity of the circle, were driven back and pursued by the fresh hunters. They turned and flew, rather than ran, in another direction; but there too they found new enemies. In this way they were alternately pursued backwards and forwards, till at length, notwithstanding the skill of the hunters, (who were merely armed with bows and arrows) they all escaped; and the party, after running for two hours, returned without having caught any thing, and their horses foaming with sweat. This chase, the greater part of which was seen from the camp, formed a beautiful scene, but to the hunters is exceedingly laborious, and so unproductive, even when they are able to worry the animal down and shoot him, that forty or fifty hunters will sometimes be engaged for more than half a day, without obtaining more than two or three antelopes."

The prong-horn is found in the vicinity of Carlton House during the summer, and is usually called a *goat* by the Canadians. The Creek Indians call them *apestachoekoos*. Lewis and Clarke saw the animal very frequently during their journey to the mouth of the Columbia River, though they were fewer on the plains of Columbia than on the eastern side of the Rocky Mountains.

Great numbers of these animals were seen by Lewis and Clarke in the month of October, near Carp Island, in the Missouri, where large flocks of them were driven into the water by the Indians.— The men were ranged along the shore so as to prevent the escape of the antelopes, and fired upon them, and sometimes the boys went into the river and killed them with sticks. Fifty-eight of the antelopes were killed by the Indians during the time they were observed by our travellers. They were then migrating from the plains east of the Missouri, where they spend the summer, towards the mountains, where they subsist on leaves and shrubbery during the winter: in the spring they resume their migrations.

The Mandan Indians capture the prong-horn antelopes by means of a pound similar to that described in the account of the rein-deer.

The following description is given by Dr. Richardson, from a recent specimen:—"The male is furnished with short, black, roundish, tapering horns, arched inwards, turning towards each other, with their points directed backwards, each horn having a single short branchlet projecting from the middle. The winter coat consists of coarse, round, hollow hairs, like those of the moose. The neck, back and legs are yellowish brown; the sides are reddish white; the belly and chest is white, with three white bands across the throat. The hairs on the occiput and back of the neck are long and tipped with black, forming a short erect mane. There is a black spot behind each cheek which exhales a strong goat-like odour. The tail is short; on the rump there is a large spot of pure white. The dimensions of the animal were as follows:—from the nose to the root of the tail four feet; height of the fore shoulder three feet; that of the hind quarter the same. Girth behind the fore-legs three feet; girth before the fore-legs two feet ten inches. The female is smaller than the male, having straighter horns, with rather a protuberance than a prong. She is also deficient in the black about the neck.

Genus VI. Goat; *Capra*, L.

Fr. Chèvre.   *Germ.* Ziege.

GENERIC CHARACTERS.

The outline of the front is rather straight, or slightly concave; there is no muzzle nor sub-ocular sinuses; the interspace of the nostrils is naked; the horns are turned upwards and outwards. The body is slender, the tail short, and the limbs somewhat robust. The teats are two in number. The hair is of two sorts; the exterior is long, or very long and smooth, forming a beard beneath the chin. Sometimes there are cuticular appendages hanging from the inferior surface of the neck. The testes are contained in a very large scrotum.

Dental System.

32 Teeth: 12 Upper { 12 Molar. 20 Lower { 8 Incisive 12 Molar.

Species I.—*Rocky Mountain Goat.*

Capra Montana; Ord.

Ovis Montana; Ord. Guthrie's Geography, Philad. ed. 292, 309. 1815. Journ. Acad. Nat. Sciences, i. part 1. p. 8.
Rupicapra Americana: Blainville. *Antilope Americana:* ibid. Bullet. de la Société Philomathique, p. 80.
Antilope Lanigera: Smith, Trans. Linnæan Society, xiii. pl. 4.

[*Commonly called Rocky Mountain Sheep.*]

This animal, concerning which very little is known, is stated by Major Long, in his communication to the Philadelphia Agricultural Society, to inhabit the portion of the Rocky Mountains situated between the forty-eighth and sixty-eighth parallels of north latitude. By Lewis and Clarke it was observed as low as forty-five degrees north. They are in great numbers about the head waters of the north fork of Columbia river, where they furnish a principal part of the food of the natives. They also inhabit the country about the sources of Marais or Muddy River, the Saskatchawan and Athabasca. They are more numerous on the western than on the eastern slope of the Rocky Mountains, but are very rarely seen at any distance from the mountains, where they appear to be better suited to live than elsewhere. They frequent the peaks and ridges during summer, and occupy the valleys in winter. They are easily obtained by the hunters, but their flesh is not much valued, as it is musty and unpleasant; neither do the traders considers their fleece of much worth. The skin is very thick and spongy, and is principally used for the purpose of making moccasins.

The Rocky Mountain goat is nearly of the size of a common sheep, and has a shaggy appearance in consequence of the protrusion of the long hair beyond the wool, which is white and soft. Their horns are five inches long and one in diameter, conical, slightly curved backward, and projecting but little beyond the wool of the head. The horns and hoofs are black.

The first indication of this animal was given by Lewis and Clarke, and it is much to be regretted that so little is still known of the manners and habits of this species. The only specimen preserved entire, that we know of, is that figured by Smith in the Linnæan Transactions, from which the figure in our plate is taken. The fineness of the wool of this animal may possibly hereafter induce persons who have it in their power to make some exertions to introduce this species among our domestic animals. It is said that the fleece of this goat is as fine as that of the celebrated shawl goat of Cashmere.

Genus VII. Sheep; *Ovis*, L.

GENERIC CHARACTERS.

The outline of the face is arched, or convex, and the mouth has no muzzle; the ears are pointed, and of middling length; the horns, which are transversely wrinkled, large and triangular, are twisted laterally into a spiral, and have an osseous core, of a cellular or cancellar structure. The limbs are slender, and covered with uniform short hair; the tail is short, curved downwards, or pendulous. Neither subocular sinus, beard, nor inguinal pores, exist in this genus.

Dental System.

32 Teeth: { 12 Upper { 12 Molar.
           { 20 Lower { 8 Incisive
                      { 12 Molar.

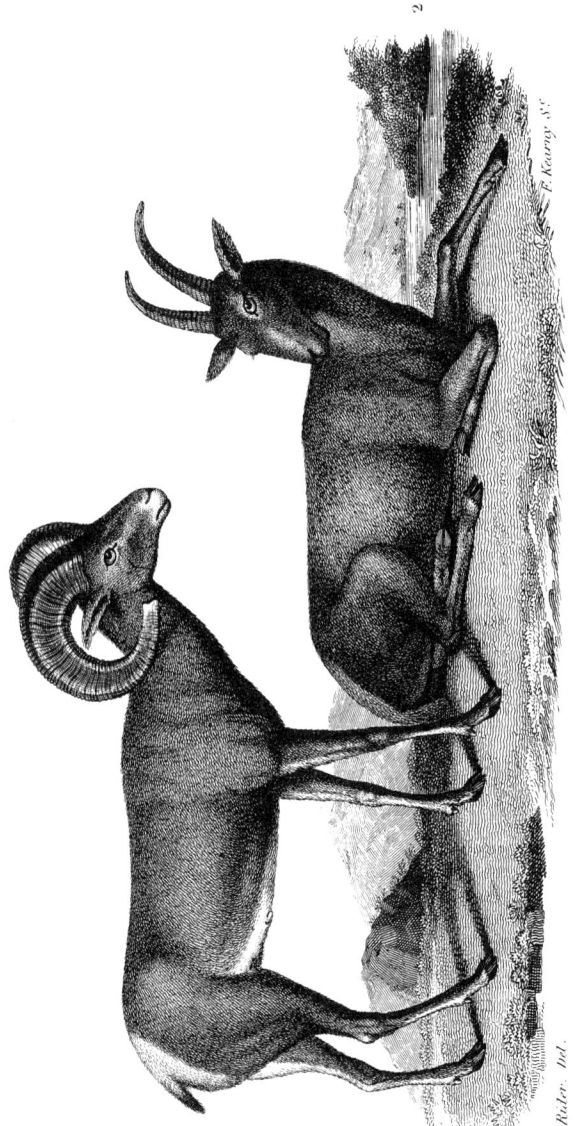

Argali. 1. Male. 2. Female.

Species I. *The Argali.*

Ovis Ammon: L.

Ovis fera Sibiricæ: vulgo, Argali dicta; Pall. Spicil. Zool. fasc. xi. pl. 1.
Monflor Argali: Shaw, Gen. Zool. ii. part 2. pl. 201.
Ovis Montana: Geoff. Ann. du Mus. ii. pl. 60.
Big Horn: Lewis and Clarke, i. 144.
Monflor D'Amerique: Desm. Mam.

The Argali is found in Northern Asia, and Eastern Siberia, whence it appears gradually to have passed into North America by crossing the ice, where the continents are separated but by a narrow strait. In America it inhabits the Rocky Mountains, in about the fiftieth degree of north latitude, and extends along the Rocky Mountain range into California. In these mountains the argali are seen in troops containing twenty or thirty, feeding upon the most precipitous parts of the ground, and leaping with wonderful activity, and at great distances, from rock to rock.

The spring of the year and autumn are said to be the sexual seasons of this species, during which period the males acquire the same disposition to fight with each other as we have described in treating of the deer.

Two specimens of the argali, a male and female, were brought in by Lewis and Clarke, and may be seen in the Philadelphia Museum, where they are preserved. The engraving will give a good idea of these animals, though the specimens just mentioned, from which the drawing was made, are much injured by time and exposure to the dust.

The male has very large horns, which arise quite near to the eyes, curve at first backwards, then bend forwards, and have their points turning upwards and outwards. These horns are triangular at the lower part, have their broadest surface forwards, and are deeply wrinkled thence for the half of their length; the superior part is smoother. The ears are straight, broad and pointed. The tail is quite short.

The colour of the argali, during summer, is of a grayish fawn, generally having along the back a deeper yellow or reddish line. Around the root of the tail, upon the buttocks, there is a spot of the same colour. The belly and inner surface of the limbs have a pale brownish or dirty white colour. The winter pelage is of a deeper reddish tint above, while the mouth, under part of the throat and belly, are nearly white.

The horns of the female are slender, when compared with those of the male, being almost straight and little wrinkled.

The dimensions of a large male killed by Mr. M'Gillivray in 1800 were these:*—from the nose to the base of the tail five feet; length of the tail four inches; girth of the body four feet; height of the body three feet eight inches. The horn was three feet and a-half long, and was one foot three inches in circumference at base.

It is much to be regretted that we are not better

---

* See his paper in the New York Medical Repository, vol vi. 238.

acquainted with the peculiar history of this animal drawn up by some one who has studied it in its native wilds; more especially as this species is said to be the source whence all the varieties of our domestic sheep are descended, an opinion which the form and proportions of its body seem to confirm, but one which would scarcely be imagined, if we relied upon the condition of the hair or wool by which the wild animal is covered.

END OF VOLUME II.

AMERICAN
Natural History

BY

JOHN D. GODMAN M.D.

Philadelphia.

CAREY, LEA & CAREY CHESTNUT STREET

1828

AMERICAN

NATURAL HISTORY.

VOLUME III.

PART I.—MASTOLOGY.

BY JOHN D. GODMAN, M.D.

PROFESSOR OF NATURAL HISTORY IN THE FRANKLIN INSTITUTE OF PENN-
SYLVANIA; ONE OF THE PROFESSORS OF THE PHILADELPHIA MUSEUM;
MEMBER OF THE AMERICAN PHILOSOPHICAL SOCIETY; OF THE PHILADEL-
PHIA ACADEMY OF NATURAL SCIENCES, &C.

PHILADELPHIA:
CAREY, LEA & CAREY—CHESTNUT STREET.

1828.

Eastern District of Pennsylvania, to wit:

BE IT REMEMBERED, That on the first day of February, in the fifty-second year of the Independence of the United States of America, A. D. 1828, P. H. Nicklin, of the said district, hath deposited in this office the title of a book, the right whereof he claims as proprietor, in the words following, to wit:

"American Natural History. Vol. III. Part I. Mastology. By John D. Godman, M. D. Professor of Natural History in the Franklin Institute of Pennsylvania; one of the Professors of the Philadelphia Museum; Member of the American Philosophical Society; of the Philadelphia Academy of Natural Sciences, &c.

In conformity to the Act of the Congress of the United States entitled, " An act for the Encouragement of Learning, by securing the copies of maps, charts, and books, to the authors and proprietors of such copies, during the times therein mentioned."—And also to the act, entitled, " An act supplementary to an act, entitled, " An act for the encouragement of learning, by securing the copies of maps, charts, and books, to the authors and proprietors of such copies during the times therein mentioned," and extending the benefits thereof to the arts of designing, engraving, and etching historical and other prints."

D. CALDWELL,
Clerk of the Eastern District of Pennsylvania.

AMERICAN NATURAL HISTORY.

CHAPTER I.

Genus Ox; *Bos;* L.

Fr. Bœuf. *Germ.* Ochs.
Sp. Buly. *Ital.* Bove.

GENERIC CHARACTERS.

The head is large, having a straight outline; large ears and eyes; a large muzzle and long smooth tongue. The subocular sinuses do not exist. The body is of large size, supported upon strong legs. A fold of skin depends below the neck, called the dewlap. The tail is frequently long and terminates in a brush; in some species it is of a middling length. The horns are conical, smooth and simple, variously curved, though often turned laterally with the points upwards.

Dental System.

32 Teeth: { 12 Upper { 12 Molar
           { 20 Lower { 8 Incisive
                      { 12 Molar.

Species I.—*The Bison.*

Bos Americanus Gmel.

Taurus Mexicanus: Hernand. Mex. 587.   Tauri Vaccæque, Ibid.
　Anim. p. 10.
The Buffalo: Catesby, Carol. 28 tab. 20.
Bœuf Sauvage: Dupratz, Louisiane, ii. 66.
American Bull: Penn. Quad. pl. ii, fig. 2.

[Commonly called Buffaloe.]

From other species of the ox kind, the Bison is well distinguished by the following peculiarities. A long shaggy hair clothes the fore part of the body, forming a well marked beard beneath the lower jaw, and descending behind the knee in a tuft. This hair rises on the top of the head in a dense mass, nearly as high as the extremities of the horns. Over the forehead it is closely curled, and matted so thickly as to deaden the force of a rifle ball, which either rebounds, or lodges in the hair, merely causing the animal to shake his head as he heavily bounds along.

The head of the bison is large and ponderous, compared with the size of the body; so that the muscles for its support, necessarily of great size, give great thickness to the neck, and by their origin from the prolonged dorsal vertebral processes form the peculiar projection called the *hump*. This hump is of an oblong form diminishing in height as it recedes, so as to give considerable obliquity to the line of the back.

Bison.

The eye of the bison is small, black, and brilliant; the horns are black and very thick near the head, whence they curve upwards and outwards, rapidly tapering towards their points. The outline of the face is somewhat convexly curved, and the upper lip, on each side being papillous within, dilates and extends downwards, giving a very oblique appearance to the lateral gape of the mouth, in this particular resembling the ancient architectural bas-reliefs representing the heads of oxen.

The physiognomy of the bison is menacing and ferocious, and no one can see this formidable animal in his native wilds, for the first time, without feeling inclined to attend immediately to his personal safety. The summer coat of the bison differs from his winter dress, rather by difference of length than by other particulars. In summer, from the shoulders backwards, the hinder parts of the animal are all covered with a very short fine hair, that is as smooth and as soft to the touch as velvet. The tail is quite short and tufted at the end, and its utility as a fly-brush is necessarily very limited. The colour of the hair is uniformly dun, but the long hair on the anterior parts of the body is to a certain extent tinged with yellowish or rust colour. These animals, however, present so little variety in regard to colour, that the natives consider any remarkable difference from the common appearance as resulting from the immediate interference of the Great Spirit.

Some varieties of colour have been observed, although the instances are rare. A Missouri trader informed the members of Long's exploring party, that he had seen a greyish white bison, and a yearling calf, that was distinguished by several white spots on the side, a star or blaze in the forehead, and white fore feet. Mr. J. Doughty, an interpreter to the expedition, saw in an Indian hut a very well prepared bison head with a star on the front. This was highly prized by the proprietor, who called it his *great medicine*, for, said he " the herds come every season to the vicinity to seek their white faced companion."

In appearance the bison cow bears the same relation to the bull, that is borne by the domestic cow to her mate. Her size is much smaller, and she has much less hair on the fore part of her body. The horns of the cow are much less than those of the bull, nor are they so much concealed by the hair. The cow is by no means destitute of beard, but though she possesses this conspicuous appendage, it is quite short when compared with that of her companion.

From July to the latter part of December the bison cow continues fat. Their breeding season begins towards the latter part of July and continues until the beginning of September, and after this month the cows separate from the bulls in distinct herds and bring forth their calves in April. The calves rarely separate from the mother before they are one year old, and cows are frequently seen accompanied by calves of three seasons.

The flesh of the bison is somewhat coarser in its fibre than that of the domestic ox, yet travellers are unanimous in considering it equally savoury as an article of food, we must, however, receive the opinions of travellers on this subject, with some allowance for their peculiar situations, being frequently at a distance from all other food and having their relish improved by the best of all possible recommendations in favour of the present viands—hunger. It is with reason, however, that the flesh is stated to be more agreeably sapid, as the grass upon which these animals feed is short, firm and nutritious, being very different from the luxuriant and less saline grass produced on a more fertile soil. The fat of the bison is said to be far sweeter and richer, and generally preferable to that of the common ox. The observations made in relation to the bison's flesh, when compared with the flesh of the domestic ox, may be extended to almost all wild meat, which has a peculiar flavour and raciness that renders it decidedly more agreeable than that of tame animals, although the texture of the flesh may be much coarser and the fibre by no means as delicate.

Of all the parts of the bison that are eaten, the hump is the most famed for its peculiar richness and delicacy; because when cooked it is said very much to resemble marrow. The Indian mode of cooking the hump is to cut it out from the vertebræ, after which the spines of bone are taken out, the denuded portion is then covered with skin, which is finally sewed to the skin covering the hump. The hair is then singed and pulled off, and the whole mass is put in a hole dug in the earth for its reception, which has been previously heated by a strong fire in and over it the evening previous to the day on which it is to be eaten. It is then covered with cinders and earth about a foot deep, and a strong fire made over it. By the next day at noon it is fit for use. The tongues and marrow bones are also highly esteemed by the hunters. To preserve the flesh for future use the hunters and Indians cut it into thin slices and dry it in the open air, which is called *jerking;* this process is speedily finished, and a large stock of meat may thus be kept for a considerable length of time.

From the dried flesh of the bison the fur traders of the north west prepare a food which is very valuable on account of the time it may be preserved without spoiling, though it will not appear very alluring to those who reside where provisions are obtained without difficulty. The dried bison's flesh is placed on skins and pounded with stones until sufficiently pulverized. It is then separated as much as possible from impurities, and one third of its weight of the melted tallow of the animal is poured over it. This substance is called *pemmican*, and being packed firmly in bags of skin of a convenient size for transportation, may be kept for one year without much difficulty, and with great care, perhaps two years.

During the months of August and September the flesh of the bison bull is poor and disagreeably flavoured; they are however much more easily killed, as they are not so vigilant as the cows, and sometimes allow the hunter to come up with them without much difficulty. Lewis & Clarke relate that once approaching a large herd, the bulls would scarcely move out of their way and as they came near, the animals would merely look at them for a moment, as at something new, and then quietly resume their grazing.

The general appearance of the bison is by no means attractive or prepossessing, his huge and shapeless form, being altogether devoid of grace and beauty. His gait is awkward and cumbrous, although his great strength enables him to run with very considerable speed over plains in summer, or in winter to plunge expeditiously through the snow.

The sense of smelling is remarkably acute in this animal, and it is remarked by hunters that the odour of the white man is far more terrifying to them than that of the Indian. From the neighbourhood of white settlements they speedily disappear: this, however, is very justly accounted for by Mr. Say, who attributes it to the impolitic and exterminating warfare, which the white man wages against all unsubdued animals within his reach.

As an exemplification of the peculiar strength of

their sense of smelling, we may here relate a circumstance mentioned by Mr. Say, in that valuable and highly interesting work, Long's Expedition to the Rocky Mountains, to which we are under continual obligations. These we are the more happy to acknowledge, because we are well acquainted with the solicitude of the gentlemen composing that expedition, to diffuse, as widely as possible, the knowledge of American Natural History.

The exploring party were riding through a dreary and uninteresting country, which at that time was enlivened by vast numbers of bisons, who were moving, in countless thousands, in every direction. As the wind was blowing fresh from the south, the scent of the party was wafted directly across the river Platte, and through a distance of eight or ten miles, every step of its progress was distinctly marked by the terror and consternation it produced among the bisons. The instant their atmosphere was infected by the tainted gale, they ran as violently as if closely pursued by mounted hunters, and instead of fleeing from the danger, they turned their heads towards the wind, eager to escape this terrifying odour. They dashed obliquely forward towards the party, and plunging into the river, swam, waded, and ran with headlong violence, in several instances breaking through the Expedition's line of march, which was immediately along the left branch of the Platte. One of the party, (Mr. Say himself,) perceiving from the direction taken by the bull who led the extended column, that he would emerge from the low river bottom at a point where the precipitous bank was deeply worn by much travelling, urged his horse rapidly forward, that he might reach this station in order to gain a nearer view of these interesting animals. He had but just reached the spot when the formidable leader, bounding up the steep, gained the summit of the bank with his fore-feet, and in this position, suddenly halted from his full career, and fiercely glared at the horse which stood full in his path. The horse was panic-struck by this sudden apparition, trembled violently from fear, and would have wheeled and taken to flight, had not his rider exerted his utmost strength to restrain him; he recoiled, however, a few feet and sunk down upon his hams. The bison halted for a moment, but urged forward by the irresistible pressure of the moving column behind, he rushed onward by the half-sitting horse. The herd then came swiftly on, crowding up the narrow defile. The party had now reached the spot, and extended along a considerable distance; the bisons ran in a confused manner, in various directions, to gain the distant bluffs, and numbers were compelled to pass through the line of march. This scene, added to the plunging and roaring of those who were yet crossing the river, produced a grand effect, that was heightened by the fire opened on them by the hunters.

To the Indians and visiters of the western regions the bison is almost invaluable; we have mentioned that they supply a large part of the food used by the natives, and covering to their tents and persons, while in many parts of the country there is no fuel to be obtained but the dried dung of this animal. The Indians always associate ideas of enjoyment with plenty of bison, and they frequently constitute the skull of one of them, their "Great Medicine." They have dances and ceremonies that are observed previous to the commencement of their hunting.

The herds of bison wander over the country in search of food, usually led by a bull most remarkable for strength and fierceness. While feeding, they are often scattered over a great extent of country, but when they move in mass they form a dense almost impenetrable column, which, once in motion, is scarcely to be impeded. Their line of march is seldom interrupted even by considerable rivers, across which they swim without fear or hesitation, nearly in the order that they traverse the plains. When flying before their pursuers, it would be in vain for the foremost to halt, or attempt to obstruct the progress of the main body, as the throng in the rear still rushing onward, the leaders must advance, although destruction awaits the movement. The Indians take advantage of this circumstance to destroy great quantities of this favourite game, and, certainly, no mode could be resorted to more effectually destructive, nor could a more terrible devastation be produced, than that of forcing a numerous herd of these large animals, to leap together from the brink of a dreadful precipice, upon a rocky and broken surface, a hundred feet below.

When the Indians determine to destroy bison in this way, one of their swiftest footed and most active young men is selected, who is disguised in a bison skin, having the head, ears, and horns adjusted on his own head, so as to make the deception very complete, and thus accoutred, he stations himself between the bison herd and some of the precipices, that often extend for several miles along the rivers. The Indians surround the herd as nearly as possible, when, at a given signal, they show themselves and rush forward with loud yells. The animals being alarmed, and seeing no way open but in the direction of the disguised Indian, run towards him, and he, taking to flight, dashes on to the precipice, where he suddenly secures himself in some previously ascertained crevice. The foremost of the herd arrives at the brink—there is no possibility of retreat, no chance of escape; the foremost may for an instant shrink with terror, but the crowd behind, who are terrified by the approaching hunters, rush forward with increasing impetuosity, and the aggregated force hurls them succesively into the gulf, where certain death awaits them.

It is extremely fortunate that this sanguinary and wasteful method of killing bisons is not very frequently resorted to by the savages, or we might expect these animals in a few years to become almost entirely extinct. The waste is not the only unpleasant circumstance consequent on it; the air for a long time after, is filled with the horrible stench arising from the putrefying carcases not consumed by the Indians after such an extensive and indiscriminate slaughter. For a very considerable time after such an event, the wolves and vultures feast sumptuously and fatten to tameness on the disgusting remains, becoming so gentle and fearless, as to allow themselves to be approached by the human species, and even to be knocked down with a stick, near places where such sacrifices of bison have been made. Lewis & Clarke bestowed the name of *Slaughter River* on one of the tributaries of the Mississippi, in consequence of the precipices along its sides, having been used by the Indians for this mode of killing the bison.

A better and more common way of killing bison is that of attacking them on horseback. The Indians, mounted and well armed with bows and arrows, encircle the herd and gradually drive them into a situation favourable to the employment of the horse. They then ride in and single out one, generally a female, and following her as closely as possible, wound her with arrows until the mortal blow is given, when they go in pursuit of others until their quivers are exhausted. Should a wounded bison attack the hunter, he escapes by the agility of his horse, which is usually well trained for the purpose. In some parts of the country, the hunter is exposed to a considerable danger of falling, in consequence of the numerous holes made in the plains by the badger.

When the hunting is ended and a sufficiency of game killed, the squaws come up from the rear to skin and dress the meat, a business in which they have acquired a great degree of dexterity, as they can, with very inferior instruments, butcher a bison with far more celerity and precision than the white hunters.

If a bison is found dead, without an arrow in the body, or any particular mark attached, it becomes the property of the finder, so that a hunter may expend his arrows to no purpose when they fall off, after wounding or fairly perforating the animal. That the Indians do frequently send their arrows through the body of this animal is well attested by a great number of witnesses. In Long's expedition to the sources of St. Peters' river, it is related that Wahnita, a distinguished chief of the Sioux, has been seen to drive his arrow through the body of one bison, and sufficiently deep into the body of a second to inflict a deadly wound.

When the ice is breaking up on the rivers in the spring of the year, the dry grass of the surrounding plains is set on fire, and the bison are tempted to cross the river in search of the young grass that immediately succeeds the burning of the old. In the attempt to cross, the bison is often insulated on a large cake of ice that floats down the river. The savages select the most favourable points for attack, and as the bison approaches, the Indians leap with wonderful agility over the frozen ice, to attack him, and as the animal is necessarily unsteady, and his footing very insecure on the ice, he soon receives his death wound and is drawn triumphantly to the shore.

The Cree Indians make a bison-pound, by fencing a circular space of about a hundred yards in diameter. The entrance is banked up with snow sufficiently high to prevent the animals from retreating after they have once entered. For about a mile on each side of the road leading to the pound, stakes are driven into the ground at nearly equal distances of about twenty yards, which are intended to look like men, and to deter the animals from endeavouring to break through the fence. Within fifty or sixty yards of the pound, branches of trees are placed between the stakes to screen the Indians who lie down behind them, to wait for the approach of the bison. The mounted hunters display the greatest dexterity in this sort of chase, as they are obliged to manœuvre around the herd in the plains so as to urge them into the road-way, which is about a quarter of a mile broad. When this is effected, the Indians raise loud shouts, and pressing closely on the animals, terrify them so much, that they rush heedlessly forwards towards the snare. When they have advanced as far as the men who are lying in ambush, they also show themselves in-

THE BISON.

creasing the consternation of the bison by shouting violently and firing their guns. The affrighted animals have no alternative but to rush directly into the pound, where they are quickly despatched by guns or arrows. In the centre of one of these pounds, there was a tree on which the Indians had hung strips of bison flesh and pieces of cloth, as tributary or grateful offerings to the Great Master of life. They occasionally place a man in the tree to sing to the presiding spirit as the bisons advance. He is obliged to remain there until all the animals that have entered the pound are killed.*

The Omawhaw Indians hunt the bison in the following manner. The hunters who are in advance of the main body on the march, employ telegraphic signals from an elevated position, to convey a knowledge of their discoveries to the people. If they see bisons, they throw up their robes in a peculiar manner as a signal for a halt. The hunters then return as speedily as possible to camp, and are received with some ceremony on their approach. The chiefs and magicians are seated in front of the people, puffing smoke from their pipes, and thanking the Master of life with such expressions as "thanks Master of life, thank you Master of life, here is smoke, I am poor, hungry, and want to eat." The hunters then draw near the chiefs and magicians, and in a low tone of voice inform them of

---

* See Franklin's Exp. p. 112.

their discovery; when questioned as to the number, they reply by holding up some small sticks in a horizontal direction, and compare one herd at a certain distance with this stick, and another with that, &c.

An old man or crier then harangues the people, informing them of the company, exhorting the women to keep a good heart, telling them that they have endured many hardships with fortitude, and that their present difficulties are ended, as on the morrow the men will go in pursuit of the bisons and bring them certainly a plenty of meat.

Four or five resolute warriors are appointed at the council of chiefs, held the evening previous, to preserve order among the hunters on the following day. It is their business, with a whip or club, to punish those who misbehave, on the spot, or whose movements tend to frighten the game before all are ready, or previously to their arrival at the place whence they are to sally forth.

The next morning all the men, not superannuated, depart at an early hour, generally mounted and armed with bows and arrows. The superintendants or officers above mentioned accompany the swiftly moving cavalcade, on foot, armed with war clubs, the whole preceded by a footman bearing a pipe. When they come in sight of the herd the hunters talk kindly to their horses, using the endearing names of father, brother, uncle, &c., begging them not to fear the bisons, but to run well and keep close, taking care at the same time not to be gored by them.

Having approached the herd as closely as they suppose the animal will permit without alarm, they halt, that the pipe bearer may perform the ceremony of smoking, which is thought necessary to success. The pipe is lighted, and he remains a short time with his head inclined, and the stem of the pipe extended towards the herd. He then puffs the smoke towards the bisons, the heavens, the earth, and the cardinal points successively. These latter are distinguished by the terms sun-rise, sun-set, cold country, and warm country.

This ceremony ended, the chief gives the order for starting. They immediately separate into two bands, which wheeling to the right and left, make a considerable circuit with a view to enclose the herd at a considerable interval between them. They then close upon the animals and every man endeavours to signalize himself by the number he can kill.

It is now that the Indian exhibits all his skill in horsemanship and archery, and when the horse is going at full speed, the arrow is sent with a deadly aim and great velocity into the body of the animal behind the shoulder, where, should it not bury itself to a sufficient depth, he rides up and withdraws it from the side of the wounded and furious animal. He judges by the direction and depth of the wound, whether it be mortal, and when the deadly blow is inflicted, he raises a triumphant shout to prevent others from engaging in the pursuit, and dashes off to seek new objects for destruction, until his quiver is exhausted or the game has fled too far.

Although there is an appearance of much confusion in this engagement, and the same animal receives many arrows from different archers before he is mortally wounded or despatched, yet as every man knows his own arrows, and can estimate the consequences of the wounds he has inflicted, few quarrels ever occur as to the right of property in the animal. A fleet horse well trained, runs parallel with the bison at the proper distance, with the reins thrown on his neck, turns as he turns, and does not lessen his speed until the shoulder of the animal is presented, and the mortal wound has been given; then by inclining to one side the rider directs him towards another bison. Such horses are preserved exclusively for the chase and are very rarely subjected to the labour of carrying burdens.*

The effect of training, on the Indian horses, is well shown in a circumstance related by Lewis and Clarke. A serjeant had been sent forward with a number of horses, and while on his way, came up with a herd of bisons. As soon as the loose horses discovered the herd, they immediately set off in pursuit, and surrounded the bisons with almost as much skill as if they had been directed by riders. At length the sergeant was obliged to send two men

---

* Say, Long's Exp. to Rocky Mountains, v. 2.

forward to drive the bisons from the route before they were able to proceed.

The skins of the bison furnish the Indians and Whites with excellent robes, for bedding, clothing, and various purposes. These are most usually the skin of cows, as the hide of the bull is too thick and heavy to be prepared in the way practised by the squaws, which is both difficult and tedious. This consists in working the hide, moistened with the brains of the animal, between the hands, until it is made perfectly supple, or till the thick texture of the skin is reduced to a porous and cellular substance. These robes form an excellent protection from rain, when the woolly side is opposed to it, and against the cold when the woolly surface is worn next the skin. But when these robes are wet, or for a considerable time exposed to moisture, they are apt to spoil and become unpleasant, as the Indian mode of dressing has no other effect than to give a softness and a pliancy to the leather. On these robes the Indians frequently make drawings of their great battles and victories; a great variety of such painted robes are to be seen in the Philadelphia Museum. The hair of the bison has been used in the manufacture of a coarse cloth, but this fabric has never been extensively employed.

We have already adverted to the great numbers of these animals which live together. They have been seen in herds of three, four, and five thousand, blackening the plains as far as the eye could view.

Some travellers are of opinion that they have seen as many as eight or ten thousand in the same herd, but this is merely a conjecture. At night it is impossible for persons to sleep near them who are unaccustomed to their noise, which from the incessant lowing and roaring of the bulls, is said very much to resemble distant thunder. Although frequent battles take place between the bulls, as among domestic cattle, the habits of the bison are peaceful and inoffensive, seldom or never offering to attack man or other animals, unless outraged in the first instance. They sometimes, when wounded, turn on the aggressor, but it is only in the rutting season that any danger is to be apprehended from the ferocity and strength of the bison bull. At all other times, whether wounded or not, their efforts are exclusively directed towards effecting their escape from their pursuers, and at this time it does not appear that their rage is provoked particularly, by an attack on themselves, but their unusual intrepidity is indiscriminately directed against all suspicious objects.

We shall conclude this account of bison, by introducing the remarks of John E. Calhoun, Esq.,* relative to the extent of country over which this animal formerly roved and which it at present inhabits.

---

* Long's Exp. to the source of the St. Peter's river, ii. p. 28.

The buffaloe was formerly found throughout the whole territory of the United States, with the exception of that part which lies east of Hudson's river and Lake Champlain, and of narrow strips of coast on the Atlantic and the Gulf of Mexico. These were swampy and had probably low thick woods. That it did not exist on the Atlantic coast is rendered probable, from the circumstance that all the early writers whom Mr. Calhoun has consulted on the subject, and they are numerous, do not mention them as existing then, but further back. Thomas Morton, one of the first settlers of New England, says, that the Indians " have also made description of great heards of well growne beasts, that live about the parts of this lake," Erocoise, now Lake Ontario, " such as the christian world, (untile this discovery,) hath not bin made acquainted with. These beasts are of the bignesse of a cowe, their flesh being very good foode, their hides good lether, their fleeces very useful, being a kind of wolle, as fine almost as the wolle of the beaver, and the salvages do make garments thereof;" he adds, " It is tenne yeares since first the relation of these things came to the eares of the English."* We have introduced this quotation, partly with a view to show that the fineness of the buffalo wool, which has caused it within a few years, to become an object of

---

* New English Canaan, by Thomas Morton, Amsterdam, 1637, p. 98.

commerce, was known as far back as Morton's time; he compares it with that of the beaver and with some truth; we were shown lower down on Red river, hats that appeared to be of a very good quality; they had been made in London with the wool of the buffaloe. An acquaintance on the part of Europeans with the animal itself, can be referred to nearly a century before that: for in 1532, Guzman met with buffalo in the province of Cinaloa.* De Laet says, upon the authority of Gomara, when speaking of the buffalo in Quivira, that they are almost black, and seldom diversified with white spots.† In his history written subsequently to 1684, Hubbard does not enumerate this animal among those of New England. Purchas informs, us that in 1613 the adventurers discovered in Virginia, " a slow kinde of cattell as bigge as kine which were good meate."‡ From Lawson, we find that great plenty of buffaloes, elks, &c., existed near Cape Fear river and its tributaries;§ and we know that some of those who first settled the Abbeville district in South Carolina, in 1756, found the buffaloe there. De Soto's party, who traversed East Florida, Georgia, Alabama, Mississippi, Arkansa Territory, and Louisiana, from 1539 to 1543, saw no buffaloe,

---

* De Laet, Americæ utriusque Descriptio, Lugd. Batav. anno 1633, lib. 6. cap. 6.
† Idem, lib. 6, cap. 17.   ‡ Purchas ut supra, p. 759.
§ Lawson ut supra, p. 48, 115 &c.

they were told that the animal was north of them; however, they frequently met with buffalo hides, particularly when west of the Mississippi; and Du Pratz, who published in 1758, informs us that at that time the animal did not exist in lower Louisiana. We know however of one author, Bernard Romans, who wrote in 1774, and who speaks of the buffalo as a benefit of nature bestowed upon Florida. There can be no doubt that the animal approached the Gulf of Mexico, near the Bay of St. Bernard; for Alvar Nunez, about the year 1535, saw them not far from the coast; and Joater, one hundred and fifty years afterwards, saw them at the Bay of St. Bernard. It is probable that this Bay is the lowest point of latitude at which this animal has been found east of the Rocky Mountains. There can be no doubt of their existence west of those mountains, though Father Venegas does not include them among the animals of California, and although they were not seen west of the mountains by Lewis and Clarke, nor mentioned by Harmon and Mackenzie as existing in New Caledonia, a country of immense extent, which is included between the Pacific Ocean, the Rocky Mountains, the territory of the United States, and the Russian possessions, on the northwest coast of America. Yet their existence at present on the Columbia, appears to be well ascertained, and we are told that there is a tradition among the natives, that shortly before the visit of our enterprising explorers, destructive fires had raged over the prairies and driven the buffalo east of the mountains. Mr. Dougherty, the very able and intelligent sub-agent, who accompanied the expedition to the Rocky Mountains, and who communicated so much valuable matter to Mr. Say, asserted that he had seen a few of them in the mountains, but not west of them. It is highly probable that the buffalo ranged on the western side of the Rocky Mountains, to as low a latitude as on the eastern side. De Laet says, on the authority of Henera, that they grazed as far south as the banks of the river Yaquimi.* In the same chapter this author states, that Martin Perez had, in 1591, estimated the province of Cinaloa, in which this river runs, to be three hundred leagues from the city of Mexico. This river is supposed to be the same, which, on Mr. Tanner's map of North America, (Philadelphia, 1822,) is named Hiaqui, and situated between the 27th and 28th degrees of north latitude. Perhaps, however, it may be the Rio Gila, which empties itself in latitude 32°. Although we may not be able to determine with precision, the southern limit of the roamings of the buffalo west of the mountains, the fact of their existence there in great abundance, is amply settled by the testimony of De Laet, on the authority of Gomara, l. 6, c. 17, and of Purchas, p. 778. Its limits to the north are

---

* " Juxta Vaquimi fluminis ripas tauri vaccæque et prægrandes cervi pascuntur," ut supra lib. 6 cap. 6.

not easier to determine. In Hakluyts' collection we have an extract of a letter from Mr. Anthonie Parkhurst, in 1578, in which he uses these words; in the Island of Newfoundland there " are mightie beastes, like to camels in greatnesse, and their feete cloven. I did see them farre off, not able to discerne them perfectly, but their steps shewed that their feete were cloven and bigger than the feete of camels. I suppose them to be a kind of buffes, which I read to bee in the countrys adjacent and very many in the firme land."* In the same collection, p. 689, we find, in the account of Sir Humphrey Gilbert's voyages, which commenced in 1583, that there are said to be in Newfoundland, " buttolfles, or a beast, it seemeth by the tract and foote, very large in the manner of an oxe." It may, however, be questioned whether these were not musk oxen, instead of the common buffalo or bison of our prairies. We have no authority of any weight, which warrants us in admitting that the buffalo existed north of Lakes Ontario, Erie, &c. and east of Lake Superior. From what we know of the country between Nelson's River, Hudson's Bay, and the lower Lakes, including New South Wales and Upper Canada, we are inclined to believe that the buffalo never abounded there, if indeed any were

---

* The principal navigations, voyages, and discoveries of the English nation, &c. by Richard Hakluyt, London, 1589, p. 676.

ever found north of the lakes. But west of Lake Winnepeck, we know that they are found as far north as the 62nd degree of north latitude. Capt. Franklin's party killed one on Salt river, about the 60th degree. Probably they are found all over the prairies which are bounded on the north by a line commencing at the point at which the 62nd degree meets the base of the Rocky Mountains, and running in a south easterly direction, to the southern extremity of Lake Winnepeck, which is but very little north of the 50th degree; on the Sardatchawan, buffalo are very abundant. It may be proper to mention here, that the small white buffalo, of which Mackenzie makes frequent mention, on the authority of the Indians, who told him that they lived in the mountains, is probably not the bison; for Lewis and Clarke inform us, that the Indians designated by that name the mountain sheep.* It is probable that west of the Rocky Mountains the buffalo does not extend far north of the Columbia. At present it is scarcely seen east of the Mississippi, and south of the St. Lawrence. Governor Cass's party found in 1819, buffalo on the east side of the Mississippi, above the falls of St. Anthony: every year this animal's rovings are restricted. In 1822, the limit of its wanderings down the St. Peter, was Great Swan Lake (near Camp Crescent.)

---

* Vol. ii. p. 325.

Drawn by C.A.Lesueur. Eng.d by G.B.Ellis

Musk Ox.

Species II.—*The Musk Ox.*

Bos Moschatus Gmel.

Musk Ox: Penn. Quad. i. 31. Ibid, Arct. Zool. 3 vol. i. 8.
Musk Ox: Hearne, Journey &c. 8vo. 135.
Bœuf Musqué: Buff. Hist. Naturelle Suppl. vi.
Ovibos* Musqué: Blainv. Nouv. Bullet. de la Soc. Philom.
Musk Ox: Parry's Voyage. i. 202.

[*Called Mathek-Mongsoo, or Ugly Moose, by the Crees, Uming Mak, by the Esquimaux.*]

To civilized man, the extreme northern regions may appear cheerless and uninviting, because they are subjected to the almost unrelenting influ-

---

* Mr. De Blainville proposed to establish a new genus, to be called *Ovibos* or Sheep-ox, of which the Musk-ox is the first species. His *generic* distinctions are drawn from the resemblance between the outline of the front of the musk-ox and that of the sheep, and from the absence of the muzzle or smooth naked surface, between the nostrils, and upon the upper lip. This division, though as well founded as that which separates *Capra* from *Ovis*, we conceive to be altogether unnecessary, as the characters are not more than sufficient to establish a *specific* difference. In regard to the muzzle, nothing is said in the text of Parry's work, though it is very distinctly represented in the plate, which is said to be very accurate, and which we have copied; as the common descriptions of the musk-ox, have mostly been taken from dried skins, it is possible, that the absence of the muzzle has been stated too hastily.

ence of wintry skies. Yet we have already seen that they are the favourite resorts of multitudes of animals, varying in size, characters and habits, from the Lemming to the Moose. A species remains to be described, which, of these forbidding regions prefers the most barren and desolate parts, and is found in the greatest abundance in the rugged and scarcely accessible districts lying nearest the North Pole. This species, so far from being condemned to a life of extreme privation and suffering, appears to derive as much enjoyment from existence, as those which feed in more luxuriant pastures, or bask in the genial rays of a summer sun.

In destining the musk ox to inhabit the domains of frost and storm, nature has paid especial attention to its security against the effects of both; first, by covering its body with a coat of long, dense hair, and then, by the shortness of its limbs, avoiding the exposure that would result from a greater elevation of the trunk. The projection of the orbits of the eyes, which is very remarkable in this species, is thought by PARRY to be intended to carry the eye clear of the large quantity of hair required to preserve the warmth of the head.

Although some few items relative to this animal are to be gathered from the works of the recent explorers of the Northern Regions, it is to HEARNE, that we are almost exclusively indebted for the Natural History of the musk ox, as we have already been for that of most of the animals inhabiting the same parts of this continent. This excellent and accurate observer travelled, in the years '69, '70, 71, and '72, and it is only to be regretted that he did did not write down all he knew in relation to the northern animals. He appears to have frequently thought that what was so familiarly known to him, would not be of much interest to others, and has thus withheld knowledge that few individuals can have a similar opportunity of gaining. Notwithstanding this, he has anticipated all the recent explorers in every essential observation.

HEARNE states that he has seen many herds of musk oxen in the high northern latitudes, during a single day's journey, and some of these herds contained from eighty to a hundred individuals, of which number a very small proportion were bulls, and it was quite uncommon to see more than two or three full grown males, even with the largest herds. The Indians had a notion that the males destroyed each other in combating for the females, and this idea is somewhat supported by the warlike disposition manifested by these animals during their sexual season. The bulls are then so jealous of every thing that approaches their favourites, that they will not only attack men or quadrupeds, but will run bellowing after ravens or other large birds that venture too near the cows.

Musk oxen are found in the greatest numbers within the arctic circle; considerable herds are occasionally seen near the coast of Hudson's bay, throughout the distance from Knapp's Bay to Wager Water. They have in a few instances been seen as low down as lat. 60° N. Capt. Parry's people killed some individuals on Melville Island, which were remarkably well fed and fat. They are not commonly found at a great distance from the woods, and when they feed on open grounds they prefer the most rocky and precipitous situations. Yet, notwithstanding their bulk and apparent unwieldiness, they climb among the rocks with all the ease and agility of the goat, to which they are quite equal in sureness of foot. Their favourite food is grass, but when this is not to be had, they readily feed upon moss, the twigs of willow, or tender shoots of pine.*

The appearance of the musk ox is singular and imposing, owing to the shortness of the limbs, its broad flattened crooked horns, and the long dense hair which envelopes the whole of its trunk, and hangs down nearly to the ground. When full

---

* It is singular and well worthy of observation, that the dung of the musk ox, though so large an animal, is not larger than, and, at the same time, is so nearly of the shape and colour of that of the Alpine Hare, that the difference is not easily distinguished except by the Indians, though the quantity generally indicates the animal to which it belonged. In the country adjacent to the Coppermine river, long ridges of this dung, together with that of deer and other animals were seen by Hearne. Similar appearances were observed by Parry on several of the North Georgian Islands.

grown, the musk ox is ten hands and a half high, according to Parry, and as large as the generality, or at least the middling size of English black cattle; but their legs, though large, are not so long; nor is their tail longer than that of a bear, and like the tail of that animal it always bends downwards and inwards, so that it is entirely hid by the long hair of the rump and hind quarters. The hunch on their shoulders is not large, being little more in proportion than that of a deer. Their hair is in some parts very long, particularly on the belly, sides and hind quarters; but the longest hair about them, particularly the bulls, is under the throat, extending from the chin to the lower part of the chest, between the forelegs; it there hangs down like a horse's mane inverted, and is full as long.*

---

* "Mr. Dragge says in his voyage, vol. 2, p. 260, that the musk ox is lower than a deer, but larger as to belly and quarters; which is very far from the truth. They are of the size I have here described them, and the Indians always estimate the flesh of a full grown cow to be equal to three deer. I am sorry also to be obliged to contradict my friend Mr. Graham, who says that the flesh of this animal is carried on sledges to Prince of Wales' Fort, to the amount of three or four thousand pounds annually. To the amount of near one thousand pounds may have been purchased from the natives in some particular years, but it more frequently happens that not an ounce is brought one year out of five, and in fact, all that has ever been carried to Prince of Wales' Fort, has most assuredly been killed out of a herd

The winter coat of the musk ox is formed of two sorts of hair, which is generally of a brownish red, and in some places of a blackish brown colour; the external being long, coarse, and straight, and the internal, fine, soft and woolly. The outer hair is so long that it hides the greater part of the limbs, causing them to look disproportionately short. As the summer comes on, the short woolly hair is gradually shed, but the summers are so short in these high latitudes, that the woolly coat commences growing almost immediately after the old coat is shed, so that the entire winter coat is completed by the return of the cold weather.

From the shortness of the limbs and the weight of the body, it might be inferred that the musk ox could not run with any speed, but it is stated by **Parry**, that although they run in a hobbling sort of canter that makes them appear as if every now and then about to fall, yet the slowest of these musk oxen can far outstrip a man. When disturbed and hunted, they frequently tore up the ground with their horns, and turned round to look at their pursuers, but never attempted to make an attack.

The month of August is the season in which the musk bulls are the most disposed to combat, as they

---

that has been accidentally found within a moderate distance of the settlement, perhaps within a hundred miles; which is only thought a step by an Indian." Hearne, 136. (The fort he mentions, was destroyed by the French in 1782.)

then fight furiously with each other for the females, and are jealous of the approach of every thing, as already stated. The cows calve about the end of May or the beginning of June; the calves are frequently whitish, but more commonly marked by a white patch or saddle upon the back.

The musk oxen killed on Melville island during Parry's visit, were very fat, and their flesh, especially the heart, although highly scented with musk, was considered very good food. When cut up it had all the appearance of beef for the market. Hearne says that the flesh of the musk ox does not at all resemble that of the bison, *(Bos Americanus)* but is more like that of the moose, and the fat is of a clear white tinged with light azure. The young cows and calves furnish a very palatable beef, but that of the old bulls is so intolerably musky, as to be excessively disagreeable. A knife used in cutting up such meat, becomes so strongly scented with this substance, as to require much washing and scouring before it is removed.* Musk ox flesh when dried, is considered by hunters and Indians to be very good. "In most parts of Hudson's Bay it is known by the name of Kew-hagon, but amongst the Northern Indians it is called Achees." The weight of

---

* Moschus iste glandulis juxta præputium positis efformari videtur; ibi materia fusca, concreta, fortissime moschi odorans inventa est.

the musk ox, according to Parry, is about 700 lbs. that of the head and hide is 130lbs.

The horns of the musk ox are employed for various purposes by the Indians and Esquimaux, especially for making cups and spoons. From the long hair growing on the neck and chest, the Esquimaux make their musquitoe wigs, to defend their faces from those troublesome insects. The hide of the musk ox makes good soles for shoes, and is much used by the natives for this purpose.

During the months of August and September the musk oxen extend their migrations to the North Georgian and other islands bordering the northern shores of the continent. By the first of October they have all left the islands and moved towards the south. By Franklin's Expedition, they were not seen lower than 66° N. though, as we have before stated from Hearne, they are occasionally seen as low as 60°.

CHAPTER II.

Order **VIII.** Cete; *Cetaceous Animals.*

Cetaceous animals in general appearance and in mode of living, bear a considerable resemblance to fish, with which they are popularly confounded; but by all the details of their conformation, their manner of respiration and the nourishment of their offspring, they are entitled to rank in the first class of animals, although at the inferior extremity of the scale.

In these creatures the head is joined to the trunk by so short and thick a neck, as to appear continuous with the body, and this large neck is in the greater number capable of very little, if any motion, owing to the consolidation of several of the slender cervical vertebræ. The trunk of the body gradually decreases until it terminates in a thick tail, which ends in a horizontal cartilaginous fin, and when used by the animal in effecting its forward motion, is moved up and down, never laterally.

The anterior extremities or arms, although in all respects analogous to those of the higher orders of animals, have the bones shortened, flattened and enveloped in a tendinous membrane, so as to be effectually converted into fins. The posterior extremities or limbs are entirely wanting.

The brain is large and well developed. The bone containing the organ of hearing, or internal ear, is separated from the rest of the head, being attached thereto by ligament alone. The orifice of the external ear is very small and destitute of external appendage. The teats, two in number, are either pectoral or abdominal.

CHAPTER III.

Family I. Sirenia; *Herbivorous Cetacea.*

This family is distinguished especially by the vegetable diet of the animals belonging to it, which is indicated by their flat grinding teeth. The head is not very large, and has always a short and obtuse snout, at the extremity of which, the external openings of the nostrils are situated, notwithstanding they pass through the bones of the head from the superior part. The mouth is garnished with long bristles or whiskers, and the teats are situated upon the chest.

The anterior extremities, though compressed, are still sufficiently free to allow them to be used for the purpose of carrying any thing by holding it against the body, the young, for instance, being thus held by the mother. The tail is not very large, but is powerful. These animals swim with great facility, and as they are able to raise the anterior parts from the water, so as to form a considerable angle with the trunk, it is considered as highly probable that the various fables of sirens, tritons and mermaids may have originated from an imperfect observation of their actions.

It must be admitted that the members of this family, present little in their general appearance to excite attention, unless it be their huge and almost shapeless bodies; but their internal structure, actions and habitudes, afford very ample scope for interesting observations, and philosophical inquiry; as it would not be easy, from any previous knowledge, to believe that merely herbivorous animals would be found inhabiting the ocean, conformed in all respects, so as closely to approach in external appearance to fish, and yet in all the characters of teeth, mode of feeding and digestive organs, to bear a very marked resemblance to herbivorous land quadrupeds.

CHAPTER IV.

GENUS I.—LAMANTIN: *Manatus*, C.

GENERIC CHARACTERS.

The head is small and conical with a broad snout, and rather small mouth; the eyes are placed high up between the extremity of the snout and the openings leading to the ears, which are very small and hardly visible. The spine is composed of seven very short cervical, seventeen dorsal, two lumbar, and twenty-two caudal vertebræ. The ribs are seventeen in number. In addition to the shoulder blade, arm and forearm, the lamantins have all the wrist or carpal bones, with the single exception of the pisiform, the phalanges of the thumb are wanting, and the corresponding metacarpal bone terminates in a point. All the other digits have three phalanges. The stomach has several cavities, the cœcum two branches, and the colon is very large; in all which circumstances they strongly resemble the pachydermatous land animals, along with which they have been considered by some naturalists.* The surface of the body is entirely destitute of hair.

Dental System.

34 Teeth: { 18 Upper { 2 Incisive. 16 Molar. } 16 Lower { 16 Molar. } }

In the upper jaw; in young individuals two small pointed incisive teeth are found, somewhat similar to those of the morse. There are no canines. The eight molars resemble each other; they have a general square form, and all present two transverse eminences, formed of three tubercles, separated from each other by a deep groove: they all have three divergent roots, one internal, the other two external. They increase gradually, but almost imperceptibly, in size from the first to the last.

In the lower jaw, neither incisive nor canine teeth are ever found, and the molars resemble those of the upper jaw, except in having a spur posteriorly, or a third eminence much smaller than the others. These teeth have two roots, one in front,

---

* Blainville at first arranged them with the unguligrada, and subsequently with the gravigrada, as the Elephants, &c. See Ranzani, Elem. di Zoologia, ii. parte iii. p. 670.

Fig. 1.

Male Narwal, or Unicorn.
15 ft. in length.

Fig. 2.

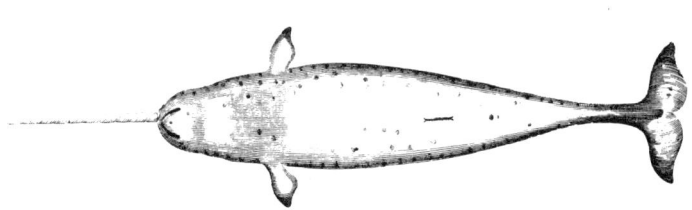

Under side view of the same Narwal.

Fig. 3.

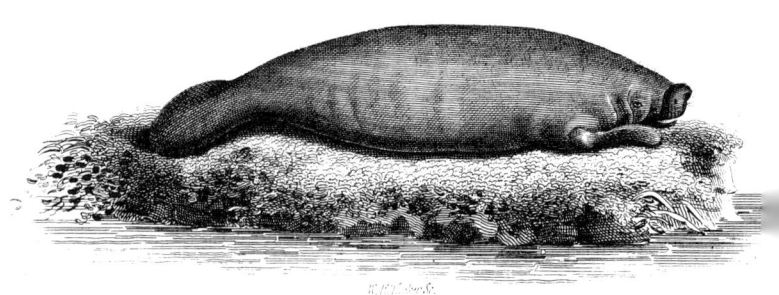

Lamantin.

the other behind, at first simple, but enlarged, and are bifurcated at their extremity.

In their reciprocal position, the eminences on one side, correspond to the grooves and intervals of the teeth on the opposite side, and to judge by the preservation of the crests of the eminences, it appears that these teeth are used more for triturating, than for crushing or bruising the food.

Species I.—*American Lamantin.*

Manatus Americanus.

Trichecus Manatus, L. Systema Naturæ.
Manate ou Vache Marine: Dampier, voyage i. 46. Sloane, Jamaica, ii. 329, La Condamine, voyage, 154.
Manati Phocæ genus, Clus. exot: 132.
Lamantin ou Manaty: Du Tertre, Hist. gen. des. Antilles.
Grand Lamantin des Antilles, Buff. Hist. Naturelle xiii. 377, C. Ann, du Museum xviii. 282. pl. 19. Ibid. Ossem. Foss iv; Desm. Nouv. Dict. d'Hist. Nat. xviii. 213 pl. G. 9.
Manatee, or Sea Cow: Bartram, Travels in Florida, 231.

If the reader infer from the number of authorities prefixed to this article, that the history of the species is amply or satisfactorily known, he will fall into an error, which a very little experience in books of travels, or systematic works of natural history would serve to correct. Indeed, as a general rule, the number of references affixed, is in an inverse proportion to the amount of knowledge concerning the animal treated of, and it not unfrequently happens that the mere mention of the name of a species, is all that occurs in a book quoted with all the formality of title and page. Bartram, for instance, who travelled in the country where the lamantin is most commonly found, gives the whole amount of his observation in nine unsatisfying lines. Other observers, who have enjoyed equally good opportunities, have contented themselves with a mention of the animal, taking it for granted that no other information was desirable.

Of this species we know little or nothing, but what is given by CUVIER in the scientific works above quoted, and from the observations made by Du Tertre in his history of the Antilles.

The general figure of the lamantin is rather elliptical and elongated. Its head is shaped like a simple truncated cone, and terminates in a thick and fleshy snout, semi-circular at its extremity, and pierced at its upper part, by two small semilunar nostrils, directed forwards. The edge of the upper lip is tumid, furrowed in the middle, and provided with thick and stiff whiskers. The lower lip is narrower and shorter than the upper, and the opening of the mouth is small. The eyes are situated towards the upper part of the head, at the same distance from the snout, as the angle of the lips. The ears are very small, scarcely perceptible, and placed at the same distance from the eyes, that the latter are from the snout.

The neck is not distinguishable by any diminution or difference in size from the head and trunk, and the latter does not diminish except from the umbilicus, whence it rapidly decreases, until it spreads out and becomes flattened, forming an oblong tail with a broad, thin, and seemingly truncated extremity. The tail forms about a fourth of the length of the animal.

The arm bones which sustain the fins are more separated from the body, than those of the Delphinus, and have digits more distinguishable through the integuments. The edges of this fin have four flat and rounded nails, which do not extend beyond the membrane, the nail of the thumb being deficient. The skin is of a gray colour, is slightly shagreened, and has upon it a few scattered hairs, which are more numerous than elsewhere about the angles of the lips, and the palmar surface of the fins.

The full grown lamantin is from fifteen to twenty feet in length, by eight in circumference, and weighs several thousand pounds.

Du Tertre states that the sight of the lamantin is very feeble, but this defect is compensated by the extreme acuteness of its hearing. In these respects it closely resembles the seal. After having satisfied its hunger by feeding on the sea grass or fucus, which constitute its principal nourishment, it delights to sleep upon the marshy grounds in shallow water, where it lies with the snout elevated above the water.

When the lamantin is discovered in this situation, the following mode of securing it, is resorted to; three, or at most, four persons get into a canoe, which is managed by the man in the bow, who moves his paddle from right to left, without lifting it from the water, so as to impel the canoe swiftly and without noise. The harpooner sits on a board placed across the canoe in the forward part, and the third person is placed in the middle of the boat to manage the line attached to the end of the harpoon.

The canoe is then swiftly paddled towards the sleeping animal, the men observing the most profound silence. When within three or four paces of the lamantin, the harpoon is suddenly struck into its body. The most violent efforts are then made by the wounded animal, which leaps up and springs forward with great force, making the sea foam, by the celerity of its movements. Tired, at length, with fruitless efforts to escape, and weakened by loss of blood, the lamantin stops short, is again wounded by other harpoons, and after a few more unavailing struggles, yields its liberty and life together.

The female lamantin brings forth her two young, which follow her closely, and are very certainly captured, if the mother be killed. The flesh of the lamantin is considered an excellent article of diet, and has, at former periods, furnished a large part of the subsistence of the inhabitants of St. Christophers, Guadaloupe, and Martinique. This flesh has the taste of veal, but is more solid, and covered in various parts with two or three inches of fat, which is used for the same purposes that lard is commonly employed for. It is so good that many persons melt the fat, and eat the oil upon bread instead of butter. When salted, the flesh of the lamantin loses its flavour and becomes very dry and hard.

The name of manati, (subsequently changed to lamantin,) is said to have been originally given to this animal by the Spaniards, in consequence of its short anterior extremities, which were regarded as hands.

CHAPTER V.

Genus II.—Steller; *Stellerus*, C.

Rytina, Ill. *Trichecus*, Gmel. *Manatus*, Stell.

GENERIC CHARACTERS.

The head is blunt, joined to the body by a short indistinct neck, and has no external ears; the eyes are defended by a sort of cartilaginous membrane instead of lids; the nostrils are situated at the extremity of the snout; both upper and lower lips are vertically divided. The anterior extremities end in flippers similar to those of the sea-turtle. The caudal fin is very broad, crescent shaped, and terminates on each side by a large point. The skin is hairless, but is defended by an uncommonly thick epidermis, composed of fibres perpendicular to the true skin. The stomach is simple; the intestines upwards of 400 feet in length; the cœcum of vast size, and divided into huge pouches.

Dental System.

4 Teeth: { 2 Upper { 2 Molar.
         { 2 Lower { 2 Molar.

These teeth are not set in the jaw by roots, but are affixed by nervous and vascular connexions.* The grinding surface is very rough, being hollowed so as form many tortuous canals.

Species I.—*Boreal Steller.*†

Stellerus Borealis: Desm.

Manatus: Stell. Act. Petrop. Com. Nov. ii. 294.
Trichecus Manatus: V. Borealis, Gmel.
Whale-tailed Manati: Penn. Arct. Zool.
Whale-tailed Trichecus: Shaw. Gen. Zool.
Rytina Stelleri: Desm. N. Dict. D'Hist. Nat. xxix. 575
Stellerus Borealis: Ibid. Mammalogie sp. 752, p. 510.

The only detailed account of the manners and habits of this singular animal, is that originally given by *Steller* in the transactions of the Imperial Academy of Sciences of St. Petersburg, in 1749. From his valuable paper, which contains numerous highly interesting observations on other animals, we have translated the following faithfully observed facts.‡

---

* Resembling in this respect the duckbill animal of New Zealand &c.

† Cuvier named the genus formed for this species, in honour of Steller; we use his name for the common appellation, because we wish to avoid confounding this animal with the *Manati* by using Pennant's term.

‡ This paper is entitled " De Bestiis marinis auctore

An unfortunate accident gave me an opportunity of observing the manners and habits of these animals, daily, before the door of our hut. They delight in the shallow sandy places, near the shores of the sea, and are very fond of frequenting the mouths of brooks and little rivers, being allured by the sweetness of the running water; they always go in troops, the half grown and young occupy the front in feeding, but are solicitously enclosed on the flanks and rear, so as to be always kept in the midst of the troop. When the tide is high, they come so close to the shore, that I have not only frequently touched them with a staff or lance, but have placed my hand upon their backs. If struck with some force, they did nothing more than move a little farther off, and in a short time forgetting the injury, they would return. Entire families commonly live together, a male with a single female, and a small quite young cub. They appeared to me to be monogamous, bringing forth at any time of the year, but most generally in the autumn, as I should judge from the young about that time; hence as I have observed them, most especially to couple in the spring. I have concluded, that they bear their young for more than a year, and do not bring forth but one cub at a birth. I never observed more than one cub in company with the mother.

Georgio Wilhelmo Stellero." Vide Nov. Comm. Acad. Scien. Imper. Petropolitanœ, tom. 2, p. 289, 294, et seq.

These most voracious animals are almost incessantly feeding, and on account of this greediness, have their heads nearly always under water, being very little solicitous concerning life and safety. A boat or a man may go into the midst of a troop, and one may be selected and secured by a hook without difficulty. All this must be done while they are feeding, as at the end of every four or five minutes, they raise their nostrils above water, and blow out the air with a small quantity of water, making a noise like the snorting of a horse. While grazing, they move slowly forwards, one foot after the other, and thus in part placidly swim, and partly walk, like oxen or sheep browsing. The half of the body, that is, the back and sides, always project above the water while they are feeding, and the gulls alight thereon for the purpose of picking up the parasite animals with which their hides are much infested, just as the ravens alight on the backs of hogs to catch their lice.

They do not devour all the sea-weeds indifferently, but chiefly two or three species of Fucus or Kelp, of which, when these animals have remained a day or so in one vicinity, large heaps of the roots and stalks are thrown ashore by the waves. Having gorged themselves fully, some of them sleep upon their backs, at some distance from the shore, lest they be left aground by the tide. They are frequently killed by the floating ice in winter, which especially occurs if the waves are blown forcibly upon the rocks, among which these beasts are entangled and killed. In the winter they become so thin, that in addition to the back bones, all the ribs may be counted through the skin.*

They are caught with a large iron hook, whose extremity resembles the fluke of an anchor; the other end has a ring to which a strong rope is secured. A strong man takes this hook, and in company with four or five others goes into a boat, which is slowly rowed towards the herd. The bearer of the hook stands in the prow of the boat, and as soon as he comes near enough, strikes it into one of the animals. Thirty or more persons on shore then get hold of the rope and drag the struggling victim towards the land. Those in the boat make themselves fast to the beast with another rope; and so fatigue it by repeated blows, until it becomes quiet, and then is despatched by spears, knives, and other weapons, and drawn on shore. Some cut huge pieces out of the living animal, which only provoked it to vibrate its tail, and struggle with its fore limbs so violent

---

* " Vernali tempore more humano coeunt, ac præcipue circa vesperam tranquillo mari; antequam vero congrediuntur præludia multa venerea præmittunt; fœmella placide natat hinc et inde in mari, mas vero semper sequitur; hunc fœmella tam diu multis gyrps et meandris eludit, donec moræ ulterioris ipsa inpatiens, velut delassata ac coacta, se in dorsum resupinet, quo facto mas furiose superveniens libidinis tributum solvit, ac ambo in mutuos amplexus ruunt." *Steller* ut supra.

ly, as to cause large pieces of the cuticle to fly off; it breathed strongly, and as if sighing. When wounded in the back, the blood spirts as high as if from a fountain; but this did not occur as long as the head was retained under water, but as soon as it was raised for the purpose of breathing, the blood gushed forth, because the lungs, lying next the back, being wounded, whenever the air was inhaled, it forced out the blood more freely.

When one of these animals is hooked, he begins to move more impetuously, whereupon the herd and those which are near are set in motion, and endeavour to assist the captive. Some of them strive to upset the boat, others endeavour to break the rope, or by blows with their tails, try to disengage the hook, which they sometimes successfully accomplish. It is a very curious trait in their character, that their conjugal love is exceedingly great: when the male is hooked, the female, after having in vain struggled to set him free, and been herself struck frequently, would nevertheless follow her companion to the shore, and would sometimes unexpectedly approach the dead body, by darting forwards like an arrow. One morning, when we came down very early to cut up the flesh and carry it home, we found the male near the body of his mate, and he remained near, even until the third day afterwards, when I went down to the shore alone, for the purpose of cleaning some of the intestines.

This animal has no voice, nor utters any sound, merely breathing forcibly, and when wounded, as if by sighs. Its organs of sight and hearing are of slight power, as they are almost always submerged and appear to be little employed.

We have stated in the specific description, as well as the generic characters, the peculiar structure of the skin of the steller. It will perhaps be still more satisfactory to the reader to have the more detailed description of it from the original observer, whose statement we subjoin.

The hide of this animal is black, rough, wrinkled, knotty, hard, tough, and destitute of hair, the epidermis being an inch thick, and scarcely to be penetrated by an axe or an iron hook. When cut transversely, this cuticle resembles ebony, both in polish and colour. The skin is smooth on the back; from the neck, to the tail fin, it has nothing but superficial circular wrinkles: the sides, however, are exceedingly knotty, having many prominent acetabula, especially about the head, bearing an unpleasant resemblance to mushrooms.

The cuticle above described, is like a crust surrounding the body, and appears to be composed of mere tubes. These tubes are placed perpendicularly to the true skin, and may be separated from each other in their length. The inferior part of each tube which is implanted in the skin, is rounded, convex and bulbous; hence, a portion of cuticle torn off, appears tuberculous like Spanish hide, while the subjacent skin presents the appearance of numerous little pits, which the cuticular bulbs had occupied. As these tubuli lie very closely together, are tough, moist and tumid, when the skin is cut horizontally they do not appear, but present a smooth surface like the pared hoof of an animal; if pieces be dried in the sun, they crack perpendicularly and may be broken like bark, at which time, this tubular structure, is perfectly obvious. Through these tubes a mucous matter flows, especially upon the sides, and about the head, and in smaller quantities upon the back. When this beast lies upon the shore for some hours, the back becomes dry, but the head and sides remain moist. The use of this singular cuticle appears to be, 1st. to preserve them from being destroyed by being thrown against the ice in winter, or the rocks at all times, and 2dly, that the vital heat may not be too much dissipated in summer, by excessive transpiration, or altogether extinguished by the cold of winter. They do not, like other animals and fish, retreat to the depths of the sea, but always expose half of their bodies to the air, while feeding.

The cuticle about the head, eyes, ears, and mammæ, and under the arms, wherever it is knotty, is attacked and infested by insects. It often happens that they perforate the cuticle and wound the true skin, in which case, large and thick warts are formed.

The true skin is about the sixth of an inch thick, is soft, white, very strong, and similar to the skin of the whale.

CHAPTER VI.

FAMILY II.—CETÆ; *Piscivorous Cetaceous Animals.*

This family is distinguished from the preceding, by the construction which has procured for all its members the name of *blowers*, in reference to the manner in which they expel the water taken in along with their food from the nasal openings. The membrane lining the nostrils being thus continually exposed to torrents of salt water, has very little, if any sensibility as an organ of smelling.

They have a pyramidal larynx, or windpipe, which is extended to the posterior opening of the nostrils, through which the air is admitted to the lungs without requiring the head and mouth to be raised above the water. Their glottis is altogether plain, and their voice is reduced to a simple bellowing. The body is destitute of hair, but is covered by a thick, smooth skin, beneath which is a great thickness of strong cellular substance containing a large quantity of oil.

The teats are situated near the anus; the fins are of no use except in swimming. Two small bones situated in the flesh near the extremity of the digestive canal, are the only vestiges of inferior extremities. Some have a dorsal fin, which is tendinous and not connected with the skeleton.

Some of these animals have conical teeth, all of the same sort, arranged along the edges of the jaws; others have only horny layers, projecting from the roof of the mouth, well known by the name of *baleen*, though generally and inaccurately called *whalebone*.

The eyes, which are flat anteriorly, have a very thick and solid sclerotica or external coat. The tongue is covered by soft and thick integument. The stomach has from five to seven distinct pouches, and instead of one spleen, there are several, which are small and lobular.

Tribe I.—Delphinus; L. *Dolphin Proper;* C.

Having teeth in both jaws, always simple and almost always conical. They have the mouth formed in advance of the head, by a sort of beak, smaller than the rest of the head. They are destitute of cœcum.

CHAPTER VII.

Section I.—*Size of the head bearing the ordinary proportion to that of the body.*

Genus?—Dolphin; *Delphinus;* L.

GENERIC CHARACTERS.

The form of the head is very various; there is but one, semilunar, external orifice to the nostrils, which is situated upon the crown of the head; the trunk of the body is elongated; the tail fin is large, bifurcated and horizontal.

Dental System.

The teeth of the pisciverous cetæ scarcely differ from each other, except in number, all appearing to have the same form. They are conical and slightly hooked; only the larger species have larger teeth than the smaller, and when their series are numerous, the anterior and posterior are smaller than those in the middle. None of them have the alveolar processes divided, nor multiplex roots; the dental capsule remains for a long time free at the base; but these teeth are not always growing, as the

Dolphin.

capsule is eventually obliterated. Then another event occurs; ossification of the jaws takes place within the alveoles, and as the teeth are not opposed to each other, and no force retains them in their places, they are soon thrust out and disappear. This explains the very variable number of teeth we find in dolphins of the same species, and still more so in those of different species. Thus, not having observed between the teeth of dolphins any essential difference of form, and their differences of number not being determinate, we have nothing but the form of their heads from which to establish the generic differences.

Species I.—*The true Dolphin.**

Delphinus Delphis; L.

Le Dauphin: Bonnaterre; Cetol. 20, pl. x, fig. 2.
Dauphin Vulgaire: Desm. Mammal. sp. 758, p. 514.

[*Called Grampus, Porpess, Herringhog, Dolphin.*].

Hitherto the subjects of our study have been inhabitants of grassy plains, or shady forests; the margins of gentle streams, or the outlets of mighty

---

* We call this the *true Dolphin* to distinguish it from the *fish* called dolphin by sailors, (the *coryphœna purpuris*) and because this species is the dolphin so celebrated in various ancient poems and fables, to which we shall hereafter refer.

rivers: we now turn our attention to creatures whose most congenial dwelling is in the bosom of the ocean.

So admirably are the beings, of which we are now to treat, adapted to an aquatic life, that they present a similarity of appearance to, and are most commonly confounded with *fish*, though this resemblance extends no farther than to the general figure of their bodies, and the modification of structure which fits their extremities for swimming. Language can scarcely convey an idea of the velocity with which they dart through the water, seeming rather to fly than to swim; resembling an arrow impelled by a powerful bow, barely long enough in sight to allow a conviction of its having passed. Of their wonderful celerity of movement, and remarkably playful disposition, we have recently enjoyed many excellent opportunities of observation. Once in particular, on a beautifully clear day, when the sea was so strongly illuminated by the sun as to render objects visible at almost any depth, and our vessel was sailing swiftly before a strong breeze, several of these animals appeared to vie with each other in showing how poor was her speed, compared with their own. As the little troop were merrily gamboling at a short distance from the vessel's side, one of the number would dart immediately in advance of her bow, and swimming with his utmost velocity, would disappear in a straight line before her, and (as the depth at which he swam was not more than three feet,) would in a minute or two be seen returning to the crew of his comrades, as if in triumph. This was repeated many times, and most probably by different individuals. These dolphins accompanied us for a considerable distance, and all their actions appeared indicative of the most playful and frolicsome disposition.*

They frequently, however, are seen sailing along with a slow and measured motion, just appearing at the surface, by elevating the crown of the head and then diving short, so as to make their bodies describe the arch of a small circle, exposing themselves to view only from the crown of the head, to a short distance behind the dorsal fin. Occasionally

---

* " On the 20th of October, 1763, a hundred of these animals approached within pistol shot of our vessel, and appeared to have come expressly for our diversion. They made singular bounds into the air; several of them in their caperings leaped three or four feet above the water, and turned over and over several times, like professed tumblers. They go almost always in troops, and swim as if arranged in battle array: they appear to move in search of the wind. We have always remarked that they swim towards the point whence the wind arises." *Dom. Pernethy*, Hist. d'un voyage aux îles Malouines, i. p. 97, &c.

" I have seen one playing around the vessel while she was going at the rate of two leagues an hour; the sailors said that it foretold a squall; in fact, one came on at midnight." *St. Pierre, voyage à l'île de France*, p. 52.

In the instance above mentioned, witnessed by the author, the vessel was moving at the rate of eight miles an hour.

a troop of them may be seen scudding along, rising in this manner in quick succession, as if anxious each to get in advance of the other: while again, a single individual may be observed successively rising and falling in the same way, as if engaged in the act of catching a prey.

In this way, shoals of dolphins may be seen almost every day, and at any hour feeding or sporting in the bay and rivers near the city of New York, where we have sometimes enjoyed an opportunity of observing from the wharf, a large shoal of them moving down the Hudson river with the tide. Some plunging along as if in haste, others apparently at play, and others very slowing rising to the surface for breath, and as gradually disappearing, allowing their dorsal fin to remain for a considerable time above the surface.

From the month of May until towards the end of Autumn, the true dolphins frequent the bays and salt-water rivers of our country, in great numbers. They are most numerous, and are best observed, during the run of the herring and shad, upon which they doubtless feast abundantly; they appear gradually to diminish in number, as these fish retire from the rivers and coast, though a small party may be occasionally seen very late in the season.

During the month of June, the actions of this animal appeared very different from what we have noticed at any other time. They swam in pairs, remained for a longer time at the surface, and seemed to be borne along by the tide rather than urged forwards by their own volition. They moved in half circles, lying rather upon their sides, and occasionally lashing the water into foam with their tails —then both disappearing, one in a few seconds would rise at a little distance as if pursued—make a short leap above the surface of the water, and on falling, again commence the same kind of semicircular movement above described, accompanied by the other. We never observed them to show the beauful inferior surface of their bodies at any other period, or to raise the tail fin above the water. But at this season, the whole inferior surface of the body on one side was frequently visible, and the tail occasionally whirled in air, and brought down with great force.

We would have inferred that these movements belonged to their ordinary gambols, was it not for the fact that they all appeared to be paired off, and almost all the pairs seen at this time were similarly occupied. With the exception of a lapse of about three weeks in the month above mentioned, we have never observed them to act in the same manner. During the period referred to, we spent a part of every day in observing them, and have repeatedly been within eight or ten feet of the spot where they were sporting. Occasionally we have watched them for hours, until the force of the tide swept them far beyond our view.

The appearance of a shoal of these animals, at sea, moving in the same direction, is considered by experienced mariners as an indication of an approaching storm, which very certainly follows their appearance. Falconer, in his beautiful poem of the Shipwreck, thus describes such a circumstance:

> " Now to the north from burning Afric's shore,
> A troop of porpoises their course explore;
> In curling-wreaths they gambol on the tide,
> Now bound aloft, now down the billow glide:
> Their tracks awhile the hoary waves retain
> That burn in sparkling trails along the main—
> These fleetest coursers of the finny race,
> When threatening clouds th' ethereal vault deface,
> Their route to leeward still sagacious form,
> To shun the fury of the approaching storm."
>
> Canto II. § II.

Relative to the breeding season of the dolphin, we have no information sufficiently exact to be relied on. We have seen them in Long Island Sound during the month of August, and the first part of September, accompanied by suckers, varying in size, and from eighteen inches to two feet or more in length. In swimming, or rather in plunging, as heretofore described, the sucker apparently rested on the lateral or humeral fin of the parent, as it always was seen as if adhering to the same place by the side of the parent, in all the movements made in ascending or descending.

THE TRUE DOLPHIN.

A full grown dolphin measures about six feet six inches in length, from the tip of the mouth to the end of the tail, and from the end of the beak to the angles of the mouth the distance is ten and a half inches; and measuring from the same point to the breathing-hole, thirteen inches. The eyes are placed almost precisely on the same line with the angle of the mouth, and are ten and a half inches distant from each other. The lateral or humeral fins are nine and a half inches long, and four broad. The dorsal fin, measured along its anterior edge, is ten inches high; measured along its base in the direction of the back, it is eight inches. The tail, measured at the extremity of its two lobes, is fourteen inches broad.

The body of the true dolphin is nearly oval, having the dorsal fin to curve backwards at its summit: the beak being flattened and pointed, and containing in both jaws a range of rounded, rather pointed teeth, which project from the sockets about three twelfths of an inch. These teeth are all arranged so as to interlock; and appear to vary according to the sex and age of the animal.* The head of this animal does not project over the beak, but is gradually enlarged backwards until in a line with the back. The line from the under surface of the lower jaw, is continued with very little change of

---

* From 84 to 95 upper, from 84 to 95 lower teeth, all of which are regarded as molars.

direction till opposite the dorsal fin, where the body is thickest. From a hand's breadth or more behind the dorsal fin, the body rapidly diminishes to the tail, and on the lower surface, the same rapid diminution takes place from beyond the anus. The tail is composed of two strong lobes terminating in points, which give a beautiful crescent shape, to the extremity of this powerful instrument for swimming.

In examining the anatomical details of this animal, one can not fail to be struck with the singular appearance of the bones forming the lateral fins. We find a shoulder blade, an arm bone, bones of the forearm, wrist and fingers, all so modified as, when covered by the skin, to resemble nothing so much as a fin, yet so obviously analogous to the same bones in the human subject, or in other animals, as to be recognised almost at a glance. The construction of the blow-hole, or breathing apparatus, is also well worthy the peculiar attention of the observer who may have an opportunity of examination, on account of its remarkable excellence of adaptation, and the delicacy of its construction.

The colour of the true dolphin, is of a black or blackish green on the upper surface, and on the inferior parts, of a light gray or whitish. There is below the eye, on each side, a whitish ray or blaze extending towards the humeral fins.

Species II.—*Gladiator Dolphin.*

Delphinus Gladiator.

Schwerdt fisch: Anderson, Island, p. 155.
Poisson à Sabre: Pages voyage au Pole nord, ii. p. 142.
Delphinus dorsi pinna altissima, & Muller, Zool. Dan. Prod. p. 8. n. 57.
Delphinus Maximus: Olafsen, Voy. en Islande.
Dauphin Epée de Mer: Bonnat. Cetol, sp. 5. p. 23.
Dauphin Gladiateur: Lacep. Hist. Nat. des Cetacés, 302. pl. 5, fig. 3.
Dauphin Espadon: Desm. Mammal. sp. 773. p. 517.

This dolphin was first described by Anderson in the work above quoted, and we have very little knowledge of it, except what is derived from him. The head of the gladiator is not highly arched above, but is gradually tapering towards the snout, which is short and as if truncated. Its mouth is furnished with small pointed teeth. Its most remarkable characteristic is a dorsal fin, which is three or four feet high, by eighteen inches wide at its base, slender towards its summit, and recurved towards the tail. Mariners believe that this fin is employed by this dolphin in attacking the whale; but Anderson states " that it is rather the mouth of the animal that is dangerous. As they commonly swim in small troops, they attack the whale in a body, and tear off great pieces of his flesh, until becoming excited to a certain degree, he thrusts out his tongue, when they immediately fasten on this organ and devour it, and finally, gaining access to his mouth, they destroy the life of the animal." Bonnaterre remarks, that the same author thinks with reason that the cetaceous animals called *Killers*, on the coasts of New England, are of this species. In fact, they have both jaws furnished with teeth which interlock, have on their backs a fin four or five feet high; swim in troops, and attack in a body young whales, just as a pack of dogs attack a bull. Some seize the whale by the tail to prevent him from using this weapon of defence, while the others attack and bite him about the head, until the unwieldy creature becomes fatigued, and thrusts forth the tongue as already stated.

The gladiator dolphin is found on the coasts of Spitzbergen, in Davis's straits, and on the New England coasts. The Chevalier Pagès, on his voyage towards the north pole, found them about the 79th degree of latitude. "The sabre-fish," says he, "are also found among the ice, but they rarely quit the frozen climate near the pole. They are from twenty-three to twenty-five feet long; they are black, and carry their sabre perpendicularly upon the back; this sabre is curved backwards, and is about four feet long. I have seen whales fly from them with the utmost celerity, and I have seen others deeply scarred by the weapon of this warlike animal."*

---

* Op. cit. apud Bonnaterre Cetologie ut supra.

Species III.—*The Sea-swine.*

Delphinus Phocœna; L.

Delphinus Phocœna: Briss. Régne An. 371, No. 2.
Dauphin Marsouin: Lacep. Cet. 284, pl. 13, fig. 2.
Dauphin Marsouin: Bonnaterre, Cétol, p. 18, sp. 1.

[*Called by the English* Porpus: *by the Dutch* Bruinvisch; *in German* Meerschwein; *by the French* Marsouin.]

We give an account of this species in this place, because authors are in the habit of ranking it among those which frequent our coasts. But we have not been able at any time to learn that the sea-swine has been seen in our waters, or that any other species than those already mentioned are known to our fishermen. The dolphin first described, or true dolphin, is the species universally known by the name of *porpus** in this country, and is at once distinguishable from the European porpus, or sea-swine, by its elongated flattened beak; the latter having a blunt snout, without any beak. If it ever is seen on our coasts, it must be very rare, as we have been thus far unable to find any one who has seen it, except in the seas bordering the shores of Europe. We therefore introduce the description of

---

* *Porpus* is a corruption of *porc-poisson*, as the French *marsouin* is of the German *meerschwein*.

the species from Bonnaterre, in order that those who have opportunities of observation, may be able to compare the animals and settle the question.

The body of the sea-swine is conical, having a triangular fin on the back. The snout is pointed, and the teeth rounded, trenchant, and enlarged at their summits. The body of this animal is round, thick, and tapering towards the tail. The head represents an obtuse cone, which is much arched above the orbits of the eyes: it thence gradually tapers down to the snout. The eyes are situated opposite the opening of the mouth: the pupil is black, surrounded with a white iris. Both jaws are nearly of the same length; the lower one being armed on each side with a range of small teeth, which are slender at the base, flattened, trenchant, and rounded at the summit; varying in number from fifty to fifty-five. Behind the eye is a small round hole, an inch in diameter; this is the ear. The blow-hole is situated upon the summit of the head, in the centre of the interval between the eyes and angle of the mouth. The nostrils are placed between the blow-holes and the extremity of the muzzle. The lateral fins are placed upon the borders of the lower surface of the body: the dorsal fin occupies the middle of the back. Behind this fin the back is flattened and raised in the middle by a projection which extends entirely to the caudal fin. On the part of the belly corresponding to the dorsal fin, there is a depression which conceals the sexual organs. The vent is equi distant from the depression and the caudal fin. The caudal fin is formed of two lobes rounded at the points, and slightly grooved. The colour of the whole superior surface is of a blackish blue: the belly whitish. The length from six to eight feet.

This animal, which is considered an excellent swimmer, habitually carries the head and tail curved downwards, and in consequence, shows nothing but the back when it comes to the surface to breathe; but as soon as it is dead, it extends itself in a straight line. They feed on small fish, which they pursue with inconceivable swiftness. The seaswine are almost always seen in troops, especially in their sexual season, which is in the month of August. It is then common to see ten or fifteen males in pursuit of a single female, and they press on with so much ardour, that they are often stranded on the beach before they are aware of their situation. The young are carried ten months; only one is brought forth at a time. An embrion, extracted from the mother by Klein, was about twenty-one inches long. The young one constantly follows the mother until weaned.

The flesh of this animal is oily and disagreeable. The Laplanders, Greenlanders, and others eat it. In some parts of the world they are killed for the sake of their oil and skins.

Fabulous History of the Dolphin.

Few animals have occupied a more distinguished place in the writings of historians and poets, than the dolphin, whose actual habits and manners we have just examined. From Herodotus, the father of Greek historians, down to a comparatively recent period, we find a succession of wonderful incidents related, originating either from the most inaccurate observation of fact, or from the wildest extravagance of fancy.

It will be an amusing, and we hope not an uninstructive lesson, to trace some of the most remarkable of these stories, both to show how far the human mind may permit itself to be misled, and to set at rest, by exposing their futility, such recitals as the beauty and excellence of poetry tend to perpetuate as probable or true. Investigations of this kind may in some degree lessen the pleasure derived from works of fiction, but the advantage of being possessed of the truth, far outweighs the temporary gratification arising from an indulged imagination.

The most ancient of the accounts given of the docility and friendly disposition of the dolphin, is that related of Arion, by Herodotus in his first book. As this may be considered the source from which most of the others have sprung, we subjoin a free translation from the venerable historian.

"Periander was king of Corinth. The Corinthians relate, (and the Lesbians agree with them,) that during his life-time, a great prodigy occurred. They say that Arion was carried from Methymna to Tænarus upon the back of a dolphin: as a musician, he was second to none, as a dithyrambist, we know he was the first, composing, giving rules and teaching at Corinth. Having spent much of his time with Periander, he desired to visit Sicily; having acquired much wealth there he wished to go back to Corinth, for which purpose he hired a vessel from some Corinthians at Tarentum. When at sea, they conspired to throw him overboard, that they might share his money; which having learned, he earnestly besought them to take his wealth, but spare his life. But they, unmoved by his entreaties, ordered him either to kill himself, and perchance obtain a burial on shore, or to cast himself instantly into the sea.

"Driven to despair, Arion besought them, since it was thus determined, that they would allow him to sing, standing completely arrayed on the deck. They promised him that he might die singing, and pleased that they should hear the most excellent of musicians, they removed from the poop towards the middle of the ship. Arion, clad in his professional robes,* took his lyre, and, standing on the deck, ran

---

* Ἐν τῇ σκευῇ πάσῃ is rendered by Dalzell, "completely arrayed," as we have given the phrase in the second para-

through the Orthian measure.* Having concluded, he threw himself into the sea, and they sailed on from Corinth. But Arion, sitting on a dolphin's back, was conveyed to Tænarus, &c. &c. These things are still told by the Corinthians and Lesbians. There is a small brazen votive tablet of Arion near Tænarus, of a man mounted upon a dolphin's back.†

---

graph. Murphy says it was not his ordinary dress he wished to die in, but one peculiar to him as a musician.

* So called, because sung at the highest and strongest pitch of the voice. " Ορθιον enim Græci dicunt quod arduum est, et quam altissima voce elevatum.—*Gesnerus.*

† Herodotus, εκ ςησ Κλειουσ.—This story did not escape the biting irony of Lucian, whose talent for ridicule has rarely been surpassed. He has a dialogue between Neptune and the very dolphin who bore Arion in safety to Tænarus, and makes him repeat Herodotus's story, as "having heard the whole of it while swimming round the ship." Lucian also accounts for the fabled attachment of the Dolphin to the human race, by making this one remind Neptune that they were changed from *men* to dolphins by Bacchus. Ovid relates the transformation in his third book of Metamorphoses, where Bacchus himself, in the semblance of his companion Acœtes, is the speaker:—

> " At Lybis obstantes dum vult obvertere remos,
> In spatium resilire manus breve vidit; et illas
> Jam non esse manus jam primas posse vocari.
> Alter ad intortos cupiens dare brachia funes
> Corpore desiluit; *falcata novissima cauda est
> Qualia dimidiæ sinuantur cornua lunæ.*"

OF THE DOLPHIN.

The sagacious and judicious Plutarch not only repeats this story, but introduces Gorgias in the conversation of the seven wise men, as saying, that he knew Arion, before he landed from off the dolphin's back, because he had on the dress he had worn at the public games. Aulus Gellius repeats the story from Herodotus, as it was originally told, and Ovid perpetuates it in verse.*

"Numerous examples, (says Aristotle,) of the gentleness and mild manners of dolphins are related. About Tarentum, Caria and other places, they tell of their love and regard for boys. A dolphin having been wounded near Caria, a troop of dolphins, it is said, gathered in the port, until the fish-

---

* Ille metu vacuus, mortem non deprecor inquit;
  Sed liceat sumta pauca referre lyra.
Dant veniam, ridentque moram: capit ille coronam,
  Quæ possit crines, Phœbe, decere tuos.
Induerat Tyrio bis tinctam murice pallam:
Reddidit icta suos pollice chorda sonos;
Flebilibus veluti numeris canentia dura
Trajectus penna tempora cantat olor.
Protinus in medias ornatus desilit undas,
Spargitur impulsa cærula puppis aqua.
Inde, fide majus tergo Delphina recurvo
Se memorant oneri supposuisse novo.
Ille sedens citharamque tenet, pretiumque vehendi
Cantat, et æquoreas carmine mulcet aquas.
Di pia facta vident: astris Delphina recepit
Jupiter, et stellas jussit habere novem.

erman set his prisoner at liberty, when they all went off. A larger dolphin likewise always accompanies them as a guard. A troop of dolphins of larger and smaller size, were once seen, and, at no great distance behind them, two dolphins appeared, bearing up the body of a young dead dolphin on their backs, by swimming beneath it, as if induced by pity lest it should be devoured by some beast.*

Œlian relates in the third chapter of his eighth book, that Ceranus, the Parian, purchased the freedom of some dolphins caught by Byzantine fishermen, and afterwards sailed towards his own country in a Milesian vessel of fifty oars. His vessel was cast away in the strait of Paros, but these dolphins which he had set at liberty, came in time to save their deliverer, and landed him on a promontory, subsequently called Cerania, in honour of him; at his death, he requested to be interred at that place; thither the dolphins went to pay their benefactor merited funeral honours.

Leonidas of Byzantium, narrates (in Œlian's 2d book, ch. 6.) that a man and his wife of Pleroselene, taught a dolphin to eat from their hands, and accustomed their son to be very familiar with the animal, which very regularly frequented the harbour of the town, appearing to regard it as his home. When old enough to take care of himself, he sought

---

* Aristoteles de Animalibus Historiæ, lib. ix. cap. 35.

his subsistence at sea, and brought a share of his success in fishing daily to his friends. The parents had given the same name to the dolphin and their son. When the boy sat upon a projecting rock, and called his friend, the dolphin immediately hastened towards him, testifying his pleasure by his frolicsome movements. This connection between the boy and dolphin, occasioned a great deal of rumour, and was very profitable to the parents.*

The younger Pliny, however, exceeds all these wonders, by the following recital. A scholar, named Hippus, in the time of Augustus, who attended a class at Puzzoli, was in the habit of going daily along the shores of Baia, and about mid-day, of stopping and throwing pieces of bread into the water to a dolphin. If the youth called the dolphin at any time, he would immediately come, and after eating his bread, would offer his back for the use of his friend who would mount thereon, and he would swim with him to Puzzoli, and afterwards carry him back in the same manner. This friendly intercourse was maintained for several years; but

---

* There is nothing improbable in the dolphin's obedience to a certain call from one accustomed to supply it with food. Animals of very inferior rank to the dolphin, may be taught as much. The improbability, is in the gratitude of the animal, evinced by the offer of part of his fish.

the boy dying, the afflicted animal came frequently to the accustomed place, remained there sorrowful and wretched, and finally died of grief!*

The reasons for believing the present species to be *the* dolphin of the poets, are the following: first, it is the only dolphin which is known habitually to frequent the coasts, or to visit the deep bays which extend far inland. The sea-swine (meerschwein, marsouin, *Delphinus Phocœna,*) have no beak extending beyond the arched part of the head, and as they are seldom seen except in the full sea, are not likely to have afforded much opportunity to the ancients for examination. That they were well acquainted with our dolphin, we have the most excellent evidence, in the figure of the one which accompanies the statue of the Venus de Medicis. Although the usual poetical licence has been taken by the sculptor, of placing the animal resting on the underjaw and neck, with its body and tail raised in fanciful undulations, from the great resemblance of the head and beak to those of the dolphin we have been examining, in conjunction with the circumstances of its habits, numbers and familiarity with the bays and rivers of almost all the world, we are persuaded of the identity of the species frequenting our waters, with that to which all the ancient fables relate.

---

* See Pliny, lib. ix, cap. viii.

We have thought it unnecessary to bring the fabulous history of the dolphin down to a later period than that of Pliny, as all the subsequent stories appear to be variations of the same. It is impossible, however, not to feel sorry that some modern works of great authority and usefulness, continue to interweave so much of what is barely *possible*, with the little that is attested in regard to this and other animals, as to give an air of fable to the whole. The following from the "Nouveau Dictionnaire d'Histoire Naturelle," may serve as an instance.

"The dolphins form among themselves a sort of society; they defend those of the troop that may be attacked, and utter frightful cries, in order to induce the aggressors to release them. The little dolphins are placed in the middle of the troop; the large and most robust at its head: they all preserve their order like a battalion of soldiers; they swim each in their ranks; the females compose the rear guard, and urge on the stragglers."*

This is not the only passage of the kind, that might be selected from the article on the dolphin, in the same work. If the time shall ever arrive, when the facts of natural history are given, without admixture with fable, the world will be more rapidly and satisfactorily advanced in improvement than can possibly be hoped for, so long as imagination

---

* Virey op. citato.

is permitted to usurp the place of truth. The latter, like perfect beauty, is unsusceptible of adornment, and is always more admirable in its simplicity than any fiction, however ingeniously contrived or gorgeously ornamented.

CHAPTER VIII.

Genus Narwal; *Monodon*; L.

GENERIC CHARACTERS.

General form of the body similar to that of the dolphin; a single spiracle or blowhole on the superior part of the head; mouth small; no teeth within the mouth, one long spiral tusk growing from the intermaxillary bone; dorsal crest or spine, instead of a fin. The eyes and ears small.*

Species I.—*The Narwal.*

Monodon Monoceros; L.

Monodon Narwhal: Fabricius, Faun. Greenl. 29.
Narwhal oder einhorn: Anderson, Island. 225.
Narwhal: Bonnat. Cétol: 10.
Narwhal Vulgaire: Lacep. Hist. Nat. des Cétacés. 142.
Narwhal, or Unicorn of the Whalers: Scoresby, Arct. Regions, i. 486.
  Ibid. Voyage to Greenland, 129.

[*Commonly called Sea-Unicorn.*]

The narwal is an inhabitant of the arctic seas, and consequently is seldom seen, except by the ad-

---

* " Penis vaginatus; mammæ lactantes binæ et genitalia feminarum sub abdomine; pone illa anus."   *Bonnat.*

venturous mariners, who seek the spoils of the whale amid the perils of polar ice and storms. Fortunately, however, some few of these, incited by hopes of gain to visit those forbidding regions, have been well qualified to make accurate scientific observations, and owing to their zealous industry, we have actually less to desire concerning the animals found in the icy seas, than in relation to many others, almost within the reach of every observer.

Among the individuals to whom science is most deeply indebted, the name of Scoresby must ever stand conspicuous; few persons have contributed so largely to the advancement of natural history, while engaged in ordinary commercial pursuits, and still fewer have effected the object so well under any circumstances. His mind appears to have been one of that rare, but amiable composition, in which genius, talent, energy and sound common sense, are blended in such just proportion, as to be capable of operating at all times, and upon all materials, to the greatest possible advantage. From his valuable researches we shall derive almost all the observations which remain to be made upon the cetaceous animals, claiming for ourselves no other merit than that of having collected and arranged them.

The vertebral column of the narwal is about twelve feet long; there are seven cervical, twelve dorsal, and thirty-five lumbar and caudal vertebræ, being in all fifty-four; twelve of which enter the tail and extend to within an inch of its extremity. The spina marrow runs through all the vertebræ, from the head of the fortieth, but does not penetrate the forty-first. The spinous processes diminish in length from the fifteenth lumbar vertebræ, until it is scarcely perceptible at the nineteenth. Large processes, attached to two adjoining vertebræ, and arising from the inferior surface of the bodies of the vertebræ, commence between the thirtieth and thirty-first, and terminate between the forty-second and forty-third. There are twelve ribs, six true and six false, on each side, which are slender for the size of the animal. The sternum is heart-shaped, with the broadest part anteriorly. Two of the false ribs, on each side. are joined by cartilages to the sixth true rib, the others are detached.

The narwal, when full grown, measures from thirteen to sixteen feet in length, exclusive of the tusk, and at the thickest part, which is two feet behind the fins, the circumference is about eight or nine feet. The part of the body anterior to the fins and head, are paraboloidal; the middle portion of body is almost cylindrical, the posterior portion, to within three or four feet of the tail, is somewhat conical; thence, a ridge commencing both at the back and belly, the section becomes first an ellipse, and then a rhombus at the junction of the tail. The perpendicular diameter, at a distance of twelve or fourteen inches from the tail, is about one foot, the transverse diameter is about seven inches. The back and belly ridges, run half way or more across the tail;

the edges of the tail run in the same way along the body, and form ridges on the sides of the rump. Posterior to a very slight elevation at the spiracle, the outline of the back forms a regular curve; the belly appears to rise, or is contracted near the vent, and expands to an obvious bump, about two feet anterior to the genitals. The back appears depressed and flat three or four feet posterior to the neck.

The head forms about one seventh of the whole length of the animal, being small, blunt, and round. The mouth is small, and incapable of much extension, having a wedge-shaped underlip. The eyes are only one inch in their largest diameter, and are placed on a line with the opening of the mouth, at about thirteen inches from the snout. The opening of the ear, situated six inches behind the eye, on the same horizontal line, is of the diameter of a small knitting needle. The scull of the narwal, like the dolphin, &c. is concave above, and sends forth a large, flat, wedge shaped process in front, which affords sockets for the tusks. There is upon this process a bed of fat extending horizontally to the thickness of ten or twelve inches, and eight or ten perpendicularly. To this fat, the roundness of the head is owing, and according to the quantity present, is the prominence of the front, and the variation of the facial angle, from 60 to 90 degrees.

The spiracle or blowhole is situated immediately over the eyes, and is a single semicircular opening about three and a half inches in diameter and one inch and a half in length. It expands immediately within the skin into a sac or air vessel, six or eight inches wide, and extending laterally and forward, into two cavities, one on each side; the extremities of which, are about twelve inches apart. These contain some mucous matter; the lining of the whole sac is a thin, greenish, black membrane. At the posterior extremity of the sac, the blowholes are seen, divided there, into two distinct canals in the skull. They are closed by a valve resembling a hare lip, one lobe of which covers each canal. This valve in the narwal, does not, (as in the whale), enter the canal in the skull, but merely closes down upon it. It, however, effectually excludes the sea-water from the lungs, whatever be the pressure; it becomes, in fact, firmer and closer, in proportion as the weight of water is increased. The valve is about six inches wide and is closed and opened by two radiated muscles. It is detached from the skull beneath, about six inches towards the snout. In consequence of this separation, the valve is sufficiently free, and has room enough in the adjoining sac to be drawn upward and forward, so as to expose the breathing canals, or falling upon them like the valve of a pump-box, to secure them against the entrance of water. The two lobes of the valve are connected by a fleshy septum, slightly attached to the cartilaginous part of the bony partition between the blow-holes in the skull.

The fins are twelve or fourteen inches long, and six or eight broad, and placed at one fifth of the length of the animal from the snout. Where fixed to the body, the fin is elliptical, its longest axis lying longitudinally, so that when the fin is elevated to the swimming position, it is horizontal, the point or tip is bent upwards or towards the back, consequently, when the fin is in the swimming position, it is concave above, and convex below, the thick edge forward and the thin edge towards the tail. The fin being horizontal, is evidently designed to balance the animal, while the tail, which is from fifteen to twenty inches long, and three or four feet broad, is the chief organ of motion, and is also used in turning. That the fins are not commonly used either for swimming or turning, appears probable from repeated observations made with a telescope from the mast head. The fins were always seen, steadily extended, and when the animal changed its direction, the tail was bent suddenly and obliquely to one side, and then slowly brought back, so that the progressive motion and change of direction were produced by the same effort; the fin at the same time remaining motionless.

The general colour of the young narwal is blackish gray on the back, variegated with numerous darker spots running into each other and forming a dusky black surface, paler and more open spots of gray on a white ground at the sides, disappearing altogether about the middle of the belly. In the elder animals, the ground is wholly white, or yellowish white, with dark gray or blackish spots of different degrees of intensity. These spots are of a roundish or oblong form: on the back, where they seldom exceed two inches in diameter, they are the darkest and most crowded together, yet with intervals of pure white among them. On the sides, the spots are fainter, smaller and more open. On the belly they are extremely faint and few, and being in considerable surfaces, are not distinguishable. A close patch of brownish black, without any white, is often found on the upper part of the neck, just behind the blowhole: the external part of the fins is also generally black at the edges, but grayish about the middle. The superior side of the tail is also blackish around the edges: but in the middle, gray with black curvilinear streaks, on a white ground, forming semicircular figures on each lobe. The inferior surfaces of the fins and tail are similar to the upper, only much paler coloured, the middle of the fins being white, and of the tail a pale gray. The sucker narwals are almost uniformly of a bluish gray or slate colour. Very old individuals become almost white.

The skin of the narwal, resembles that of the whale, except that it is thinner. The cuticle is about as thick as writing paper; the rete mucosum three eighths or three tenths of an inch thick; the cutis thin, but strong and compact on the outer side.

We may next consider the most remarkable peculiarity which distinguishes this animal; the long spiral tooth or tusk, which has obtained for it the name of Unicorn. This tusk grows from the left side of the head, and is sometimes nine or ten feet long. Egede, in his description of Greenland, describes this tusk as being fourteen or fifteen feet long. It projects from the inferior part of the upper jaw, and points forward and slightly downward, being parallel in direction to the roof of the mouth. It is spirally striated from right to left, nearly straight, and tapers to a round blunt point. It is of a yellowish white colour, and consists of a compact kind of ivory, and is usually hollow from the base to within a few inches of the point. A tusk of the average length, five feet, is about two inches and a half in diameter at the base; one inch and three fourths in the middle, and about three eighths within an inch of the end. In such a tusk there are five or six turns of the spiral, extending from the base to within six or seven inches of the point. Beyond this, the end is not striated, but smooth, clean, and white; the striated part is usually gray and dirty. The tusk is commonly covered with a greasy blackish brown incrustation over the greatest part of its surface; the under part and a few inches of the point, are kept quite clear and polished by some use which prevents the adherence of the matter just mentioned. A horn externally of seven feet in length, is bedded about fifteen or sixteen inches in the skull. All the male narwals, killed by Scoresby, excepting one, had tusks of from three to seven feet in length, projecting from the left side of the head.

In addition to this external tusk, peculiar to the male,* there is another on the right side of the head about nine inches long, imbedded in the skull. In females as well as in young males, in which the tooth does not appear externally, the rudiments of two tusks are generally found in the upper jaw. These are entirely solid, and are placed back in the substance of the skull, about six inches from its most prominent part. These rudiments of tusks are eight or nine inches long, both in the male and female; in the former they are smooth, tapering, and terminate at the root with an oblique truncation; in the latter they have an extremely rough surface, and finish at the base with a large irregular knob placed towards one side, which gives the tusks something of the form of pocket-pistols. Two or three instances have occurred of male narwals having

---

* Scoresby, in his Greenland voyage, killed a female narwal, having an external horn, four feet three inches long; twelve inches of which were imbedded in the skull. It had also a milk tusk, as is usual, nine inches long, which was of a conical form and obliquely truncated at the thicker end, and without the knob found in many of the milk tusks. The horn was on the left side of the head, and the spiral was *dextrorsal.*

been taken, which had two external tusks. This is a rare circumstance, and it rarely or never occurs that an external horn is found on the right side.

What purpose this singular and formidable tusk can serve, is not easily to be determined. It is not essential to the defence of the animal, or else the young and a vast majority of the females would be left unprotected. It has been suggested, that it is employed by the animal in piercing thin ice for the convenience of rising to respire, and that it is occasionally employed in killing prey. But nothing has yet been observed, sufficient to enable us to draw any positive conclusion on the subject.

The food of the narwal appears to be principally molluscous animals, such as the cuttle-fish &c., but judging by the materials occasionally found in their stomachs, more substantial food is frequently devoured by them. In the stomach of one examined by Scoresby, besides the beaks and other remains of cuttle-fish, there was part of the spine of a *pleuronectes*, or flat-fish, probably a small turbot; fragments of the spine of a *gadus;* the backbone of a *raia,* with nearly a whole skate, *raia-batis,* which was two feet three inches long, and one foot eight inches broad. That an animal having no teeth except the external tusk, a small mouth, and a tongue incapable of protrusion, should be able to swallow a fish nearly three times as great as the width of its own mouth, is really surprising. Scoresby inclines to the opinion, that the skates had been pierced with the horn, and killed before they were swallowed by the narwal, as it is otherwise very difficult to conceive how an animal so large as the skate, would allow itself to be sucked down the throat of a smooth-mouthed animal, having no means of crushing or detaining it.

The narwal is a harmless animal, of an active disposition, and swims with considerable swiftness. When at the surface, for the sake of respiring, these animals frequently lie motionless for several minutes, with their heads and backs just appearing above water. Occasionally, numerous small herds are seen together, each herd generally consisting of individuals of the same sex.

The narwal is sometimes shot with a rifle, kept for that purpose in the *crow's-nest* of the whaling-ships. When harpooned, the narwal dives as swiftly, but not so deeply as the common whale. It commonly descends about two hundred fathoms, and then returns to the surface, where it is soon killed with lances.

The whole body of the narwal is covered by a layer of blubber immediately beneath the skin, which is from two to three inches thick, and yields a considerable quantity of fine oil. The Greenlanders and Esquimaux employ the whole animal to various uses. The flesh is eaten, the oil burned in their lamps, the intestines wrought into lines and dresses, and the tusks are used for spears &c. It

THE NARWAL.

is said that the king of Denmark has a magnificent and valuable throne made entirely of narwal tusks.

The following are the dimensions of a male narwal, killed by Scoresby near Spitzbergen in 1817.

	Feet	Inches
Length, exclusive of the tusk,	15	0
——— from the snout to the eyes,	1	$1\frac{1}{2}$
————————————— fins,	3	1
————————————— backridge,	6	0
————————————— vent,	9	9
Circumference $4\frac{1}{2}$ inches from snout,	3	5
——————— at the eyes and blowhole,	5	$3\frac{1}{2}$
——————— just before the fins,	7	5
——————— at the forepart of backridge,	8	5
——————— at the vent	5	8
Tusk, length externally,	5	$0\frac{1}{2}$
——————— diameter at base,	0	$2\frac{1}{4}$
Blowhole length $1\frac{1}{2}$ inch. breadth,	0	$3\frac{1}{4}$
Tail do 14 do	3	$0\frac{1}{2}$
Fins do 13 do	0	$7\frac{1}{2}$

Heart weighed 11 pounds. Temperature of the blood an hour after death, 97°.

A fine specimen of the tusk or horn of the narwal may be seen in the Philadelphia Museum.

CHAPTER X.

Section II.—*Size of the head disproportioned to that of the body.*

Genus Cachalot; *Physeter:* L.

GENERIC CHARACTERS.

The head in these animals is of huge size, forming a third, or even half of their entire length. The upper is broad, high, destitute of corneous fringes and teeth, or having short teeth, almost entirely concealed within the gums. The lower jaw is elongated, narrow, and armed with thick conical teeth, which fit into corresponding depressions in the upper jaw. The spiracles are placed at or near the extremity of the superior part of the snout. There is a dorsal fin in some species, in others merely an eminence. In the superior parts of the head there are large cavities, circumscribed by cartilaginous partitions, and communicating with different parts of the body by particular canals. These are filled with an oil that becomes fixed and crystallized on cooling, and is the well known substance spermaceti.

The teeth are ovoid and recurved; externally they somewhat resemble ivory, internally they are softer, and ash coloured. They are commonly about six inches long, and three in circumference at the base, and are thought to become larger and more recurved as the animal grows. The upper jaw has as many alveolar depressions as there are teeth in the lower, but what is most remarkable, is, that in the interstices separating these depressions, are to be found about twenty small teeth, horizontally placed, and raised about one-twentieth of an inch above the gum. These teeth are acutely pointed, and present a flat, even, and oblique surface, filling the intervals separating the alveoles. This oblique surface is all that is seen of them, the other parts of these teeth being imbedded in the gum.*

Species I.—*The Spermaceti Cachalot.*

Physeter Macrocephalus.

Le Grand Cachalot; Bonnat. Cetol. 12.
Cachalot Macrocéphale; Desm. Mam. 524, p. 790.
Cachalot Macrocéphale; Lacep. Hist. Nat. des Cétacés, pl. 10.

The spermaceti cachalot is found in greatest abundance in the Pacific Ocean, where large numbers of them are annually killed by the American and other whalers, for the sake of their oil and spermaceti.

The spermaceti cachalot is gregarious, and herds

* See Desmarest's Mammalogie; Bonnaterre Cétologie; Sibbald Phalainologia nova.

are frequently seen containing two hundred or more individuals. Such herds, with the exception of two or three old males, are composed of females, who appear to be under the direction of the males. The males are distinguished by the whalers as *bulls;* the females they call *cows*. The bulls attack with great violence, and inflict dreadful injuries upon other males of the species, which attempt to join their herd. These animals live separately, while young, according to their age and sex. The young and half grown males are found by themselves; the old *cows* protect the young females. When the young bulls attain sufficient strength, they venture into a herd under the protection of some old bulls, an intrusion that is said to produce a severe contest, by which they succeed in gaining admittance to, or are driven from the herd.

The mode of attacking these animals is as follows:—Whenever a number of them are seen, four boats, each provided with two or three lines, two harpoons, four lances and a crew of six men, proceed in pursuit, and, if possible, each boat strikes or " fastens to" a distinct animal, and each crew kill their own. When engaged in distant pursuit, the harpooner generally steers the boat, and in such cases the proper boat steerer occasionally strikes, but the harpooner mostly kills it. If one cachalot of a herd is struck, it commonly takes the lead and is followed by the rest. The one which is struck, seldom descends far under water, but generally swims off with great rapidity, stopping after a short course, so that the boat can be drawn up to it by the line, or be rowed sufficiently near to lance it. In the agonies of death, the struggles of the animal are truly tremendous, and the surface of the ocean is lashed into foam by the motions of the fins and tail. Tall jets of blood are discharged from the blowholes, which show that the wounds have taken mortal effect, and seeing this, the boats are kept aloof, lest they should be dashed to pieces by the violent efforts of the victim.

When a herd is attacked in this way, ten or twelve of the number are killed; those which are only wounded are rarely captured. After the cachalot is killed, the boats tow it to the side of the ship, and if the weather be fine, and other objects of chase in view, they are again sent to the attack.

The separation of the blubber from the animal, or " flensing," is sometimes done differently from the manner used in the polar whaling. A strap of blubber is cut in a spiral direction, and being raised by tackles, turns the cachalot round as on an axis, until nearly all the blubber is stripped off. The material contained within the head, consisting of spermaceti mixed with oil, being in a fluid state while warm, is taken out of large cachalots in buckets, while the animal remains in the water; but in smaller ones, the part of the head containing the spermaceti, is hoisted upon deck before the cavity is opened.

THE SPERMACETI CACHALOT.

The substances taken from the head, congealing as soon as cold, the compound is thrown in its crude state into casks, and is purified at the end of the voyage on shore. The oil is reduced from the blubber shortly after it is on board, in " try works," with which the ships engaged in this business are always provided. There are two coppers in the try works, placed side by side, near the fore hatch. These, with their furnaces and casing of brickwork, occupy a space of five or six feet in length, by eight or nine in breadth, (or fore and aft—and athwart ship,) and four or five feet in height. The cavity of the brick arches sustaining the coppers and furnaces, forms a water cistern, so that while the fire is burning, the deck is secured from injury by the changing of the water in the cistern twice or thrice in every watch. As the oil is extracted it is thrown into coolers, whence, after about twenty-four hours, it is transferred to casks. At first the coppers are heated with wood, but afterwards the cracklings or fritters of the blubber, which still contain some oil, are employed as fuel, and produce a fierce fire. About three tons of oil are commonly obtained from a large cachalot of this species; from one to two tons are procured from a small one. A cargo, produced from one hundred cachalots, may be from 150 to 200 tons of oil, besides the spermaceti, &c.

CHAPTER IX.

Genus—Whale; *Balæna;* L.

GENERIC CHARACTERS.

Whales possess no true teeth; the upper jaw resembles the keel of a vessel, or the roof of a house reversed. It is furnished on each side with transverse horny layers of a peculiar substance, called *Baleen,* which at the edges are split into long slender fringes. The spiracles or blowholes are separated, and placed about the middle of the superior part of the head. Some species have a dorsal fin; others merely a prominence.

Species I.—*The Whale.*

Balæna Mysticetus. L.

Φαλαινα Arist. An. 1. c. v. III. c. xvi. Μυςτικητος, ib. III. c. x. Æl. Hist. an. v. c. iv.
Hvalfisch; Egede Greenland, 48.
La Baleine Franche; Bonnat. Cétol. 1.
The Common or Greenland Whale; Scoresby Arct. Regions, i. 449.

In attempting to describe a creature so gigantic and surpassing in strength as the whale, we deeply feel the want of expressions suitable to our purpose, and vainly endeavour to remove this difficulty

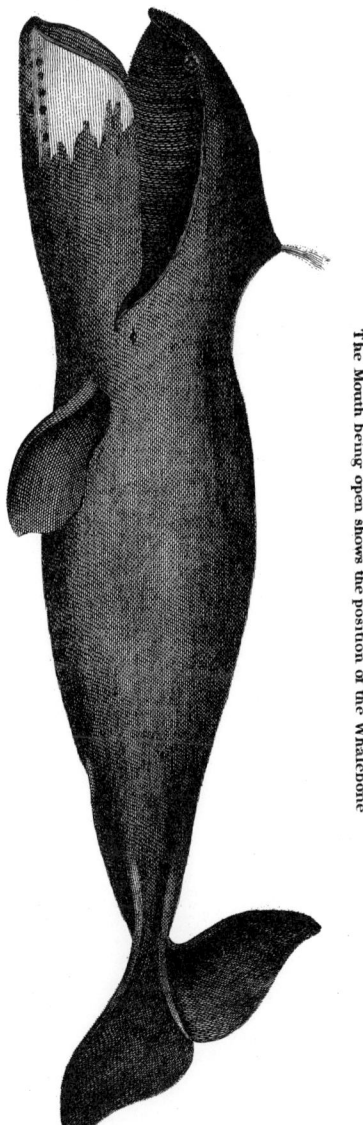

BALÆNA MYSTICETUS, or COMMON WHALE.
58 Feet long
The Mouth being open shows the position of the Whalebone

Scale, one-tenth of Inch to a Foot.

by resorting to comparisons scarcely less inadequate, or conveying at best but vague and unsatisfactory ideas. The sublime in magnitude among organized and animated beings, the whale is adapted in all his attributes to the fathomless and illimitable waters he is destined to inhabit: contrasted with other animals, his strength as far transcends their greatest exertions, as the irresistible heavings of the mighty deep exceed the harmless rippling of a sylvan stream. It is only by successive approaches and detailed examination, that we can arrive at a proper conception of this animal, and, therefore, the statements which are freest from attempts to emulate by ambitious style the magnitude of the subject, will lead us to the most satisfactory conclusions.

Having never personally enjoyed opportunities of studying the whale in his native floods, and having derived all that we know in relation thereto, from Scoresby, we should deem it injustice to the reader to give this account in any other language than that of the original. We do this without reluctance, as our object is to convey the most accurate knowledge, rather than to produce a work exclusively of our own composition, and because we believe that where an original observer is competent to express what he has seen, his remarks must have a force and value far greater than can be imparted by another, however great may be his command of language, or his felicity of expression. All that follows in relation to the whale, is selected from the different works of the accurate and philosophical Scoresby.

The Whale.

This valuable and interesting animal, generally called the whale by way of eminence, is the object of our most important commerce to the polar seas—is productive of more oil than any other of the cetacea, and being less active, slower in its motion, and more timid than any other of the kind, of similar or nearly similar magnitude, is more easily captured.

Large as the size of the whale certainly is, it has been much over-rated; for such is the avidity with which the human mind receives communications of the marvellous, and such the interest attached to those researches, which describe any remote and extraordinary production of nature, that the judgment of the traveller receives a bias, which, in cases of doubt, induces him to fix upon that extreme point in his opinion, which is calculated to afford the greatest surprise and interest. Hence, if he perceives an animal remarkable for its minuteness, he is inclined to compare it with something still more minute: if remarkable for its bigness, with something fully larger. When the animal inhabits an element where he can not examine it, or is seen under any circumstance which prevent the possibility of his determining its dimensions, his decision will certainly be in that extreme which excites the most interest. Thus a mistake in the size of the whale would easily be made; and there is every probability of such an error having been committed two or three centuries back, from which period some of our present dimensions have been derived, when we know that whales were usually viewed with superstitious dread, and their magnitude and powers in consequence, highly exaggerated. Besides, errors of this kind having a tendency to increase, rather than to correct one another, from the circumstance of each writer on the subject, being influenced by a similar bias; the most gross and extravagant results are at length obtained. Thus authors, we find, of the first respectability in the present day, give a length of 80 or 100 feet, or upwards, to the mysticetus, and remark with unqualified assertion, that when the captures were less frequent, and the animals had sufficient time to attain their full growth, specimens were found of 150 to 200 feet in length, or even longer; and some ancient naturalists, indeed, have gone so far, as to assert that whales had been seen of above 900 feet in length.

But whales in the present day are by no means so bulky. Of 332 individuals, in the capture of which, I have been personally concerned, no one I believe exceeded 60 feet in length; and the largest I ever measured, was 58 feet from one extremity to the other, being one of the longest to appearance, which I ever saw. An uncommon whale, which was caught near Spitsbergen, about twenty years ago, the whalebone of which measured almost fifteen feet, was not, I understand, so much as 70 feet in length; and the longest actual measurement that I have met with, or heard of, is given by Sir Charles Giesecke, who informs us, that in the Spring of 1813, a whale was killed at Godharn, of the length of 67 feet; these however are very uncommon instances. I therefore conceive that 60 feet may be considered as the size of the largest animals of this species, and 65 feet in length as a magnitude which very rarely occurs.

Yet I believe that whales now occur of as large dimensions as at any former period, since the commencement of the whale fishery. This point I endeavoured to prove, from various historical records, in a paper, read before the Wernerian Society, on the 19th day of December, 1818, and since inserted in the Edinburgh Philosophical Journal, No. 1. p. 83.

In this paper, I brought forward the authorities of Zorgdrager, the writer of an account of the whale fishery, and one of the early superintendents of the Dutch northern fisheries, together with opinions or remarks of Captain Anderson, Gray, Heley, and others, who were among the earliest of the English whalers, which satisfactorily prove, that the average and largest produce of a whale in oil, was not greater near two hundred years ago, than it is at the present time; and to these are added the testimonies of Captain Jenkinson and Edge, as to the length of the whale, which likewise corresponds pretty nearly with the measurements I have myself made.

Jenkinson, in his voyage to Russia, performed in 1557, saw a number of whales, some of which, by estimation, were 60 feet long, and are described as being "very monstrous." Edge, who was one of the Russia Company's chief and earliest whale fishers, having been ten years to Spitsbergen, prior to the year 1625, calls the whale "a sea beaste of hughe bigness, about 65 foot long, and 35 foot thick," having whalebone ten or eleven feet long, (a common size at present) and yielding about 100 hogsheads of oil; and in a descriptive plate, accompanying Captain Edge's paper on the fishery, published by Purchas in 1625, is a sketch of a whale, with this remark subjoined—"a whale is ordinarily about 60 foot long."

Hence, I conceive, we may satisfactorily conclude that whales of as large size are found now, as at any former period, since the Spitsbergen fishery was discovered; and I may also remark, that where any respectable authority affords actual measurement exceeding 70 feet, it will always be found that the specimen referred to, was not one of the mysticetus kind, but of B. Physalis or the B. Musculus animals, which considerably exceed in length any of the common whales that I have either heard of, or met with.

When fully grown, therefore, the length of the whale may be stated as varying from 50 to 65, and rarely, if ever, reaching 70 feet; and its greatest circumference from 30 to 40 feet. It is thickest a little behind the fins, or in the middle between the anterior and posterior extremes of the animal; from whence it gradually tapers, in a conical form, towards the tail, and slightly towards the head. Its form is cylindrical from the neck to within ten feet of the tail, beyond which, it becomes somewhat quadrangular, the greatest ridge being upwards, or on the back, and running backward nearly across the middle of the tail. The head has somewhat of a triangular shape. The under part, the arched outline of which is given by the jaw bones, is flat, and measures 16 to 20 feet in length, and 10 to 12 feet in breadth. The lips, extending 15 or 20 feet in length, and five or six in height, and forming the cavity of the mouth, are attached to the under jaw, and rise from the jawbones, at an angle of about 80 degrees, having the appearance, when viewed in front, of the letter U. The upper jaw, including the crown bone or skull, is bent down at the extremity, so as to shut the front and upper parts of the cavity of the mouth, and is overlapped by the lips in a squamous manner at the sides.

When the mouth is open, it presents a cavity as large as a room, and capable of containing a merchant ship's jolly boat, full of men, being six or eight feet wide, ten or twelve feet high, (in front) and fifteen or sixteen feet long.

The fins, two in number, are placed between one third and two-fifths of the animal, from the snout, and about two feet behind the angle of the mouth;

they are from seven to nine feet in length, and four or five in breadth. The part by which they are attached to the body is somewhat elliptical, and about two feet in diameter; the side which strikes the water is nearly flat. The articulation being spherical, the fins are capable of motion in any direction; but, from the tension of the flesh and skin below, they can not be raised above the horizontal position. Hence, the account given by some naturalists, that the whale supports its young by its fin on its back, must be erroneous. The fins after death are always hard and stiff; but in the living animal, it is presumed, from the nature of the internal structure, that they are capable of considerable flexion. The whale has no dorsal fin. The tail, comprising in a single surface 80 or 100 square feet, is a formidable instrument of motion and defence. Its length is only five or six feet; but its width is from 18 to 24 or 26 feet. Its position is horizontal. In its form it is flat and semilunar; indented in the middle; the two lobes somewhat pointed, and turned a little backward. Its motions are rapid and universal; its strength immense.

The eyes are situated in the sides of the head, about a foot, obliquely, above and behind the angle of the mouth. They are remarkably small, in proportion to the bulk of the animal's body, being little larger than those of an ox. The whale has no external ear; nor can any orifice for the admission of sound be discovered until the skin is removed.

THE WHALE.

On the most elevated part of the head, about sixteen feet from the anterior extremity of the jaw, are situated two blow-holes, or spiracles, consisting of two longitudinal apertures, six or eight inches in length. These are the proper nostrils of the whale; a moist vapour, mixed with mucous, is discharged from them when the animal breathes; but no water accompanies it, unless an expiration of the breath be made under the surface.

The mouth, in place of teeth, contains two extensive rows of fins or whalebone, which are suspended from the sides of the crown bone. These series of fins are generally curved longitudinally, although they are sometimes straight, and give an arched form to the roof of the mouth. They are covered immediately by the lips attached to the lower jaw, and enclose the tongue between their lower extremities, each series, or " side of bone," as the whale fishers term it, consists of upward of 300 laminæ;* the longest are near the middle, from whence they gradually diminish away to nothing, at each extremity; fifteen feet is the greatest length of the whalebone; but ten or eleven feet is the average size, and thirteen feet is a magnitude seldom met with. The greatest breadth, which is at the gum, is ten or twelve inches. The laminæ, composing the two series of bone, are ranged side by side two-thirds of an inch apart, (thickness of the

---

* In a very small whale the number was 316 or 320.

blade included,) and resemble a frame of saws in a saw-mill, the interior edges are covered with a fringe of hair, and the exterior edges of every blade, excepting a few at each extremity of the series, is curved and flattened down, so as to present a smooth surface to the lips. In some whales a curious hollow on one side, and ridge on the other, occurs in many of the central blades of whalebone, at regular intervals of six or seven inches. May not this irregularity, like the rings in the horn of the ox, which they resemble, afford an intimation of the age of the whale? if so, twice the number of running feet in the longest lamina of whalebone, in the head of a whale not full grown, would represent its age in years. In the youngest whales, called suckers, the whalebone is only a few inches long; when the length reaches six feet or upwards, the whale is said to be *size*. The colour of the whalebone is brownish black, or bluish black. In some animals it is striped longitudinally with white. When newly cleaned, the surface exhibits a fine play of colour. A large whale sometimes affords a ton and a half of whalebone. If the " sample blade," that is, the largest lamina in the series, weigh seven pounds, the whole produce may be estimated at a ton; and so on in proportion. The whalebone is inserted into the crown bone, in a sort of rabbit. All the blades in the same series are connected together by the gum, in which the thick ends are inserted. This substance (the gums) is white, fibrous, tender, and tasteless; it cuts like cheese. It has the appearance of the interior or kernel of the cocoa nut. The tongue occupies a large portion of the cavity of the mouth: and the arch formed by the whalebone, is capable of protrusion, being fixed from root to lip, to the fat extending between the jaw bones.

A slight beard, consisting of a few short scattered white hairs, surmounts the anterior extremity of both jaws.

The throat is remarkably straight.

Two paps in the female, afford the means of rearing the young. They are situated on the abdomen, one on each side of the pudendum, and are two feet apart. They appear not to be capable of protrusion, beyond the length of a few inches. In the dead animal they are always found retracted.

The milk of a whale, resembles that of a quadruped, in its appearance. It is said to be rich and well flavoured. The vent is about six inches behind the pudendum of the female: but in the male, it is further back.

The colour of the mysticetus is velvet black, gray, (composed of dots of blackish brown on a white ground,) and white with a tinge of yellow. The back, most of the upper jaw, and part of the lower jaw, together with the fins and tail, are black. The tongue, the lower part of the under jaw and lips, sometimes a little of the upper jaw, at the extremity, and a portion of the belly are white; and the eye-lids, the junction of the tail with the body, a portion in the axillæ of the fins, &c. are gray. I have seen whales, that were all over piebald. The older animals contain the most gray and white; under size whales, are altogether of a bluish black, and suckers of pale bluish or bluish gray colour.

The skin of the body is slightly furrowed, like the water-lines on coarse laid paper. On the tail-fins, &c. it is smooth. The cuticle, or that part of the skin which can be pulled off in sheets, after it has been a little dried in the air, or particularly in frost, is not thicker than parchment. The rete mucosum in adults, is about three fourths of an inch in thickness over most parts of the body; in suckers nearly two inches; but on the under side of the fins, on the inside of the lips, and on the surface of the tongue, it is much thinner. This part of the integuments is generally of the same colour throughout its thickness. The fibres, of which it is composed, are perpendicular to the surface of the body: under this lies the true skin, which is white and tough. As it imperceptibly becomes impregnated with oil, and passes gradually into the form of blubber, its real thickness can not easily be stated. The most compact part, perhaps, may be a quarter of an inch thick.

Immediately beneath the skin, lies the blubber or fat, encompassing the whole body of the animal, together with the fins and tail. Its colour is yellowish white, yellow or red. In the very young animals, it is always yellowish white. In some old animals it resembles in colour the substance of the salmon. It swims in water. Its thickness all round the body, is eight or ten or twenty inches, varying in different parts as well as in different individuals. The lips are composed almost entirely of blubber, and yield from one to two tons of pure oil each. The tongue is chiefly composed of a soft kind of fat, that affords less oil than any other blubber; in the centre of the tongue, and towards the root, the fat is intermixed with fibres of a muscular substance. The under jaw, excepting the two jaw bones, consists almost wholly of fat, and the crown bone possesses a considerable coating of it; the fins are principally blubber, tendons and bones, and the tail possesses a thin stratum of blubber. The oil appears to be retained in the blubber in minute cells, connected together by a strong reticulated combination of tendinous fibres. These fibres being condensed at the surface, appear to form the substance of the skin. The oil is expelled when heated, and in a great measure discharges itself out of the *henks*, whenever putrefaction in the fibrous parts of the blubber takes place. The blubber and the whalebone are the parts of the whale, to which the attention of the fisher is directed. The flesh and bones, excepting occasionally the jaw bone, are rejected. The blubber, in its fresh state, is without any unpleasant smell, and it is not until after the termination of the voyage, when the cargo is unstowed, that a Greenland ship becomes disagreeable.

THE WHALE.

Four tons of blubber, by measure, generally affords three tons of oil,* but the blubber of a sucker contains a very small portion. Whales have been caught that afforded nearly thirty tons of pure oil, and whales yielding twenty tons of oil, are by no means numerous. The quantity of oil, yielded by a whale, generally bears a certain proportion to the length of its longest blade of whalebone.

The average quantity is expressed in the following table.†

Length of whalebone in feet.	1	2	3	4	5	6	7	8	9	10	11	12
Oil yielded in tons.	$1\frac{1}{2}$	$2\frac{1}{4}$	$2\frac{3}{4}$	$3\frac{1}{4}$	4	5	$6\frac{1}{2}$	$8\frac{1}{2}$	11	$13\frac{1}{2}$	17	21

Though this statement, on the average, be exceedingly near the truth, yet exceptions sometimes occur. A whale of $2\frac{1}{2}$ feet bone, for instance, has been known to produce near ten tons of oil, and another of twelve feet bone only nine tons. Such instances, however, are very uncommon.

---

* The ton or tun of oil, is 252 gallons, wine measure; it weighs, at temperature 60°, 1933lb. 12oz. 14dr. avoirdupois.

† This table is somewhat different from that given in Wernerian Memoirs, (vol. 1. p. 582,) an increased number of observations having enabled me to improve it.

A stout whale of sixty feet in length, is of the enormous weight of seventy tons; the blubber weighs about thirty tons, the bones of the head, whalebone, fins, and tail, eight or ten; carcass thirty or thirty-two.

The flesh of the young whale is of a red colour; and when cleared of fat, broiled and seasoned with pepper and salt, does not eat unlike coarse beef; that of the old whale, approaches to black, and is exceedingly coarse. An immense bed of muscles, surrounding the body, is appropriated chiefly to the movements of the tail. The tail consists principally of two reticulated beds of sinewy fibres, compactly interwoven, and containing very little oil. In the central bed, the fibres run in all directions; in the other, which encompasses the central one in a thinner stratum, they are arranged in regular order. These substances are extensively used, particularly in Holland, in the manufacture of glue.

Most of the bones of the whale are very porous, and contain large quantities of fine oil. The jaw bones, which measure twenty to twenty-five feet in length, are often taken care of, principally on account of the oil that drains out of them, when they come into a warm climate. When exhausted of oil, they readily swim in water. The external surface of the most porous bones is compact and hard; the ribs are pretty nearly solid; but the crown bone is almost as much honey-combed as the jaw bones. The number of ribs, according to Sir Charles Giesecke, is thirteen on each side. The bones of the fins are analogous, both in proportion and number to those of the fingers of the human hand. From this peculiarity of structure, the fins have been denominated by Dr. Fleming, " swimming paws." The posterior extremity of the whale, however, is a real tail; the termination of the spine, or os coccygis, running through the middle of it, almost to the edge.

As the whale is flensed while afloat, with nearly the whole of the carcass under water, few opportunities of examining its anatomical structure occur. The smallest animals of the species, mere cubs or " suckers," may indeed be hoisted on deck; and it is in such cases only that I have had a chance of inspecting them entirely out of the water. One of these having been taken, the head was hoisted aboard in a mass, and the body, when stripped of the fat, was so small as to be quite within the power of the tackles. Some new facts, respecting the anatomy of the whale, arose out of the investigation of this, and another of the species, killed in the summer of 1821, which I shall attempt to describe. The following measurements and weight, it must be observed, all refer to a sucking whale, that at the time of capture, was under maternal protection, but the other details in general may be considered as applying to the whole species of the Balæna Mysticetus.

This whale, though a " sucker," was nineteen feet in length, and fourteen feet five inches in circumference, at the thickest part of the body. The external skin, consisting of cuticle and rete mucosum, was on the body an inch and three quarters thick, being about twice the thickness of the same membranes in a full grown animal. The blubber, on an average, was five inches in thickness. The largest of the whalebone measured only twelve inches; about one half of which was imbedded in the gum. The external part of these fringes, not exceeding six inches in length, did not seem sufficient to enable the little whale yet to catch, by filtration out of the sea, the shrimps and other insects on which the animal, in a more advanced stage, is dependent for its nourishment: maternal assistance and protection, therefore, appeared to have been essential for its support. The muscles about the neck, appropriated to the movements of the jaws, formed a bed, if extended, of nearly five feet broad, and a foot thick. The central part of the diaphragm was two inches in thickness. The two principal arteries in the neck (the carotid,) were so large as to admit a man's hand and arm.

The brain lies in a small cavity in the upper and back part of the skull. The cavity included within the *pia mater*, exclusive of the foramen magnum, measured only eight inches by five. The upper part of the brain lies very near the surface of the skull. The convolutions of the cortical substance lie in beautiful fringed folds, attached to the medullary portion, which is white, as in the human brain. The general appearance of the brain is not unlike that of the other mammalia, but its smallness is remarkable. The quantity of brain in a human subject of 140 or 160 pounds weight, is, according to Haller, 4 pounds; in this whale, of 11,200 pounds, or seventy times the weight of a man, the brain was only 3 pounds 12 ounces. According to Cuvier, the brain in man varies from one thirty-first to one twenty-second part of his weight;* whereas, in this animal, the proportion of brain was only a three thousandth part.

The heart, which is of an oblong form, much compressed, resembles in colour and substance, the heart of an ox. The breadth of it, in this specimen, was 29 inches, the height 12, the thickness 9, and the weight of it 64lbs. Diameter of the aorta about 6 inches.

Large as the whale is in bulk, the throat is but narrow. In this animal the diameter of the œsophagus, when fully distended, was scarcely $2\frac{1}{2}$ inches, with difficulty admitting my hand.

The epiglottis is a beautiful valve, formed almost like the termination of the proboscis of an elephant. Though the larynx in the whale has a free communication with the mouth, as in quadrupeds, yet the mysticetus does not appear to have any voice. In

---

* Leçons d'Anat. Comp. ii. p. 149. The proportion the human brain bears to the weight of the body, appears to be, on an average, less than is stated by Cuvier. According to Haller, the proportion in a man of 160lb. weight is one-fortieth; in a man of 140lbs., one-thirty-fifth, in a child six years old, one-twenty-second.

other cetacea, however, this is not always the case; some of the dolphins, in particular, having been heard to emit a shrill sound, which in the beluga may be heard before the animal arises to the surface of the water.*

The external blowholes or spiracles, were, in the sucking whale, four inches in length; in the full grown animal, they form two curved slits, above ten inches long. In passing downward through the blubber, the blowholes, which at the surface are nearly longitudinal, as in the annexed figure, *a, a,*

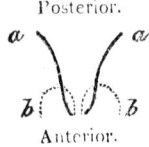

twist into a semicircular and transverse position, in the form of the dotted line b, b, then penetrating the skull, they proceed backward and downward in two conical parallel canals, until they open near the back of the under part of the skull, where they inosculate and form a single membranous sac, within a few inches of the epiglottis. The first impression of each blowhole on the upper part of the skull, is marked as in the following cut, (representing the upper surface of the anterior part of the whale's skull, the skin and fat being removed,) by an oblong cavity, *b, b,*

---

* Captain Parry's Voyage for the discovery of a North West passage, p. 35.

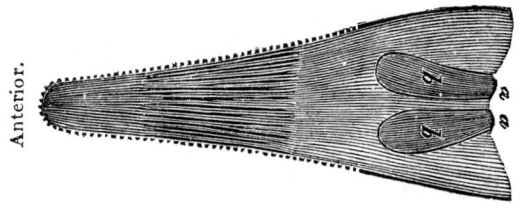

which is the seat of a muscular substance attached by its anterior extremity to the surface of the skull, and also attached, by its posterior and inferior extremity, to the interior of the skull, at some depth in the blowing canal, *a, a*. The part of this muscle that penetrates the bony canal, is of a conical form, the apex downward, or within, represented at *b,* in the annexed figure of a vertical section of the skull;

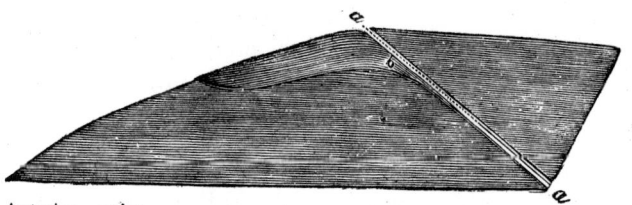

Anterior portion.

so that, when this interior portion contracts, the muscular cone *b,* is drawn tight into the orifice, and completely closes the breathing canal *a, a;* while, on the other hand, the action of the external part of the muscle draws the conical plug forward and upward, and affords a free passage for the air in respiration. This beautiful structure it is, (aided, perhaps, by the epiglottis,) that enables the animal, under the immense pressure to which it is sometimes exposed, to exclude the sea-water from its lungs. This pressure, under some depths to which the whale is known to descend, is upwards of a ton upon every square inch; yet, so far from the water being forced down the spiracles, the enormous load serves only more effectually to press down, and close the valves that defend the passages to the lungs.

The whale has no external ear, and the opening of the passage to this organ is so small as not to be easily discovered. In the sucking whale, it was only one-sixth of an inch in diameter. An elegant contrivance appears in the meatus auditorius externus for protecting the ear against pressure from without. It consists of a little plug, like the end of the finger, inserted into a corresponding cavity, in the midst of the canal, by a slight motion of which the opening can either be effectually shut for the exclusion of the sea-water, or opened for the admission of sound.

In the sucking whale, the skull or crown bone was six feet in length, from the anterior extremity to the condyles. In a full grown animal, in which the whalebone was ten feet four inches, the length of the skull, measured along the upper and convex side of the curve, was twenty feet eight inches, the cavity on the crown of the same, occupied by the muscular valve of the blow-holes, was 14 inches wide, and 24 inches long.

The whale being very nearly of the same specific gravity as sea-water, (some few individuals sinking, and others barely floating when dead) the weight may be calculated with considerable precision. The body of the whale may be divided into three segments, forming tolerably regular geometric solids. First; the *head* a parabolic conoid, which in the sucking-whale is four feet in diameter, and five and a half feet in height; its solid contents about thirty-four and a half cubic feet. Secondly; the middle segment, extending from the head to the thickest part of the body: this is a frustum of a cone in the sucking-whale, three feet in length, and four to five feet in diameter, producing a solid content of forty-eight cubic feet. Thirdly; the posterior segment, extending from the greatest circumference to the tail: this segment is a paraboloid or parabolic conoid, with its smaller end truncated. Its length in the sucking-whale is eight feet; its diameters one and five feet; and its solid contents eighty-one and a half cubic feet. And to these products may be added about ten cubic feet, the estimated bulk of the fins and tail, which make an amount of 174 cubic feet; this sum, divided by 35, the number of cubic feet of sea-water in the Greenland ocean, in a ton weight, gives the weight of the animal five tons within a cubic foot.

One of the largest mysticete, of sixty feet in length, the head twenty feet in length, by twelve feet in diameter, the middle section six feet by thir teen diameter, the third section twenty-six feet in length, by twelve and two feet diameter, will appear (if calculated the same way with an allowance of five tons for the fins and tail) to be of the prodigious weight of 114 tons! But as the last section is somewhat more slender than the body to which it is referred, this calculation may be a little in excess.

The largest animals of this species may, however, I conceive, be safely stated at a hundred tons in weight; and an ordinary full grown animal at seventy tons.

The most useful and ennobling view of natural history is, unquestionably, that which gives us the most exalted conceptions of the wisdom, power, and goodness of the Creator. And the branch of this science, that is in the highest degree calculated to assist us in tracing "the works of Nature up to Nature's God," is probably the physiology of animals. In every genus of animals we discover peculiar marks of adaptation for their economy or mode of life, and an endless variety of inimitable contrivances for accomplishing this adaptation.

The whale, which is a mammiferous animal, and closely allied, in its anatomical structure to the class of quadrupeds, affords, in the modification of the parts and principles of land animals, for applying them to a tribe inhabiting the sea, a great number of those striking displays of wisdom and power, the very contemplation of which is calculated to elevate in no inconsiderable degree, our conceptions of the Great Supreme. The mysticetus feeds on the smallest insects; its capacious mouth, with the vast fringes of whalebone, which is a most admirable filter, enables it to receive some tons of water at a mouthful, and to separate every substance from it, of the size of a pin's head and upwards. The physalis feeds on herrings, mackarel and other fishes of a similar kind; its whalebone therefore is shorter, stronger, and less compact than that of the mysticetus, and the filter formed by it less perfect:

As the whale must rise to the surface of the sea to breathe, its tail is placed horizontally, to enable it to ascend and descend more quickly; and its nostrils, or blowholes, instead of being placed at the snout, are generally on the most elevated part of the head, that they may be readily lifted clear of the water.

When the whale descends to the depths of the ocean, it becomes exposed to an enormous pressure from the superincumbent water. This pressure is sufficient to force the water through the pores of the hardest wood; yet it is effectually resisted by the skin of the whale, though it is remarkably soft and flexible. To exclude the water from the lungs, which would occasion suffocation if admitted, the blowholes are defended by the peculiar valves that have been already described.

The variety discovered in the structure of whales, is by no means one of the least interesting parts of their physiology. In other classes of animals, whose habits are similar, we often find, that each organ is the same as the corresponding one, in almost all the species of the same genus, or even of the same order; excepting when their peculiar habits, or necessities, require a modification of the general structure or principle. But in whales, as if it were intended not only to exhibit the matchless wisdom of the Creator, but, to show that his resources are unlimited, the structure of the breathing canals is varied in the different genera of cetaceous animals, and a number of contrivances alike extraordinary, equally beautiful, and equally efficient, are adapted for performing the same office.

THE WHALE.

TABLE of the comparative dimensions of six Mysticete, from my own measurements.

	Ft. in.	Ft. in.	Ft. in.	Ft. in.	Ft. in.	Ft. in.
Longest blade of whalebone,	1 0	6 0		11 2	11 6	13 7
Extreme length,	17 0	28 0	51 0	50 0	58 0	52 0
Length of the head,	5 0	8 6	16 0	15 6	19 0	20 0
Breadth of under jaw,				9 6	12 0	
Length from tip of lip to fin,	5 6	10 0		18 0		
— to greatest circumference,	7 0	18 6		24 0		
Circumference at the neck,	10 0	20 0		31 6	35 0	34 0
Greatest circumference,	12 0	15 6		34 0		
Circumference by the genitalia,	9 0		6 6	19 6		
— near the tail,	2 11	4 0	7 0	6 4	8 6	9 0
Fin length,	2 3		4 0	4 0	5 6	
— breadth,	1 3		5 0	5 6	6 0	6 0
Tail length,			20 0	17 6	24 6	20 10
— breadth,	4 9	8 2	15 6	15 0	18 6	19 6
Lip length,						6 2
— breadth,						
Produce in oil (tons)	1	4	16	16	19	24
Sex,	F.	M.		F.		M.

The whale seems dull of hearing. A noise in the air, such as that produced by a person shouting, is not noticed by it, though at the distance only of a ship's length; but a very slight splashing in the water in calm weather excites its attention, and alarms it.

Its sense of seeing is acute, whales are observed to discover one another in clear water, when under the surface, at an amazing distance. When at the surface, however, they do not see far.

They have no voice; but in breathing or blowing, they make a very loud noise. The vapour they discharge is ejected to the height of some yards, and appears at a distance, like a puff of smoke. When the animals are wounded, it is often stained with blood; and, on the approach of death, jets of blood are sometimes discharged alone. They blow strongest, densest, and loudest, when "running." When in a state of alarm, or when they first appear at the surface, after being a long time down, they respire or blow about four or five times a minute.

The whale being somewhat lighter than the medium in which it swims, can remain at the surface of the sea, with its "crown," in which the blowholes are situated, and a considerable extent of the back, above water, without any effort or motion. To descend, however, requires an exertion. The proportion of the whale that appears above water, when alive, or when recently killed, is probably not a twentieth part of the animal; but within a day after death, when the process of putrefaction commences, the whale swells to an enormous size, until at least a third of the carcass appears above water, and sometimes the body is burst by the force of air generated within.

By means of the tail principally, the whale advances through the water. The greatest velocity is produced by powerful strokes against the water, impressed alternately upward and downward; but a slower motion, it is believed, is elegantly produced, by cutting the water laterally and obliquely downward, in a manner similar to that in which a boat is forced along, with a single oar, by the operation of skulling. The fins are generally stretched out in an horizontal position; their chief application seems to be, the balancing of the animal, as the moment life is extinct, it always falls over on its side, or turns upon its back. They appear also to be used in bearing off their young, in turning, and giving a direction to the velocity produced by the tail.

Bulky as the whale is, and inactive, or indeed clumsy as it appears to be, one might imagine that all its motions would be sluggish, and its greatest exertions productive of but little celerity. The fact, however, is the reverse. A whale extended motionless at the surface of the sea, can sink in the space of five or six seconds or less, beyond the reach of its human enemies. Its velocity along the surface, or perpendicularly, or obliquely downward, is the same. I have observed a whale descending after I had harpooned it, to the depth of 400 fathoms, with the average velocity of seven or eight miles per hour. The usual rate at which whales swim, however, even when they are on their passage from one situation to another, seldom exceeds four miles an hour; and though, when urged by the sight of any enemy, or alarmed by the stroke of a harpoon, their extreme velocity may be at the rate of eight or nine miles an hour; yet we find this speed never continues longer than for a few minutes, before it relaxes to almost one half; hence, for the space of a few minutes, they are capable of darting through the water, with the velocity almost of the fastest ship under sail, and of ascending with such rapidity as to leap entirely out of the water. This feat they sometimes perform as an amusement apparently, to the high admiration of the distant spectators; but to the no small terror of the inexperienced fishers, who even under such circumstances, are often ordered, by the fool-hardy harpooner, to " pull away," to the attack. Sometimes, the whales throw themselves into a perpendicular posture, with their heads downwards, and rearing their tails on high in the air, beat the water with awful violence. In both these cases, the sea is thrown into foam, and the air filled with vapours: the noise in calm weather is heard to a great distance; and the concentric waves, produced by the concussions on the water, are commuicated abroad to a considerable extent. Sometimes the whale shakes its tremendous tail in the air, which, cracking like a whip, resounds to the distance of two or three miles.

When it retires from the surface, it first lifts its head, then plunging it under water, elevates its back, like the segment of a sphere, deliberately rounds it away towards the extremity, throws its tail out of the water, and then disappears.

In their usual conduct, whales remain at the surface to breathe, about two minutes, seldom longer; during which time, they "blow" eight or nine times, and then descend for an interval usually of five or ten minutes, but sometimes, when feeding, fifteen or twenty. The depth to which they commonly descend, is not known, though, from the eddy occasionally observed on the water, it is evidently at times, only trifling. But when struck, the quantity of line they sometimes take out of the boats, in a perpendicular descent, affords a good measure of the depth. By this rule, they have been known to descend to the depth of an English mile, and with such velocity, that instances have occurred, in which whales have been drawn up by the line attached, from a depth of 700 or 800 fathoms, and have been found to have broken their jaw-bones, and sometimes crown-bone, by the blow struck against the bottom. Some persons are of opinion, that whales can remain under a field of ice, or at the bottom of the sea in shallow water, when undisturbed, for many hours at a time. Whales are seldom found sleeping, yet, in calm weather, among ice, instances occasionally occur.

The food of the whale consists of various species of actiniæ, cliones, sepiæ, medusæ, caneri, and helices, or, at least, some of these genera are always to be seen, wherever any tribe of whales is found stationary and feeding. In the dead animals, however, in the very few instances, in which I have been enabled to open their stomachs, squillæ or shrimps, were the only substances discovered. In the mouth of a whale just killed, I once found a quantity of the same kind of insect.

When the whale feeds, it swims with considerable velocity below the surface of the sea, with its jaws widely extended. A stream of water consequently enters its capacious mouth, and along with it large quantities of water insects; the water escapes again at the sides; but the food is entangled and sifted, as it were, by the whalebone, which, from its compact arrangement, and the thick internal covering of hair, does not allow a particle the size of the smallest grain to escape.

There does not seem to be sufficient dissimilarity in the form and appearance of the mysticete found in the polar seas, to entitle them to a division into other species; yet such is the difference observed in the proportions of these animals, that they may be well considered as sub-species or varieties. In some of the mysticete, the head measures four tenths of the whole length of the animal; in others, scarcely three tenths; in some the circumference is upwards of seven tenths of the length, in others less than six tenths, or little more than one half.

The sexual intercourse of whales, is often observed about the latter end of summer; and females, with cubs or suckers along with them, being most commonly met with in the spring of the year, the time of their bringing forth, it is presumed, is in February or March; and their period of gestation about nine or ten months. In the latter end of April, 1811, a sucker was taken by a Hull whaler, to which the funis umbilicalis was still attached. The whale has one young at a birth. Instances of two being seen with a female are very rare. The young one, at the time of parturition, is said to be at least ten, if not fourteen feet in length. It goes under the protection of its mother for probably a year, or more; or until, by the evolution of the whalebone, it is enabled to procure its own nourishment. Supposing the criterion before mentioned, of the notches in the whalebone being indicative of the number of years growth, to be correct, then it would appear that the whale reaches the magnitude called *size,* that is, with a six feet length of whalebone, in twelve years, and attains its full growth at the age of twenty or twenty-five. Whales, doubtless, live to a great age. The marks of age are, increase in the quantity of gray colour in the skin, and a change to a yellowish tinge of the white parts about the head; a decrease in the quantity of oil yielded by a certain weight of blubber; an increase of hardness in the blubber, and in the thickness and strength of the ligamentous fibres of which it is partly composed.

The maternal affection of the whale, which, in other respects, is apparently a stupid animal, is striking and interesting, the cub, being insensible to danger, is easily harpooned; when the tender attachment of the mother is so manifested as not unfrequently to bring her within the reach of the whalers. Hence, though a cub is of little value, seldom producing above a ton of oil, and often less, yet it is sometimes struck as a snare for its mother. In this case she joins it at the surface of the water, whenever it has occasion to rise for respiration; encourages it to swim off; assists its flight, by taking it under her fin, and seldom deserts it while life remains. She is then dangerous to approach; but affords frequent opportunities for attack. She loses all regard for her own safety, in anxiety for the preservation of her young; dashes through the midst of her enemies; despises the danger that threatens her; and even voluntarily remains with her offspring, after various attacks on herself, from the harpoons of the fishers. In June, 1811, one of my harpooners struck a sucker, with the hope of its leading to the capture of the mother. Presently she arose close by the "fast boat," and seizing the young one, dragged about a hundred fathoms of line with remarkable force and velocity. Again she arose to the surface: darted furiously to and fro; frequently stopped short, or suddenly changed her direction, and gave every possible intimation of extreme agony. For a length of time she continued thus to act, though closely pursued by the boats; and, inspired with courage and resolution by the concern for her offspring, seemed regardless of the danger which surrounded her. At length one of the boats approached so near that a harpoon was hove at her. It hit, but did not attach itself. A second harpoon was struck; this also failed to penetrate; but a third was more effectual, and held. Still she did not attempt to escape; but allowed other boats to approach; so that, in a few minutes, three more harpoons were fastened; and, in the course of an hour afterwards, she was killed.

There is something extremely painful in the destruction of a whale, when thus evincing a degree of affectionate regard for its offspring, that would do honour to the superior intelligence of human beings; yet the object of the adventure, the value of the prize, the joy of the capture, can not be sacrificed to feelings of compassion. Whales, though often found in great numbers together, can scarcely be said to be gregarious; found most generally solitary, or in pairs, excepting when drawn to the same spot, by the attraction of an abundance of palatable food, or a choice situation of the ice.

The superiority of the sexes, in point of numbers, seems to be in favour of the male. Of 124 whales which have been taken near Spitsbergen, in eight years, in ships commanded by myself, 70 were males, and 54 were females, being in the proportion of five to four nearly. The mysticetus occurs most abundantly in the frozen seas of Greenland and Davis's Strait—in the bays of Baffin and Hudson—in the sea to the northward of Behring's Strait, and along some parts of the northern shores of Asia, and probably America. It is never met with in the German Ocean, and rarely within 200 leagues of the British coast; but along the coasts of Africa and South America, it is met with periodically in considerable numbers, In these regions it is attacked and captured by the Southern British and American Whalers, as well as by some of the people inhabiting the coasts, to the neighbourhood of which it resorts. Whether this whale is precisely of the same kind as that of Spitzbergen and Greenland, is uncertain, though it is evidently a mysticetus. One striking difference, possibly the effect of situation and climate, is, that the mysticetus found in southern regions is often covered with barnacles, (Lepas diadema, &c.) while those of the Arctic seas are free from these shell-fish.

It would be remarkable, if an animal like the whale, which is so timid that a bird alighting upon its back sometimes sets it off in great agitation and terror, should be wholly devoid of enemies. Besides man, who is doubtless its most formidable adversary, it is subject to annoyance from sharks, and it is also said from the narwal, sword-fish, and thresher. With regard to the narwal, I am persuaded that this opinion is incorrect, for so far from its being an enemy, it is found to associate with the whale in the greatest apparent harmony, and its appearance, indeed, in the Greenland sea is hailed by the fishers, the narwal being considered as the harbinger of the whale. But the sword-fish and thresher, (if such an animal there be) may possibly be among the enemies of the whale, notwithstanding I have never witnessed their combats; and the shark is known certainly to be an enemy, though perhaps not a very formidable one. Whales indeed flee the seas where it abounds, and evince by marks occasionally found on their tails, a strong evidence of their having been bit by the shark. A living whale may be annoyed, though it can scarcely be supposed to be ever overcome by the shark; but a dead whale is an easy prey, and affords a fine banquet to this insatiable creature.

The whale, from its vast bulk, and variety of products, is of great importance in commerce, as well as in the domestic economy of savage nations; and its oil and whalebone are of extensive application in the arts and manufactures. A description of its most valuable products, and of the uses to which they are applied, being included in the account of the whale fishery, which follows, it will only be necessary, in this place, to mention the purposes to which parts and products, not now objects of commerce, are or might be applied.

Though to the refined palate of a modern European, the flesh of a whale, as an article of food, would be received with abhorrence, yet we find that it is considered by some of the inhabitants of the northern shores of Europe, Asia, and America, as well as those on the coasts of Hudson's Bay, and Davis's strait, as a choice and staple article of subsistence. The Esquimaux eat the flesh and fat of the whale, and drink the oil with greediness. Indeed, some tribes, who are not familiarized with spiritous liquors, carry along with them in their canoes, in their fishing excursions, bladders filled with oil, which they use in the same way, and with a similar relish, that a British sailor does a dram.* They also eat the skin of the whale raw, both adults and children; for it is not uncommon, when the females visit the whale-ships, for them to help themselves to pieces of skin, preferring those with which a little blubber is connected, and to give it as food to their infants suspended on their backs, who suck it with apparent delight.

Blubber, when pickled and boiled, is said to be very palatable; the tail, when parboiled and then fried, is said to be not unsavoury, but even agreeable eating; and the flesh of young whales, I know from experiment, is by no means indifferent food.

Not only is it certain that the flesh of the whale is now eaten by savage nations, but it is also well

---

* Ellis's voyage to Hudson's Bay, p. 233.

authenticated that, in the 12th, 13th, 14th, and 15th centuries, it was used as food by the Icelanders, the Netherlanders, the French, the Spanish, and probably by the English. M. S. B. Noel, in a tract on the whale fishery,* informs us, that about the 13th century, the flesh, particularly the tongue of whales, was sold in the markets of Bayonne, Cibourre and Beariz, where it was esteemed as a great delicacy, being used at the best tables; and even so late as the 15th century, he conceives, from the authority of Charles Etienne, that the principal nourishment of the poor in Lent, in some districts of France, consisted of the flesh and fat of the whale.

Besides forming a choice eatable, the inferior products of the whale are applied to other purposes by the Indian and Esquimaux of Arctic countries, and with some nations are essential to their comfort, some membranes of the abdomen are used for an upper article of clothing, and the peritoneum, in particular, being thin and transparent, is used instead of glass in the windows of their huts; the bones are converted into harpoons and spears, for striking the seal, or darting at the sea-birds, and are also employed in the erection of their tents, and with some tribes, in the formation of their boats; the sinews are divided into filaments, and used as thread, with which they join the seams of their

---

* Memoire sur " l'Antiquité de la Pêche de la Baleine par les nations Européennes."

coats and tent cloths, and sew with great taste and nicety the different articles of dress they manufacture; and the whalebone and other superior products, so valuable in European markets, have also their uses among them.

I shall conclude this account of the mysticetus, with a sketch of some of the characters which belong generally to cetaceous animals.

Whales are viviparous: they have but one young at a time, and suckle it with teats. They are furnished with lungs, and are under the necessity of approaching the surface of the water at intervals to respire in the air. The heart has two ventricles and two auricles. The blood is warmer than in the human species; in a narwal that had been an hour and a half dead, the temperature of the blood was 97°; and in a mysticetus recently killed 102°. All of them inhabit the sea. Some of them procure their food by means of a kind of sieve, composed of two fringes of whalebone; these have no teeth. Others have no whalebone, but are furnished with teeth. They all have two lateral or pectoral fins, with concealed bones like those of a hand; and a large flexible horizontal tail, which is the principal member of motion. Some have a kind of dorsal fin, which is an adipose or cartilaginous substance, without motion. This fin, varying in form, size, and position, in different species, and being in a conspicuous situation, is well adapted for a specific distinction. The appearance and dimensions of the whalebone and teeth, especially the former, are other specific characteristics. All whales have spiracles or blowholes, some with one, others with two openings, through which they breathe; some have a smooth skin all over the body; others have rugæ or sulci about the region of the thorax and on the lower jaw. And all afford, beneath the integuments, a quantity of fat or blubber, from whence a useful and valuable oil, the train oil of commerce, is extracted.

Species II.—*The Razor-back.*

Balæna Physalis; L.

Balænoptera Gibbar: La Cepède.

This is the longest animal of the whale tribe; and probably, the most powerful and bulky of created beings. It differs from the mysticetus, in its form being less cylindrical, and its body longer and more slender; in its whalebone being shorter; its produce in blubber and oil being less; in its colour being of a bluer tinge; in its fins being more in number, in its breathing or blowing being more violent; in its speed being greater; in its actions being quicker and more restless, and in its conduct being bolder.

The length of the physalis is about 100 feet; its greatest circumference 30 or 35. The body is not cylindrical, but is considerably compressed on the side, and angular at the back. A transverse section near the fins is an oblong, and at the rump a rhombus. The longest lamina of whalebone measures about four feet; it affords ten or twelve tons of blubber. Its colour is a pale bluish black, or dark bluish gray, in which it resembles the sucking mysticetus. Besides the two pectoral fins, it has a small horny protuberance, or rayless and immoveable fin, on the extremity of the back. Its blowing is very violent, and may be heard in calm weather, at the distance of about a mile. It swims with a velocity at the greatest of about twelve miles an hour. It is by no means a timid animal, yet it does not appear to be revengeful or mischievous. When closely pursued by boats, it manifests little fear, and does not attempt to outstrip them in the race; but merely endeavours to avoid them by diving or changing its direction. If harpooned, or otherwise wounded, it then exerts all its energies, and escapes with its utmost velocity, but shows little disposition to retaliate on its enemies, or to repel their attacks by engaging in a combat. Though at a distance the physalis is sometimes mistaken by the whalers for the mysticetus; yet its appearance and actions are so different, that it may be generally distinguished. It seldom lies quietly on the surface of the water when blowing, but usually has a velocity of four or five miles an hour; and when it descends, it very rarely throws its tail in the air, which is a very general practice with the mysticetus.

The great speed and activity of the physalis, render it a difficult and dangerous object of attack; while the small quantity of inferior oil it affords, makes it unworthy the general attention of the fishers. When struck, it frequently drags the fast boat with such speed through the water, that it is liable to be carried immediately beyond the reach of assistance, and soon out of sight of both boats and ship. Hence the striker is under the necessity of cutting the line, and sacrificing his employer's property, for securing the safety of himself and companions. I have made different attempts to capture one of these formidable creatures. In the year 1818, I ordered a general chase of them, providing against the danger of having my crew separated from the ship, by appointing a rendezvous on the shore, not far distant, and preparing against the loss of much line, by dividing it at 200 fathoms from the harpoon, and affixing a buoy to the end of it. Thus arranged, one of these whales was shot, and another struck. The former dived with such impetuosity, that the line was broken by the resistance of the buoy, as soon as it was thrown into the water, and the latter was liberated within a minute by the the division of the line, occasioned, it was supposed, by its friction against the dorsal fin. Both of them escaped. Another physalis was struck by one of my inexperienced harpooners, who mistook it for a mysticetus. It dived obliquely with such velocity, that 480 fathoms of line were withdrawn from the boat in about a minute of time. This whale was also lost by the breaking of the line.

The following observations on this animal have been derived from different persons who have had opportunities of examining it when dead.

Length of a physalis found dead in Davis's Strait 105 feet, greatest circumference about 38. Head small, compared with that of the common whale; fins long and narrow; tail about twelve feet broad, finely formed; whalebone about four feet in length, thick, bristly and narrow; blubber six or eight inches thick, of indifferent quality; colour bluish black on the back, and bluish gray on the belly; skin smooth, excepting about the side of the thorax, where longitudinal rugæ or sulci occur. The physalis occurs in great numbers in the Arctic seas, especially along the edge of the ice, between Cherie Island and Nova Zembla, and also near Jan Mayen. Persons trading to Archangel have often mistaken it for the common whale. It is seldom seen among much ice, and seems to be avoided by the mysticetus; as such, the whale fishers view its appearance with painful concern. It inhabits most generally in the Spitzbergen quarter, the parallels of from 70 to 76 degrees, but in the months of June, July, and August, when the sea is usually open, it advances along the land to the northward as high as the 80th degree of latitude. In open seasons it is seen near the headland at an earlier period. A whale, probably of this kind, 101 feet in length, was stranded on the banks of the Humber, about the middle of September, 1750.

SPECIES III.—*The Broad-nosed Whale.*

Balæna Musculus; L.

Balænoptera Rorqual: La Cepède.

This species of whale frequents the coasts of Scotland, Ireland, and Norway, &c. and is said to feed principally upon herrings. Several characters of the musculus very much resemble those of the physalis, though I believe there is an essential difference between the two animals; the musculus being shorter, having a larger head and mouth, and rounder under jaw, than the physalis. Several individuals, apparently of this kind, have been stranded or killed on different parts of the coast of the United Kingdom. One, 52 feet in length, was stranded near Eyemouth, June 19th, 1752. Another, nearly 70 feet in length, ran ashore on the coast of Cornwall, on the 18th, of June, 1797. Three were killed on the northwest coast of Ireland, in the year 1762, and two in 1763; one or two have been killed in the Thames, and one was embayed and killed in

Baltic sound, Shetland, in the winter of 1817–18; some remains of which I saw. This latter whale, was 82 feet in length, the jaw bones were 31 feet long, the longest lamina of whalebone about three feet long. Instead of hair at the inner edge and at the front of each blade of whalebone, it had a fringe of bristly fibres; and it was stiffer, harder, and more horny in its texture than common whalebone. This whale produced only about five tons of oil, all of it of an inferior quality, some of it viscid and bad. It was valued altogether, expenses of removing the produce and extracting the oil deducted, at no more than 60*l.* Sterling. It had the usual sulci about the thorax, and a dorsal fin.

In its blowing, swimming, and general action, as well as in its appearance in the water, the musculus very much resembles the physalis, from which, indeed, while living, it can scarcely be distinguished.

Species IV.—*The Finner.*

Balæna Boops: L.

Balænoptera Jubartes; La Cepède.

Length about 46 feet; greatest circumference of the body about 20 feet; dorsal protuberance or fin, about two feet and a half high; pectoral fin, four or five feet long, externally, and scarcely a foot broad; tail about three feet deep, and ten broad; whalebone about 300 laminæ on each side, the longest about 18 inches in length; the under jaw about 15 feet long, or one third of whole length of the animal; sulci about two dozen in number; two external blowholes; blubber on the body, two or three inches thick; under the sulci none.

In the Memoirs of the Wernerian Society, a description of a whale, corresponding in its dimensions, at least, with the Balæna Boops, has been given to the public by Mr. P. Neill, Edinburgh.* This whale was stranded on the banks of the Forth, near Alloa, and had been considerably mutilated before Mr. Neill had an opportunity of examining it. It is considered by him, a Balæna Rostrata. From his valuable paper, part of the above description is taken, which differs so much from a Rostrata noticed below, particularly in its larger dimensions, and in the greater proportion which the head bears to the body, that it would appear to belong either to the Balæna Boops or to an undescribed species. From the inaccuracy of the sketches of almost all the whales hitherto figured, the naturalist is rather plagued than assisted by them. As such, the figures given by La Cepède and others, can scarcely be of any service, in determining the species of this whale.

---

* Vol. 1. p. 201.

Species V.—*The Beaked Whale.*

Balæna Rostrata; L.

Balænoptera Acuto Rostrata: La Cepéde.

This is the last and the smallest of the whalebone whales with which I am acquainted. An animal of this species was killed in Scalpa Bay, November 14, 1808. Its length was $17\frac{1}{2}$ feet, circumference 20 feet, length from the snout to the dorsal fin $17\frac{1}{2}$ feet, from the snout to the pectoral fin 5 feet, from the snout to the eye $3\frac{1}{2}$ feet, and from the snout to the blowholes 3 feet. Pectoral fins two feet long and seven inches broad; dorsal fin 15 inches long by 9 inches high, tail 15 inches long by $4\frac{1}{2}$ feet broad. Largest whalebone about six inches. Colour of the back black; of the belly glossy white; and of the grooves of the plicæ, according to Mrs. Traill, who saw it on the beach in Scalpa Bay, a sort of flesh colour.

The Rostrata is said to inhabit principally the Norwegian seas, and to grow to the length of 25 feet. One of the species was killed near Spitzbergen, in the year 1813, some of the whalebone of which I now have in my possession. It is thin, fibrous, of a yellowish white colour, and semi-transparent, almost like lantern horns. It is curved like a scymetar, and fringed with white hair on the convex edge and point. Its length is 9 inches; greatest breadth $2\frac{1}{4}$.

THE WHALE FISHERY.

Observations on the Fishery of different latitudes and seasons, and under different circumstances of Ice, Wind, and Weather.

It is not yet ascertained, what is the earliest period of the year, in which it is possible to fish for whales. The danger attending the navigation, amidst massive drift ice in the obscurity of night, is the most formidable objection against attempting the fishery before the middle of the month of April, when the sun, having entered the northern tropic, begins to enlighten the Polar regions throughout the twenty-four hours. Severity of frost, prevalence of storms, and frequency of thick weather, arising from snow and frost rime, are the usual concomitants of the spring of the year; and these, when combined with the darkness incident to night, a tempestuous sea, and crowded ice, must probably produce as high a degree of horror in the mind of the navigator, who is unhappily subjected to their distressful influence, as any combination of circumstances which the imagination can present. Some ships have sailed to the northward of the seventy-eighth degree of latitude, before the close of the month of March; but I am not acquainted with a single instance, where the hardy fishers have, at this season, derived any compensation for the extraordinary dangers to which they were exposed. In the course of the month of April, on certain occasions, considerable progress has been made in the fishery, notwithstanding the frequency of storms. At the first stage of the business, in *open* seasons, the whales are usually found in most abundance on the borders of the ice, near Hackluyt's Headland, in the latitude of 80°. A degree or two farther south, they are sometimes seen, though not in much plenty; but in the 76th degree, they sometimes occur in such numbers, as to present a tolerable prospect of success in assailing them. Some rare instances have occurred, wherein they have been seen on the edge of the ice, extending from Cherry Island to Point-look-out, in the early part of the season.

In the year 1803, the fishery of April was considerable in the latitude of 80°; in 1813, many whales were seen in the same latitude; but the weather being tempestuous in an almost unprecedented degree, but few were killed; and in the intermediate years, the fishery was never general in this month, and but seldom begun at all before the commencement of May. In 1814, the fishery commenced before the middle of April, and some ships derived uncommon advantage from an early arrival. In 1815, some ships were near Spitzbergen in March, and fished in the first week of April in the latitude of 80°, where a great number of whales were seen. Accompanying the ice in its drift, along the coast to the southward, the same *tribe* of whales were seen in the latitude of 78°, about the middle and end of the month, and a considerable number were killed. In 1816, fish were seen in 80°, in the same month, but few killed, on account of the formation of bay ice upon the sea. In 1817, the weather was very tempestuous in April, and scarcely any whales were killed; and in 1818, the fishery of this month was inconsiderable.

Grown fish are frequently found at the edge, or a little within the edge of the loose ice, in the 79th degree of north latitude, in the month of May; and small whales of different ages at fields, and sometimes in bays of the ice in the 80th degree.

Usually, the fish are most plentiful in June; and on some occasions they are met with in every degree of latitude from 75° to 80°. In this month, the large whales are found in every variety of situation; sometimes in open water, at others in the loose ice, or at the edges of fields and floes, near the main impervious body of ice, extending towards the coast of *West* Greenland. The smaller animals of the species are, at the same time, found farther to the south, than in the spring, at floes, fields, or even among loose ice, but most plentiful about fields or floes, at the border of the main western ice, in the latitude of 78 or 78½ degrees.

In July, the fishery generally terminates, sometimes at the beginning of the month, at others, though more rarely, it continues throughout the greater part of it. Few small fish are seen at this season. The large whales, when plentiful, are found occasionally in every intermediate situation, between the open sea and the main ice, in one direction, and between the latitudes of 75° and 79° in the other, but rarely as far north as 80°.

The parallel of 78 to $78\frac{1}{2}$ degrees, is, on the whole, the most productive fishing station. The interval between this parallel and 80°, or any other situation more remote, is called the " northward," and any situation in a lower latitude than 78°, is called the " southward."

Though the 79th degree affords whales in the greatest abundance, yet the 76th degree affords them, perhaps, more generally. In this latter situation, a very large kind of the mysticetus is commonly to be found throughout the season, from April to July inclusive. Their number, however, is not often great; and as the situation in which they occur is unsheltered, and, consequently, exposed to heavy swells, the southern fishery is not much frequented.

The parallel of 77° to $77\frac{1}{2}°$, is considered a " dead latitude," by the fishers, but occasionally it affords whales also.

From an attentive observation of facts, it would appear, that different tribes of the mysticetus inhabit different regions, and pursue different routes on their removal from the places where first seen.

These tribes seem to be distinguished by a difference of age or manners, and in some instances, apparently by a difference of species, or sub-species. The whales seen in the spring in the latitude of 80°, which are usually full grown animals, disappear generally by the end of April; and the place of their retreat is unknown. Those inhabiting the regions of 78°, are of a mixed size. Such as resort to fields in May and the beginning of June, are generally young animals; and those seen in the latitude of 76°, are almost always of the very largest kind. Instances are remembered by some aged captains, wherein a number have been taken in the *southward* fishing stations, which were astonishingly productive of oil. It is probable, that the difference in the appearance of the heads, or the difference of proportion existing between the heads and bodies of some mysticete, are distinguishable of a difference in the species, or sub-species. Those inhabiting southern latitudes, have commonly long heads and bodies, compared with their circumference, moderately thick blubber and long whalebone; those of the mean fishing latitude, that is 78°—79°, have more commonly short broad heads, compared with the size of the body. In some individuals, the head is at least one-third of the whole length of the animal, but in others scarcely two-sevenths. Hence, it is exceedingly probable, that the whales seen early in April, in the latitude of 80°, are a peculiar tribe, which do not re-appear during the remainder of the season; and that those inhabiting the latitude of 78° and of 76°, are likewise distinct tribes.

Notwithstanding, if we descend to particulars, the great variety and uncertainty which appear in the nature of the situations preferred by the whales, and the apparent dissimilarity observed in their habits, it is probable, that, were the different tribes distinguished, we should find a much greater degree of similarity in their choice of situation and in their general habits than we are at present able to trace.

Annoyed as the whales are by the fishers, it is not surprising that they sometimes vary their usual places of resort, and it is not improbable, were they left undisturbed for a few years, but that they might return to the bays and sea-coasts of Spitzbergen and its neighbouring islands, as was formerly the custom with certain tribes, at the commencement of this fishery. We are doubtless in a great measure indebted to the necessity they are under, of performing the function of respiration in the air, at stated intervals, for being able to meet with them at all; though the coast of Spitzbergen may possibly possess powerful attraction to the mysticete, by affording them a greater abundance of palatable food than the interior western waters, covered perpetually by the ice. From this necessity of respiring in the air, we may account for their appearance in the open sea in the early part of the spring. The ice at this season, connected by the winter's frost, is so consolidated, as to prevent the whales from breathing among it, excepting

WHALE-FISHERY. 151

within so much of its confines as may be broken by the violence of the sea in storms. After the dissolution of the continuity of the ice, by north, northwest, or west winds, they find sufficient convenience for respiration in the interior, and often retreat thither to the great disadvantage of the whalers. In such cases, if the formation of bay ice, or the continuity of the border of the heavy ice, prevents the ships from following, the whales completely escape their enemies, until the relaxation of the frost permits an entrance.

It is not uncommon, however, for an adult tribe of whales, to resort partially to the open sea, between the latitudes of 76° and 79°, during the months of May and June, and, though more rarely, during the early part of July, when, at length, they suddenly betake themselves to the ice, and disappear altogether.

The systematical movements of the whales receive additional illustration from many well known facts. Sometimes a large tribe, passing from one place to another, which, under such circumstances, is denominated a " run of fish," has been traced in its movements in a direct line from the south towards the north, along the seaward edge of the western ice, through a space of two or three degrees of latitude; then it has been ascertained to have entered the ice, and penetrated to the north-westward, beyond the reach of the fishers. In certain years, it is curious to observe, that the whales commence a simultaneous retreat throughout the whole fishing limits, and all disappear within the space of a very few days. On such occasions it has often happened, that not a single whale has been seen by any individual belonging to the whole Greenland fleet, after perhaps the middle of June, but more commonly after the first or second week in July, notwithstanding many of the fleet may have cruised about in the fishing region for a month afterwards. In the year 1813, whales were found in considerable numbers in the open sea, during the greater part of the fishing season, but in the greatest abundance about the end of June and beginning of July. On the 6th of July, they departed into the ice, and were followed by the fishers; several were killed during the three succeeding days, but they wholly disappeared after the 9th. Notwithstanding, several ships cruised "the country," for some weeks afterwards, in all navigable directions, through an extent of four degrees of latitude, and penetrated the ice as far as the main western body, in different parallels, it does not appear that a single whale was caught, and as far as I was able to learn, but one was seen, and this individual was observed to be rapidly advancing towards the north-west. I do not mention this as an uncommon circumstance, because a similar case occurs frequently, but as a single illustration of the foregoing observation.

When the fishery for the season, in the opinion of the British whalers, has altogether ceased, it appears from the observation of the Dutch,* that it may frequently be recommenced in the autumn, at the verge of the most northern waters, near Hackluyt's Headland. They consider the fish which then appear as the same tribe that are seen in this place in the spring of the year, and enter the ice, immediately after it opens in the north. On the recommencement of the frost, they instinctively return to prevent themselves being enclosed so far within the ice, as to occasion suffocation from the freezing up of the openings through which they might otherwise breathe.

This tribe are supposed by the Dutch to be really inhabitants of the sea adjoining West Greenland; that they always retreat thither whenever the state of the ice will admit, and only appear within the observation of the fishers, when the solidity of the ice prevents their attaining those favourite situations, where they probably find the most agreeable food.†

The whales, of lower latitudes, however, whose food lies near the eastern margin of the main ice, when they enter the ice in May and June, seem to exhibit an intention of evading their pursuers; for in whatever manner they may retreat for a while, they frequently return to the same or other similar place,

---

* Beschryving der Walvisvangst, vol. 1, p. 52.

† Beschryving, &c. vol. 1. p. 53.—As I have never seen whales in this situation in the autumn myself, I give the information entirely on the authority of the work here quoted.

accessible to the fishers. But after the month of July, this tribe also penetrates so deeply into the ice, that it gets beyond the reach of its enemies.

Experience proves, that the whale has its favourite places of resort, depending on a sufficiency of food, particular circumstances of weather, and particular positions and qualities of the ice. Thus, though many whales may have been seen in open water, when the weather was fine, after the occurrence of a storm, perhaps not one is to be seen. And, though fields are sometimes the resort of hundreds of whales, yet, whenever the loose ice around separates entirely away, the whales quit them also. Hence fields seldom afford whales in much abundance, excepting at the time when they first " break out," and become accessible; that is, immediately after a vacancy is made on some side by the separation of adjoining fields, floes, or drift ice. Whales, on leaving fields which have become exposed, frequently retire to other more obscure situations in a west or northwest direction; but occasionally they retreat no further than the neighbouring drift ice, from whence they sometimes return to the fields at regular intervals of six, twelve or twenty-four hours.

Whales are rarely seen in abundance in the large open space of water, which sometimes occurs amidst fields and floes, nor are they commonly seen in a very open pack, unless it be in the immediate neighbourhood of the main western ice. They seem to have a preference for close packs and patches of ice; and for fields under certain circumstances; for

WHALE-FISHERY. 155

deep bays or *bights,* and sometimes for clear water situations; occasionally for detached streams of drift ice; and most generally, for extensive sheets of bay ice. Bay ice is a very favourite retreat of the whales, so long as it continues sufficiently tender to be conveniently broken, for the purpose of respiration. In such situations, whales may frequently be seen in amazing numbers, elevating and breaking the ice with their *crowns,** where they are observed to remain much longer at rest than when seen in open water, or in the clear interstices of the ice, or indeed in almost any other situation.

Description of the boats and principal instruments used in the capture of the whale.

Whale-boats are, of course, peculiarly adapted for the occupation they are intended to be employed in. A well constructed " Greenland boat," possesses the following properties. It floats lightly and safely on the water,—is capable of being rowed with great speed, and readily turned round,—it is of such capacity that it carries six or seven men, seven or eight hundred weight of whale-lines, and various other materials, and yet retains the necessary properties of safety, buoyancy, and speed, either in smooth water, or where it is exposed to a considerable sea. Whale-boats being very liable to receive damage, both from whales and ice, are al-

---

* The eminence on the head of the whale, in which the blow-holes are situated, is thus called.

ways *carver-built*,—a structure which is easily repaired. They are usually of the following dimensions. Those called " six oared boats," adapted for carrying seven men, six of whom, including the harpooner, are rowers, are generally 26 to 28 feet in length, and about five feet nine inches in breadth. Six men boats, that is, with five rowers and a steersman, are usually 25 to 26 feet in length, and about five feet six inches in breadth. And " four oared boats," are usually twenty-three to twenty-four in length, and about five feet three inches in breadth. The main breadth of the two first classes of boats is at about three-sevenths of the length of the boat reckoned from the stem; but, in the last class, it is necessary to have the main breadth within one-third of the length of the boat from the stem. The object of this is, to enable the smaller boat to support, without being dragged under water, as great a strain on the lines as those of a larger class; otherwise, if such a boat were sent out by itself, its lines would be always liable to be lost before any assistance could reach it. The five oared or six men boat, is that which is in most general use; though each fishing ship generally carries one or two of the largest class. These boats are now commonly built of fir-boards, one-half or three-fourths of an inch thick, with timbers, keel, gunwales, stem and sternpost of oak. An improvement in the timbering of whale-boats has lately been made, by sawing the timber out of very straight grained oak, and bending them to the required form, after being made supple, by the application of steam, or immersion in boiling water. This improvement, which renders the timbers more elastic, than when they are sawn out of crooked oak, at the same time makes the boat stronger and lighter. Though the principle has long been acted upon in clincher-built boats, with ash timbers, the application to carver-built whale-boats, is, I believe, new. The bow and stern of Greenland boats, are both sharp, and, in appearance, very similar; but the stern forms a more acute angle than the bow. The keel has some inches depression in the middle, from which the facility of turning is acquired.

The instruments of general use in the capture of the whale, are the harpoon and lance.

The harpoon (fig. 4.) is an instrument of iron, of about three feet in length. It consists of three conjoined parts, called the "socket," "shank," and "mouth," the latter of which includes the barbs or "withers." This instrument, if we except a small addition to the barbs, and some enlargement of dimensions, maintains the same form in which it was originally used in the fishery two centuries ago. At that time, the mouth or barbed extremity was of a triangular shape, united to the shank in the middle of one of the sides; and this being scooped out on each side of the shank, formed two simple flat barbs. In the course of last century, an improvement was made, by adding another small barb, resembling the beard of a fish-hook, within each of the former withers, in a reverse position. The two principal withers, in the present improved harpoon, measure about eight inches in length and six in breadth; the shank is eighteen inches to two feet in length, and four-tenths of an inch in diameter, and the socket, which is hollow, swells from the size of the shank to near two inches diameter, and is about six inches in length. Now, when the harpoon is forced by a blow into the fat of the whale, and the line is held tight, the principal withers seize the strong ligamentous fibres of the blubber, and prevent it from being withdrawn; and in the event of its being pulled out, so far as to remain entangled by one wither only, which is frequently the case, then the little reverse barb, or "stop wither," as it is called, collecting a number of the same reticulated sinewy fibres, which are very numerous near the skin, prevents the harpoon from being shaken out by the ordinary motions of the whale. The point and exterior edges of the barbs of the harpoon, are sharpened to a rough edge, by means of a file. This part of the harpoon is not formed of steel, as it is frequently represented, but of common soft iron; so that when blunted, it can be readily sharpened by a file, or even by scraping it with a knife. The most important part in the construction of this instrument, is the shank. As this part is liable to be forcibly and sudddenly extended, twisted and bent, it requires to be made of the softest and most pliable iron. That kind which is of the most approved tenacity, is made of old horse-shoe nails or *stubs*, which are formed into small rods, and two or three of these welded together; so that should a flaw happen to occur in any one of the rods, the strength of the whole might still be depended on. Some manufacturers enclose a quantity of stub-iron in a cylinder of best foreign iron, and form the shank of the harpoon out of a single rod. A test sometimes used for trying the sufficiency of a harpoon, is to wind its shank round a bolt of inch iron, in the form of a close spiral, then to unwind it again, and put it into a straight form. If it bears this without injury in the *cold* state, it is considered as excellent. The breaking of a harpoon is of no less importance than the value of a whale, which is sometimes estimated at more than 1000*l.* sterling.

Next in importance to the harpoon, is the lance, (fig. 5.) which is a spear of iron of the length of six feet. It consists of a hollow socket six inches long, swelling from half an inch, the size of the shank, to near two inches in diameter, into which is fitted a four feet stock or handle of fir; a shank five feet long, and half an inch in diameter; and a mouth of steel, which is made very thin, and exceedingly sharp, seven or eight inches in length, and two or $2\frac{1}{2}$ in breadth.

These two instruments, the harpoon and lance, with the necessary apparatus of lines, boats, and oars, are all that are essential for capturing the whale. But besides these instruments, so successfully used in the whale-fishery, there is likewise an auxiliary weapon which has, at different periods, been of some celebrity. This is the harpoon-gun,

It is well calculated to facilitate the capture of whales under particular circumstances, particularly in calm clear weather, when the fish are apt to take the alarm, whenever the boats approach within fifteen or twenty yards of them. The harpoon-gun was invented in the year 1731, and used, it seems, by some individuals with success. Being, however, difficult, and somewhat dangerous in its application, it was laid aside for many years. It has, however, subsequently been highly improved, and rendered capable of throwing a harpoon near forty yards with effect; yet, on account of the difficulty and address requisite in the management of it, and loss of fish, which, in unskilful hands it has been the means of occasioning, together with some accidents which have resulted from its use,—it has not been so generally adopted as might have been expected.

In its present improved form, the harpoon-gun consists of a kind of swivel, having a barrel of wrought iron, 24 to 26 inches in length, of 3 inches exterior diameter, and $1\frac{7}{8}$ inches bore. It is furnished with two locks, which act simultaneously, for the purpose of diminishing the liability of the gun missing fire. Fig. 1. is a representation of the harpoon-gun; and fig. 2. and 3. show the form of the harpoon which is fired from it. The shank of this harpoon is double, terminating in a cylindrical knob, fitting the bore of the gun. Between the two parts of the shank is a wire ring, to which is attached the line. Now, when the harpoon is introduced into the barrel of the gun, the

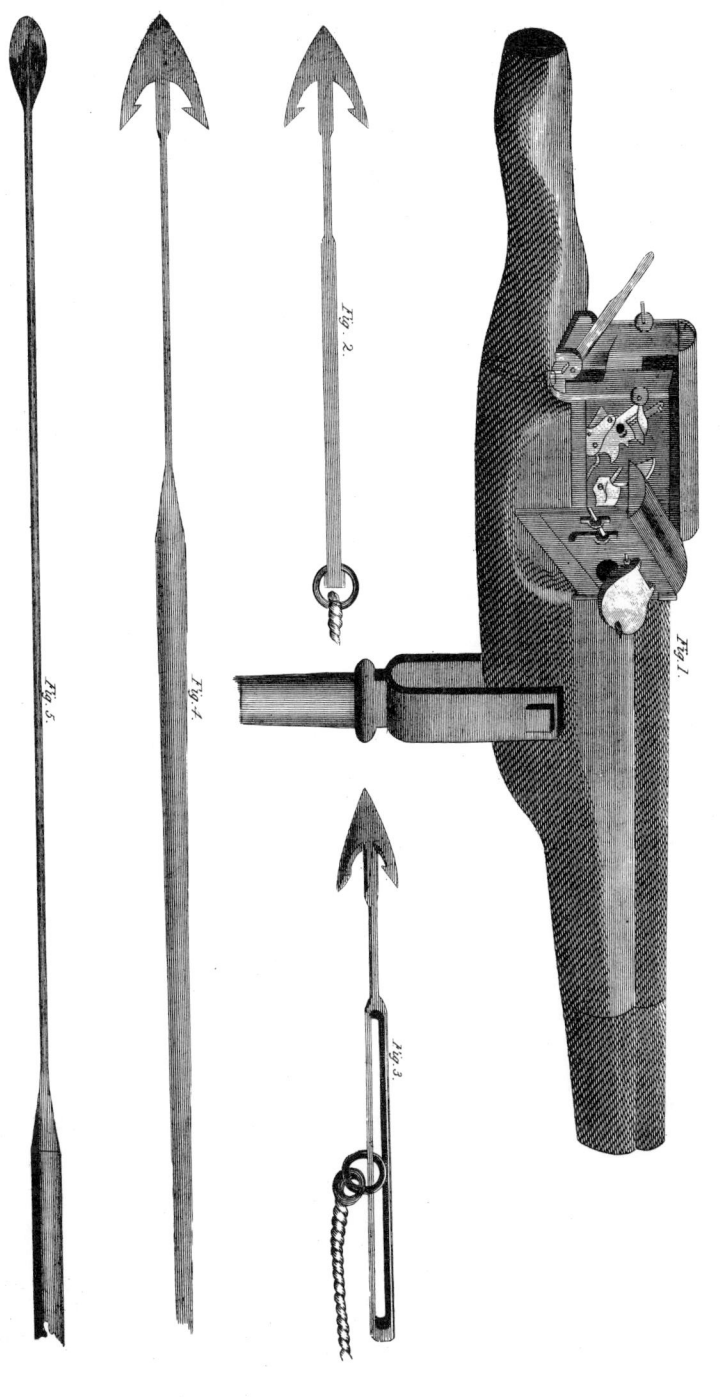

WHALE-FISHERY.

ring, with the attached line, remains on the outside near the mouth of the harpoon; but the instant that it is fired, the ring flies back against the cylindrical knob. Some harpoons have been lately made with a single shank, similar to the common "hand-harpoon," but swelled at the end to the thickness of the bore of the gun. The whale line closely spliced round the shank, is slipped towards the mouth of the harpoon, when it is placed in the gun, and when fired, is prevented from disengaging itself, by the size of the knob at the end.

Proceedings on Fishing Stations.

On fishing stations, when the weather is such as to render the fishing practicable, the boats are always ready for instant service. Suspended from davits or cranes by the side of the ship, and furnished with the requisite implements, two boats at least, the crews of which are always in readiness, can, in a general way, be manned and lowered into the water, within the space of one minute of time.

Wherever there is a probability of seeing whales, when the weather and situation are such, as to present a possibility of capturing them, the crow's-nest*, is generally occupied by the master, or some

---

* The crow's-nest, is an apparatus placed on the main-topmast, or top-gallant-mast head, as a kind of watch tower, for the use of the master, or officer of the watch, in the fishing seas, for sheltering him from the wind, when en-

one of the officers, who, commanding from thence an extensive prospect of the surrounding sea, keeps an anxious watch for the appearance of a whale; assisted by a telescope, he views the operations of any ship which may be in sight at a distance; and occasionally sweeps the horizon with his glass, to extend the limited sphere of vision, in which he is able to discriminate a whale with the naked eye, to an area vastly greater. The moment that a fish is seen, he gives notice to the " watch upon deck," part of whom leap into a boat, are lowered down, and push off towards the place. If the fish be large, a second boat is immediately despatched to the support of the other. When the whale again appears, two boats row towards it with their utmost speed; and though they may be disappointed in all their attempts, they generally continue the pursuit, until the fish either takes the alarm, and escapes them, or they are recalled by signal to the ship. When two or more fish, appear at the same time in different situations, the number of boats, sent in pursuit, is commonly increased; and when the whole of the

---

gaged in piloting the ship, through crowded ice, or for obtaining a more extensive view of the sea around, when looking out for whales. In difficult situations, a master's presence at the mast-head is sometimes required for many hours in succession, when the temperature of the air is from 10° to 20° degrees below the freezing point, or more. It is therefore necessary for the preservation of his health, as well as his comfort, that he should be sheltered from the gale.

boats are sent out, the ship is said to have "a loose fall."

During fine weather, in situations where whales are seen, or where they have recently been seen, or where there is a great probability of any making their appearance, a boat is generally kept in readiness, manned and afloat. If the ship sails with considerable velocity, this boat is towed by a rope astern; but when the ship is pretty still, whether moored to ice, laid to, or sailing in light winds, the "bran boat," as it is called, often pushes off to a little distance from the ship. A boat on watch, commonly lies still in some eligible situation. with all its oars elevated out of the water, but in readiness in the hands of the rowers for immediate use.

The harpooner and boat steerer, keep a careful watch on all sides, while each of the rowers looks out in the direction of his oar. In field fishing, the boats approach the ice with their sterns, and are each of them fastened to it, by means of a boat-hook, or an iron spike with a cord attached, either of which is held by the boat-steerer, and is slipped or withdrawn, the moment a whale appears. There are several rules observed in approaching a whale, as precautions, to prevent, as far as possible, the animal from taking the alarm. As the whale is dull of hearing, but quick of sight, the boat-steerer always endeavours to get behind it; and, in accomplishing this, he is sometimes justified in taking a circuitous route. In calm weather, where guns are not used, the greatest caution is necessary, before a whale can be reached; smooth careful rowing is always requisite, and sometimes sculling is practised.

When it is known that a whale seldom abides longer on the water than two minutes, that it generally remains from five to ten or fifteen minutes under water;* that in this interval it sometimes moves through a space of half a mile or more,— and that the fisher has very rarely, any certain intimation of the place in which it will reappear;— the difficulty and address, requisite to approach sufficiently near, during its short stay on the surface, to harpoon it, will readily be appreciated. It is, therefore, a primary consideration with the harpooner always to place his boat as near as possible to the spot, in which he expects the fish to rise, and he conceives himself successful in the attempt when the fish "comes up within a start," that is, within the distance of about 200 yards. In all cases when a whale that is pursued, has but once been seen, the fisher is considerably indebted to what is called chance for a favourable position. But when the whale has been twice seen, and its change of place, if any, noticed, the harpooner makes the best use of the intimation derived from his observation on its

---

* Before I had particularly minuted the time, that a whale stays on the surface, and remains below, I believed each interval, and especially the former, was much greater than it really is.

apparent motion, and places his boat accordingly; thus he anticipates the fish in its progress, so that when it rises to the surface, there is probability of its being within the favourable precincts of a start.

A whale moving forward at a small distance beneath the surface of the sea, leaves a sure indication of its situation, in what is called an " eddy," having somewhat the resemblance of the " wake," or track of a ship, and in fine calm weather, its change of position is sometimes pointed out by the birds, many of which closely follow it when at the surface, and hover over it when below, whose keener vision can discover it, when it is totally concealed from human eyes. By these indications many whales have been taken.

Whenever a whale lies on the surface of the water, unconscious of the approach of its enemies, the hardy fisher rows directly upon it; and an instant before the boat touches it, buries his harpoon in its back. But if, while the boat is yet at a little distance, the whale should indicate his intention of diving, by lifting his head above its common level, and then plunging it under water, and raising his body until it appear like the large segment of a sphere,—the harpoon is thrown from the hand, or fired from a gun, the former of which, when skilfully practised, is efficient at the distance of eight or ten yards, and the latter at the distance of thirty yards, or upward. The wounded whale, in the surprise and agony of the moment, makes a convulsive effort to escape. Then is the moment of danger. The boat is subjected to the most violent blows from its head, or its fins, but particularly from its ponderous tail, which sometimes sweeps the air with such tremendous fury, that both boat and men are exposed to one common destruction.

The head of the whale is avoided, because it cannot be penetrated with the harpoon; but any part of the body, between the head and tail, will admit of the full length of the instrument, without danger of obstruction. The harpoon, therefore, is always struck into the back, and generally well forward towards the fins, thus affording the chance, when it happens to drag and plough along the back, of retaining its hold during a longer time, than when struck in closer to the tail.

The moment that the wounded whale disappears, or leaves the boat, a jack or flag, elevated on a staff, is displayed; on sight of which, those on watch in the ship, give the alarm, by stamping on the deck, accompanied by a simultaneous and continued shout of " a fall,"* at the sound of this, the sleeping crew are roused, jump from their beds, rush upon deck, with their clothes tied by a string in

---

* The word fall, as well as many others used in the fishery, is derived from the Dutch language. In the original it is written val, implying jump, drop, fall, and is considered as expressive of the conduct of the sailors, when manning the boats, on an occasion requiring extreme dispatch.

their hands, and crowd into the boats, with a temperature of zero. Should a fall occur, the crew would appear upon deck, shielded only by their drawers, stockings, and shirts, or other habiliments in which they sleep. They generally contrive to dress themselves, in part at least, as the boats are lowered down; but sometimes they push off in the state in which they rise from their beds, row away towards the "fast boat," and have no opportunity to clothe themselves for a length of time afterwards. The alarm of a "fall," has a singular effect on the feelings of a sleeping person, unaccustomed to the whale-fishing business. It has often been mistaken as a cry of distress. A landsman in a Hull ship, seeing the crew, on an occasion of a fall, rush upon deck, with their clothes in their hands, when there was no appearance of danger, thought the men were all mad; but, with another individual the effect was totally different. Alarmed with the extraordinary noise; and still more so, when he reached the deck, with the appearance of all the crew seated in the boats in their shirts, he imagined the ship was sinking. He therefore endeavoured to get into a boat himself, but every one of them being fully manned, he was always repulsed. After several fruitless endeavours to gain a place among his comrades, he cried out, with feelings of evident distress, "what shall I do?—Will none of you take me in?"

The first effort of a "fast-fish," or whale that has been struck, is to escape from the boat, by sinking under water. After this, it pursues its course directly downward, or re-appears at a little distance, and swims with great celerity, near the surface of the water, towards any neighbouring ice, among which it may obtain an imaginary shelter; or it returns instantly to the surface, and gives evidence of its agony, by the most convulsive throes, in which its fins and tail are alternately displayed in the air, and dashed into the water with tremendous violence. The former behaviour, however, that is, to dive towards the bottom of the sea, is so frequent, in comparison of any other, that it may be considered as the general conduct of a fast fish.

A whale struck near the edge of any large sheet of ice, and passing underneath it, will sometimes run the whole of the lines out of the boat, in the space of eight or ten minutes of time. This being the case, when the " fast-boat" is at a distance, both from the ship and from any other boat, it frequently happens that the lines are all withdrawn before assistance arrives, and, with the fish, entirely lost. In some cases, however, they are recovered. To retard, therefore, as much as possible, the flight of the whale, it is usual for the harpooner, who strikes it, to cast one, two, or more turns of line round a kind of post called a bollard; which is fixed within ten or twelve inches of the stem of the boat, for the purpose. Such is the friction of the line, when running round the bollard, that it frequently envelopes the harpooner in smoke; and if the wood were not repeatedly wetted, would probably set fire to the boat. During the capture of one whale, a groove is sometimes cut in the bollard, near an inch in depth; and, were it not for a plate of brass, iron, or a block of lignum-vitæ, which covers the top of the stem where the line passes over, it is apprehended that the action of the line on the material of the boat, would cut it down to the water's edge, in the course of one season of successful fishing. The approaching distress of a boat, for want of line, is indicated by the elevation of an oar, in the way of a mast, to which is added a second, a third, or even a fourth, in proportion to the nature of the exigence. The utmost care and attention are requisite, on the part of every person in the boat, when the lines are running out; fatal consequences having been sometimes produced by the most trifling neglect. When the line happens " to run foul," and can not be cleared on the instant, it sometimes draws the boat under water; on which, if no auxiliary boat, or convenient piece of ice be at hand, the crew are plunged into the sea, and are obliged to trust to the buoyancy of their oars, or to their skill in swimming, for supporting themselves on the surface. To provide against such an accident, as well as to be ready to furnish an additional supply of lines, it is usual, when boats are sent in pursuit, for two to go out in company; and when a whale has been struck, for the first assisting boat which approaches, to join the fast-boat, and to stay by it until the fish re-appears. The other boats, likewise, make towards the one carrying a flag, and surround it at various distances, awaiting the appearance of the wounded whale.

On my first voyage to the whale-fishery, such an accident, as above alluded to, occurred. A thousand fathoms of line were already out, and the fast-boat was forcibly pressed against the side of a piece of ice. The harpooner, in his anxiety to retard the flight of the whale, applied too many turns of the line round the bollard, which, getting entangled, drew the boat beneath the ice. Another boat, providentially, was at hand, into which the crew, including myself, who happened to be present, had just time to escape. The whale, with near two miles length of line, was, in consequence of the accident, lost, but the boat was recovered. On a subsequent occasion, I underwent a similar misadventure, but with a happier result; we escaped with a little wetting into an accompanying boat, and the whale was afterwards captured, and the boat with its lines recovered.

When fish have been struck by myself, I have on different occasions estimated their rate of descent. For the first 300 fathoms, the average velocity was usually after the rate of eight to ten miles per hour. In one instance, the third line of 120 fathoms was run out in 61 seconds; that is at the rate of $8\frac{1}{6}$ English miles, or $7\frac{1}{8}$ nautical miles per hour. By the motions of the fast-boat, the simultaneous movements of the whale are estimated. The auxiliary boats, accordingly, take their stations about the situation where the whale, from these motions, may reasonably be expected to appear.

The average stay under water, of a wounded whale, which steadily descends after being struck, according to the most usual conduct of the animal, is about 30 minutes. The longest I ever observed was 56 minutes; but in shallow water, I have been informed, it has sometimes been known to remain an hour and a half at the bottom after being struck, and yet has returned to the surface alive. The greater the velocity, the more considerable the distance to which it descends; and the longer the time it remains under water, so much greater in proportion is the extent of its exhaustion and the consequent facility of accomplishing its capture. Immediately that it reappears, the assisting boats make for the place with their utmost speed, and as they reach it, each harpooner plunges his harpoon into its back, to the amount of three, four, or more, according to the size of the whale, and the nature of the situation. Most frequently, however, it descends for a few minutes after receiving the second harpoon, and obliges the other boats to await its return to the surface, before any further attack can be made. It is afterwards actively plied with lances, which are thrust into its body, aiming at its vitals. At length, when exhausted by numerous wounds and the loss of blood, which flows from the huge animal in copious streams, it indicates the approach of its dissolution, by discharging from its "blowholes," a mixture of blood along with the air and mucus which it usually expires, and finally jets of blood alone. The sea, to a great extent around, is dyed with its blood, and the ice, boats, and men, are sometimes drenched with the same. Its track is likewise marked by a broad pellicle of oil, which exudes from its wounds, and appears on the surface of the sea. Its final capture is sometimes preceded by a convulsive struggle, in which, its tail, reared, whirled, and violently jerked in the air, resounds to the distance of miles. In dying, it turns on its back or on its side; which joyful circumstance is announced by the capturers with the striking of their flags, accompanied by three lively huzzas!

The remarkable exhaustion observed in the first appearance of a wounded whale at the surface, after a descent of 700 or 800 fathoms perpendicular. does not depend on the nature of the wound it has received; for a hundred superficial wounds received from harpoons, could not have the effect of a single lance penetrating the vitals, but is the effect of the almost incredible pressure to which the animal must have been exposed. The surface of the body of a large whale, may be considered as comprising an area of 1540 square feet. This, under the common weight of the atmosphere only, must sustain a pressure of 3,104,640 pounds, or 1386 tons. But at the depth of 800 fathoms, where there is a column of water equal in weight to about 154 atmospheres, the pressure on the animal must be equal to 211,200 tons.* This is a degree of pressure of which we can have but an imperfect conception. It may assist our comprehension, however, to be informed, that it exceeds in weight sixty of the largest ships of the British navy when manned, provisioned, and fitted for a six months cruise.

Every boat fast to a living whale carries a flag, and the ship to which such boats belong, also wears a flag, until the whale is either killed or makes its escape. These signals serve to indicate to surrounding ships the exclusive title of the " fast ship," to the entangled whale, and to prevent their interference, excepting in the way of assistance, in the capture.

A very natural inquiry connected with this subject, is, what is the length of time requisite for cap-

---

* From experiments made with sea-water taken up near Spitzbergen, I find that 35 cubical feet weigh a ton. Now supposing a whale to descend to the depth of 800 fathoms or 4800 feet, which, I believe, is not uncommon, we have only to divide 4800 feet, the length of the column of water pressing upon the whale, by 35 feet, the length of a column of sea-water a foot square, weighing a ton, the quotient 137 1-7, shows the pressure per square foot upon the whale, in tons; which multiplied by 1540, the number of square feet of surface exposed by the animal, affords a product of 211,200 tons, besides the usual pressure of the atmosphere.

turing a whale? This is a question which can only be answered indirectly; for I have myself witnessed the capture of a large whale, which has been effected in twenty-eight minutes; and have also been engaged with another fish which was lost, after it had been entangled about sixteen hours. Instances are well authenticated, in which whales have yielded their lives to the lances of active fishers, within the space of fifteen minutes from the time of being struck; and in cases when fish have been shot with a harpoon-gun, in a still shorter period; while other instances are equally familiar and certain, wherein a whale having gained the shelter of a pack or compact patch of ice, has sustained or avoided every attack upon it, during the space of forty or fifty hours. Some whales have been captured when very slightly entangled with a single harpoon, while others have disengaged themselves, though severely wounded with lances, by a single act of violent and convulsive distortion of the body, or tremendous shake of the tail, from four or more harpoons; in which act, some of the lines have been broken with apparent ease, and the harpoons, to which other lines were attached, either broken or torn out of the body of the vigorous animal. Generally, the speedy capture of a whale depends on the activity of the harpooners, the favourableness of situation and weather, and, in no inconsiderable degree, on the peculiar conduct of the whale attacked. Under the most favourable circumstances, namely, when the fishermen are very active, the ice very open, or the sea free from ice and the weather fine,—the average length of time occupied in the capture of a whale, may be stated as not exceeding an hour.* The general average, including all sizes of fish, and all circumstances of capture, may probably be two or three hours.

The method practised in the capture of whales, under favourable circumstances, is very uniform with all the fishers, both British and foreigners. The only variation observable in the proceedings of the different fishers, consisting in the degree of activity and resolution displayed, in pursuance of the operations of harpooning and lancing the whale, and in the address manifested in improving by any accidental movement of the fish, which may lay it open to an effectual attack,—rather than in any thing different or superior in the general method of conducting the fishery. It is true, that with some the harpoon-gun is much valued, and used with advantage, while with others, it is held in prejudiced aversion; yet, as this difference of opinion affects

---

* Twelve large whales, taken in different voyages, memoranda of whose capture I have preserved, were killed, on an average, in 67 minutes. The shortest time expended in the taking of one of the twelve whales, was 28 minutes, the longest time 2 hours. One of these whales we believed, descended 670 fathoms perpendicular; another 720; and a third 750, one descended 1400 fathoms obliquely, and another 1600 fathoms.

only the first attack and entanglement of the whale, the subsequent proceedings with all the fishers, may still be said to be founded on equal and unanimous principles. Hence, the mode described in the preceding pages, of conducting the fishery for whales under favourable circumstances, may be considered as the general plan pursued by all the fishers of all the ports of Britain, as well as those of the nations who resort to Spitzbergen. Neither is there any difference in the plan of attack, or mode of capture between fish of large size, and those of lesser growth; the proceedings are the same, but, of course, with the smaller whales less force is requisite; though it sometimes happens, that the trouble attached to the killing of a very small whale, exceeds that connected with the capture of one of the largest individuals. The progress or flight of a large whale can not be restrained; but that of an under size fish may generally be confined within the limits of 400 to 600 fathoms of line. A full grown fish generally occupies the whole, or nearly the whole, of the boats belonging to one ship in its capture; but three, four, or sometimes more small fish, have been killed at the same time, by six or seven boats. It is not unusual for small whales to run downward, until they exhaust themselves so completely, that they are not able to return to the surface, but are suffocated in the water. As it is requisite that a whale that has been drowned should be drawn up by the line, which is a tedious and troublesome operation, it is usual to guard against such an event by resisting its descent with a light strain on the line, and also by hauling upon the line, the moment its descent is stopped, with a view of irritating the wound, and occasioning such a degree of pain, as may induce it to return to the surface, where it can be killed and secured without further trouble. Seldom more than two harpoons are struck into an under size whale.

The ease with which some whales are subdued, and the slightness of the entanglement by which they are taken, is truly surprising; but with others it is equally astonishing, that neither line nor harpoon, nor any number of each, is sufficiently strong to effect their capture. Many instances have occurred where whales have escaped, from four, five, or even more harpoons, while fish, equally large, have been killed through the medium of a single harpoon. Indeed, whales have been taken in consequence of the entanglement of a line, without any harpoon at all; though, when such a case has occurred, it has evidently been the result of accident. The following instances are in point.

A whale was struck from one of the boats of the ship Nautilus, in Davis's Straits. It was killed, and as is usual after the capture, it was disentangled of the line connected with the "first fast-boat," by dividing it at the splice of the foreganger, within eight or nine yards of the harpoon. The crew of the boat from which the fish was first struck, in the meantime were employed in heaving in the lines, by means of a winch fixed in the boat for the purpose, which they progressively effected for some time. On a sudden, however, to their great astonishment, the lines were pulled away from them, with the same force and violence, as by a whale when first struck. They repeated their signal, indicative of a whale being struck; their shipmates flocked towards them, and while every one expressed a similar degree of astonishment with themselves, they all agreed that a fish was fast to the line. In a few minutes, they were agreeably confirmed in their opinion, and relieved from suspense, by the rising of a large whale close by them, exhausted with fatigue, and having every appearance of a fast-fish. It permitted itself to be struck by several harpoons at once, and was speedily killed. On examining it after death, for discovering the cause of such an interesting accident, they found the line, belonging to the above mentioned boat, in its mouth, where it was still firmly fixed by the compression of its lips. The occasion of this happy and puzzling accident, was therefore solved;—the end of the line, after being cut from the whale first killed, was in the act of sinking in the water; the fish in question, engaged in feeding, was advancing with its mouth wide open, and accidentally caught the line between its extended jaws;—a sensation so utterly unusual as that produced by the line, had induced it to shut its mouth and grasp the line, which was the cause of its alarm, so firmly between its lips, as to produce the effect just stated. This circumstance took place many years ago, but a similar one occurred in the year 1814.

A harpooner, belonging to the Prince of Brazil, of Hull, had struck a small fish. It descended, and remained for some time quiet, and at length appeared to be drowned. The strain on the line being then considerable, it was taken to the ship, with a view of heaving the fish up. The force requisite for performing this operation, was extremely various; sometimes, the line came in with ease, at others, a quantity was withdrawn with great force and rapidity. As such, it appeared evident that the fish was yet alive. The heaving, however, was persisted in, and after the greater part of the lines had been drawn on board, a dead fish appeared at the surface, secured by several turns of the line round its body. It was disentangled with difficulty, and was confidently believed to be the whale they had struck. But when the line was cleared from the fish, it proved to be merely the "bight," for the end still hung perpendicularly downward. What was then their surprise to find that it was still pulled away with considerable force. The capstern was again resorted to, and shortly afterwards, they hove up, also dead, the fish originally struck, with the harpoon still fast. Hence it appeared, that the fish first drawn up, had got accidentally entangled with the line, and in its struggles to escape, had still further involved itself, by winding the line repeatedly round its body. The first fish entangled, as was suspected, had long been dead; and it was this lucky interloper, that occasioned the jerks and other singular effects observed on the line.

Alterations produced in the manner of conducting the Fishery, by peculiar Circumstances of Situation and Weather.

Hitherto I have only attempted to describe the method adopted for the capture of whales, under favourable circumstances, such as occur in open water or amongst open ice in fine weather; as, however, this method is subject to various alterations, when the situation and circumstances are peculiar, I shall venture a few remarks on the subject.

1. *Pack-fishing.*—The borders of close packs of drift ice are frequently a favourite resort of large whales. To attack them in such a situation, subjects the fisher to great risk in his lines and boats, as well as uncertainty in effecting their capture. When a considerable swell prevails on the borders of the ice, the whales, on being struck, will sometimes recede from the pack, and become the prize of their assailers; but most generally flee to it for shelter, and frequently make their escape. To guard against the loss of lines as much as possible, it is pretty usual either to strike two harpoons from different boats at the same moment, or to bridle the lines of

a second boat upon those of the boat from which the fish is struck. This operation consists in fixing other lines to those of the fast-boat at some distance from the harpoon, so that there is only one harpoon and one line immediately attached to the fish, but the double strength of a line from the place of their junction to the boats. Hence, should fish flee directly into the ice, and proceed to an inaccessible distance, the two boats, bearing an equalstra in on each of their lines, can at pleasure draw the harpoon, or break the single part of the line immediately connected with it, and in either case, secure themselves against any considerable loss.

When a pack, for its compactness, prevents boats from penetrating, the men travel over the ice, leaping from piece to piece, in pursuit of the entangled whale. In this pursuit, they carry lances with them and sometimes harpoons, with which, whenever they can approach the fish, they attack it, and if they succeed in killing it, they drag it towards the exterior margin of the ice, by means of the line fastened to the harpoon with which it was originally struck. In such cases, it is generally an object of importance to sink it beneath the ice; for effecting which purpose, each lobe of the tail is divided from the body, excepting a small portion of the edge, from which it hangs pendulous in the water. If it still floats, bags of sand, kedges or small cannon, are suspended by a block on the bight of the line, wherewith the buoyancy of the dead whale is usually overcome. It then sinks, and is easily hauled out by the line into the open sea.

To particularize all the variety of pack fishing, arising from winds and weather, size of the fish, state and peculiarities of the ice, &c. would require more space than the interest of the subject, to general readers, would justify. I shall, therefore, only remark, that pack-fishing is, on the whole, the most troublesome and dangerous of all others;—that instances have occurred of fish having been entangled during 40 or 50 hours, and have escaped after all;—and that other instances are remembered, of ships having lost the greater part of their stock of lines, several of their boats, and sometimes, though happily, less commonly, some individuals of their crews.

2. *Field-fishing.*—The fishery for whales, when conducted at the margin of those wonderful sheets of solid ice, called fields, is, when the weather is fine, and the refuge for ships secure, of all other situations which the fishery of Greenland presents, the most agreeable and sometimes the most productive. A fish struck at the margin of a large field of ice, generally descends obliquely beneath it, takes four to eight lines from the fast-boat, and then returns exhausted to the edge. It is then attacked in the usual way, with harpoons and lances, and is easily killed. There is one evident advantage in field-fishing, which is this: When the fast-boat lies at the edge of a firm unbroken field, and the

line proceeds in an angle beneath the ice, the fish must necessarily arise somewhere in a semicircle, described from the fast-boat as a centre, with a sweep not exceeding the length of the lines out; but most generally it appears in a line extending along the margin of the ice, so that the boats, when dispersed along the edge of the field, are effectual and as ready for promoting the capture as twice the number of boats or more, when fishing in open situations; because, in open situations, the whale may arise any where within a circle, instead of a semicircle, described by the length of the lines withdrawn from the fast-boat; whence, it frequently happens, that all the attendant boats are disposed in a wrong direction, and the fish recovers its breath, breaks loose, and escapes before any of them can secure it by a second harpoon. Hence, when a ship fishes at a field with an ordinary crew, and six or seven boats, two of the largest fish may be struck at the same time with every prospect of success; while the same force attempting the capture of two at once in an open situation, will, not unfrequently, occasion the loss of both. There have indeed been instances of a ship's crew, with seven boats, striking at a field, six fish at the same time, and of success in killing the whole. Generally speaking, six boats at a field are capable of performing the same execution as near twice that number in open situations. Besides, fields sometimes afford an opportunity of fishing, when in any other situation there can be little or no chance of success, or, indeed, when to fish elsewhere is utterly impracticable. Thus calms, storms, and fogs, are great annoyances in the fishery in general, and frequently prevent it altogether; but at fields the fishery goes on under any of these disadvantages. As there are several important advantages attending the fishery at fields, so, likewise, there are some serious disadvantages, chiefly relating to the safety of the ships engaged in the occupation. The motions of fields are rapid, various, and unaccountable, and the power with which they approach each other, and squeeze every resisting object, immense,—hence occasionally vast mischief is produced, which it is not always in the power of the most skilful and attentive master to foresee and prevent.

Such are the principal advantages and disadvantages of fields of ice to the whale-fishers. The advantages, however, as above enumerated, though they extend to large floes, do not extend to small floes, or to such fields, how large soever they may be, as contain tracks or holes, or are filled up with thin ice on the interior. Large and firm fields are the most convenient, and likewise the most advantageous for the fishery; the most convenient, because the whales, unable to breathe beneath a close extensive field of ice, are obliged to make their appearance again above water among the boats on the look out; and they are most advantageous, because not only the most fish commonly resort to them, but a greater number can be killed with less force, and in a short space of time, than in any other situation. Thin fields, or fields full of holes, being by no means advantageous to fish by, are usually avoided, because a "fast-fish," retreating under such a field, can respire through the holes in the centre as conveniently as on the exterior; and a large fish usually proceeds from one hole to another, and if determined to advance can not possibly be stopped. In this case, all that can be done is, to break the line or draw the harpoon out. But when the fish can be observed "blowing," in any of the holes in a field, the men travel over the ice and attack it with lances, pricking it over the nose, to endeavour to turn it back. This scheme, however, does not always answer the expectations of the fishers, as frequently the fear of his enemies acts so powerfully on the whale, that he pushes forward to the interior to his dying moment. When killed, the same means are used as in pack-fishing, to sink it, but they do not always succeed; for the harpoon is frequently drawn out, or the line broken in the attempt. If, therefore, no attempt to sink the fish avails, there is scarcely any other practicable method of making prize of it, (unless when the ice happens to be so thin that it can be broken with a boat, or a channel readily cut in it with an ice saw,) than cutting the blubber away, and dragging it piece by piece across the ice to the vessel, which requires immense labour and is attended with vast loss of time. Hence, we have a sufficient reason for avoiding such situations whenever fish can be found elsewhere. As connected with this subject, I can not pass over a circumstance which occurred within my own observation, and which excited my highest admiration.

On the 8th of July, 1813, the ship Esk lay by the edge of a large sheet of ice, in which were several thin parts, and some holes. Here a fish being heard blowing, a harpoon, with a line connected to it, was conveyed across the ice, from a boat on guard, and the harpooner succeeded in striking the whale, at the distance of 350 yards from the verge. It dragged out ten lines, (2400 yards,) and was supposed to be seen blowing in different holes in the ice. After some time, it happened to make its appearance on the exterior, when a harpoon was struck at the moment it was on the point of proceeding again beneath. About a hundred yards from the edge, it broke the ice where it was a foot in thickness, with its crown, and respired through the opening. It then determinately pushed forward, breaking the ice as it advanced, in spite of the lances constantly directed against it. It reached at length a kind of bason in the field, where it floated on the surface of the water, without any incumbrance from ice. Its back being fairly exposed, the harpoon, struck from the boat on the outside, was observed to be so slightly entangled, that it was ready to drop out. Some of the officers lamented this circumstance, and expressed a wish that the harpoon were better fast, observing, at the same time, that if it should slip out, the fish would either be lost, or they would be under the necessity of flensing it where it lay, and of dragging the pieces of blubber over the ice to the ship; a kind and degree of labour, every one was anxious to avoid. No sooner was the wish expressed, and its importance made known, than one of the sailors, a smart and enterprising fellow, stept forward and volunteered his services to strike it better in. Not at all intimidated by the surprise which was manifested in every countenance, by such a bold proposal, he pulled out his pocket-knife, leapt upon the back of the living whale, and immediately cut the harpoon out. Stimulated by this courageous example, one of his companions proceeded to his assistance. While one of them hauled upon the line and held it in his hands, the other set his shoulder against the extremity of the harpoon, and though it was without a stock, he contrived to strike it again into the fish more effectually than it was at first; the fish was in motion before they finished. After they got off its back, it advanced a considerable distance, breaking the ice all the way, and survived this uncommon treatment, ten or fifteen minutes. This admirable act was an essential benefit. The fish fortunately sunk spontaneously, after being killed, on which it was hauled out to the edge of the ice by the line, and secured without further trouble. It proved a stout whale, and a very acceptable prize.

When a ship approaches a considerable field of ice, and finds whales, it is usual to moor to the leeward side of it, from which the adjoining ice usually first separates. Boats are then placed on watch, on each side of the ship, and stationed at intervals of 100 or 150 yards from one another, along the edge of the ice. Hence, if a fish arises any where between the extreme boats, it seldom escapes unhurt. It is not uncommon for a great number of ships to moor to the same sheet of ice. When the whale fishery of the Hollanders was in a flourishing state, above 100 sail of ships might sometimes be seen moored to the same field of ice, each having two or more boats on watch. The field would, in consequence, be so nearly surrounded with boats, that it was almost impossible for a fish to rise near the verge of the ice, without being within the limits of a start of some of them.

3. *Fishing in crowded ice, or in open packs.*—In navigably open drift ice, or among small detached streams and patches, either of which serve in a degree to break the force of the sea, and to prevent any considerable swell from arising, we have a situation, which is considered as one of the best possible for conducting the fishery in; consequently, it comes under the same denomination as those favourable situations, in which I have first attempted to describe the proceedings of the fishers in killing the whale. But the situation I now mean to refer to, is, when the ice is crowded and nearly close; so

close, indeed, that it scarcely affords room for boats to pass through it, and by no means sufficient space for a ship to be navigated among it. This kind of situation occurs in somewhat open packs, or in large patches of crowded ice, and affords a fair probability of capturing a whale, though it is seldom accomplished without a considerable degree of trouble. When the ice is very crowded, and the ship can not sail into it with propriety, it is usual to seek out for a mooring to some large mass of ice, if such can be found, extending two or three fathoms or more, under water. A piece of ice of this kind, is capable not only of holding the ship "head-to-wind," but also to windward of the smaller ice. The boats then set out in chase of any fish which may be seen; and when one happens to be struck, they proceed in the capture in a similar manner as when in more favourable circumstances, excepting so far as the obstruction which the quality and arrangement of the ice may offer, to the regular system of proceeding. Among crowded ice, for instance, the precise direction pursued by the fish is not easily ascertained, nor can the fish itself be readily discovered on its first arrival at the surface, after being struck, on account of the elevation of the intervening masses of ice, and the great quantity of line it frequently takes from the fast-boat. Success in such a situation, depends on the boats being spread widely abroad, and on a judicious arrangement of each boat: on a keen look out on the part of the harpooners in the boat, and on their occasionally taking the benefit of a hummock of ice, from the elevation of which the fish may sometimes be seen "blowing" in the interstices of the ice; on pushing or rowing the boats with the greatest imaginable celerity, towards the place where the fish may have been seen; and, lastly, on the exercise of the highest degree of activity and despatch in every proceeding.

If these means be neglected, the fish will generally have taken his breath, renewed its strength, and removed to some other quarter, before the arrival of the boats; and it is often remarked, that if there be one part of the ice more crowded or more difficult of access than another, it commonly retreats thither for refuge. In such cases, the sailors find much difficulty in getting to it with their boats, having to separate many pieces of ice before they can pass through between them. But when it is not practicable to move the pieces, and when they can not travel over them, they must either drag the boats across the intermediate ice, or perform an extensive circuit, before they can reach the opposite side of the close ice, into which the whale has retreated.

A second harpoon, in this case, as indeed in all others, is a material point. They proceed to lance whenever a second harpoon is struck, and strike more harpoons as the auxiliary boats progressively arrive at the place.

When the fish is killed, it is often at a distance from the ship, and so circumstanced, that the ship can not get near it. In such cases, the fish must be towed by the boats to the ship; an operation which, in crowded ice, is most troublesome and laborious.

4. *Bay-ice fishing.*—Bay-ice constitutes a situation, which, though not particularly dangerous, is yet, on the whole, one of the most troublesome in which whales are killed. In sheets of bay-ice, the whales find a very effectual shelter; for so long as the ice will not "carry a man," they can not be approached with a boat, without producing such a noise, as most certainly warns them of the intended assault. And if a whale, by some favourable accident, were struck, the difficulties of completing the capture are always numerous, and sometimes prove insurmountable. The whale having free locomotion beneath the ice, the fishers pursue it under great disadvantage. The fishers can not push their boats towards it but with extreme difficulty; while the whale, invariably warned by the noise of their approach, possesses every facility for avoiding its enemies. In the year 1813, I adopted a new plan of fishing in bay-ice, which was attended with the most fortunate result. The ship under my command (the Esk of Whitby) was frozen into a sheet of bay-ice, included in a triangular space, formed by massive fields and floes. Here a number of small whales were seen sporting around us, in

every little hole or space in the bay-ice, and occasionally they were observed to break through it, for the purpose of breathing. In various little openings, free of ice, near the ship, few of which were twenty yards in diameter, we placed boats; each equipped with a harpoon and lines, and directed by two or three men. They had orders to place themselves in such a situation, that if a fish appeared in the same opening, they could scarcely fail of striking it. Previous to this, I provided myself with a pair of ice shoes, consisting of two pieces of thin deal, six feet in length, and seven in breadth. They were made very thin at both ends; and, in the centre of each, was a hollow place exactly adapted for the reception of the sole of my boot, with a loop of leather for confining the toes. I was thus enabled to retain the ice shoes pretty firmly to my feet, when required, or, when I wished it, of disengaging them in a moment. Where the ice was smooth, it was easy to move in a straight line; but, in turning, I found a considerable difficulty, and required some practice before I could effect it, without falling. I advanced, with tolerable speed, when the ice was level on the surface, by sliding the shoes alternately forward, but when I met with rough hilly places, I experienced great inconvenience. When, however, the rough places happened to consist of strong ice, which generally was the case, I stepped out of my ice shoes, until I reached a weaker part. Equipped with this apparatus, I travelled safely over ice which had not been frozen above twenty-four hours, and which was incapable of supporting the weight of the smallest boy in the ship.

Whenever a fish was struck, I gave orders to the harpooner, in running the lines, to use every means of drowning it; the trouble of hauling it up, under the circumstances in which the ship was placed, being a matter of no consideration. This was attempted, by holding a steady tight strain on the line, without slacking it or jerking it unnecessarily, and by forbearing to haul at the line when the fish was stopped. By this measure, one fish, the stoutest of the three we got, was drowned. When others were struck, and the attempt to drown them failed, I provided myself with a harpoon; and, observing the direction of the line, travelled towards the place where I expected the fish to rise. A small boat was launched, more leisurely, in the same direction, for my support; and whenever the ice in my track was capable of supporting a man, assistance was afforded me in dragging the line. When the wounded fish appeared, I struck my harpoon through the ice, and then, with some occasional assistance, proceeded to lance it, until it was killed. At different times the fish rose beneath my feet, and broke the ice on which I stood; on one occasion, when the ice was fortunately more than usually strong, I was obliged to leave my ice-shoes and skip off. In this way we captured three fish, and took their produce on board, while several ships near us made not the least progress in the fishery. After they were killed, we had much trouble in getting them to the ship, but as we could not employ ourselves to advantage in any other way, we were well satisfied with the issue. This part of the business, however, I could not effect alone, and all hands who were occasionally employed in it, broke through the ice. Some individuals broke in two or three times, but no serious accident ensued. As a precaution, we extended a rope from man to man, which was held in the hands of each in their progress across the ice, and which served for drawing those out of the water who happened to break through. Sometimes ten or a dozen of them would break in at once; but so far was such an occurrence from exciting distress, that each of their companions indulged a laugh at their expense, notwithstanding they, probably, shared the same fate a minute or two afterwards. The shivering tars were, in general, amply repaid for the drenching they had suffered, by a dram of spirits, which they regularly received on such occasions. I have seen instances, indeed, of sailors having voluntarily broken through the ice, for the mere purpose of receiving the usual precious beverage.

5. *Fishing in Storms.*—Excepting in situations sheltered from the sea by ice, it would be alike useless and presumptuous to attempt to kill whales during a storm. Cases, however, occur, wherein

fish that were struck during fine weather, in winds which do not prevent the boats from plying about, remain entangled, but unsubdued, after the commencement of a storm. Sometimes the capture is completed, at others the fishers are under the necessity of cutting the lines, and allowing the fish to escape. Sometimes, when they have succeeded in killing it, and in securing it during the gale with a hawser to the ship, they are enabled to make a prize of it on the return of moderate weather; at others, after having it to appearance secured, by means of a sufficient rope, the dangerous proximity of a lee pack constrains them to cut it adrift and abandon it, for the preservation of their vessel. After thus being abandoned, it becomes the prize of the first who gets possession of it, though it be in the face of the original capturers. A storm commencing while the boats are engaged with an entangled fish, sometimes occasions serious disasters. Generally, however, though they suffer the loss of the fish, and perhaps some of their boats and materials, yet the men escape with their lives.

6. *Fishing in Foggy Weather.*—The fishery in storms, in exposed situations, can never be voluntary, as the case only happens when a storm arises subsequent to the time of a fish being struck; but in foggy weather, though occasionally attended with hazard, the fishery is not altogether impracticable. The fogs which occur in the icy regions in June and July, are generally dense and lasting. They

are so thick, that objects can not be distinguished at the distance of 100 or 150 yards, and frequently continue for several days without attenuation. To fish with safety and success, during a thick fog, is, therefore, a matter of difficulty, and of still greater uncertainty. When it happens that a fish conducts itself favourably, that is, descends almost perpendicularly, and on its return to the surface remains nearly stationary, or moves round in a small circle, the capture is usually accomplished without hazard or particular difficulty; but when, on the contrary, it proceeds with any considerable velocity in a horizontal direction, or obliquely downwards, it soon drags the boats out of sight of the ship, and shortly so confounds the fishers in the intensity of the mist, that they lose all traces of the situation of their vessel. If the fish, in its flight, draws them beyond the reach of the sound of a bell, or a horn, their personal safety becomes endangered; and if they are removed beyond the sound of a cannon, their situation becomes extremely hazardous, especially if no other ships happen to be in the immediate vicinity. Meanwhile, whatever may be their imaginary or real danger, the mind of their commander must be kept in the most anxious suspense until they are found; and whether they may be in safety, or near perishing with fatigue, hunger and cold, so long as he is uncertain, his anxiety must be the same. Hence it is, that feelings excited by uncertainty are frequently more violent and distressing than those produced by the actual knowledge of the truth.

Keen and vigilant observance of the direction pursued by the whale, on the part of the persons engaged in the chase, and a corresponding observance of the same by their commanders, can be the only means within the power of each party of securing the ship and boats from being widely separated, without knowing what course to pursue for re-uniting them. Much depends on the people employed in the boats using every known means to arrest the progress of the fish in its flight, by attacking it with the most skilful, active, and persevering efforts, until it is killed; and then, as speedily as may be, of availing themselves of the intimation they may possess relative to the position of the ship, for the purpose of rejoining her. But as their knowledge of the direction of their movements generally depends on the wind, unless they happen to be provided with a compass, and have attentively marked their route by its indications, any change in the direction of the wind, must be attended with serious consequences.

Anecdotes illustrative of Peculiarities in the Whale Fishery.

On the 25th of June, 1812, one of the harpooners belonging to the Resolution, of Whitby, under my command, struck a whale by the edge of a small floe of ice. Assistance being promptly afforded, a second boat's lines were attached to those of the fast-boat; in a few minutes after the harpoon was discharged. The remainder of the boats proceeded to some distance, in the direction the fish seemed to have taken. In about a quarter of an hour the fast-boat, to my surprise, again made a signal for lines. As the ship was then within five minutes sail, we instantly steered towards the boat, with the view of affording assistance, by means of a spare boat we still retained on board. Before we reached the place, however, we observed four oars displayed in signal order, which, by their number, indicated a most urgent necessity for assistance. Two or three men were, at the same time, seated close by the stern, which was considerably elevated, for the purpose of keeping it down,—while the bow of the boat, by the force of the line, was drawn down to the level of the sea,—and the harpooner, by the friction of the line round the bollard, was enveloped in smoky obscurity. At length, when the ship was scarcely 100 yards distant, we perceived preparations for quitting the boat. The sailors' *pea*-jackets were cast upon the adjoining ice,—the oars were thrown down,—the crew leaped overboard,—the bow of of the boat was buried in the water,—the stern rose perpendicular, and then majestically disappeared. The harpooner having caused the end of the line to be fastened to the iron ring at the boat's stern, was the means of its loss;* and a *tongue* of the ice, on which was a depth of several feet of water, kept the boat, by the pressure of the line against it, at such a considerable distance as prevented the crew from leaping upon the floe. Some of them were, therefore, put to the necessity of swimming for their preservation, but all of them succeeded in scrambling upon the ice, and were taken aboard of the ship a few minutes afterwards. I may here observe, that it is an uncommon circumstance for a fish to take more than two boats' lines in such a situation;—none of our harpooners, therefore, had any scruple in leaving the fast-boat, never suspecting, after it had received the assistance of one boat, with six lines or upwards, that it would need any more.

Several ships being about us, there was a possibility that some person might attack and make a prize of the whale, when it had so far escaped us, that we no longer retained any hold of it; as such, we set all sail the ship could safely sustain, and worked through several narrow and intricate channels in the ice, in the direction I observed the fish had retreated. After a little time, it was de-

---

* " Giving a whale the boat" as the voluntary sacrifice of a boat is termed, is a scheme not unfrequently practised by the fisher, when in want of line. By submitting to this risk, he expects to gain the fish, and still has the chance of recovering his boat and its materials. It is only practised in open ice or at fields.

scried by the people in the boats, at a considerable distance to the eastward; a general chase immediately commenced, and in the space of an hour three harpoons were struck. We now imagined the fish was secure, but our expectations were premature. The whale resolutely pushed beneath a large floe that had recently been broke to pieces by the swell, and soon drew all the lines out of the second fast-boat, the officer of which, not being able to get any assistance, tied the end of his line to a hummock of ice, and broke it. Soon afterwards, the other two boats, still *fast*, were dragged against the broken floe, when one of the harpoons drew out. The line of only one boat, therefore, remained fast to the fish, and with six or eight lines out, was dragged forward into the shattered floe with astonishing force. Pieces of ice, each of which was sufficiently large to have answered the purpose of mooring a ship, were wheeled about by the strength of the whale; and such was the tension and elasticity of the line, that whenever it slipped clear of any mass of ice, after turning it round, into the space between any two adjoining pieces, the boat and its crew flew forward through the creek, with the velocity of an arrow, and never failed to launch several feet upon the first mass of ice that it encountered.

While we scoured the sea, around the broken floe with the ship, and while the ice was attempted in vain by the boats, the whale continued to press forward in an easterly direction towards the sea.

At length, when fourteen lines (about 1680 fathoms) were drawn from the fourth fast-boat, a slight entanglement of the line broke it at the stem. The fish then again made its escape, taking along with it a boat and twenty-eight lines. The united length of the lines was 6720 yards, or upwards of $3\frac{3}{4}$ English miles; value, with the boat, above 150 pounds sterling.

The obstruction of the sunken boat, to the progress of the fish, must have been immense; and that of the lines likewise considerable; the weight of the lines alone, being 35 hundred weight.

So long as the fourth fast-boat, through the medium of its lines, retained its hold of the fish, we searched the adjoining sea with the ship in vain; but, in a short time after the line was divided, we got sight of the object of pursuit, at the distance of near two miles to the eastward of the ice and boats, in the open sea. One boat only with lines, and two empty boats, were reserved by the ship. Having, however, fortunately fine weather, and a fresh breeze of wind, we immediately gave chase under all sails; though, it must be confessed, with the insignificant force by us, the distance of the fish, and the rapidity of its flight considered, we had but very small hopes of success. At length, after pursuing it five or six miles, being at least nine miles from the place where it was struck, we came up with it, and it seemed inclined to rest after its extraordinary exertions. The two dismantled or empty boats having been furnished with two lines each, (a very inadequate supply,) they, together with the one in a good state of equipment, now made an attack upon the whale. One of the harpooners made a blunder; the fish saw the boat, took the alarm, and again fled. I now supposed it would be seen no more; nevertheless, we chased nearly a mile in the direction I imagined it had taken, and placed the boats, to the best of my judgment, in the most advantageous situations. In this case we were extremely fortunate. The fish rose near one of the boats, and was immediately harpooned. In a few minutes, two more harpoons entered its back, and lances were plied against it with vigour and success. Exhausted by its amazing exertions to escape, it yielded itself at length to its fate, received the piercing wounds of the lances without resistance, and finally died without a struggle. Thus terminated with success, an attack upon a whale, which exhibited the most uncommon determination to escape from its pursuers, seconded by the most amazing strength of any individual whose capture I ever witnessed. After all, it may seem surprising, that it was not a particularly large individual; the largest lamina of whalebone only measuring nine feet six inches, while those affording twelve feet bone are not uncommon.* The quantity of line

---

* It has been frequently observed, that whales of this size are the most active of the species; and those of very large growth are, in general, captured with less trouble.

withdrawn from the different boats engaged in the capture, was singularly great. It amounted, altogether, to 10,440 yards,* or nearly six English miles. Of these, thirteen new lines lost, together with the sunken boat; the harpoon connecting them to the fish, having dropt out before the whale was killed.

After having taken a large circuit with the ship Esk in the open sea in search of whales, we saw two or three individuals, when at the distance of about twenty miles from the middle hook of the Foreland.† The weather was fine and no ice in sight. A boat was despatched towards one of the fish we saw, which was immediately struck. The men

---

* The following is a correct statement of the quantity of lines withdrawn from each of the fast-boats, viz.

			Yards.
From the first fast-boat 13 new lines, (the whole of which, together with the boat, were lost,) harpoon drew,			3120
From the second fast-boat $6\frac{1}{4}$ lines; line broke,			1560
——— third do $3\frac{1}{2}$ lines; harpoon drew,			840
——— fourth do 14 lines; line broke,			3360
——— fifth do $\frac{1}{2}$ line; harpoon drew,			120
——— sixth do $2\frac{1}{2}$ lines, do do			600
——— seventh do $2\frac{1}{2}$ lines, do do			600
——— eighth do 1 line, do do			240
		Total in yards	10,440

† Charles Island, lying parallel to the west side of Spitzbergen, is usually denominated the Foreland; the middle Hook is a remarkable ridge of mountains near the middle of the Island.

were already considerably fatigued, having been employed immediately before in the arduous operation hereafter to be described, called *making off;* but, of course, proceeded in the boats to the chase of the fast fish. It made its appearance before they all had left the ship. Three boats then approached it, unluckily at the same moment. Each of them so incommoded the other, that no second harpoon could be struck. The fish then took the alarm and ran off towards the east, at the rate of about four miles per hour; some of the boats gave chase and others took hold of the fast-boat and were towed by it to windward. When two boats, by great exertions on the part of their crews, had got very near the fish, and the harpooners were expecting every moment to be able to strike it, it suddenly shifted its course under water, and in a few minutes discovered itself in a southerly direction, at least half a mile from any boat. It then completed a circuit round the fast-boat, with the sweep of nearly a mile as a radius, and though followed in its track by the boats, it dived before any of them got near it, and evaded them completely. When it appeared again, it was at least half a mile to windward of any of them, and then continued arduously advancing in the same direction. At various times during the pursuit, the boats having the most indefatigable crews, reached the fish within ten or fifteen yards, when, apparently aware of their design, it immediately sunk and changed its course; so that it invariably made its next appearance in a quarter where no boats were near.

The most general course of the whale being to windward, it soon withdrew all the boats many miles from the ship, notwithstanding our utmost efforts, under a press of sail to keep near them.

After six or seven hours pursuit without success, the sky became overcast, and we were suddenly enveloped for some time in the obscurity of a thick fog. In this interval the boats were all moored to the fast-boat, the men being fearful of being dispersed; but on the disappearing of the fog, the pursuit was recommenced with renewed vigour. Still the harpooners were not able to succeed. They were now convinced of the necessity of using every measure to retard the flight of the fish. For this purpose they slacked out nine lines, a weight in air of 11cwt., while the crew of the fast-boat endeavoured farther to retard its progress, by holding their oars firmly in the water, as if in the act of backing the boat astern. But this plan did not succeed. They then lashed two or three boats with their sides to the stern of the fast-boat, and these were dragged broadside first, with little diminished velocity for some time. But the fish at length feeling the impediment, suddenly changed its course, and again disappointed the people in two of the boats which had got extremely near it.

Several times the harpooners seized their weapons and were on the point of launching them at the fish, when in an instant it shot from them with singular velocity and disappeared. In this way the chase was continued for fourteen hours, when the fish again turned to leeward. But the men, exhausted by such continued exertion, together with the hard labour to which they had been previously subjected, at the same time being without meat or drink, and sparingly sheltered from the inclemency of the weather by clothes drenched in oil, were incapacitated from taking advantage of the only chance they had ever had of success from the beginning of the chase.

By this time we had reached the boats with the ship. The wind had increased to a gale, and a considerable sea had arisen. We had no hope therefore of success. As however we could not possibly recover the lines at this time, stormy as the weather was, we applied a cask as a buoy to support them, and moored an empty boat having a jack flying in it, to the cask with the intention of keeping near it during the storm, and with the expectation of recovering our lines, and a faint hope likewise of gaining the fish after the termination of the gale. The boat was then abandoned. We made an attempt to keep near the boat with the ship, but the increasing force of the gale, drove us in spite of every effort to leeward. On the first cessation we made all sail, and plyed towards the boat; succeeded in finding it, recovered boat and line, but lost the whale.

On the 28th of May, 1817, the Royal Bounty, of Leith, Captain Drysdale, fell in with a great number of whales in the latitude of 77° 25′ N., and longitude 5° or 6° E. Neither ice nor land was in sight, nor was there supposed to be either the one or the other within 50 or 60 miles. A brisk breeze of wind prevailed, and the weather was clear. The boats were therefore manned and sent in pursuit. After a chase of about five hours, the harpooner commanding a boat, who, with another in company, had rowed out of sight of the ship, struck one of the whales. This was about 4 A. M. of the 29th. The captain supposing, from the long absence of the two most distant boats, that a fish had been struck, directed the course of the ship towards the place where he had last seen them, and about 8 A. M. he got sight of a boat which displayed the signal for being fast. Some time afterwards, he observed the other boat approach the fish, a second harpoon struck, and the usual signal displayed. As, however, the fish dragged the two boats away with considerable speed, it was mid-day before any assistance could reach them. Two more harpoons were then struck, but such was the vigour of the whale, that although it constantly dragged through the water from four to six boats, together with 1600 fathoms of line, which it had drawn out of the different boats, yet it pursued its flight nearly as fast as a boat could row; and such was the terror that it manifested on the approach of its enemies, that whenever a boat passed beyond its tail, it invariably dived. All their endeavours to lance it, were therefore in vain. The crews of the loose boats, being unable to keep pace with the fish, caught hold of and moored themselves to the fast-boats, and for some hours afterwards, *all hands* were constrained to sit in idle impatience, waiting for some relaxation in the speed of the whale. Its most general course had hitherto been to windward, but a favourable change taking place, enabled the ship, which had previously been at a great distance, to join the boats at 8 P. M. They succeeded in taking one of the lines to the ship, which was made fast to the ship, with a view of retarding its flight. They then furled the top-gallant-sails, and lowered the top-sails; but after supporting the ship a few minutes head to wind, the wither of the harpoon *upset*, or twisted aside, and the instrument was disengaged from its grasp. The whale immediately set off to windward, with increased speed, and it required an interval of three hours before the ship could again approach it. Another line was then taken on board which immediately broke. A fifth harpoon had previously been struck, to replace the one which was pulled out, but the line attached to it was soon afterwards cut. They then instituted various schemes for arresting the speed of the fish, which occupied their close attention nearly twelve hours. But its velocity was yet such, that the master, who had himself proceeded to the attack, was unable to approach sufficiently near to strike a harpoon. After a long chase, however, he succeeded in getting hold of one of the lines which the fish dragged after it, and of fastening another line to it. The fish then fortunately turned towards the ship, which was at a considerable distance to leeward. At 4 P. M. of the 30th, 36 hours after the fish was struck, the ship again joined the boats; when, by a successful manœuvre, they secured two of the fast-lines on board. The wind blowing a moderately brisk breeze, the top-gallant sails were taken in, the courses hauled up, and the top-sails clewed down; but notwithstanding the resistance a ship thus situated must necessarily offer, she was towed by the fish directly to windward, with the velocity of at least one and a half to two knots, during an hour and a half. And then, though the whale must have been greatly exhausted, it beat the water with its fins and tail in so tremendous a way, that the sea around was in a continual foam, and the most hardy of the sailors scarcely dared to approach it. At length, about 8 P. M., after 40 hours of almost incessant, and for the most part fruitless exertions, this formidable and astonishingly vigorous animal was killed. The capture and the flensing occupied 48 hours. The fish was 11 feet bone (the length of the longest laminæ of whalebone;) and its produce filled 47 butts, or 23½ ton casks with blubber.*

---

* This interesting occurrence was communicated to me

Excepting when it has young under its protection, the whale generally exhibits remarkable timidity of character. A bird perching on its back alarms it. The fisher, however, is sometimes liable to danger from its fury.

The Aimwell, while cruising in the Greenland seas, in the year 1810, had boats in chase of whales. One of them was harpooned. But instead of sinking immediately, on receiving the wound, as is the most usual manner of the whale, this individual only dived for a moment, and then rose again beneath the boat, struck it in the most vicious manner with its fins and tail, stove it, upset it, and then disappeared. The crew, seven in number, got on the bottom of the boat, but the unequal action of the line, which for some time remained entangled with the boat, rolled it occasionally over, and thus plunged the crew repeatedly into the water. Four of them, after each immersion, recovered themselves, and clung to the boat; but the other three, one of whom was the only person acquainted with the art of swimming, were drowned before assistance could arrive. The four men in the boat being rescued, the attack was renewed, and two more harpoons struck. But the whale, irritated instead of being enervated by its wounds, recommenced its furious

---

by the *late* Captain of the Royal Bounty, in a letter containing the account of the transaction, as inserted in his log book.

conduct. The sea was in a foam. Its tail and fins were in awful play; and in a short time, harpoon after harpoon drew out; the fish was loosened from its entanglement, and escaped.

On the 3rd of June, 1811, a boat from the Resolution, commanded at the time by myself, put off in pursuit of a whale, and was rowed upon its back. At the moment that it was harpooned, it struck the side of the boat a violent blow with its tail, the shock of which threw the boat steerer to some distance into the water. A repetition of the blow, projected the harpooner and line manager in a similar way. One of the men regained the boat, but as the fish immediately sunk, and drew the boat away from the place, his two companions in misfortune were soon left far beyond the reach of assistance. The harpooner, though a practised swimmer, felt himself so bruised by a blow he had received on the chest, that he was totally incapacitated from giving the least support to his fellow sufferer. The ship being happily near, a boat arrived to their succour, at the moment when the line manager, who was unacquainted with the art of swimming, was on the point of sinking to rise no more. The fish, after a close pursuit, was subdued.

A large whale, harpooned from a boat belonging to the same ship, became the subject of a general chase. Being myself in the first boat which approached the fish, I struck my harpoon at arm's length, by which we fortunately evaded a blow which appeared to be aimed at the boat. Another boat then advanced, and another harpoon was struck, but not with the same result, for the stroke was returned by a tremendous blow from the fish's tail. The boat was sunk by the shock, and at the same time whirled round with such velocity, that the boat steerer was precipitated into the water, on the side next the fish, and was accidentally carried down to a considerable depth by its tail. After a minute or two he arose, and was taken up, along with his companions, into my boat. A similar attack was made on the next boat which came up; but the harpooner, being warned of the prior conduct of the fish, used such precaution, that the blow, though equal in strength, took effect only in an inferior degree. The activity and skill of the lancers soon overcame this designing whale, and added its produce to the cargo of the ship. Such intentional mischief on the part of the whale, it must be observed, is a somewhat rare occurrence.

Proceedings after a Whale is Killed.

Before a whale can be flensed, as the operation of taking off the fat and whalebone is called, some preliminary measures are requisite. These consist in securing the fish to a boat, cutting away the attached whale-lines, lashing the fins of the whale together, and towing it to the ship.

The first operation performed on a dead whale, is to secure it to a boat. This is easily effected, by lashing it with a rope, passed several times through two holes pierced in the tail, to the boat's bow. The more difficult operation of freeing the whale from the entanglement of the lines, is then attempted. As the whale, when dead, always lies on its back, or on its side, the lines and harpoons are generally far under water. When they are seen passing obliquely downward, they are hooked with a grapnel, pulled to the surface and cut. But when they hang perpendicularly, or when they can not be seen, they are discovered by a process, called " sweeping a fish."

While this is in progress, the men of the other boats, having first lashed the tail to a boat, are employed in lashing the fins together across the belly of the whale. I have observed two or three curious circumstances connected with these operations, which I shall venture to mention.

On one occasion, I was myself engaged in the capture of a fish, upon which, when to appearance dead, I leaped, cut holes in the fins, and was in the act of reeving a rope through them, when the fish sunk beneath my feet. As soon as I observed that the water had risen above my knees, I made a spring towards a boat at the distance of three or four yards from me, and caught hold of the gunwale. Scarcely was I on board before the fish began to move forward, turned from its back upon its belly, reared its tail aloft, and began to shake it with such prodigious violence, that it resounded through the air to the distance of two or three miles. After two or three minutes of this violent exercise, it ceased, rolled over upon its side, and died.

In the year 1816, a fish was to all appearance killed. The fins were partly lashed, and the tail on the point of being secured, and all the lines excepting one were cut away, the fish meanwhile lying as if dead. To the astonishment and alarm, however, of the sailors, it revived, began to move, and pressed forward in a convulsive agitation; soon after, it sunk in the water to some depth, and then died. One line remained attached to it, by which it was drawn up and secured. A fish being properly secured, is then " taken in tow," that is, all the boats join themselves in a line, by ropes always carried for the purpose, and unite their efforts in rowing towards the ship. The course of the ship, in the mean time, is directed towards the boats, but in calms, or when the ship is moored to the ice, at no great distance, or when the situation of the fish is inconvenient or inaccessible, the ship awaits the approach of the fish.

The fish having reached the ship, is taken to the *larboard* side, arranged and secured for flensing. For the performance of this operation, a variety of knifes and other instruments is requisite.

Towards the stern of the ship, the head of the fish is directed; and the tail, which is first cut off, sent abreast of the fore chains. The smallest or posterior part of the whale's body, where the

WHALE-FISHERY.

tail is united, is called the rump, and the extremity or anterior part of the head is drawn in an opposite direction by means of the nose tackles. Hence, the body of the fish is forcibly extended. The right-side fin, being next the ship, is lashed upward towards the gunnel. A band of blubber, two or three feet in width, encircling a fish's body, and lying between the fins and the head, being the fat of the neck, or what corresponds to the neck in other animals, is called the *kent*, because by means of it the fish is turned over or *kented*. In the commencement of this band of fat or kent is fixed the lower extremity of a combination of powerful blocks, called the *kent purchase*. Its upper extremity is fixed round the head of the main mast, and its *fall* or rope, is applied to the windlass, drawn tight, and the upper surface of the fish rising several inches above the water. The enormous weight of a whale prevents the possibility of raising it more than one fourth, or one fifth part out of the water, except, indeed, when it has been some days dead, in which case it swells in consequence of air generated by putrefaction, until one third of its bulk, appears above the surface; the fish then lying belly upwards, extended and well secured, is ready for the operation of flensing.

Process of Flensing.

After the whale is properly secured along side of the ship, the harpooners, having their feet armed with spurs, to prevent them from slipping, descend upon the fish. Two boats, each of which is under the guidance of one or two boys, attend upon them, and serve to hold all their knives, and other apparatus. Thus provided, the harpooners, directed by the specksioner,* divide the fat into oblong pieces, or "slips," by means of "blubber spades" and "blubber knives;" then, affixing a "speck-tackle" to each slip, flay it progressively off, as it is drawn upwards. The speck-tackles, which are two or three in number, are rendered effective by capstern winches, or other mechanical powers. The flensers commence with the belly and under jaw, being the only part then above water. The blubber, in pieces of half a ton each, is received on deck, and divided into portable, cubical, or oblong pieces, containing near a solid foot of fat, and passed down between decks, when it is packed in a receptacle provided for it in the hold, or other suitable place, called the flens-gut, where it remains until further convenience.

All the fat being taken away from the belly, and the right fin removed, the fish is then turned round on its side by means of the kent, which, by the power of the windlass, readily performs this office.

---

* The name of this officer was introduced by the Dutch, and is derived from the word *speck*, which, in their language, is applied to the fat of the whale, as well as to that of other animals.

The upper surface of fat is again removed, together with the left fin, and after a second kenting, one of the "lips" is taken away, by which the whalebone of one side of the head, now lying nearly horizontal, is exposed. The fish being a little further turned, the whalebone of the left side is dislodged by the use of "bone hand-spikes," "bone knives," and "bone spades." These constitute what are called "bone geer," and are used, with the assistance of speck tackles, for taking up the whalebone in one mass. On its arrival on deck, it is split with bone wedges into "junks," containing five to ten blades each, and stowed away. A further kenting brings the fish's back upward, and the next exposes the second side of bone. As the fish is turned round, every part of the blubber becomes successively uppermost and is removed. At length, when the whole of the blubber, whalebone, and jaw bones have been taken on board, the kent, which now appears a slip of perhaps 30 feet in length, is also separated, together with the rump rope, and nose tackle, on which, the carcass being at liberty, generally sinks in the water and disappears.

When sharks are present, they generally help themselves very plentifully, during the progress of the flensing; but they often pay for their temerity with their lives. Fulmars pay close attendance in immense numbers. They seize the fragments occasionally disengaged by the knife, while they are swimming in the water: but most of the other gulls, who attend on the occasion, take their share on the wing. The burgomaster is decidedly master of the feast. Hence every bird is obliged to relinquish the most delicious morsel, when the burgomaster descends to claim it.

When despatch is seconded by ability, the operation of flensing can be performed on a whale, affording 20 or 30 tons of blubber, in the space of three or four hours. Flensing in a swell is a most difficult and dangerous undertaking, and when the swell is considerable, it is commonly impracticable. No ropes or blocks are capable of bearing the jerk of the sea. The harpooners are annoyed by the surge, and repeatedly drenched in water, and are likewise subject to be wounded by the breaking of ropes, or hooks, or tackles, and even by strokes from each other's knives. Hence, accidents in this kind of flensing are not uncommon. The harpooners not unfrequently fall into the fish's mouth, when it is exposed by the removal of a surface of blubber; where they might easily be drowned, but for prompt assistance.

Some years ago, I was witness of a circumstance in which a harpooner was exposed to the most imminent risk of his life at the conclusion of a flensing process, by a very curious accident. The harpooner stood on one of the jaw bones of the fish, with a boat by his side. In this situation, while he was in the act of cutting the kreng* adrift, a boy inadvertently

---

* The carcass, after being flensed, is so called.

stuck the point of the boat-hook, by which he usually held the boat, through the ring of the harpooner's spur, and in the same act, seized the jaw bone of the fish with the hook of the same instrument. Before this was discovered, the kreng was set at liberty, and began instantly to sink. The harpooner then threw himself towards the boat, but being firmly entangled by the foot, he fell into the water. Providentially he caught the gunwale of the boat with both hands; but overpowered by the force of the sinking kreng, he was on the point of relinquishing his grasp, when some of his companions got hold of his hands while others threw a rope round his body. The carcass of the fish was now suspended entirely by his body, which was consequently so dreadfully extended, that there was some danger of his being drawn asunder. But such was his terror of being taken under water, that notwithstanding the excruciating pain he suffered, he constantly cried to his companions, to "haul away the rope." He remained in this dreadful state until means were adopted for hooking the kreng with a grapnel, and drawing it back to the surface of the water.

Process of Making Off.

When a fish is caught, or sometimes when there is a good prospect of success in the fishery, even before a fish is caught, the centre of the ship's hold is disencumbered of a few of its casks, to be in readiness for the reception of the blubber. The cavity thus made, together with all the space between decks which can conveniently be appropriated to the same purpose, receives the name of the *flens-gut*. Now, when the flens-gut is filled with blubber, or when, no fish having been seen, a favourable opportunity of leisure is presented, the operation of *making off** is generally commenced. This consists of freeing the fat from all extraneous substances, especially the muscular parts, and the skin; then cutting it into small pieces, and putting it into casks through the bunghole. Before the process of making off can, however, be commenced, several preparatory measures are necessary. The ship must be moored to a convenient piece of ice, or placed in an open situation, and the sails so reduced as to require no further attention in the event of bad weather occurring. The hold of the ship must be cleared of its superstructure of casks, until the " ground tier," or lowest stratum of casks. is exposed; and the ballast water must be " started," or pumped out of all the casks that are removed upon deck, as well as out of those in the ground tier, which are first prepared for the reception of the blubber. In " breaking out the hold," it is

---

* The expression " making off," seems to be derived from the word *afmaaken* of the Dutch, signifying to finish, adjust, or complete, referring to the nature of the operation, as a concluding, finishing, or adjusting process.

not necessary to lay open more of the ground tier at a time, than three or four casks extended in length.

The water which is discharged from the casks in the hold, provided they have been before in use, gives out a great quantity of a strong disagreeable vapour, consisting probably of sulphuretted and phosphuretted hydrogen, with a mixture of other gaseous fluids, produced by the decomposition of the oleaginous, and other animal substances, left in the casks after former voyages. This decomposition seems to be encouraged, if not wholly produced, by the action of the water on the animal matter; because the same casks, if bunged close, when empty, give out but a small quantity of gas and that of inferior pungency. The gas proceeding from oily casks, having contained water, resembles, in some degree, though vastly more pungent, the gas evolved by "bilge water," or the stagnant water which rests among the timbers of a very tight ship. The gas discharged from oily casks, is usually stronger and more abundant, in proportion as the water from which it is disengaged, has been a longer time in the casks. A considerable quantity of it is generated in the space of three or four months. This gas blackens metals, even gold, restores some metallic oxides, is disagreeable in respiration, and affects the eyes of the persons employed in the hold, where it is most abundant, so as to occasion ophthalmic inflammation, and frequently temporary blindness.

While the line-managers, together with the "skeeman,"* the cooper, and perhaps a few others, are employed in breaking out the hold, the rest of the crew on the deck arrange all the variety of apparatus used for the preparation of the blubber, before it is put into the casks. Of this apparatus, the most considerable part is the "speck-trough," with its appendages. It consists of a kind of oblong box or chest, about twelve feet in length, 1¾ feet in breadth, and 1½ feet in depth. The speck-trough is fixed upon the deck, as nearly as possible over the place where the casks are to be filled in the hold. A square hole, made in its bottom, is placed either over the nearest hatch-way to the scene of operation, or upon a corresponding hole cut in the deck.

The speck-trough is then secured, and its lid turned backward into a horizontal position; in which position it is supported on one side by its hinges, and on the other by screw props or pillars; or it is altogether rested upon several little stools. The surface of the lid, which thus placed, forms a level table, is then covered with blocks of whales' tail, from end to end. This substance, from its sinewy and elastic nature, makes excellent "chopping blocks," and preserves the "chopping knives" from injury, when used for dividing the blubber

---

* The officer who has the direction of operations conducting in the hold.

upon it. Into the square hole in the bottom of the speck-trough is fitted an iron-frame, to which is suspended a canvass tube or "hose," denominated a *lull*. The lull is open at both ends. Its diameter is about a foot, and its length sufficient to reach from the deck to the bottom of the hold. To the middle, or towards the upper part of the lull, is attached a "pair of nippers," consisting of two sticks fastened together by a kind of hinge at one end, and capable of being pressed together at the other. The nippers being passed across the body of the lull, and their detached extremities brought together, they embrace it so closely, that nothing can pass downward while they remain in this position; but when, on the other hand, the nippers are extended, the lull forms a free channel of communication between the speck-trough and the hold.

Every thing being now in readiness, the blubber, as it is thrown out of the flens-gut, undergoes the following several operations. It is received upon deck by the "krengers," whose office is to remove all the muscular parts, together with such spongy or fibrous fat, as is known by experience to produce very little oil. When these substances, which go under the general denomination of kreng, are included among the blubber in the casks, they undergo a kind of fermentation, and generates such a quantity of gas, as sometimes to burst the containing vessels, and occasion the loss of their contents. From the krengers, the blubber passes to the harpooners. Each of these officers, provided with a blubber-knife, or a strand knife, places himself by the side of a "closh," which is an upright fixed in the deck, from the top whereof, project several sharp spikes. An attendant, by means of a pair of "hand hooks," or a "pick haak," then mounts a piece of blubber upon the spikes of the closh, and the harpooner slices off the skin. From the skinners, the blubber is passed into an open space called the bank, prepared as a depositary, in front of the speck-trough and it is then laid upon the "chopping blocks," as wanted. It now falls under the hands of the boat steerers, who armed with "chopping knives," are arranged in a line by the side of the chopping-blocks, with the speck-trough before them. Thus prepared, they divide the blubber, as it is placed on their blocks, into oblong pieces, not exceeding four inches in diameter, and push it into the speck-trough intended for its reception, And, finally, the blubber falls under the direction of the line managers stationed in the hold, who receive it into tubs, through the medium of the lull; and pass it, without any instrument but their hands, into the casks through their bung-holes. The casks being closely filled, are then securely bunged up.

When the ground tier casks, as far as they have been exposed, are filled, the second tier of casks is "stowed" upon it, and likewise filled with blubber, together with the third tier casks when necessary. As in this progressive manner, when fish can be had in sufficiency, all the hold is filled, and likewise the space between decks,—it is evident, that the process of making-off must be tedious, disagreeable and laborious. Fifty men, actively employed, can prepare and pack about three tons of blubber in an hour; though, more frequently, they are contented with making-off little more than half that quantity.*

When a ship, which makes a successful fishing, is deficient in casks, the remaining vacancies adapted for the reception of the cargo, are filled with "blubber in bulk," that is, the blubber, in large pieces as it is taken off the whales, is laid skin downward, upon the highest tier of casks, and over this, stratum after stratum, until the vacancies are filled. A little salt is usually scattered over the surface of each stratum of blubber, which assists in preserving the animal fibre, and in preventing the discharge of the oil. Blubber in bulk, notwithstanding every precaution, however, generally loses much of its oil.

A quick passage homeward, with cool weather and smooth sea, are favourable for its preservation,

---

* The operation of making-off was always, in the earlier ages of the fishery, performed on shore; and even so recently as the middle of last century, it was customary for ships to proceed into a harbour, and there remain so long as this process was going on.

but under the influence of opposite circumstances, it becomes greatly reduced.

Process of boiling Blubber, or extracting Oil.

The blubber, which is originally in the state of firm fat, is found, on arrival in a warm climate, to be in a great measure resolved into oil. The casks containing the blubber are conveyed by a mechanical apparatus to the top of a wooden cistern, called the *starting-back,* capable of containing from 3 to 6 or 10 tons, into which their contents are started through the bung-holes. When the copper or boiler, which is a vessel of about the same capacity as the starting-back, is properly cleansed, the contents of the starting-back, on lifting a clough at the extremity, or turning a stop-cock, fall directly into the copper, one edge of which is usually placed beneath. The copper is filled within two or three inches of the top, a little space being requisite to admit of the expansion of the oil when heated; and then a brisk fire is applied in the furnace, and continues until the oil begins to boil. This usually takes place in less than two hours. Many of the fritters or fenks (the refuse) float on the surface of the oil before it is heated, but after it is boiled off, the whole, or nearly so, subside to the bottom. From the time the copper begins to warm, until it is boiled off, or ceases to boil, its contents must be incessantly stirred by means of a pole armed with a kind of broad blunt chisel, to prevent the fenks from adhering to the bottom or sides of the vessel. When once the contents of the copper boil, the fire in the furnace is immediately reduced, and shortly afterwards altogether withdrawn. Some persons allow the copper to boil an hour, others during two or three hours. The former practice is supposed to produce finer or paler oil, the latter a greater quantity. Supposing the copper to be filled at four in the morning, it is generally brought to boil by half past five, and boiled off at half past six or seven. It then stands to cool or subside, until about two in the afternoon, when the bailing process commences. A back or cooler having been prepared for the reception of the oil, by putting into it a quantity of water,* for the double purpose of preventing the heat of the oil from warping or rending the back, and for receiving any impurities which it may happen to hold in suspension; a wooden spout, with a large square box-like head, which head is filled with brush-wood or broom, that it may act as a filter, is then placed along, from the " copper-head "† to the cooler, so as to form a communication between the two. The oil in the copper being now separated from the fenks, water, and other impuri-

---

* Some persons dispense with the water, believing that it promotes rancidity in the oil.

† The platform built around the edge of the copper, is called the *copper-head*.

ties, all of which have subsided to the bottom, is, in a great measure, run off through the pipe communicating with the cooler, and the remainder is carefully lifted in copper or tin ladles, and poured upon the broom in the spout, from whence it runs into the same cooler, or any other cooler, at the pleasure of the " boilers."* Besides oil and fenks, the blubber of the whale likewise affords a considerable quantity of watery liquor, produced probably from the putrescence of the blood, on the surface of which, some of the fenks, and all the greasy animal matter called *footje* or *footing*, float, and upon the top of these the oil. Great care, therefore, is requisite, on approaching these impure substances, to take the oil off by means of shallow tinned iron or copper ladles, called *skimmers*, without disturbing the refuse, and mixing it with the oil. There must always, however, be a small quantity towards the conclusion, which is a mixture of oil and footing; such is put into a cask or other suitable vessels by itself, and when the grossy part has thoroughly subsided, the most pure part is skimmed off, and becomes fine oil, and the impure is allowed to accumulate by itself in another vessel, where, in the end, it affords " brown oil."

The refuse now left in the copper, is *bailed* into

---

* The men employed in extracting oil are thus denominated.

a tunnel or spout, which conveys it into the fenk-back, where it remains as long as the capacity of the vessel will admit; a portion of brown oil, which is constantly found rising to the surface, being, in the meantime, occasionally skimmed off.

A few years ago, my father instituted a process for reducing blubber into oil, by the use of steam; and a similar process has been adopted in Hull, and other ports, and applied to the extraction of oil, with considerable advantage.

From a ton, or 252 gallons by measure, of blubber, there generally arises from 50 to 65 gallons of refuse, whereof the greater part is a watery fluid. The constant presence of this fluid, which boils at a much lower temperature than the oil, prevents the oil itself from boiling, which is, probably an advantage, since, in the event of the oil being boiled, some of the finest and most inflammable part, would fly off in the form of vapour; whereas, the principal part of the steam, which now escapes, is produced from the water.

Some persons make a practice of adding a quantity of water, amounting, perhaps, to half a ton, to the contents of each copper, with the view of weakening or attenuating the viscid impurities contained in the blubber, and thus obtaining a finer oil; others consider the quantity of watery fluid, already in the blubber, as sufficient for producing every needful effect.

Each day, immediately after the copper is emptied, and while it is yet hot, the men employed in the manufacture of the oil, having their feet defended by strong leathern or wooden shoes, descend into it, and scour it out with sand and water, until they restore the natural surface of the copper, wherever it is discoloured. This serves to preserve the oil from becoming high coloured,* which will always be the case, when proper cleanliness is not observed.

The starting-back being previously filled with blubber, its contents are again transferred into the copper, and the fire is applied as before. This is generally accomplished by four, or half past four o'clock in the afternoon. The copper again boils by half an hour after five or six, and is boiled off by seven or eight in the evening. The men employed in this service, consisting of about six persons, alternately watch in the night by couples. Those on watch, commence about two in the morning to empty the copper, which done, they again fill it from the starting-back, which is always made ready the night before. Thus the process goes on, until the whole cargo is finished.

By means of three coolers, severally capable of containing at least twice the quantity of oil produc-

---

* The palest coloured oil is most esteemed by buyers, and is supposed to be the best; simply, perhaps, because it seems to have been manufactured with care, and appears to be free from any admixture of brown or black oil, produced from the fenk-back, or found in the hold of the ship.

ed from one boiling of blubber in the copper, each can be allowed, in turn, to stand undisturbed upwards of twenty-four hours. Thus, while one is in the act of being filled, the other stands to cool and settle, and the third is drawn off. If the backs be twice this size, or four times the capacity of the copper, every one will require two days to be filled by one copper, and after being filled, may subside during two or three days undisturbed. Even two backs in number, of this capacity, would admit of an interval of twenty four hours each, after being filled, before it would be necessary to begin to empty it. Thus prepared and cooled, the oil is in a marketable state, and requires only to be transferred from the coolers into casks for convenience of conveyance to any part of the country. Each of the coolers, it has been observed, is furnished with a stop-cock, beneath which there is a platform adapted for receiving the casks, when they are filled, with great ease, by the introduction of a leathern tube, extending from the orifice of the stop-cock into the bung-hole.

At the conclusion of the process of boiling, each vessel's cargo manufactured on the premises, the backs are completely emptied of their contents. To effect this, water is poured in, until the lower part of the stratum of oil rises to within a few lines of the level of the stop-cock, and permits the greater part of the oil to escape. The quantity left, amounts, perhaps, to half an inch, or an inch in depth. To recover this oil without waste requires a little address. A deal-board, in length a little exceeding the breadth of the cooler, is introduced at one end, a little diagonally, and placed edgewise in its contents. The ends of the board being covered with flannel, when pressed forcibly against the two opposite sides of the cooler, prevent the oil from circulating past. The board is then advanced slowly forward, towards the part of the back, where the stop-cock is placed; and in its progress, (the ends being kept close to the side of the cooler, and the upper edge a little above the surface of the oil,) all the oil is now collected by the board, while the water has a free circulation beneath it. When the oil accumulates to the depth of the board, its further motion is suspended, until the oil, thus collected, is drawn off. Another similar board is afterwards introduced at the farthest extremity of the cooler, and passed forward in the same manner, whereby the little oil which escapes the first is collected. Now the remnant, which still refuses to run off by the orifice of the stop-cock, being collected in a corner, is taken up by *skimmers;* and the footing or sediment which appears at the last, is disposed of in the same way as the footing from the copper, until the oil it contains rises to the surface and can be removed.

In most of the out-ports, the oil is generally deposited in casks, in which it remains until it is disposed of by the importers. In London, however, and in some concerns in Hull and other ports, the speculators in the whale fishery are provided with cisterns or tanks, wherein they can deposit their oil, and preserve it until a convenient time for selling, without being subject to the waste which usually takes place when it is put into casks. From these cisterns, any quantity can be drawn off at pleasure.

The smell of oil, during its extraction, is undoubtedly disagreeable; but, perhaps, not more so than the vapour arising from any other substance submitted to the action of heat when in a putrid state. The prevailing opinion, however, that a whale ship must always give out the same unpleasant smell, is quite erroneous. The fact is, that the fat of the whale, in its fresh state, has no offensive flavour whatever, and never becomes disagreeable until it is brought into a warm climate and becomes putrid; neither is a whale ship more unpleasant than any other trader, until after her cargo is opened on her arrival in port.

Description of Whalebone, and the Method of Preparing it.

Whalebone, or whale fins, as the substance is sometimes, though incorrectly named, is found in the mouth of the common Greenland whale, to which it serves as a substitute for teeth. It forms an apparatus most admirably adapted, as a filter, for separating the minute animals, on which the whale feeds, from the sea water in which they exist.

It is a substance of a horny appearance and consistence, extremely flexible and elastic, generally of a bluish black colour, but not unfrequently striped longitudinally with white, and exhibiting a beautiful play of colour on the surface. Internally it is of a fibrous texture, resembling hair; and the external surface consists of a smooth enamel, capable of receiving a good polish.

This substance, when taken from the whale, consists of laminæ, connected by what is called the gum, in a parallel series, and ranged along each side of the mouth of the animal. The laminæ are about 300 in number, in each side of the head. The length of the longest blade, which occurs near the middle of the series, is the criterion fixed on by the fishers, for designating the size of the fish. Its greatest length is about 15 feet; but an instance very rarely occurs of any being met with above $12\frac{1}{2}$ or 13 feet. Its greatest breadth, which is at the root end, is 10 or 12 inches, and its greatest thickness four-tenths or five-tenths of an inch.

The two *sides* or series of the whalebone, are connected at the upper part of the head, or crown bone of the fish, within a few inches of each other, from whence they hang downward, diverging so far as to enclose the tongue between their extremities; the position of the blades, with regard to each other, resembles a frame of saws in a saw mill; and taken altogether, they exhibit, in some measure, the form and position of the roof of a house. The smaller extremity and interior edge of each blade of bone, or the edge annexed to the tongue, are covered with a long fringe of hair, consisting of a similar kind of substance as that constituting the exterior of the bone. Whalebone is generally brought from Greenland in the same state as when taken from the fish, after being divided into portable *junks*, or pieces, comprising ten or twelve laminæ in each; but occasionally it is subdivided into separate blades, and the gum and hair removed when at sea.

One of the first importations of whalebone into England, was probably in the year 1594, when a quantity of this substance, being part of the cargo of a wrecked Biscayan ship, was picked up at Cape Breton, by some English ships, fitted out for the whale and morse fisheries, after the example of the Icelanders and Biscayans.*

This substance has been held in such high estimation, that, since the establishment of the Spitzbergen whale fishery, the British have occasionally purchased it of the Dutch, at the rate of 700*l.* per ton.† It is calculated, that at least 100,000*l. per annum* were paid to the Dutch for this article, about the years 1715 to 1721, when the price was 400*l.*‡

---

* Hakluyt's Voyages, vol. iii. p. 194.
† Macpherson's Annals of Commerce, vol. iii. p. 512.
‡ Elking's View of the Greenland Trade, &c. p. 65.

About the year 1763, the price in England was 500*l.* *per* ton; but after an extensive importation of this article from New England, the price delined to 350*l.** and subsequently as low as 50*l.* *per* ton. Of late years the price has usually been fluctuating between 50*l.* and 150*l.* *per* ton. Whalebone becomes more valuable as it increases in length and thickness.

On or near the premises where the oil is extracted, the whalebone is commonly cleaned and prepared.

The first operation, if not already done, consists in depriving it of the gum. It is then put into a cistern containing water, until the dirt upon its surface becomes soft. When this effect is sufficiently produced, it is taken out, piece by piece, laid on a plank placed on the ground, where the operator stands, and scrubbed or scoured with sand and water, by means of a broom or a piece of cloth. It is then passed to another person, who, on a plank or bench, elevated to a convenient height, scrapes the root-end where the gum was attached, until he produces a smooth surface; he or another workman, then applies a knife or a pair of shears to the edge, and completely detaches all the fringe of hair connected with it. Another person, who is generally the superintendent of the concern, afterwards re-

---

* Macpherson's Annals, vol. iii. p. 371.

ceives it, washes it in a vessel of clean water, and removes, with a bit of wood, the impurities out of the cavity of the root. Thus cleansed, it is exposed to the air and sun until thoroughly dry, when it is removed into a warehouse, or other place of safety and shelter.

Before it is offered for sale, it it usually scrubbed with brushes and hair-cloth, by which the surface receives a polish, and all dirt or dust adhering to it is removed; and finally, it is packed in portable bundles, consisting of about a hundred weight each. The *size*-bone, or such pieces as measure six feet or upward in length, is kept separate from the *under-size;* the latter being usually sold at half the price of the former. Each blade being terminated with a quantity of hair, there is sometimes a difficulty in deciding, whether some blades of whalebone are size or not. Owing to the diminished value of under-size bone, and more particularly, in consequence of the captain and some of the officers engaged in a fishing ship, having a premium on every size fish, it becomes a matter of some importance in a doubtful case, to decide this point. From a decision, which I understand has been made in a court of law, it is now a generally received rule, that so much of the substance terminating each blade, as gives rise to two or more hairs, is whalebone: though, in fact, the hair itself is actually the same substance as that of which the whalebone is composed.

APPENDIX.

A.

The Common Wild Cat; or Bay Lynx.

Felis Rufa, Guld. Penn. &c.

(See vol. 1. plate opposite page 203, fig. 2.)

It is highly probable that all the species of wild cat described under the names of *Fasciata, aurea, Montana,* &c., may be correctly referred to the present, which is the only species, (in addition to the *Felis canadensis* Geoff. Borealis; Temm.) of whose existence in this country, sufficiently satisfactory evidence is to be obtained. At least we have not been able to find any other in the cabinets of natural history to which we have had access, nor in the caravans of living American animals, frequently exhibited within our vicinity. The naturalists attached to the different exploring parties which have traversed vast extents of the American territory, have not been able, by their own efforts, nor through the aid of the Indians, to procure any species but the common wild-cat: we therefore deem it most correct to wait for additional observations, before we admit the existence of so many species as have been proposed.

The common wild-cat stands very high upon its legs, and has a short tail, which is curved upwards at its extremity; which circumstances tend to give the animal an appearance of being somewhat disproportioned. In other respects, its physiognomy reminds one strongly of the domestic cat, to which its general aspect and movements are very similar. The residence of the wild-cat, is usually in wooded districts, where it preys upon birds, squirrels, and other small animals, which are taken by surprise, according to the manner of all the animals belonging to the genus Felis.

The common wild-cat is about two feet long, and twelve or thirteen inches in circumference. The tail but little surpasses three inches in length.

The general colour of the pelage, is a deep reddish, mingled with small spots of blackish brown; the inferior parts of the body and throat, as well as the inferior surface of the tail are white, or whitish. Numerous small, nearly vertical streaks of black, are to be observed on the front between the ears, and down to the space surrounding the orbits, which are encircled by a clear pale, red, or whitish fur. There are small pencils of hairs to the tips of the ears.

B.

We believe the observation was first made by MITCHILL, that the opossum, *Didelphis Virginiana*, is never found to inhabit the country north of the Hudson, and we have been informed by a scientific friend, who has devoted an especial degree of attention to the subject, that from repeated researches and inquiries, he is satisfied that the observation of our distinguished countryman above named, is correct. This fact appears the more singular, when it is remembered, that numbers of the species are found along the southern banks of the river, and it is well known, that in other parts of the country, rivers of nearly equal size offer no barrier to the diffusion of this species. It is a curious and interesting inquiry to determine the causes of this limitation; in the present state of our knowledge, we have no satisfactory explanation to offer.

In relation to the generation of the opossum, considering that this work is destined for general readers, we have deemed it advisable to omit what we have prepared on this subject, and to make the facts we have been enabled to collect and observe, the subject of a paper to be published in a work exclusively devoted to students of natural science.

D.

Wistar's Fossil Elk.

(Vol. 2. fig. * and † in plate opposite page 197.)

The late distinguished professor Wistar published an account of some fossil skulls, (exhumed at Big-Bone Lick, Kentucky, by Gen. Clark, one of the enterprising explorers of the western regions,) which were presented to the American Philosophical Society, by Jefferson. Among other descriptions, is the following of the head of one of the largest species of the genus Cervus.

The breadth of the skull, at its narrowest part, is 4.75 inches. The depth, from the margin of the occipital surface to the most distant part of the great foramen of the occipital bone, is 5.25 inches. From the superior surface, immediately posterior to the base of the horns, to the body of the sphenoid bone, immediately under it, 4.7 inches. The length of the cranium, from the centre of the space between the horns, to the projection of the occipital bone, is 6.37 inches.

Dr. Wistar compared this skull with that of the American Elk, *Cervus Caradensis,* and the Rein-Deer *C. Tarandus,* and concluded, that it more nearly resembled the Elk, at the occiput, though differing from it greatly in the position and projection of the horns. According to the measurements and comparisons of Dr. Wistar, this Elk

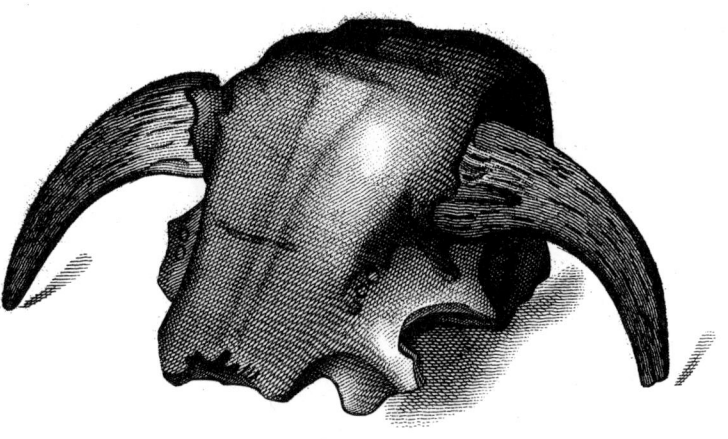

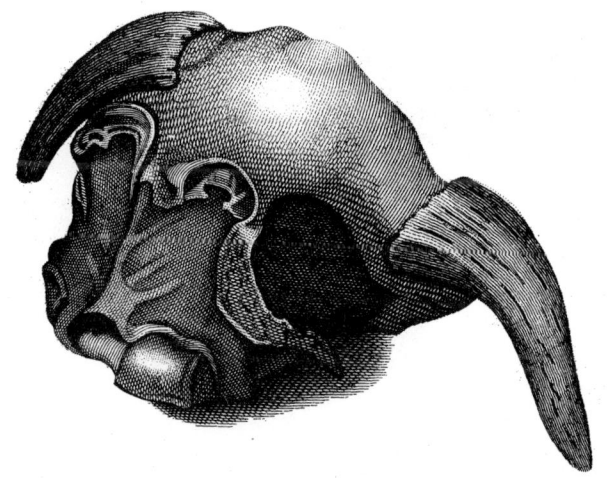

Wistar's Fossil Ox.

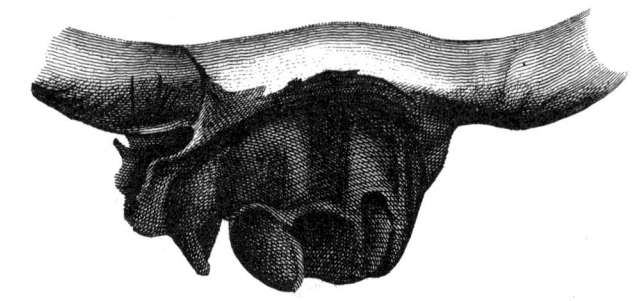

Common Ox.

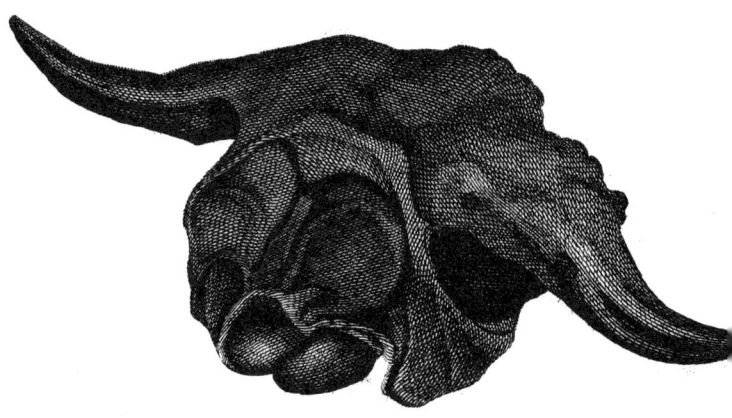

Bison

C.W. Peale Del.

F. Kear

must have been much larger than either the American Elk, or the Rein-Deer. The species is doubtless entirely extinct.*

The name of "*Americanus*" has been given to this species.

F.

Wistar's Fossil Ox.

In the paper above referred to, Dr. Wistar described the fossil skull of an ox, obtained from the same locality, which he considered as nearly allied to the Bison, Bos Americanus. The most remarkable peculiarity of this skull, is the projection or convexity of that portion of the facial or frontal surface between the horns. The accompanying plate gives a front and back view of this skull. The species has been named "*Bombifrons*."

Great Fossil Ox.

The portion of the skull, and nucleus of the horn, belonging to the valuable cabinet of the American

---

* See American Philosophical Trans. vol. 1. new series, p. 377. In the figures we have transferred from Wistar's plate, the posterior and superior view of the skull is marked with a *; the profile view with a †.

Philosophical Society, from which specimen, the annexed accurate drawing was made by M. LE SUEUR, was first described and figured in the annals of the museum, by Cuvier, and subsequently in his great work on Fossil Bones. The drawing renders any detailed description unnecessary. The nucleus of the horn, measures twenty-eight inches in circumference. Though nothing but the fragment here represented is preserved, there can be no doubt but that the animal was of great size and belonged to a species which is utterly extinct. The species has been named " *Catifrons.*"

Dekay's Fossil Ox.

We must refer the reader to the 2d vol. of the Annals of the Lyceum of New York, for the full description of the fragments of this skull, and the comparisons instituted by DR. DEKAY to determine the species.

Dr. Dekay considers that none of the Genus Bos, now to be found in this country, have crania in the slightest degree resembling this specimen. It was thrown out at the eruption caused by an earthquake in 1812, which entirely destroyed the town of New Madrid, on the Mississippi.

Dr. Dekay proposes to designate the species of Fossil crania to which he refers those of Pallas and Ozeretskovsky, by the name of Bos Pallasii, and the

Great Joint Ox.

New Madrid fragment he refers provisionally to the same.

Mitchill's Fossil Walrus.

Dr. Mitchill has received from the shores of Long Island, a very interesting skull belonging to a species of the genus *Trichecus*. This skull is agatised and in a fine state of preservation. It has been referred to the examination of a committee of the Lyceum, and their report will be found in the 2d volume of the annals of that excellent Institution.

CONCLUSION.

This work has been delayed by uncontrollable circumstances, for a much longer period than was anticipated. It is not now the time to offer any apology for the manner in which the undertaking has been accomplished. In reviewing what we have done, it is easy to perceive that much may be hereafter improved. These, and all other deficiencies will no doubt be indicated by those who interest themselves in the execution of such performances. We shall certainly profit by their suggestions, whether made in a spirit of candour or malevolence.

We have been as *original* as it was possible to be, in such a work, unless the whole business of the author's life, had been the collection of materials. The observations we have had an opportunity of making from living nature, we fear not to have compared with those made by any other individual. Wherever we have been obliged to compile, we have anxiously endeavoured to approximate the truth, and have faithfully acknowledged the aid obtained from different sources.

CONCLUSION.

It has been our intention to render this study pleasing and intelligible, more than to discuss minutia of classification; to give the *Natural History*, instead of the nomenclature of American animals; to impart information to those seeking for knowledge, rather than to prepare a book for such as consider themselves the founders of systems and settlers of moot points in philosophy. If we have accomplished nothing more, we have rendered it much easier for our successors to attempt the composition of a better work, having saved them the toil of examining a vast number of books, to glean the detached observations worthy of being brought together.

"Reader, I have given thee an Account of my
"intendments and endeavours in this Performance;
"and if it hath, (as I am too conscious to myself, it
"often hath,) happened, that I have any where fail-
"ed of my design; if in a long and tedious Work,
"I have, thro' inadvertency, streights of time, and
"hurry sometimes of other business, made any balk,
"and committed mistakes, let thy humanity excuse
"the humane infirmities of Thine, and his Coun-
"try's Faithful Servant,"

JOHN D. GODMAN.

GENERAL SYNOPSIS OF MAMMALIA

INHABITING NORTH AMERICA.

BY CHARLES L. BONAPARTE.

Mammalia are vertebrated, warm blooded, viviparous animals; suckling their young; breathing by lungs which float freely in the chest, imperforated; the heart is bilocular and biauricular.

In the present state of science, they form the first class of the first type of the animal kingdom.

GENERAL DIVISIONS,

Or, view of the natural families of the system, adopted in classifying the North American Mammalia.

SUB-CLASS 1. QUADRUPEDA.

Limbs four, obvious: head separated from the body by the intervention of a neck.

SECTION I.

Unguiculata; nails covering only the tips of the digits.
§ Three kinds of teeth.

ORDER I. PRIMATES.

Mammæ 2, pectoral: penis free: anterior limbs terminated by hands.

TRIBE I. BIMANA.

FAMILY 1. Bimana. Anterior limbs only, terminated by hands: body vertical, plantigrade.

Tribe II. Quadrumana.

The four limbs terminated by hands.

Family 2. Simiæ. Resembling man; 4 incisive teeth in each jaw.

Family 3. Lemurini. Resembling carnivorous animals; incisors varying in number, shape and situation; nostrils at the tip of the snout.

Family 4. Dermoptera. Digits of the anterior limbs moderate, robust, all furnished with compressed incurved nails; connecting membrane pilous.

ORDER II. Cheiroptera.

Mammæ 2, pectoral; penis free: limbs connected by a membrane formed for flying.

Family 5. Cheiroptera. Digits of the anterior limbs excessively elongated, comprised in an expansion of the naked membrane of the flanks, thumb free, but not opposable.

ORDER III. Feræ.

Mammæ abdominal, numerous; penis attached to the belly; limbs free, formed for walking; the anterior not terminating by hands.

Family 6. Insectivora. Plantigrade; no carnivorous teeth; false molars acute; 3 or 4 tuberculous grinders on each side of both jaws; from one to six incisors.

Family 7. Carnivera. Last molar, at least, tuberculous; 2 strong canine, and six incisive teeth above and below.

* Plantigrada. ** Digitigrada.

Family 8. Marsupialia. Females with a pouch; both sexes furnished with marsupial bones: hind thumb destitute of nail: opposable; sometimes wanting.

* 2 canines and several small incisors above and below.

** No canine below—at least 6 incisors above.

APPENDIX.

ORDER IV. Pinnipedia.

Mammæ abdominal; penis attached to the belly: feet very short, covered by a skin formed for swimming, the posterior turned backward.

Family 9. Pinnipedia.

§§ Not more than two kinds of teeth.

ORDER V. Glires.

No canine teeth; incisive 2 below, 2, 4, or 6 above; 22 molar at most; jaws moving horizontally.

* Females with a pouch; both sexes with marsupial bones.

Family 10. Marsupialia.(1) Incisive 2 or 6 above.

** No pouch, no marsupial bones.

† Clavicles distinct omnivorous.

Family 11. Murina.

†† Clavicles rudimental. Herbivorous.

Family 12. Aculeata. Skin covered with prickles; upper incisors 2; toes 4–5.

Family 13. Duplicidentata. Skin covered with hair; upper incisors 4, (6 in young subjects;) toes 5–4.

Family 14. Subungulata. Skin covered with hair; upper incisors 2; molars 16; posterior toes 3 or 5, but lateral each side, very small.

* 5 toed.   ** 4–3 toed.

ORDER VI. Bruta.

No canine nor incisor teeth; (except in one genus in which there are 4 below;) from 14 to 98 molars, or none; nails enveloping the extremities of the digits, almost hoop shaped.

---

(1) We scatter the marsupial animals, as naturally they should be separated: their resemblance being merely of analogy and not of affinity, two things often confounded in natural history.

APPENDIX.

Family 15. Tardigrada. All having teeth; 18 molars at most; no incisors; snout short; limbs much elongated.

Family 16. Effodientia. Some edentous; some having incisors; molars from 26 to 98; snout elongated; limbs well proportioned to the body.

* Incisors and molars. ** Molars. *** No teeth at all.
‡ Ungulata. ‡‡ Vermilinguia.

Section II. Ungulata.

Nails hoof-shaped, covering the last phalanges of the digits: no clavicles; the fore-arm always in a state of pronation.

ORDER VII. Pecora.

Rarely three kinds of teeth; no incisors above; feet didactyle, with two hoofs; the metacarpal and metatarsal bones united; four stomachs; ruminating.

Family 17. Cavicornia. No canine teeth; both sexes having permanent horns, composed of a solid nucleus, growing from the frontal bones, and of an elastic thin case.

* Lacrymatories; nucleus entirely solid. ** No lacrymatories; nucleus of the horns cellular.

Family 18. Devexa. No canine teeth; both sexes with permanent solid horns covered by a skin.

Family 19. Capreoli. No canine teeth; in general the males only having caducous solid, branched horns, covered at least for a time by a hirsute skin.

Family 20. Tylopoda. With canine teeth; hornless.

ORDER VIII. Bellua.

Generally three kinds of teeth; stomach simple; or divided into several pouches, but not for rumination.

Family 21. Solidungula. Feet apparently monodactyle.

Family 22. Fissipedes. Toes 3 or 4, but in the intermediaries approximated; others 4–3 toed.

* Toes 4–3.   ** Toes 4–4   *** Toes 2–2.

Family 23. Pachydermata. Feet pentadactyle or tridactyle, the other digits being rudimental; digits only perceived externally, &c.

* Pentadactyla, (Proboscida.)   ** Tridactyla.

Sub-class II. Bipeda.

No hind limbs; (merely indicated by bones.) Fore limbs fins; neck not distinct from the body; body pisciform, terminating in a cartilaginous horizontal fin-shaped tail.

(Live in the water; have no external ears, nor hair on the body.

ORDER IX. Cete.

Family 24. Sirenia. Mammæ pectoral; no blow-holes.

Family 25. Hydraula. Mammæ inguinal; with blow-holes.

ANALYTICAL TABLE

OF THE

NORTH AMERICAN GENERA.

ORDER PRIMATES.

FAMILY BIMANA.

GENUS 1. Homo.

ORDER CHEIROPTERA.

FAMILY CHEIROPTERA.

GENUS 2. Vespertilio.

ORDER FERÆ.

FAMILY INSECTIVORA.

GENUS 3. Sorex. Ears short, rounded.
GENUS 4. Scalops. No external ears: snout simple.
GENUS 5. Condylura. No external ears: snout stellated.

FAMILY CARNIVORA.

* *Plantigrada.* *Treading on the whole sole of the foot.*

GENUS 6. Ursus. Seven molar on each side: tail short: no anal odoriferous follicules.
GENUS 7. Procyon. Six molars on each side: tail very long, pilous: no anal follicules.
GENUS 8. Meles. Five molars on each side· tail short, pilous: an anal pouch filled with fetid unctuous substance.
GENUS 9. Gulo. Five molars above, six below on each side: tail moderate or short: two folds of the skin near the anus, but no anal pouch.

APPENDIX. 255

** *Digitigrada. Treading on the extremities of their digits.*

a. Only one tuberculous behind the upper carnivorous tooth: body much elongated, vermiform: feet short.

Genus 10. Mustela. Toes cleft: tail moderate and bushy.
Genus 11. Mephitis. Toes cleft: tail long and bushy or wanting.
Genus 12. Lutra. Toes palmated.

b. Two tuberculous behind the upper carnivorous tooth.

Genus 13. Canis. Feet 5—4 toed; nails not retractile: tongue smooth.

c. No small tooth behind the inferior large molar.

Genus 14. Felis. Feet 5—4 toed: nails retractile: tongue prickly.

FAMILY MARSUPIALIA.

Genus 15. Didelphis.

ORDER PINNIPEDIA.

FAMILY PINNIPEDIA.

Genus 16. Phoca. Both jaws furnished with incisive and canine teeth.
Genus 17. Trichecus. No incisors nor canine below; superior canine greatly prolonged below the lower jaw.

ORDER GLIRES.

FAMILY MURINI.

Genus 18. Castor. Feet five toed, anterior cleft, posterior palmated; tail wide, depressed, thick, oval, naked and scaly.
Genus 19. Fiber. Feet five toed, anterior simple, posterior furnished with stiff bristles replacing the membrane; tail compressed, linear, scaly, with scattered bristles.
Genus 20. Arvicola. Feet simple; tail cylindrical, hairy, grinders without radicles.
Genus 21. Neotoma. Feet simple; tail cylindrical, hairy: grinders with profound radicles, and with small marked triangles.
Genus 22. Sigmodan. Feet simple; cylindrical, hairy: molars in each jaw, six, subequal, with radicles, and with deep, alternate folds towards the summit.
Genus 23. Mus. Feet simple; tail cylindrical, subnaked, scaly, with scattered hairs.
Genus 24. Gerbillus. Hind feet very long, five toed, each furnished with a distinct metatarsal bone; tail elongated, more or less bushy, but without tuft at tip.

Genus 25. Arctomys. Feet and tail short; nails robust; inferior incisive subulate.

Genus 26. Sciurus. Hind feet turned towards each other; nails very sharp; tail long and bushy; inf. incisive much compressed.

Genus 27. Pteromys. Tail long and bushy; skin of the flanks extended between the fore and hind limbs.

FAMILY ACULEATA.

Genus 28. Hystrix.

FAMILY DUPLICIDENTATA.

Genus 29. Lepus. Hind limbs very long: ears very long: tail short.

ORDER PECORA.

FAMILY CAVICORNIA.

* Nucleus of the horns solid.

Genus 30. Antelope.

** Nucleus of the horns cellular.

Genus 31. Ovis. Tail destitute of terminal tuft.

Genus 32. Bos. Tail ending in a bushy tuft.

FAMILY CAPREOLI.

Genus 33. Cervus.

ORDER CETE.

FAMILY SIRENIA.

Genus 34. Manatus. Body oblong, ending in an oval, horizontal fin; pectoral fins furnished with rudiments of nails.

Genus 35. Stellerus. Body elongated, ending in a crescent shaped fin; no rudiment of nails.

FAMILY HYDRAULA.

Genus 36. Delphinus. Head proportioned: teeth.

Genus 37. Monodon. Head proportioned: no teeth.

Genus 38. Physeter. Head exceedingly disproportioned: teeth.

Genus 39. Balæna. Head exceedingly disproportioned: teeth cartilaginous, or rather cartilages instead of teeth.

INDEX.

	Vol.	Page
Antelope	ii.	320
Arctomys	ii.	98
Monax	ii.	100
Empetra	ii.	108
Franklinii	ii.	109
Richardsonii	ii.	111
Tridecemlineatus	ii.	112
Ludovicianus	ii.	114
Parryii	ii.	120
Argali	ii.	329
Arvicola	ii.	63
Xanthognatus	ii.	65
Riparius	ii.	67
Hispidus	ii.	68
Floridanus	ii.	69
Badger	i.	176
American	i.	179
Balæna	iii.	98
Mysticetus	iii.	ib.
Physalis	iii.	137
Musculus	iii.	141
Boops	iii.	142
Rostrata	iii.	144
Bat	i.	48
Carolina	i.	67
New York	i.	68
Hoary	i.	ib.
Arcuated	i.	70
Subulate	i.	71
Bear	i.	109
Brown	i.	113
American or Black	i.	114
Grizzly	i.	131

INDEX.

	Vol.	Page
Bear Polar	i.	143
Beaver	ii.	19
Fabulous History	ii.	38
Description of	ii.	55
Belluæ	ii.	202
Bison	iii.	4
Bos	iii.	3
Americanus	iii.	4
Moschatus	iii.	29
Bruta	ii.	173
Cachalot	iii.	93
Spermaceti	iii.	94
Canis	i.	232
Familiaris	i.	243
Lupus	i.	255
Latrans	i.	260
Nubilus	i.	265
Lycaon	i.	267
Lagopus	i.	268
Argentatus	i.	274
Fulvus	i.	276
Cinereo-Argentatus	i.	280
Velox	i.	282
Capra	ii.	326
Montana	ii.	ib.
Carnivora	i.	107
Amphibia	i.	305
Castor	ii.	19
Fiber	ii.	21
Cat	i.	285
Common wild	iii.	239
Cervus	ii.	271
Alcis	ii.	274
Tarandus	ii.	283
Canadensis	ii.	294
Macrotis	ii.	304
Virginianus	ii.	306
Cete (Order)	iii.	37
(Family)	iii.	56
Cheiroptera	i.	48
Condylura	i.	97
Cougar	i.	291
Deer	ii.	271
Black-Tail	ii.	304
Common	ii.	306

INDEX.

	Vol.	Page
Delphinus (Tribe)	iii.	57
Genus	iii.	58
Delphis	iii.	59
Gladiator	iii.	67
Phocœna	iii.	69
Didelphis	ii.	4
Virginiana	ii.	7
Digitigrada	i.	191
Dog	i.	232
Domestic	i.	243
Newfoundland	i.	254
Dolphin Proper	iii.	57
(Genus)	iii.	58
True	iii.	59
Gladiator	iii.	67
Elk	ii.	294
Wistar's Fossil	iii.	242
Elephant	ii.	253
Fossil	ii.	255
Elephas	ii.	253
Primogenius	ii.	255
Ermine Weasel	i.	193
Felis	i.	285
Concolor	i.	291
Canadensis	i.	302
Feræ	iii.	239
Rufa	i.	146
Fiber	ii.	57
Zibethicus	ii.	58
Field Mouse	ii.	63
Flying Squirrel	ii.	146
Fox Arctic	i.	268
Black or Silver	i.	274
Red	i.	276
Gray	i.	280
Swift	i.	282
Gerbillus	ii.	93
Canadensis	ii.	94
Labradorius	ii.	97
Glires	ii	17
Glutton	i	184
Gulo	i.	ib.
Luscus	i.	185
Hairy Campagnol	ii.	68
Hare	ii.	155
American	ii.	157

260 INDEX.

	Vol.	Page
Hare Polar	ii.	162
Herbivorous Cetacious animals	iii.	39
Hystrix	ii.	149
Dorsata	ii.	150
Inclaviculata	ii.	149
Indian	i.	17
Insectivora	i.	73
Isatis	i.	268
Jumping Mouse	ii.	93
Labrador	ii.	97
Lamantin	iii.	41
American	iii.	43
Lemming	ii.	73
Hudson's Bay	ii.	ib.
Lemmus	ii.	ib.
Hudsonius	ii.	ib.
Lepus	ii.	155
Americanus	ii.	157
Glacialis	ii.	162
Lutra	i.	220
Brasiliensis	i.	222
Marina	i.	228
Lynx Northern	i.	302
Bay	iii.	239
Man American	i.	17
Manatus	iii.	41
Americanus	iii.	43
Marmot	ii.	98
Maryland	ii.	100
Quebec	ii.	108
Franklin's	ii.	109
Tawny American	ii.	111
Hood's	ii.	112
Prairie	ii.	114
Parry's	ii.	120
Marsh Campagnol	ii.	67
Marsupialia	ii.	3
Marten	i.	191
Pine	i.	200
Pennant's	i.	203
Mastodon	ii.	204
Gigantic	ii.	208
Meadow Mouse	ii.	65
Megatherium	ii.	173
Cuvieri	ii.	180

INDEX.

	Vol.	Page
Megatherium Jeffersonii	ii.	196
Mephitis	i.	211
Americana	i.	213
Meles	i.	176
Labradoria	i.	179
Mink	i.	206
Mole Star-nose	i.	100
Monodon	iii.	80
Monoceros	iii.	ib.
Moose	ii.	274
Morse	i.	351
Mouse Common	ii.	84
Rustic	ii.	88
Mus	ii.	76
Decumanus	ii.	78
Rattus	ii.	83
Musculus	ii.	84
Agrarius	ii.	88
Musk Ox	iii.	29
Musk Rat	ii.	57
Mustela	i.	191
Erminea	i.	193
Martes	i.	200
Pennanti	i.	203
Lutriola	i.	206
Zibellina	i.	208
Narwal	iii.	81
Opossum	ii.	4
Common	ii. 7–iii.	241
Otter	i.	220
American	i.	222
Sea	i.	228
Ovis	ii.	328
Ammon	ii.	329
Ox	iii.	3
Wistar's Fossil	iii.	243
Great Fossil	iii.	ib.
Dekay's Fossil	iii.	244
Pecora	ii.	267
Phoca	i.	306
Vitulina	i.	313
Cristata	i.	336
Barbata	i.	342
Grœnlandica	i.	343
Fetida	i.	345

INDEX.

	Vol.	Page
Phoca Ursina	i.	346
Physeter	iii.	93
Macrocephalus	iii.	94
Piscivorous Cetaceous animals	iii.	56
Plantigrada	i.	108
Porcupine Canada	ii.	149
Porpus	iii.	69
Pouched Rat	ii.	89
Proboscidia	ii.	202
Procyon	i.	161
Lotor	i.	163
Pseudostoma	ii.	89
Pteromys	ii.	146
Volucella	ii.	ib.
Rabbit	ii.	157
Raccoon	i.	161
Rat	ii.	76
Brown or Norway	ii.	78
Black	ii.	83
Rein Deer	ii.	283
Sable	i.	208
Scalops	i.	81
Sciurus	ii.	122
Vulpinus	ii.	128
Cinereus	ii.	129
Carolinensis	ii.	131
Niger	ii.	133
Macroureus	ii.	134
Grammurus	ii.	136
Quadrivittatus	ii.	137
Hudsonius	ii.	138
Rufiventer	ii.	141
Striatus	ii.	142
Lateralis	ii.	144
Seal	i.	306
Common	i.	313
Hooded	i.	336
Great	i.	342
Harp	i.	343
Fetid	i.	345
Ursine	i.	346
Sea-Swine	iii.	169
Sea-Unicorn	iii.	81
Sirenia	iii.	39
Sheep	ii.	328

INDEX.

		Vol.	Page
Shrew		i.	74
Mole		i.	81
Skunk		i.	211
Sloth		ii.	173
Giant		ii.	ib.
Cuvier's		ii.	180
Jefferson's		ii.	197
Sorex		i.	74
Parvus		i.	78
Brevicaudus		i.	79
Araneus		i.	80
Squirrel		ii.	122
Fox		ii.	128
Cat		ii.	129
Common Gray		ii.	131
Black		ii.	133
Great Tailed		ii.	134
Line do		ii.	136
Four Lined		ii.	137
Hudson's Bay		ii.	138
Red Belly		ii.	141
Ground		ii.	142
Rocky Mountain		ii.	144
Stellerus		iii.	49
Borealis		iii.	50
Steller		iii.	49
Boreal		iii.	50
Trichecus		i.	351
Rosmarus		i.	354
Tardigrada		ii.	173
Ursus		i.	109
Americanus		i.	114
Horribilis		i.	131
Maritimus		i.	143
Vespertilio		i.	51
Carolinensis		i.	67
Noveboracensis		i.	67
Pruinosus		i.	ib.
Arcuatus		i.	70
Subulatus		i.	71
Walrus		i.	354
Mitchill's Fossil		iii.	245
Weasel		i.	193
Whale		iii.	98
Razor-back		iii.	137

		Vol.	Page
Whale Broad-nosed	- - -	iii.	141
Finner	- - - -	iii.	142
Beaked	- - - -	iii.	144
Wolf Common	- - - -	i.	255
Prairie or Barking	- - -	i.	260
Dusky	- - - - -	i.	265
Black	- - - - -	i.	267
Wolverine	- - - -	i.	185
Wood-Rat	- - - -	ii.	69

ERRATUM.

Page 239, 5th line, for 203 read 302.

RAMBLES OF A NATURALIST.

BY

JOHN D. GODMAN, M. D.

PHILADELPHIA:
THOMAS T. ASH—KEY AND BIDDLE.
1833.

PREFACE.

The beautiful sketches, under the title of "The Rambles of a Naturalist," which first appeared in "The Friend," a religious and literary weekly journal of this city, were lately published collectively in *Waldie's Select Circulating Library*, and being deemed worthy of still further dissemination, they have been dressed in the present garb, and thus offered to public patronage. Though they were dashed off on the spur of the occasion, they possess all the characteristic freshness and vigour of the author's style, and were among the last effusions of his vigorous mind. They were written while he was confined to the bed of sickness, from which he was removed in a few weeks afterwards to the tomb and the series consequently interrupted. The sale of the present edition may contribute to more pious uses than mere commercial profits.

As a suitable accompaniment, the "Reminiscences" of Dr. Reynell Coates, likewise first published in "The Friend," have been appended; the whole forming a delightful pocket companion for a spring or summer ramble.

The biographical sketch is from the pen of Dr. Drake, of Cincinnati, and first appeared in the "Western Journal of the Medical and Physical Sciences."

"The great characteristics of Dr. Godman's mind," says a friend, who knew him well, "were his retentive memory, an unwearied industry and quick perception, and his capacity of concentrating all his powers upon any given object of pursuit. What he had once read or observed, he rarely, if ever, forgot. Hence it was, that although his early education was much neglected, he became an excellent linguist, and made himself master of Latin, French, and German, besides acquiring a knowledge of Greek, Italian and Spanish. He had read the best works in all these languages, and wrote with facility the Latin and French.

"His powers of observation were quick, patient, keen and discriminating; and it was these qualities that rendered him so admirable a naturalist. He came to the study of natural history as an investigator of facts, and not as a pupil of the schools; and while he regarded systems and nomenclature with perhaps too little respect, his great aim was to learn the instincts, the structure and the habits of all animated beings. This science was his favourite pursuit, and he devoted himself to it with indefatigable zeal. He has been heard to say, that in investigating the habits of the shrew mole, he walked many hundred miles. Those parts of his natural history in which he relates the results of his own observation, are among the most interesting essays on that subject in our language. This praise is due in a still greater degree to his Rambles of a Naturalist, which are not inferior in poetical beauty and vivid and accurate description, to the celebrated letters of Gilbert White on the Natural History of Selbourne. These essays were among the last productions of his pen, and were written in the intervals of acute pain and extreme debility. They form a mere sketch of what he intended, and had he lived to complete them, he would have left a work and a name of enduring popularity.

" There were few subjects of general literature, excepting the pure and mixed mathematics, with which Dr. Godman was not more or less familiar. Among other pursuits to which his attention had been turned, was the study of ancient coins, of which he had acquired a critical knowledge.

" The powers of his mind were always buoyant. His eagerness in the pursuit of knowledge seemed like the impulse of gnawing hunger and unquenchable thirst. Neither adversity nor disease could allay it, and had it pleased Providence to heal his mortal wound, and prolong his life and strength, he would have borne away the palm from all his contemporaries.

"The fine imagination and deep enthusiasm of Dr. Godman occasionally burst forth in impassioned poetry. He wrote verse and prose with almost equal facility, and had he lived and enjoyed leisure to prune the exuberance of his style, and to bestow the last polish upon his labours, he would have ranked as one of the great masters of our language, both in regard to the curious felicity, and the strength and clearness of his diction. The following specimens of his poetical compositions are selected less for their intrinsic excellence, than for the picture which they furnish of his private meditations."

A MIDNIGHT MEDITATION.

'Tis midnight's solemn hour ! now wide unfurled
Darkness expands her mantle o'er the world:
The fire-fly's lamp has ceased its fitful gleam
The cricket's chirp is hushed; the boding scream
Of the grey owl is stilled; the lofty trees
Scarce wave their summits to the failing breeze;
All nature is at rest, or seems to sleep;
'Tis thine alone, oh man ! to watch and weep !
Thine 'tis to feel thy system's sad decay,
As flares the taper of thy life away
Beneath the influence of fell disease :—
Thine 'tis to *know* the want of mental ease
Springing from memory of time misspent;
Of slighted blessings ; deepest discontent,
And riotous rebellion 'gainst the laws
Of health, truth, heaven, to win the world's applause !

Such was thy course, Eugenio, such thy hardened heart,
Till mercy spoke, and death unsheathed the dart,
Twanged his unerring bow, and drove the steel,
Too deep to be withdrawn, too wide the wound to heal;
Yet left of life a feebly glimmering ray,
Slowly to sink and gently ebb away.

—And yet, how blest am I?
While myriad others lie
In agony of fever or of pain,
With parching tongue and burning eye,
Or fiercely throbbing brain;
My feeble frame, though spoiled of rest,
Is not of comfort dispossest.
My mind awake, looks up to thee,
Father of mercy! whose blest hand I see
In all things acting for our good,
Howe'er thy mercies be misunderstood.

—See where the waning moon
Slowly surmounts yon dark tree tops,
Her light increases steadily, and soon
The solemn night her stole of darkness drops:
Thus to my sinking soul in hours of gloom,
The cheering beams of hope resplendent come,
Thus the thick clouds which sin and sorrow rear
Are changed to brightness, or swift disappear.
Hark! that shrill note proclaims approaching day;
The distant east is streaked with lines of gray;
Faint warblings from the neighbouring groves arise,
The tuneful tribes salute the brightening skies.
Peace breathes around; dim visions o'er me creep,
The weary night outwatched, thank God! I too may sleep.

Lines written under a feeling of the immediate approach of Death.

The damps of death are on my brow, the chill is in my heart,
My blood has almost ceased to flow, my hopes of life depart;
The valley and the shadow before me open wide,
But thou, Oh Lord! even there wilt be my guardian and my guide.
For what is pain, if thou art nigh its bitterness to quell?
And where death's boasted victory, his last triumphant spell?
Oh! Saviour, in that hour when mortal strength is nought,
When nature's agony comes on, and every anguished thought
Springs in the breaking heart a source of darkest woe,
Be nigh unto my soul, nor permit the floods o'erflow.
To thee! to thee alone! dare I raise my dying eyes;
Thou didst for all atone, by thy wondrous sacrifice;
Oh! in thy mercy's richness extend thy smiles on me,
And let my soul outspeak thy praise throughout eternity!

"Beneath the above stanzas is the following note: 'Rather more than a year has elapsed since the above was first written. Death is now certainly near at hand;

but my sentiments remain unchanged, except that my reliance on the Saviour is stronger.'

" This reliance on the mercies of God through Christ Jesus, became indeed the habitual frame of his mind; and imparted to the closing scenes of his life a solemnity and a calmness, a sweet serenity and a holy resignation, which robbed death of its sting, and the grave of its victory. It was a melancholy sight to witness the premature extinction of such a spirit; yet the dying couch on which genius, and virtue, and learning thus lay prostrated, beamed with more hallowed lustre, and taught a more salutary lesson than could have been imparted by the proudest triumphs of intellect. The memory of Dr. Godman, his blighted promise, and his unfinished labours, will long continue to call forth the vain regrets of men of science and learning. There are those who treasure up in their hearts as a more precious recollection, his humble faith and his triumphant death, and who can meet with an eye of pity, the scornful glance of the scoffer, and the infidel, at being told that if Dr. Godman was a philosopher, he was also a Christian."

MEMOIR

OF

DR. JOHN D. GODMAN.

Of Dr. Godman's early years, we have received a number of interesting memoranda, from his first medical preceptor, Dr. Luckey, now of Circleville, in this state. According to this gentleman, Dr. G. was born at Wilmington, in the state of Delaware. At an early period he lost his parents, and was left without patrimony, or deprived of it. Dr. Luckey first saw him in 1810, when he was fifteen years old. The doctor was, at that time, a senior student in the office of Dr. Thomas E. Bond, of Baltimore. "The office," says Dr. L., "was fitted up with taste, and boys, attracted by its appearance, would frequently drop in, to gaze on the labelled jars and drawers. Among them I discovered, one evening, an interesting lad, who was amusing himself with the manner in which his comrades pronounced the 'hard words,' with which the furniture was labelled. He appeared to be quite an adept in the Latin language. A strong curiosity soon prompted me to enquire 'Who are you?' 'Don't you recollect,' says he, 'that you visited a boy at Mr. Creery's, who had a severe attack of bilious colic?' 'I do. But what is your name, my little boy?' He was small of his age. 'My name, sir, is John D. Godman.' 'Did you study the Latin language with Mr. Creery?' 'No, he does not teach any but an English school.' 'Do you intend to prosecute your studies alone?' 'I do. And I will, if I live, make myself a Latin, Greek, and French scholar."

In the autumn of 1811, Dr. Luckey commenced the practice of medicine in Elizabethtown, Pennsylvania, and the next summer received a letter from his *protegé*, stating that he had been bound an apprentice to the printer of a newspaper. With this business, he was, from the beginning, exceedingly dissatisfied, as he evinced in his numerous letters to Dr. Luckey.

In one of these, dated July 23d, 1812, he expressed the opinion, that it was worse than " cramping his genius over a pestle and mortar"—it was " cramping it over a font of types, where there are words without ideas."

Addicted to reading, and aspiring to a more intellectual pursuit, it is not probable that our young printer was much devoted to the drudgery of the office, or performed his duties *con amore;* which may sufficiently explain the origin of the difficulties set forth in the following paragraph from a subsequent letter to the same.

"Every thing is in *statu quo* with me. The same series of oppressions, impositions and insults are still my lot to bear. But I will not bear them long. From the oldest to the youngest, master and man, all seem to have a disposition to peck at me. You will (or may be) surprised to hear that I can never make a printer. It is an erroneous opinion of some people, that no one can make a printer unless he be a scholar. On the contrary, scholars can hardly, if at all, be printers. I would not wish you to think that I count myself a scholar. On the contrary I think myself no scholar."

The following extract from another letter, dated October 23d, 1813, shows that, at this early period, young Godman was threatened with the malady which ultimately destroyed him.

"The disease for which I mentioned a recipe in my last has commenced its direful effects on my poor body. A continued pain in my breast, and at night a slow but burning fever, convince me that I am travelling down a much frequented road to the place where disease has no effect. This my friend is no phantasy. I do not say it from affectation. I feel it. I cannot believe in this disease being contagious, or I should be certain that I have caught it. I sleep with a youth who was born with it and has it fully."

In the opinion of Dr. L., the deceased, at that early period, laboured under a hypertrophy of the heart.

Through the whole of his apprenticeship, young Godman had a strong desire to study medicine, but his guardian was opposed to any change of destination. Early in the month of January, 1814, he writes to Dr. L.—

"At the suggestion of Dr. Anderson, I have determined to commence the study of chemistry, as he says it will be a great improvement to the mind, and more so, I may be enabled, the ensuing season (if I should live so long) to attend the lectures at the University (of Maryland,) and it seems to run greatly in Dr. A.'s head that I shall one day be a physician. How far this surmise may be right, time will disclose. It may indeed so happen, and should I study chemistry now, I shall not have it to do at a future period. I must, however, ask your opinion in this affair."

On the 24th of the same month, he writes to the same gentleman—

"I have read the catechetical part of Parke's Chemistry, and I can assure you I liked it not a little. But my knowledge, so far as I may obtain it, will only be theoretical."

In the same letter he sets forth his early views of the Christian religion:

"I have not ever had a fixed determination to read the works of that Modern Serpent,* nor had I determined *not* to do it; and it seems to me surprising, that a fellow

* Thomas Paine.

student of yours should recommend the perusal of such writings as Thomas Paine's.

"I had, thank heaven, before I asked you the question, and still have, the "Apology for the Bible," by the celebrated Lord Regius, of Landaff, (Bishop Watson.) There is a great comfort in the belief of that glorious doctrine of salvation, that teaches us to look to the Great Salvator for happiness in a future life; and it has always been my earnest desire, and I must endeavour to die the death of the righteous, that my last end and future state may be like his. It would be a poor hope indeed—it would be a sandy foundation for the dying soul, to have no hope but such as might be derived from the works of Bolingbroke and Paine; and how rich the consolation and satisfaction afforded by the glorious tidings of the blessed Scriptures. It is my opinion, there has never one of these modern deists died as their writings would lead us to believe; nor are but few of their writings read at the present day."

In the year 1814, when the war raged in the Chesapeake, he became a sailor under Com. Barney, and was engaged in the service at the bombardment of Fort M'Henry. Early in the next year, Dr. Luckey, captivated by his genius, and touched by his misfortunes, resolved to invite him to his house, in Elizabethtown, and afford him all the facilities in his power for studying the profession to which he aspired. It does not appear how he had rid himself of his apprenticeship; but he seems to have been at liberty to accept the doctor's generous invitation. This he did, with emotions of joy which are uttered in the following simple and affecting reply, dated April 4th, 1815.

"I have this hour received your last letter, and I can assure you, that language is inadequate to express to you my sincere, unfeigned joy, for the pleasing news you have communicated to me. Let the manner in which these lines are penned, convince you of the state of my mind at present. I was, thirty minutes before I received your letter, on the point of going to a printer, in this city, to seek employment, and, but for Providence, I should have done so. You may suppose that, as soon as I read your letter, I abandoned this intention and returned to my sister's house,* 'with fire in each eye and paper in each hand,' to answer your epistle of friendship's own dictating. I must lay this aside for a short time, till my mind becomes settled and undisturbed. I stopped at the line above, in order that I might recover a small degree of composure, in order to express myself as I ought, to so good a friend. I will certainly comply with your request, should it please God to continue my health and strength during the ensuing week. Should it please the mercy of Providence to suffer me to take up my residence with you, I shall endeavour, by the most indefatigable study and diligence, to give you the satisfaction

---

* Mrs. Stella Miller, of Baltimore.

your kindness to me deserves. I am in hopes that I shall be able to come some day in the course of the next week; but, as my journey must be a pedestrian one, I should not wish to mention a particular day."

"On the 10th of April, four days after the date of this letter, he arrived," says Dr. L., "at my house, and took up his residence in my family. He made his promise good, for in *six weeks* he had acquired more knowledge in the different departments of medical science, than most students do in a year. During this short period he not only read Chaptal, Fourcroy, Chesselden, Murray, Brown, Cullen, Rush, Sydenham, Sharp, and Cooper, but wrote annotations on each, including critical remarks on the incongruities in their reasonings. He remained with me five months, and at the end of that time, you would have imagined from his conversation, that he was an Edinburgh graduate. When he sat down to study, so completely was he absorbed by his subject, that it seemed as though the amputation of one of his limbs would scarcely withdraw his attention."

A circumstance having no connection with the relation between him and his benefactor, but involving them both, led to premature separation. One or both of them were requested by the political party to which they belonged, to deliver orations on the approaching Fourth of July. Dr. L. began at the appointed hour, and went through with his discourse, but attempts were made by the opposite party to offer insult and create disturbance;

at which our young orator became indignant ; and yielding to the impulse of his strong native feelings, not only refused to deliver what he had prepared, but resolved on returning forthwith to Baltimore. His oration was left with his preceptor, who speaks of it as not unworthy of Patrick Henry.

Departing from Elizabethtown, he returned to Baltimore, and became a pupil of Dr. Hall; and, in the succeeding autumn, began to attend the lectures in that city. His pecuniary difficulties, however, were pressing, and, in the ensuing February, 1816, he wrote to his benefactor in the following eloquent and affecting style :

" Need I then inform you how high my expectations were raised, when I commenced attending the lectures this winter—need I say I was almost certain of future competency ? Alas ! my friend, the Great Ruler of events has interposed (in order to teach me resignation to his will) this heavy disappointment. By unforeseen events—by domestic calamities, I have been compelled to relinquish the study of medicine, so long the ultimatum of all my hopes. FATHER OF ALL, THY WILL BE DONE. I have made this my motto—my consolation; and did I not daily see the truth of " *Omnia pro optimo*," I might perhaps repine. I am now in expectation of a situation with an eminent apothecary of this city, and I may be enabled, at a future period, to recommence the study of medicine."

This situation however he did not obtain.

"Let me now give you a retrospect of 'the days of my life.' Since I have returned from you, I have discovered my *real* age, in an old book of my father's, (and you would hardly suppose it,) I was 21 years old the 20th day of December, 1815. Before I was two years old I was motherless—before I was five years old I was fatherless and friendless—I have been cast among strangers—I have been deprived of property by *fraud*, that was mine by right—I have eaten the bread of misery—I have drunk of the cup of sorrow—I have passed the flower of my days in a state little better than slavery, and have arrived—at what? Manhood, poverty, and desolation. Heavenly Parent, teach me patience and resignation to thy will."

About this time he seems to have found a patron in Professor Davidge, and, on the 18th of April following, he wrote to Dr. Luckey—

"I still continue to study with Dr. Wright, (the partner of Dr. Davidge,) and provided it shall be the will of heaven, I may possibly procure admission in the course of the next year into the venerable circle of medicine."

In speaking of his perplexed and embarrassed situation, and of the mutations of fortune, he says—

"There is only one thing which points to, and affords immutable consolation, and that is, the observance of religion. Although we should be incapable of reaping enjoyment in this world, even from uninterrupted prosperity, yet we can ardently long for, and sincerely believe, we may be eternally happy in the next."

In this situation he finished his medical education. In the language of Professor Sewall*—

"Here he pursued his studies with such diligence and zeal, as to furnish, even at that early period, strong intimations of his future eminence. So indefatigable was he in the acquisition of knowledge, that he left no opportunity of advancement unimproved, and notwithstanding the deficiencies of his preparatory education, he pressed forward with an energy and perseverance, that enabled him not only to rival, but to surpass all his fellows."

He appears to have attended the lectures in the Baltimore school, through the sessions commencing in the autumns of 1816, and 1817. In the course of the last, Professor Davidge was disabled, by an accident, for several weeks, and Mr. Godman was appointed to supply his place. This, as he had been an apprentice to a trade, not three years before, in the same city, was an honourable testimony to his talents and industry, and must have been highly gratifying to his ambition, According to Professor Sewall, (*loco citato.*)

"This situation he filled for several weeks with so much propriety—he lectured with such enthusiasm and eloquence, his illustrations were so clear and happy, as

---

* Eulogy on Dr. Godman, p. 4.

to gain universal applause; and at the time he was examined for his degree, the superiority of his mind, as well as the extent and accuracy of his knowledge, were so apparent, that he was marked by the professors of the University as one who was destined at some future period to confer high honour upon the profession."

In reference to his graduation, on the 10th of February, 1818, he wrote to his friend, Dr. Luckey, in these emphatical words:

"I know not what to tell you for news, unless I tell you that I passed my graduate examination, on Saturday; (Feb. 7,) which lasted twenty minutes; and, of course, 1 have now the 'vast unbounded prospect all before me;' though 'shadows, clouds, and darkness rest upon it.' I will go to the country to practise, most probably to Frederick county."

In the United States, it is common to see young men, without preparatory education or fortune, become practitioners of medicine; but most of this class struggle into the ranks of the profession, totally unprepared; and depart from it for other pursuits, or for the grave, unknown and unhonoured by the scientific world. *Such* an admission, must not be confounded with that of young Godman; who scorned to enter the profession unqualified and unauthorised by those who guard, or ought to guard, its portals. In this respect he was a shining example; and his subsequent success should animate every friendless young man, who may engage in the study of medicine, to imitate his industry and unfaltering perseverance. By these means, if not blessed with his genius, they may prepare themselves for extensive usefulness, and earn respectability if not renown.

We come now to contemplate Dr. Godman, as a member of the profession. His first location was in the village of New Holland, on the banks of the Susquehanna; where, however, he remained but a few months. The next was on the Patapsco, near Baltimore, whence, in July, 1819, he wrote to Dr. Luckey as follows:

"My success in business has been considerable, or my practice, at least, has been as extensive as I could rationally expect." "What my success may be in the end is at present very doubtful. I still have considerable expectation of being recalled to Baltimore, in order to fill the place which I held in the University. If it so happen, I shall be much delighted, as a country life is very little, or not at all, to my taste."

In these rural situations he devoted himself to the study of nature; and, at a subsequent time, set forth the fruits of his observations in a series of papers, entitled the Rambles of a Naturalist. But his ardent temperament was little adapted to the stagnant existence of a village doctor. He thirsted for competition, and longed to engage in the rivalries which prevail among the candidates for fame. Nature seems to have urged him on. It was she who revealed to him the compass of his intellectual powers; and bid him seek a theatre commensurate with their efficiency. A different arrangement from what he had anticipated was made in the Baltimore school; he returned, however, to that city, but at length boldly resolved to fix himself in Philadelphia, and become a public teacher of anatomy and physiology.

But an unexpected event gave, for the time being, a different direction to his efforts. The writer of this article was enquiring, at that time, for a suitable person to fill the chair of surgery in the medical college of Ohio, the first session of which had just closed; and Dr. Godman was recommended. His qualifications for the first place, were expressed by Professor Gibson, then of the University of Pennsylvania, but previously a member of the Baltimore institution, in the following unequivocal and prophetic language. "In my opinion, Dr. Godman would do honour to any school in America." He was forthwith appointed; and arrived in Cincinnati the ensuing October, (1821,) in time to enter on the second session of the school.

For the practical details of such a professorship, he could not of course be well prepared, as his surgical experience was exceedingly limited; but he was learned in the institutes of the science, and his knowledge of anatomy was comprehensive, accurate and commanding. As a dissector, he was equally rapid and adroit. His lectures were well received by the class, who admired his genius, were captivated by his eloquence, and charmed with the *naiveté* of his manners.

In the course of the session, difficulties, of which he was neither the cause nor the victim, were generated in the faculty, the class was small, and the prospects of the institution overcast: under these circumstances, Dr. Godman resigned, but did not at that time return to the east.

Not long before, the author of this narrative had issued proposals for a medical journal, to be edited by the professors of the college, and obtained a number of subscribers; but the distracted state of the institution prevented the fulfilment of the design. To this enterprise, as soon as he had resigned, Dr. Godman directed his attention; and assisted by Mr. Foote, a liberal and literary bookseller in this city, in a few weeks issued the first number of the *Western Quarterly Reporter*. Thus, if not the first to project, Dr. G. had the honour of being the first to commence, a journal of medicine, in the Valley of the Mississippi. At the end of the 6th number, of a hundred pages each, the work was discontinued, for, previously to that time, its editor had returned to Philadelphia. More than three hundred pages of this periodical were from his own pen; chiefly in translations and reviews of anatomy, physiology, and medical jurisprudence.

Dr. Godman resided in our city for one year only; but in that short period he deeply inscribed himself on the public mind. The memory of his works still remains with us. In addition to writing for his medical journal, and to his practice, which was considerable for a stranger, he erected an apparatus for sulphurous fumigation, and translated and published a French pamphlet on that remedy; he read medical books, and many current works of general literature; prosecuted the study of the German and Spanish languages; and labelled the ancient coins and medals of the Western Museum. In the midst of the whole, he found time to cultivate his social relations; and every day added a new friend to the catalogue of those, who loved him for his simplicity and frankness, not less than they admired him for his genius, vivacity, and diligence. Thus, to use an idiomatic expression, he was a growing man, and might have remained with us and done well. But the hand of destiny was upon him. He had left the banks of the Patapsco, to be a public teacher: the same object had drawn him from Philadelphia to Cincinnati; and that object, at length, restored him to the great emporium of the medical sciences. Contrary to the wishes and importunities of his western friends, in the autumn of 1822, with his young family, he set off for the theatre of his future glory; which he reached in safety, though not without some of the many difficulties, at that time connected with a journey across the state of Ohio; of which, in a letter from Wheeling to one of his friends in this city, he gave a familiar account, in all respects so characteristic, that we hope to be excused for extracting it:

"We arrived last night, after a journey which exceeded

in miseries any twenty journeys I ever made in my life. Thank God, the whole has been productive of nothing worse, than some hoarseness to my wife, and a galloping consumption of my bank notes. We were thirteen days on the way, *twelve* of which gave us as heavy rains as ever poor mortals could venture to travel in; and this produced such a delightfully *soft* state of the roads, that but for the rocks, (which fortunately were not twenty feet below the surface,) we might have been extracted some thousand years hence, in a high state of preservation, to decorate Best's museum, having one of Dorfeuille's mummy labels around our necks.

"If I were one of the 'tristful travellers,' I might draw much 'matter of melancholy' from these 'misadventures,' as my friend Sancho Panza calls them. But as the blessed sun of heaven has driven forth once more in his beamy chariot, and the clouds are scattered from their long held seats, those which have loured on my mind, have also fled; and with 'a light heart,' I am once more preparing to encounter all the good or ill that God may send."

Of Dr. Godman's life and labours from this time forward, we shall say but little, as they are known to all the reading people of the United States, both in and out of the profession; and as our chief object is to present the difficulties and triumphs of his earlier years, for the benefit of our younger readers.

In Philadelphia he immediately began to lecture on

anatomy and physiology, his first and greatest objects; and succeeded so well, that, in 1826, he was called to Rutgers' College, in the city of New York, as associate of Mott and Hosack.

In 1824 he was made one of the editors, (a *working* editor,) of the Philadelphia Journal of the Medical Sciences: and continued a liberal contributor to that respectable periodical, to the last weeks of his life.

At different times he published a number of interesting and eloquent introductory lectures.

He was the writer of several elaborate analytical and critical reviews, in the American Quarterly.

At the present time, actual discoveries in anatomy are no more to be expected, yet Dr. G., with admirable skill, revealed many new connections and relations of certain parts, and described them in a volume which he entitled Anatomical Investigations.

He translated and published from the Latin, French, and German languages, a variety of papers and distinct treatises; several of them on subjects not professional, as for example, Lavasseur's Narrative of La Fayette's Visit to the United States.

He wrote critical and emendatory notes on several important English and continental works, which the booksellers of this country were about to publish.

The article of Natural History, in the Encyclopedia Americana, was exclusively confided to him, and his labours upon it ended only with his life.

He studied the Zoology of North America, both existing and fossil, and favoured us with an interesting and extended history of all its own quadrupeds, embracing a great variety of new observations.

Such were the labours of the deceased, during the seven years that he resided in Philadelphia and New York. For the whole of that period, his life was one of unmitigated toil. As far back as November, 1823, he writes to his friend Dr. Best,

" Whatever you may think of my long continued silence, it has been unavoidably produced by the incessant and laborious employments which have occupied the whole of my time."

In 1824, he writes to another friend—

" My time has been very much occupied in the various duties which devolve on me here, and I am obliged to neglect my friends, in appearance, because it is out of my power to bestow the necessary attention to correspondence."

Again, in 1825, he says to the same—

" It is needless to tell you, that I am excessively occupied, and shall be more so as the winter approaches."

In the next year we find him still in the same condition—

" If you expect news at my hands," says he to Dr. Best, " you expect in vain. My life is one monotonous round of incessant toil after bread and *fame*, that ' *certain portion of uncertain paper*.' Of my success in the bread making way, I can, thank God, speak more satisfactorily than when we last met, though still nothing to boast of."

Again in the same year he writes—

"You recollect how much and how hard I had to work, when you were here—that was nothing to what I have to do now, as vigilance and labour are incessantly demanded, not only to gain more 'reputation,' but to retain that which I have already with vast toil acquired."

In the following year, after he had removed to New York, and was there a candidate for professional business, he writes to the same friend—

"The prospects of our college are fair enough at present, but what will be the event, cannot be told until the time of trial arrives. For my own part, I am not a little sick of the life such a business occasions, and think you far better off, in a situation, where you can acquire a subsistence and respect, without the incessant worry and vexation attendant on a life of professional ambition. For my own part, I shall lay myself as much out for the profession as I can, though I fear, not the best subject for improvement in that way. My situation is such, that I am obliged to rely, in a very great degree, on my pen, and that, you will say, produces habits very little compatible with the introduction of one's self into practice, where there are so many professed bowers, scrapers, and flatterers."

In the ensuing winter he was seized with the disease of the lungs of which he finally died, and was compelled to suspend his lectures. In the following January, 1829, he speaks to the same gentleman, of his situation and labours, in these affecting words—

"My excessive exertion, and the exposure to a dreadful climate destroyed me. My lungs became diseased, and last winter, I was threatened with so rapid a decline as to force me to escape from the climate of New York, by going to the West Indies. The months of February, March, and April, my wife and I spent in the Danish Island of Santa Cruz, where I very nearly perished from my disease, though I certainly should have done so in New York. On my return to Philadelphia, in May, I took a house in Germantown, within seven miles from the city, where I have since resided. During the warm weather I was able to creep about, but since the first of the fall have been confined to a single room. My health during all this time has been in a very wretched state, and my *consumption* very obvious indeed, for I wasted to bones and lost all my strength. Until the last three weeks past, I was exceedingly low, unable to sit up, eat, or perform any function advantageously. Since the time mentioned I have greatly recovered in all respects. My cough is by no means troublesome, and I eat and sleep well. What is best of all is that I have never had hectic since leaving New York, where I was not properly prescribed for. Notwithstanding all these drawbacks, I have had my family to support, and have done so merely by my pen. This you may suppose severe enough for one in my condition, nevertheless necessity is a ruthless master. At present, that I am comparatively well, my literary occupations form my chief pleasure, and all the regret I experience is, that my strength is so inadequate to my wishes. Should my health remain as it is now I shall do very well, and I cannot but hope, since we have recently passed through a tremendous spell of cold weather without my receiving any injury. All my prospects as a public teacher of anatomy are utterly destroyed, as I can never hope, nor would I venture if I could, again to resume my labours. My success promised to be very great, but it has pleased God that I should move in a different direction."

In the following year, continuing to write for the support of his family till the last month of his existence, he was taken from them, and in him they lost their all. Twelve years of unfaltering industry, that had carried his name into all the countries where science is cultivated, had not enabled him to accumulate property; and ended by consigning him to the grave, ere he reached the noon-day of life, or had put forth, to their full extent, the vast intellectual powers, with which he was endowed. In all this, there is much more to grieve than astonish us. As a physician and surgeon, Dr. Gòdman's business was never considerable. At the very beginning of his professional career, his mind took a different direction. No human heart was ever imbued with a deeper thirst for knowledge, or warmed with a nobler love of glory. He made the former subservient to the latter; but the objects of his ambition were teaching and writing, not the practice of his profession. Perhaps, indeed, he adapted the aims of his ambition to his taste. He relished reading, writing, and lecturing, more than the practice of medicine; and sought to derive from them, that emolument, which, in this country, they seldom afford, and which can much more certainly be drawn from a close attention to the practical duties of the profession. Had he possessed a patrimony, this course would have been unexceptionable; without such a reliance, no young physician should neglect the means of acquiring professional business, at the outset of his career.

Dr. Godman was, without doubt, a man of genius; but he was not, perhaps, so much the expositor, as the historian of nature. Observing, imaginative, fluent, and graphical, he abounded less in deep and original analysis than vivid and accurate delineations. Thus his mind, like that of Lucretius, Darwin, and Good, was poetical and philosophical; and he left behind him several fugitive pieces, written chiefly in his last illness, which prove that he might have shone as the poet of nature, not less than her historian, had circumstances awakened his powers.

He possessed uncommon abilities for dissection, and was accustomed, in the presence of his class, to disentangle the structures intended for exhibition; thus showing their connections and dependences, while he described them with that clearness, animation, and eloquence, which only can render the study of anatomy attractive.

In every situation, and on every subject, his attention was active and acute, his perceptions rapid, his memory exceedingly retentive, and his ratiocination profound and analytical.

For languages, he had both taste and talents; and, succeeded in acquiring a practical knowledge of a greater number, perhaps, than any American physician who had preceded him.

The qualities of his heart harmonised with those of his head. They did honour to the profession, and inspired confidence wherever he went. To pure moral habits, and incorruptible honesty, he added that unsuspecting frankness, and all those fine and glowing sensibilities, which at once excite our respect, and win our affection.

But it is not our design to attempt an extended delineation of his character, and we shall close an article already prolonged far beyond our original intention, with his own statement of his opinions and hopes, in regard to that world of which he is now a " bright inhabitant."

In his last letter to Dr. Best, who followed him in a few months, he writes:—

"It gives me great happiness to learn that *you* have been taught, as well as myself, to fly to the Rock of Ages for shelter against the afflictions of this life, and for hopes of eternal salvation. But for the hopes afforded me, by an humble reliance on the all-sufficient atonement of our blessed Redeemer, I should have been the most wretched of men. But I trust, that the afflictions I have endured have been sanctified to my awakening, and to the regeneration of my heart and life. May we, my dear friend, persist to cling to the only sure support against all that is evil in life, and all that is fearful in death."

Thus fell from the firmament of the American profession, before he had reached his meridian splendour, one of the brightest stars which have yet risen above its horizon; but he was one only, and we may hope that his own example will contribute to place some other in the constellation.

RAMBLES OF A NATURALIST.

NO. I.

From early youth devoted to the study of nature, it has always been my habit to embrace every opportunity of increasing my knowledge and pleasures by actual observation, and I have found ample means of gratifying this disposition, wherever my place has been allotted by Providence. When an inhabitant of the country, it was sufficient to go a few steps from the door to be in the midst of numerous interesting objects; when a resident of the crowded city, a healthful walk of half an hour placed me where my favourite enjoyment was offered in abundance; and now, when no longer able to seek in fields and woods and running streams for that knowledge which cannot readily be elsewhere obtained, the recollection of my former rambles is productive of a satisfaction, which past pleasures but seldom bestow. Perhaps a statement of the manner in which my studies were pursued, may prove interesting to those who love the works of nature, and may not be aware how great a field for original observation is within their reach, or how vast a variety of instructive objects are easily accessible, even to the occupants of a bustling metropolis. To me it will be a source of great delight to spread these resources before the reader, and enable him so cheaply to participate in the pleasures I have enjoyed, as well as place him in the way of enlarging the general stock of knowledge by communicating the results of his original observations.

One of my favourite walks was through Turner's lane, near Philadelphia, which is about a quarter of a mile long, and not much wider than an ordinary street, being closely fenced in on both sides; yet my reader may feel surprised when informed that I found ample employment for all my leisure, during six weeks, within and about its precincts. On entering the lane from the Ridge road, I observed a gentle elevation of the turf beneath the lower rails of the fence, which appeared to be uninterruptedly continuous; and when I had cut through the verdant roof with my knife, it proved to be a regularly arched gallery or subterranean road, along which the inhabitants could securely travel at all hours without fear of discovery. The sides and bottom of this arched way were smooth and clean, as if much used; and the raised superior portion had long been firmly consolidated by the grass roots, intermixed with tenacious clay. At irregular and frequently distant intervals, a side path diverged into the neighbouring fields, and by its superficial situation, irregularity, and frequent openings, showed that its purpose was temporary, or had been only opened for the sake of procuring food. Occasionally I found a little gallery diverging from the main route beneath the fence, towards the road, and finally opening on the grass, as if the inmate had come out in the morning to breathe the early air, or to drink of the crystal dew which daily gemmed the close cropped verdure. How I longed to detect the animal which tenanted these galleries, in the performance of his labours! Farther on, upon the top of a high bank, which prevented the pathway from continuing near the fence, appeared another evidence of the industry of my yet unknown miner. Half a dozen hillocks of loose, almost pulverised earth were thrown up, at irregular distances, communicating with the main gallery by side passages. Opening one of these carefully, it appeared to differ little from the common gallery in size, but it was very difficult to ascertain where the loose earth came from, nor have I ever been able to tell, since I never witnessed the formation of these hillocks, and conjectures are forbidden, where nothing but observation is requisite to the decision. My farther progress was now interrupted by a delightful brook which sparkled across the road over a clear sandy bed; and here my little galleries turned into the field, coursing along at a moderate distance from the stream. I crept through the fence into the meadow on the west side, intending to discover, if possible, the animal whose works had first fixed my attention, but as I approached the bank of the rivulet something suddenly retreated towards the grass, seeming to vanish almost unaccountably from sight. Very carefully examining the point at which it disappeared, I found the entrance of another gallery or burrow, but of very different construction from that first observed. This new one was formed in the grass, near and among whose roots and lower stems a small but regular covered way was practised. Endless, however, would have been the attempt to follow this, as it opened in various directions, and ran irregularly into the field, and towards the brook, by a great variety of passages. It evidently belonged to an animal totally different from the owner of the subterranean passage, as I subsequently discovered, and may hereafter relate. Tired of my unavailing pursuit, I now returned to the little brook, and seating myself on a stone, remained for some time unconsciously gazing on the fluid which gushed along in unsullied brightness over its pebbly bed. Opposite to my seat, was an irregular hole in the bed of the stream, into which, in an idle mood, I pushed a small pebble with the end of my stick. What was my surprise, in a few seconds afterwards, to observe the water in this hole in motion, and the pebble I had pushed into it gently approaching the surface. Such was the fact; the hole was the dwelling of a stout little crayfish or fresh water lobster, who did not choose to be incommoded by the pebble, though doubtless he attributed its sudden arrival to the usual accidents of the stream, and not to my thoughtless movements. He had thrust his broad lobster-like claws under the stone, and then drawn them near to his mouth; thus making a kind of shelf; and as he reached the edge of the hole, he suddenly extended his claws, and rejected the incumbrance from the lower side, or down stream. Delighted to have found a living object with whose habits I was unacquainted, I should have repeated my experiment, but the crayfish presently returned with what might be called an armful of rubbish, and threw it over the side of his cell, and down the stream as before. Having watched him for some time while thus engaged, my attention was caught by the considerable number of similar holes along the margin and in the bed of the stream. One of these I explored with a small rod, and found it to be eight or ten inches deep, and widened below into a considerable chamber, in which the little lobster found a comfortable abode. Like all of his tribe, the crayfish makes considerable opposition to being removed from his dwelling, and bit smartly at the stick with his claws: as my present object was only to gain acquaintance with his dwelling, he was speedily permitted to return to it in peace.

Under the end of a stone lying in the bed of the stream, something was floating in the pure current, which at first seemed like the tail of a fish, and being desirous to obtain a better view, I gently raised the stone on its edge, and was rewarded by a very beautiful sight. The object first observed was the tail of a beautiful salamander, whose sides were of a pale straw colour, flecked with circlets of the richest crimson. Its long lizard like body seemed to be semitransparent, and its slender limbs appeared like mere productions of the skin. Not far distant, and near where the upper end of the stone had been, lay crouched, as if asleep, one of the most beautifully coloured frogs I had ever beheld. Its body was slender compared with most frogs, and its skin covered with stripes of bright reddish brown and grayish green, in such a manner as to recall the beautiful markings of the tiger's hide; and since the time alluded to, it has received the name of *Tigrina* from Leconte, its first scientific describer. How long I should have been content to gaze at these beautiful animals, as they lay basking in the living water, I know not, had not the intense heat made me feel the necessity of seeking a shade. It was now past 12 o'clock, I began to retrace my steps towards the city; and without any particular object moved along by the little galleries examined in the morning. I had advanced but a short distance, when I found the last place where I had broken open the gallery was *repaired.* The earth was perfectly fresh, and I had lost the chance of discovering the miner, while watching my new acquaintances in the stream. Hurrying onward, the same circumstance uniformly presented; the injuries were all efficiently repaired, and had evidently been very recently completed. Here was one point gained; it was ascertained that these galleries were still inhabited, and I hoped soon to become acquainted with the inmates. But at this time, it appeared fruitless to delay longer, and I returned home, filled with anticipations of pleasure from the success of my future researches. These I shall relate on another occasion, if such narrations as the present be thought of sufficient interest to justify their presentation to the reader.

NO. II.

On the day following my first related excursion, I started early in the morning, and was rewarded by one sight, which could not otherwise have been obtained, well worth the sacrifice of an hour or two of sleep. There may be persons who will smile contemptuously at the idea of a *man's* being delighted with such trifles; nevertheless, we are not inclined to envy such as disesteem the pure gratification afforded by these simple and easily accessible pleasures. As I crossed an open lot on my way to the lane, a succession of gossamer spider webs, lightly suspended from various weeds and small shrubs, attracted my attention. The dew which had formed during the night was condensed upon this delicate lace, in globules of most resplendent brilliance, whose clear lustre pleased while it dazzled the sight. In comparison with the immaculate purity of these dewdrops, which reflected and refracted the morning light in beautiful rays as the gossamer webs trembled in the breeze, how poor would appear the most invaluable diamonds that were ever obtained from Golconda or Brazil! How rich would any monarch be that could boast the possession of *one* such, as here glittered in thousands on every herb and spray! They are exhaled in an hour or two and lost, yet they are almost daily offered to the delighted contemplation of the real lover of nature, who is ever happy to witness the beneficence of the great Creator, not less displayed in trivial circumstances, than the most wonderful of his works.

No particular change was discoverable in the works of my little miners, except that all the places which had been a second time broken down were again repaired, showing that the animal had passed between the times of my visit; and it may not be uninteresting to observe how the repair was effected. It appeared, when the animal arrived at the spot broken open or exposed to the air, that it changed its direction sufficiently downwards to raise enough of earth from the lower surface to fill up the opening; this of course slightly altered the direction of the gallery at this point, and though the earth thrown up was quite pulverulent, it was so nicely arched as to retain its place, and soon became consolidated. Having broken open a gallery where the turf was very close, and the soil tenacious, I was pleased to find the direction of the chamber somewhat changed; on digging farther with my clasp knife, I found a very beautiful cell excavated in very tough clay, deeper than the common level of the gallery and towards one side. This little lodging-room would probably have held a small melon, and was nicely arched all round. It was perfectly clear, and quite smooth, as if much used; to examine it fully, I was obliged to open it completely. (The next day, it was replaced by another, made a little farther to one side, exactly of the same kind; it was replaced a second time, but when broken up a third time, it was left in ruins.) As twelve o'clock approached, my solicitude to discover the little miner increased to a considerable degree; previous observation led me to believe that about that time his presence was to be expected. I had trodden down the gallery for some inches in a convenient place, and stood close by, in vigilant expectation. My wishes were speedily gratified; in a short time the flattened gallery began at one end to be raised to its former convexity, and the animal rapidly advanced. With a beating heart, I thrust the knife blade down by the side of the rising earth, and quickly turned it over to one side, throwing my prize fairly into the sunshine. For an instant, he seemed motionless from surprise, when I caught and imprisoned him in my hat. It would be vain for me to attempt a description of my pleasure in having thus succeeded, small as was my conquest. I was delighted with the beauty of my captive's fur; with the admirable adaptation of his diggers or broad rose-tinted hands; the wonderful strength of his forelimbs, and the peculiar suitableness of his head and neck to the kind of life the Author of nature had designed him for. It was the shrew-mole, or *scalops canadensis*, whose history and peculiarities of structure are minutely related in the 1st volume of Godman's American Natural History. All my researches never enabled me to discover a nest, female or young one of this species. All I ever caught were males, though this most probably was a mere accident. The breeding of the scalops is nearly all that is wanting to render our knowledge of it complete.

This little animal has eyes, though they are not discoverable during its living condition, nor are they of any use to it above ground. In running round a room, (until it had perfectly learned where all the obstacles stood,) it would uniformly strike hard against them with its snout, and then turn. It appeared to me as singular that a creature which fed upon living earth worms with all the greediness of a pig, would not destroy the larvæ or maggots of the flesh fly. A shrew-mole lived for many weeks in my study, and made use of a gun case, into which he squeezed himself, as a burrow. Frequently he would carry the meat he was fed with into his retreat; and as it was warm weather, the flies deposited their eggs in the same place. An offensive odour led me to discover this circumstance, and I found a number of large larvæ over which the shrew-mole passed without paying them any attention: nor would he, when hungry, accept of such food, though nothing could exceed the eager haste with which he seized and munched earth worms. Often when engaged in observing him thus employed, have I thought of the stories told me, when a boy, of the manner in which snakes were destroyed by swine; his voracity readily exciting a recollection of one of these animals, and the poor worms writhing and twining about his jaws answering for the snakes. It would be tedious were I to relate all my rambles undertaken with a view to gain a proper acquaintance with this creature, at all hours of the day, and late in the evening, before daylight, &c. &c.

Among other objects which served as an unfailing source of amusement, when resting from the fatigue of my walks, was the little inhabitant of the brook, called the *gyrinus natator*. These merry swimmers occupied every little sunny pool in the stream, apparently altogether engaged in sport. A circumstance connected with these insects, gives them additional interest to a close observer; they are allied by their structure and nature to those nauseous vermin, the cimices (or *bedbugs*.) All of which, whether found infesting fruits or our dormitories, are distinguished by their disgusting odour. But their distant relatives, called by the boys the *water-witches* and *apple smellers*, the gyrinus natator above alluded to, has a delightful smell, exactly similar to that of the richest, mellowest apple. This peculiarly pleasant smell frequently causes the idler many unavailing efforts to secure some of these creatures, whose activity in water renders their pursuit very difficult, though by no means so much so as that of some of the long legged water spiders which walk the waters dry shod, and evade the grasp with surprising ease and celerity.

What purposes either of these racers serve in the great economy of nature, has not yet been ascertained, and will scarcely be determined until our store of *facts* is far more extensive than at present. Other and still more remarkable inhabitants of the brook, at the same time, came within my notice, and afforded much gratification in the observation of their habits.

NO. III.

In moving along the borders of the stream, we may observe, where the sand or mud is fine and settled, a sort of mark or cutting, as if an edged instrument had been drawn along, so as to leave behind it a track or groove. At one end of this line, by digging a little into the mud with the hand, you will generally discover a shell of considerable size, which is tenanted by a molluscous animal of singular construction. On some occasions, when the mud is washed off from the shell, you will be delighted to observe the beautifully regular dark lines with which its greenish smooth surface is marked. Other species are found in the same situations, which, externally, are rough and inelegant, but within are ornamented to a most admirable degree, presenting a smooth surface of the richest pink, crimson, or purple, to which we have nothing of equal elegance to compare it. If the mere shells of these creatures be thus splendid, what shall we say of their internal structure, which, when examined by the microscope, offers a succession of wonders? The beautiful apparatus for respiration, formed of a network regularly arranged, of the most exquisitely delicate texture; the foot, or organ by which the shell is moved forward through the mud or water, composed of an expanded spongy extremity, capable of assuming various figures to suit particular purposes, and governed by several strong muscles that move it in different directions; the ovaries, filled with myriads, not of eggs, but of perfect shells, or complete little animals, which, though not larger than the point of a fine needle, yet when examined by the microscope, exhibit all the peculiarities of conformation that belong to the parent; the mouth, embraced by the nervous ganglion, which may be considered as the animal's brain; the stomach, surrounded by the various processes of the liver, and the strongly acting, but transparent heart, all excite admiration and gratify our curiosity. The puzzling question often presents itself to the enquirer, why so much elaborateness of construction, and such exquisite ornament as are common to most of these creatures, should be bestowed? Destined to pass their lives in and under the mud, possessed of no sense that we are acquainted with, except that of touch, what purpose can ornament serve in them? However much of vanity there may be in asking the question, there is no answer to be offered. We cannot suppose that the individuals have any power of admiring each other, and we know that the foot is the only part they protrude from their shell, and that the inside of the shell is covered by the membrane called the mantle. Similar remarks may be made relative to conchology at large: the most exquisitely beautiful forms, colours and ornaments are lavished upon genera and species which exist only at immense depths in the ocean, or buried in the mud; nor can any one form a satisfactory idea of the object the great Author of nature had in view, in thus profusely beautifying creatures occupying so low a place in the scale of creation.

European naturalists have hitherto fallen into the strangest absurdities concerning the motion of the bivalved shells, which five minutes' observation of nature would have served them to correct. Thus they describe the upper part of the shell as the *lower*, and the *hind* part as the front, and speak of them as moving along on their rounded convex surface, like a boat on its keel; instead of advancing with the edges or open part of the shell towards the earth. All these mistakes have been corrected, and the true mode of progression indicated from actual observation, by our fellow citizen, Isaac Lea, whose recently published communications to the American Philosophical Society, reflect the highest credit upon their author, who is a naturalist in the best sense of the term.

As I wandered slowly along the borders of the run, towards a little wood, my attention was caught by a considerable collection of shells lying near an old stump. Many of these appeared to have been recently emptied of their contents, and others seemed to have long remained exposed to the weather. On most of them, at the thinnest part of the edge, a peculiar kind of fracture was obvious, and this seemed to be the work of an animal.

A closer examination of the locality showed the footsteps of a quadruped which I readily believed to be the muskrat, more especially as upon examining the adjacent banks numerous traces of burrows were discoverable. It is not a little singular that this animal, unlike all others of the larger gnawers, as the beaver, &c. appears to increase instead of diminishing with the increase of population. Whether it is that the dams and other works thrown up by men, afford more favourable situations for their multiplication, or their favourite food is found in greater abundance, they certainly are quite as numerous now, if not more so, than when the country was first discovered, and are to be found at this time almost within the limits of the city. By the construction of their teeth, as well as all the parts of the body, they are closely allied to the rat kind; though in size and some peculiarities of habit, they more closely approximate the beaver. They resemble the rat especially, in not being exclusively herbivorous, as is shown by their feeding on the uniones or muscles above mentioned. To obtain this food, requires no small exertion of their strength; and they accomplish it by introducing the claws of their fore-paws between the two edges of the shell, and tearing it open by main force. Whoever has tried to force open one of these shells, containing a living animal, may form an idea of the effort made by the muskrat:—the strength of a strong man would be requisite to produce the same result in the same way.

The burrows of muskrats are very extensive, and consequently injurious to dykes and dams, meadow banks, &c. The entrance is always under water, and thence sloping upwards above the level of the water, so that the muskrat has to dive in going in and out. These creatures are excellent divers and swimmers, and being nocturnal are rarely seen unless by those who watch for them at night. Sometimes we alarm one near the mouth of the den, and he darts away across the water, near the bottom, marking his course by a turbid streak in the stream: occasionally we are made aware of the passage of one to some distance down the current in the same way; but in both cases the action is so rapidly performed, that we should scarcely imagine what was the cause, if not previously informed. Except by burrowing into and spoiling the banks, they are not productive of much evil, their food consisting principally of the roots of aquatic plants, in addition to shellfish. The musky odour, which gives rise to their common name, is caused by glandular organs placed near the tail, filled with a viscid and powerfully musky fluid, whose uses we know but little of, though it is thought to be intended as a guide by which these creatures may discover each other. This inference is strengthened by finding some such contrivance in different races of animals, in various modifications. A great number carry it in pouches similar to those just mentioned. Some, as the musk animal, have the pouch under the belly; the shrew has the glands on the side;

the camel on the back of the neck; the crocodile under the throat, &c. At least no other use has ever been assigned for this apparatus; and in all creatures possessing it, the arrangement seems to be adapted peculiarly to the habits of the animals. The crocodile, for instance, generally approaches the shore in such a manner, as to apply the neck and throat to the soil, while the hinder part of the body is under water. The glands under the throat leave the traces of his presence, therefore, with ease, as they come in contact with the shore. The glandular apparatus on the back of the neck of the male camel, seems to have reference to the general elevation of the olfactory organs of the female; and the dorsal gland of the peccary, no doubt has some similar relation to the peculiarities of the race.

The value of the fur of the muskrat causes many of them to be destroyed, which is easily enough effected by means of a trap. This is a simple box, formed of rough boards nailed together, about three feet long, having an iron door, made of pointed bars, opening *inwards*, at both ends of the box. This trap is placed with the end opposite to the entrance of a burrow observed during the day time. In the night when the muskrat sallies forth, he enters the box, instead of passing into the open air, and is drowned, as the box is quite filled with water. If the traps be visited and emptied during the night, two may be caught in each trap, as muskrats from other burrows may come to visit those where the traps are placed, and thus one be taken going in as well as on coming out. These animals are frequently very fat, and their flesh has a very wholesome appearance, and would probably prove good food. The musky odour, however, prejudices strongly against its use ; and it is probable that the flesh is rank, as the muscles it feeds on are nauseous and bitter, and the roots which supply the rest of its food are generally unpleasant and acrid. Still we should not hesitate to partake of its flesh in case of necessity, especially if of a young animal, from which the musk bag had been removed immediately after it was killed.

In this vicinity, the muskrat does not build himself a house for the winter, as our fields and dykes are too often visited. But in other parts of the country where extensive marshes exist, and muskrats are abundant, they build very snug and substantial houses, quite as serviceable and ingenious as those of the beaver. They do not dam the water as the beaver, nor cut branches of trees to serve for the walls of their dwellings. They make it of mud and rushes, raising a cone two or three feet high, having the entrance on the south side under water. About the year 1804, I saw several of them in Worrell's marsh, near Chestertown, Maryland, which were pointed out to me by an old black man who made his living principally by trapping these animals, for the sake of their skins. A few years since I visited the marshes, near the mouth of Magerthy river in Maryland, where I was informed by a resident, that the muskrats still built regularly every winter. Perhaps these quadrupeds are as numerous in the vicinity of Philadelphia as elsewhere, as I have never examined a stream of fresh water, dyked meadow, or milldam, hereabout, without seeing traces of vast numbers. Along all the water courses and meadows in Jersey, opposite Philadelphia, and in the meadows of the neck, below the navy yard, there must be large numbers of muskrats. Considering the value of the fur, and the ease and trifling expense at which they might be caught, we have often felt surprised that more of them are not taken, especially as we have so many poor men complaining of wanting something to do. By thinning the number of muskrats, a positive benefit would be conferred on the farmers and furriers, to say nothing of the profits to the individual.

NO. IV.

My next visit to my old hunting ground, the lane and brook, happened on a day in the first hay harvest, when the verdant sward of the meadows was rapidly sinking before the keen edged scythes swung by vigorous mowers. This unexpected circumstance afforded me considerable pleasure, for it promised me a freer scope to my wanderings, and might also enable me to ascertain various particulars, concerning which my curiosity had long been awakened. Nor was this promise unattended by fruition of my wishes. The reader may recollect, that, in my first walk, a neat burrow in the grass, above ground, was observed, without my knowing its author. The advance of the mowers explained this satisfactorily, for in cutting the long grass, they exposed several nests of field mice, which, by means of these grass-covered alleys, passed to the stream in search of food or drink, unseen by their enemies, the hawks and owls. The numbers of these little creatures were truly surprising; their fecundity is so great, and their food so abundant, that were they not preyed upon by many other animals, and destroyed in great numbers by man, they would become exceedingly troublesome. There are various species of them, all bearing a very considerable resemblance to each other, and having to an incidental observer much of the appearance of the domestic mouse. Slight attention, however, is requisite to perceive very striking differences, and the discrimination of these will prove a source of considerable gratification to the enquirer. The nests are very nicely made, and look much like a bird's nest, being lined with soft materials, and usually placed in some snug little hollow, or at the root of a strong tuft of grass. Upon the grass roots and seeds these nibblers principally feed; and where very abundant, the effects of their hunger may be seen in the brown and withered aspect of the grass they have injured at the root. But under ordinary circumstances, the hawks, owls, domestic cat, weasels, crows, &c. keep them in such limits, as prevent them from doing essential damage.

I had just observed another and a smaller grassy covered way, where the mowers had passed along, when my attention was called towards a wagon at a short distance, which was receiving its load. Shouts and laughter, accompanied by a general running and scrambling of the people, indicated that some rare sport was going forward. When I approached, I found that the object of chase was a jumping mouse, whose actions it was truly delightful to witness. When not closely pressed by its pursuers, it ran with some rapidity in the usual manner, as if seeking concealment. But in a moment it would vault into the air, and skim along for ten or twelve feet, looking more like a bird than a little quadruped. After continuing this for some time, and nearly exhausting its pursuers with running and falling over each other, the frightened creature was accidentally struck down by one of the workmen, during one of its beautiful leaps, and killed. As the hunters saw nothing worthy of attention in the dead body of the animal, they very willingly resigned it to me; and with great satisfaction I retreated to a willow shade, to read what nature had written in its form for my instruction. The general appearance was mouse-like; but the length and slenderness of the body, the shortness of its fore limbs, and the disproportionate length of its hind limbs, together with the peculiarity of its tail, all indicated its adaptation to the peculiar kind of action I had just witnessed. A sight of this little creature vaulting or bounding through the air, strongly reminded me of what I had read of the great kanguroo of New Holland; and I could not help regarding our little jumper as in some respects a sort of miniature resemblance of that curious animal. It was not evident, however, that the jumping mouse derived the aid from its tail, which so powerfully assists the kanguroo. Though long and sufficiently stout in proportion, it had none of the robust muscularity which, in the New Holland animal, impels the lower part of the body immediately upward. In this mouse, the leap is principally, if not entirely effected by a sudden and violent extension of the long hind limbs, the muscles of which are strong, and admirably suited to their object. We have heard that these little animals feed on the roots, &c. of the green herbage, and that they are every season to be found in the meadows. It may perhaps puzzle some to imagine how they subsist through the severities of winter, when vegetation is at rest, and the earth generally frozen. Here we find another occasion to admire the all-perfect designs of the awful Author of nature, who has endowed a great number of animals with the faculty of retiring into the earth, and passing whole months in a state of repose so complete, as to allow all the functions of the body to be suspended, until the returning warmth of the spring calls them forth to renewed activity and enjoyment. The jumping mouse, when the chill weather begins to draw nigh, digs down about six or eight inches into the soil, and there forms a little globular cell, as much larger than his own body as will allow a sufficient covering of fine grass to be introduced. This being obtained, he contrives to coil up his body and limbs in the centre of the soft dry grass, so as to form a complete ball; and so compact is this, that, when taken out, with the torpid animal, it may be rolled across a floor without injury. In this snug cell, which is soon filled up and closed externally, the jumping mouse securely abides through all the frosts and storms of winter, needing neither food nor fuel, being utterly quiescent, and apparently dead, though susceptible at any time of reanimation, by being very gradually stimulated by light and heat.

The little burrow under examination, when called to observe the jumping mouse, proved to be made by the merry musicians. of the meadows, the field crickets, *acheta campestris.* These lively black crickets are very numerous, and contribute very largely to that general song which is so delightful to the ear of the true lover of nature, as it rises on the air from myriads of happy creatures rejoicing amid the bounties conferred on them by Providence. It is not *a voice* that the crickets utter, but a regular vibration of musical chords, produced by nibbing the nervures of the elytra against a sort of network intended to produce the vibrations. The reader will find an excellent description of the apparatus in Kirby and Spence's book, but he may enjoy a much more satisfactory comprehension of the whole, by visiting the field cricket in his summer residence, see him tuning his viol, and awakening the echoes with his music. By such an examination as may be there obtained, he may derive more knowledge than by frequent perusal of the most eloquent writings, and perhaps observe circumstances which the learned authors are utterly ignorant of.

Among the great variety of burrows formed in the grass, or under the surface of the soil, by various animals and insects, there is one that I have often anxiously and as yet fruitlessly explored. This burrow is formed by the smallest quadruped animal known to man, the minute *shrew*, which, when full grown, rarely exceeds the weight of *thirty-six grains*. I had seen specimens of this very interesting creature in the museum, and had been taught, by a more experienced friend, to distinguish its burrow, which I have often perseveringly traced, with the hope of finding the living animal, but in vain. On one occasion, I patiently pursued a burrow nearly round a large barn, opening it all the way. I followed it under the barn floor, which was sufficiently high to allow me to crawl beneath. There I traced it about to a tiresome extent, and was at length rewarded by discovering where it terminated, under a foundation stone, perfectly safe from my attempts. Most probably a whole family of them were then present, and I had my labour for my pains. As these little creatures are nocturnal, and are rarely seen from the nature of the places they frequent, the most probable mode of taking them alive would be, by placing a small mousetrap in their way, baited with a little tainted or slightly spoiled meat. If a common mouse trap be used, it is necessary to work it over with additional wire, as this shrew could pass between the bars even of a close mouse trap. They are sometimes killed by cats, and thus obtained, as the cat never eats them, perhaps on account of their rank smell, owing to a peculiar glandular apparatus on each side, that pours out a powerfully odorous greasy substance. The species of the shrew genus are not all so exceedingly diminutive, as some of them are even larger than a common mouse. They have their teeth coloured at the tips in a remarkable manner; it is generally of a pitchy brown, or dark chestnut hue, and, like the colouring of the teeth in the beaver and other animals, is owing to the enamel being thus formed, and not to any mere accident of diet. The shrews are most common about stables and cow-houses; and there, should I ever take the field again, my traps shall be set, as my desire to have one of these little quadrupeds is still as great as ever.

NO. V.

Hitherto my rambles have been confined to the neighbourhood of a single spot, with a view of showing how perfectly accessible to all, are numerous and various interesting natural objects. This habit of observing in the manner indicated, began many years anterior to my visit to the spots heretofore mentioned, and have extended through many parts of our own and another country. Henceforward my observations shall be presented without reference to particular places, or even of one place exclusively, but with a view to illustrate whatever may be the subject of description, by giving all I have observed of it under various circumstances.

A certain time of my life was spent in that part of Anne Arundel county, Md. which is washed by the river Patapsco on the north, the great Chesapeake bay on the west, and the Severn river on the south. It is in every direction cut up by creeks, or arms of the rivers and bay, into long, flat strips of land, called necks, the greater part of which is covered by dense pine forests, or thickets of small shrubs and saplings, rendered impervious to human footsteps by the growth of vines, whose inextricable mazes nothing but a fox, wild cat, or weasel, could thread. The soil cleared for cultivation is very generally poor, light, and sandy, though readily susceptible of improvement, and yielding a considerable produce in Indian corn, and most of the early garden vegetables, by the raising of which for the Baltimore market the inhabitants obtain all their ready money. The blight of slavery has long extended its influence over this region, where all its usual effects are but too obviously visible. The white inhabitants are few in number, widely distant from each other, and manifest, in their mismanagement and half indigent circumstances, how trifling an advantage they derive from the thraldom of their dozen or more of sturdy blacks, of different sexes and ages. The number of marshes formed at the heads of the creeks, render this country frightfully unhealthy in autumn, at which time the life of a resident physician is one of incessant toil and severe privation. Riding from morning till night, to get round to visit a few patients, his road leads generally through pine forests, whose aged and lofty trees, encircled by a dense undergrowth, impart an air of sombre and unbroken solitude. Rarely or never does he encounter a white person on his way, and only once in a while will he see a miserably tattered negro, seated on a sack of corn, carried by a starveling horse or mule, which seems poorly able to bear the weight to the nearest mill. The red-head woodpecker, and the flicker or yellowhammer, a kindred species, occasionally glance across his path; sometimes when he turns his horse to drink at the dark coloured branch, (as such streams are locally called,) he disturbs a solitary rufous thrush engaged in washing its plumes; or as he moves steadily along, he is slightly startled by a sudden appearance of the towhé bunting close to the side of the path. Except these creatures, and these by no means frequently seen, he rarely meets with animated objects; at a distance the harsh voice of the crow is often heard, or flocks of them are observed in the cleared fields, while now and then the buzzard, or turkey vulture, may be seen wheeling in graceful circles in the higher regions of the air, sustained by his broadly expanded wings, which apparently remain in a state of permanent and motionless extension. At other seasons of the year, the physician must be content to live in the most positive seclusion; the white people are all busily employed in going to and from market; and even were they at home, they are poorly suited for companionship. I here spent month after month, and, except the patients I visited, saw no one but the blacks; the house in which I boarded was kept by a widower, who, with myself, was the only white man within the distance of a mile or two. My only compensation was this, the house was pleasantly situated on the bank of Curtis's creek, a considerable arm of the Patapsco, which extended for a mile or two beyond us, and immediately in front of the door expanded so as to form a beautiful little bay. Of books I possessed very few, and those exclusively professional; but in this beautiful expanse of sparkling water, I had a book opened before me, which a life-time would scarcely suffice me to read through. With the advantage of a small but neatly made and easily manageable skiff, I was always independent of the service of the blacks, which was ever repugnant to my feelings and principles. I could convey myself in whatever direction objects of enquiry might present, and as my little bark was visible for a mile in either direction from the house, a handkerchief waved, or the loud shout of a negro, was sufficient to recall me, in case my services were required.

During the spring months, and while the garden vegetables are yet too young to need a great deal of attention, the proprietors frequently employ their blacks in hauling the seine; and this in these creeks is productive of an ample supply of yellow perch, which affords a very valuable addition to the diet of all. The blacks in an especial manner profit by this period of plenty, since they are permitted to eat of them without restraint, which cannot be said of any other sort of provision allowed them. Even the pigs and crows obtain their share of the abundance, as the fishermen, after picking out the best fish, throw the smaller ones on the beach. But as the summer months approach, the aquatic grass begins to grow, and this fishing can no longer be continued, because the grass rolls the seine up in a wisp, so that it can contain nothing. At this time the spawning season of the different species of sun-fish begins, and to me this was a time of much gratification. Along the edge of the river, where the depth of water was not greater than from four feet to as shallow as twelve inches, an observer would discover a succession of circular spots cleared of the surrounding grass, and showing a clear sandy bed. These spots, or cleared spaces, we may regard as the nest of this beautiful fish. There, balanced in the transparent wave, at the distance of six or eight inches from the bottom, the sunfish is suspended in the glittering sunshine, gently swaying its beautiful tail and fins; or, wheeling around in the limits of its little circle, appears to be engaged in keeping it clear of all incumbrances. Here the mother deposits her eggs or spawn, and never did hen guard her callow brood with more eager vigilance, than the sun-fish the little circle within which her promised offspring are deposited. If another individual approach too closely to her borders, with a fierce and angry air she darts against it, and forces it to retreat. Should any small, and not too heavy object be dropped in the nest, it is examined with jealous attention, and displaced if the owner be not satisfied of its harmlessness. At the approach of man she flies with great velocity into deep water, as if willing to conceal that her presence was more than accidental where first seen. She may, after a few minutes, be seen cautiously venturing to return, which is at length done with velocity; then she would take a hurried turn or two around, and scud back again to the shady bowers formed by the river grass which grows up from the bottom to within a few feet of the surface, and attains to twelve, fifteen, or more feet in length. Again she ventures forth from the depths; and if no further cause of fear presented, would gently sail into the placid circle of her home, and with obvious satisfaction explore it in every part.

Besides the absolute pleasure I derived from visiting the habitations of these glittering tenants of the river, hanging over them from my little skiff, and watching their every action, they frequently furnished me with a very acceptable addition to my frugal table. Situated as my boarding house was, and all the inmates of the house busily occupied in raising vegetables to be sent to market, our bill of fare offered little other change than could be produced by varying the mode of cookery. It was either broiled bacon and potatoes, or fried bacon and potatoes, or cold bacon and potatoes, and so on at least six days out of seven. But, as soon as I became acquainted with the habits of the sun-fish, I procured a neat circular iron hoop for a net; secured to it a piece of an old seine, and whenever I desired to dine on *fresh* fish, it was only necessary to take my skiff, and push her gently along from one sun-fish nest to another, myriads of which might be seen along all the shore. The fish, of course, darted off as soon as the boat first drew near, and during this absence the net was placed so as to cover the nest, of the bottom of which the meshes but slightly intercepted the view. Finding all things quiet, and not being disturbed by the net, the fish would resume its central station, the net was suddenly raised, and the captive placed in the boat. In a quarter of an hour, I could generally take as many in this way as would serve two men for dinner, and when an acquaintance accidentally called to see me, during the season of sun-fish, it was always in my power to lessen our dependence on the endless bacon. I could also always select the finest and largest of these fish, as while standing up in the boat, one could see a considerable number at once, and thus choose the best. Such was their abundance, that the next day would find all the nests re-occupied. Another circumstance connected with this matter gave me no small satisfaction; the poor blacks, who could rarely get time for angling, soon learned how to use my net with dexterity; and thus, in the ordinary time allowed them for dinner, would borrow it, run down to the shore, and catch some fish to add to their very moderate allowance.

NO. VI.

After the sun-fish, as regular annual visitants of the small rivers and creeks containing salt or brackish water, came the crabs in vast abundance, though for a very different purpose. These singularly constructed and interesting beings furnished me with another excellent subject for observation; and, during the period of their visitation, my skiff was in daily requisition. Floating along with an almost imperceptible motion, a person looking from the shore might have supposed her entirely adrift; for as I was stretched at full length across the seats, in order to bring my sight as close to the water as possible without inconvenience, no one would have observed my presence from a little distance. The crabs belong to a very extensive tribe of beings, which carry their *skeletons* on the *outside* of their bodies, instead of within; and of necessity the fleshy, muscular, or moving power of the body, is placed in a situation the reverse of what occurs in animals of a higher order, which have internal skeletons or solid frames to their systems. This peculiarity of the crustaceous animals and various other beings, is attended with one apparent inconvenience; when they have grown large enough to fill their shell or skeleton completely, they cannot grow farther, because the skeleton being external, is incapable of enlargement. To obviate this difficulty, the Author of nature has endowed them with the power of casting off the entire shell, increasing in size, and forming another equally hard and perfect, for several seasons successively, until the greatest or maximum size is attained, when the change or sloughing ceases to be necessary, though it is not always discontinued on that account. To undergo this change with greater ease and security, the crabs seek retired and peaceful waters, such as the beautiful creek I have been speaking of, whose clear, sandy shores are rarely disturbed by waves causing more than a pleasing murmur, and where the number of enemies must be far less in proportion than in the boisterous waters of the Chesapeake, their great place of concourse. From the first day of their arrival in the latter part of June, until the time of their departure, which in this creek occurred towards the first of August, it was astonishing to witness the vast multitudes which flocked towards the head of the stream.

It is not until they have been for some time in the creek, that the moult or sloughing generally commences. They may be then observed gradually coming closer in shore, to where the sand is fine, fairly exposed to the sun, and a short distance farther out than the lowest water mark, as they must always have at least a depth of three or four inches water upon them.

The individual having selected his place, becomes perfectly quiescent, and no change is observed during some hours but a sort of swelling along the edges of the great upper shell at its back part. After a time this posterior edge of the shell becomes fairly disengaged like the lid of a chest, and now begins the more difficult work of withdrawing the great claws from their cases, which every one recollects to be vastly larger at their extremities and between the joints than the joints themselves. A still greater apparent difficulty presents in the shedding of the sort of tendon which is placed within the muscles. Nevertheless, the Author of nature has adapted them to the accomplishment of all this. The disproportionate sized claws undergo a peculiar softening, which enables the crab, by a very steadily continued, scarcely perceptible effort, to pull them out of their shells, and the business is completed by the separation of the complex parts about the mouth and eyes. The crab now slips out from the slough, settling near it on the sand. It is now covered by a soft, perfectly flexible skin; and though possessing precisely the same form as before, seems incapable of the slightest exertion. Notwithstanding that such is its condition, while you are gazing on this helpless creature, it is sinking in the fine loose sand, and in a short time is covered up sufficiently to escape the observation of careless or inexperienced observers. Neither can one say how this is effected, although it occurs under their immediate observation; the motions employed to produce the displacement of the sand are too slight to be appreciated, though it is most probably owing to a gradual lateral motion of the body by which the sand is displaced in the centre beneath, and thus gradually forced up at the sides until it falls over and covers the crab. Examine him within twelve hours, and you will find the skin becoming about as hard as fine writing paper, producing a similar crackling if compressed; twelve hours later the shell is sufficiently stiffened to require some slight force to bend it, and the crab is said to be in *buckram*, as in the first stage it was in *paper*. It is still helpless, and offers no resistance; but at the end of thirty-six hours, it shows that its natural instincts are in action, and by the time forty-eight hours have elapsed, the crab is restored to the exercise of all his functions. I have stated the above as the periods in which the stages of the moult are accomplished, but I have often observed that the rapidity of this process is very much dependent upon the temperature, and especially upon sunshine. A cold, cloudy, raw, and disagreeable spell happening at this period, though by no means common, will retard the operation considerably, protracting the period of helplessness. This is the harvest season of the white fisherman and of the poor slave. The laziest of the former are now in full activity, wading along the shore from morning till night, dragging a small boat after them, and holding in the other hand a forked stick with which they raise the crabs from the sand. The period during which the crabs remain in the paper state is so short, that great activity is required to gather a sufficient number to take to market, but the price at which they are sold is sufficient to awaken all the cupidity of the crabbers. Two dollars a dozen is by no means an uncommon price for them, when the season first comes on; they subsequently come down to a dollar, and even to fifty cents, at any of which rates the trouble of collecting them is well paid. The slaves search for them at night, and then are obliged to kindle a fire of pine-knots on the bow of the boat, which strongly illuminates the surrounding water, and enables them to discover the crabs. Soft crabs are, with great propriety, regarded as an exquisite treat by those who are fond of such eating; and though many persons are unable to use crabs or lobsters in any form, there are few who taste of the soft crabs without being willing to recur to them. As an article of luxury they are scarcely known north of the Chesapeake, though there is nothing to prevent them from being used to considerable extent in Philadelphia, especially since the opening of the Chesapeake and Delaware canal. The summer of 1829 I had the finest soft crabs from Baltimore. They arrived at the market in the afternoon, were fried according to rule, and placed in a tin butter kettle, then covered for an inch or two with melted lard, and put on board the steam boat which left Baltimore at five o'clock the same afternoon. The next morning before ten o'clock they were in Philadelphia, and at one they were served up at dinner in Germantown. The only difficulty in the way is that of having persons to attend to their procuring and transmission, as when cooked directly after they arrive at market, and forwarded with as little delay as above mentioned, there is no danger of their being the least injured.

At other seasons, when the crabs did not come close to the shore, I derived much amusement by taking them in the deep water. This is always easily effected by the aid of proper bait; a leg of chicken, piece of any raw meat, or a salted or spoiled herring, tied to a twine string of sufficient length, and a hand net of convenient size, is all that is necessary. You throw out your line and bait, or you fix as many lines to your boat as you please, and in a short time you see, by the straightening of the line, that the bait has been seized by a crab, who is trying to make off with it. You then place your net where it can conveniently be picked up, and commence steadily but gently to draw in your line, until you have brought the crab sufficiently near the surface to distinguish him; if you draw him nearer, he will see you and immediately let go, otherwise his greediness and voracity will make him cling to his prey to the last. Holding the line in the left hand, you now dip your net edge foremost into the water at some distance from the line, carry it down perpendicularly until it is five or six inches lower than the crab, and then with a sudden turn out bring it directly before him, and lift up at the same time. Your prize is generally secured, if your net be at all properly placed;

for as soon as he is alarmed, he pushes directly downwards, and is received in the bag of the net. It is better to have a little water in the bottom of the boat to throw them into, as they are easier emptied out of the net, always letting go when held over the water. This a good crabber never forgets, and should he unluckily be seized by a large crab, he holds him over the water and is freed at once, though he loses his game. When not held over the water, they bite sometimes with dreadful obstinacy, and I have seen it necessary to crush the forceps or claws before one could be induced to let go the fingers of a boy. A poor black fellow also placed himself in an awkward situation; the crab seized him by a finger of his right hand, but he was unwilling to lose his captive by holding him over the water, instead of which he attempted to secure the other claw with his left hand, while he tried to crush the biting claw between his teeth. In doing this, he somehow relaxed his left hand, and with the other claw, the crab seized poor Jem by his under lip, which was by no means a thin one, and forced him to roar with pain. With some difficulty he was freed from his tormentor, but it was several days before he ceased to excite laughter, as the severe bite was followed by a swelling of the lip, which imparted a most ludicrous expression to a naturally comical countenance.

NO. VII.

On the first arrival of the crabs, when they throng the shoals of the creeks in vast crowds, as heretofore mentioned, a very summary way of taking them is resorted to by the country people, and for a purpose that few would suspect without having witnessed it. They use a three pronged fork or gig made for this sport, attached to a long handle; the crabber standing up in the skiff, pushes it along until he is over a large collection of crabs, and then strikes his spear among them. By this several are transfixed at once and lifted into the boat, and the operation is repeated until enough have been taken. The purpose to which they are to be applied is to feed the hogs, which very soon learn to collect in waiting upon the beach when the crab spearing is going on. Although these bristly gentry appear to devour almost all sorts of food with great relish, it seemed to me that they regarded the crabs as a most luxurious banquet; and it was truly amusing to see the grunters, when the crabs were thrown on shore for them, and were scampering off in various directions, seizing them in spite of their threatening claws, holding them down with one foot, and speedily reducing them to a state of helplessness by breaking off their forceps. Such a crunching and cracking of the unfortunate crabs I never have witnessed since; and I might have commiserated them more, had not I known that death in some form or other was continually awaiting them, and that their devourers were all destined to meet their fate in a few months in the sty, and thence through the smoke house to be placed upon our table. On the shores of the Chesapeake I have caught crabs in a way commonly employed by all those who are unprovided with boats and nets. This is to have a forked stick and a baited line, with which the crabber wades out as far as he thinks fit, and then throws out his line. As soon as he finds he has a bite, he draws the line in, cautiously lifting but a very little from the bottom. As soon as it is near enough to be fairly in reach, he quickly, yet with as little movement as possible, secures the crab by placing the forked stick across his body and pressing him against the sand. He must then stoop down and take hold of the crab by the two posterior swimming legs, so as to avoid being seized by the claws. Should he not wish to carry each crab ashore as he catches it, he pinions or *spansels* (as the fishermen call it) them. This is a very effectual mode of disabling them from using their biting claws, yet it is certainly not the most humane operation; it is done by taking the first of the sharp-pointed feet of each side, and forcing it in for the length of the joint behind the moveable joint or thumb of the opposite biting claw. The crabs are then strung upon a string or wythe, and allowed to hang in the water until the crabber desists from his occupations. In the previous article crabs were spoken of as curious and interesting, and the reader may not consider the particulars thus far given as being particularly so. Perhaps, when he takes them altogether, he will agree that they have as much that is curious about their construction as almost any animal we have mentioned, and in the interesting details we have as yet made but a single step.

The circumstance of the external skeleton has been mentioned, but who would expect an animal, as low in the scale as a crab, to be furnished with ten or twelve pair of jaws to its mouth? Yet such is the fact, and all these variously constructed pieces are provided with appropriate muscles, and move in a manner which can scarcely be explained, though it may be very readily comprehended when once observed in living nature. But, after all the complexity of the jaws, where would an inexperienced person look for their teeth? surely not in the stomach? Nevertheless, such is their situation; and these are not mere appendages, that are called teeth by courtesy, but stout regular grinding teeth, with a light brown surface. They are not only within the stomach, but fixed to a cartilage nearest to its lower extremity, so that the food, unlike that of other creatures, is submitted to the action of the teeth as it is passing *from* the stomach; instead of being chewed before it is swallowed. In some species the teeth are five in number; but throughout this class of animals the same general principle of construction may be observed. Crabs and their kindred have no brain, because they are not required to reason upon what they observe; they have a nervous system excellently suited to their mode of life, and its knots or ganglia send out nerves to the organs of sense, digestion, motion, &c. The senses of these beings are very acute, especially their sight, hearing and smell. Most of my readers have heard of crabs' eyes, or have seen these organs in the animal on the end of two little projecting knobs, above and on each side of the mouth; few of them, however, have seen the crab's ear, yet it is very easily found, and is a little triangular bump placed near the base of the feelers. This bump has a membrane stretched over it, and communicates with a small cavity, which is the internal ear. The *organ* of smell is not so easily demonstrated as that of hearing, though the evidence of their possessing the sense to an acute degree is readily attainable. A German naturalist inferred, from the fact of the nerve corresponding to the olfactory nerve in man being distributed to the antennæ, in insects, that the antennæ were the organs of smell in them. Cuvier and others suggest that a similar arrangement may exist in the crustacea. To satisfy myself whether it was so or not, I lately dissected a small lobster, and was delighted to find that the first pair of nerves actually went to the antennæ, and gave positive support to the opinion mentioned. I state this, not to claim credit for ascertaining the truth or inaccuracies of a suggestion, but with a view of inviting the reader to do the same in all cases of doubt. Where it is possible to refer to *nature* for the actual condition of facts, learned *authorities* give me no uneasiness. If I find that the structure bears out their opinions, it is more satisfactory; when it convicts them of absurdity, it saves much fruitless reading, as well as the trouble of shaking off prejudices.

The first time my attention was called to the extreme acuteness of sight possessed by these animals, was during a walk along the flats of Long Island, reaching towards Governor's Island in New York. A vast number of the small land crabs, called fiddlers by the boys (gecarcinus,) occupy burrows or caves dug in the marshy soil, whence they come out and go for some distance, either in search of food or to sun themselves. Long before I approached close enough to see their forms with distinctness, they were scampering towards their holes, into which they plunged with a tolerable certainty of escape; these retreats being of considerable depth, and often communicating with each other, as well as nearly filled with water. On endeavouring cautiously to approach some others, it was quite amusing to observe their vigilance; to see them slowly change position, and from lying extended in the sun, beginning to gather themselves up for a start should it prove necessary; at length standing up as it were on tiptoe, and raising their pedunculated eyes as high as possible. One quick step on the part of the individual approaching was enough—away they would go, with a celerity which must appear surprising to any one who had not previously witnessed it. What is more remarkable, they possess the power of moving equally well with any part of the body foremost, so that when endeavouring to escape, they will suddenly dart off to one side or the other, without turning round, and thus elude pursuit. My observations upon the crustaceous animals have extended through many years, and in very various situations; and for the sake of making the general view of their qualities more satisfactory, I will go on to state what I remarked of some of the genera and species in the West Indies, where they are exceedingly numerous and various. The greater proportion of the genera feed on animal matter, especially after decomposition has begun; a large number are exclusively confined to the deep waters, and approach the shoals and lands only during the spawning season. Many live in the sea, but daily pass many hours upon the rocky shores for the pleasure of basking in the sun; others live in marshy or moist ground, at a considerable distance from the water, and feed principally on vegetable food, especially the sugar cane, of which they are extremely destructive. Others again reside habitually on the hills or mountains, and visit the sea only once a year for the purpose of depositing their eggs in the sand. All those which reside in burrows made in moist ground, and those coming daily on the rocks to bask in the sun, participate in about an equal degree in the qualities of vigilance and swiftness. Many a breathless race have I run in vain, attempting to intercept them, and prevent their escaping into the sea. Many an hour of cautious and solicitous endeavour to steal upon them unobserved, has been frustrated by their long sighted watchfulness; and several times, when, by extreme care and cunning approaches, I have actually succeeded in getting between a fine specimen and the sea, and had full hope of driving him farther inland, have all my anticipations been ruined by the wonderful swiftness of their flight, or the surprising facility with which they would dart off in the very opposite direction, at the very moment I felt almost sure of my prize. One day, in particular, I saw on a flat rock, which afforded a fine sunning place, the most beautiful crab I had ever beheld. It was of the largest size, and would have covered a large dinner plate, most beautifully coloured with bright crimson below, and a variety of tints of blue, purple, and green above; it was just such a specimen as could not fail to excite all the solicitude of a collector to obtain. But, it was not in the least deficient in the art of self-preservation; my most careful manœuvres proved ineffectual, and all my efforts only enabled me to see enough of it to augment my regrets to a high degree. Subsequently I saw a similar individual in the collection of a resident; this had been killed against the rocks during a violent hurricane, with very slight injury to its shell. I offered high rewards to the black people if they would bring me such a one, but the most expert among them seemed to think it an unpromising search, as they knew of no way of capturing them. If I had been supplied with some powder of nux vomica with which to poison some meat, I *might* have succeeded.

NO. VIII.

The fleet running crab (cypoda pugilator,) mentioned as living in burrows dug in a moist soil, and preying chiefly on the sugar cane, is justly regarded as one of the most noxious pests that can infest a plantation. Their burrows extend to a great depth, and run in various directions; they are also, like those of our fiddlers, nearly full of muddy water, so that, when these marauders once plump into their dens, they may be considered as entirely beyond pursuit. Their numbers are so great, and they multiply in such numbers, as in some seasons to destroy a large proportion of a sugar crop, and sometimes their ravages, combined with those of the rats and other plunderers, are absolutely ruinous to the sea-side planters. I was shown, by the superintendent of a place thus infested, a great quantity of cane utterly killed by these creatures, which cut it off in a peculiar manner, in order to suck the juice; and he assured me that, during that season, the crop would be two thirds less than its average, solely owing to the inroads of the crabs, and rats, which if possible are still more numerous. It was to me an irresistible source of amusement to observe the air of spite and vexation with which he spoke of the crabs; the rats he could shoot, poison, or drive off for a time with dogs. But the crabs would not eat his poison, while sugar cane was growing; the dogs could only chase them into their holes; and if, in helpless irritation, he sometimes fired his gun at a cluster of them, the shot only rattled over their shells like hail against a window. It is truly desirable that some summary mode of lessening their number could be devised, and it is probable that this will be best effected by poison, as it may be possible to obtain a bait sufficiently attractive to ensnare them. Species of this genus are found in various parts of our country, more especially towards the south. About Cape May, our friends may have excellent opportunities of testing the truth of what is said of their swiftness and vigilance.

The land crab, which is common to many of the West India islands, is more generally known as the Jamaica crab, because it has been most frequently described from observation in that island. Wherever found, they all have the habit of living, during great part of the year, in the highlands, where they pass the day time, concealed in huts, cavities, and under stones, and come out at night for their food. They are remarkable for collecting in vast bodies, and marching annually to the sea side, in order to deposit their eggs in the sand; and this accomplished, they return to their former abodes, if undisturbed. They commence their march in the night, and move in the most direct line towards the destined point. So obstinately do they pursue this route, that they will not turn out of it for any obstacle that can possibly be surmounted. During the day time they skulk and lie hid as closely as possible, but thousands upon thousands of them are taken for the use of the table by whites and blacks, as on their seaward march they are very fat and of fine flavour. On the homeward journey, those that have escaped capture are weak, exhausted, and unfit for use. Before dismissing the crabs, I must mention one which was a source of much annoyance to me at first, and of considerable interest afterwards, from the observation of its habits. At that time I resided in a house delightfully situated about two hundred yards from the sea, fronting the setting sun, having in clear weather the lofty mountains of Porto Rico, distant about eighty miles, in view. Like most of the houses in the island, ours had seen better days, as was evident from various breaks in the floors, angles rotted off the doors, sunken sills, and other indications of decay. Our sleeping room, which was on the lower floor, was especially in this condition; but as the weather was delightfully warm, a few cracks and openings, though rather large, did not threaten much inconvenience. Our bed was provided with that indispensable accompaniment, a musquito bar or curtain, to which we were indebted for escape from various annoyances. Scarcely had we extinguished the light, and composed ourselves to rest, than we heard, in various parts of the room, the most startling noises. It appeared as if numerous hard and heavy bodies were trailed along the floor; then they sounded as if climbing up by the chairs and other furniture, and frequently something like a large stone would tumble down from such elevations with a loud noise, followed by a peculiar chirping note. What an effect this produced upon entirely inexperienced strangers, may well be imagined by those who have been suddenly waked up in the dark, by some unaccountable noise in the room. Finally, these invaders began to ascend the bed; but happily the musquito bar was securely tucked under the bed all around, and they were denied access, though their efforts and tumbles to the floor produced no very comfortable reflections. Towards daylight they began to retire, and in the morning no trace of any such visitants could be perceived. On mentioning our troubles, we were told that this nocturnal disturber was only Bernard the Hermit, called generally the soldier crab, perhaps from the peculiar habit he has of protecting his body by thrusting it into an empty shell, which he afterwards carries about, until he outgrows it, when it is relinquished for a larger. Not choosing to pass another night quite so noisily, due care was taken to exclude Monsieur Bernard, whose knockings were thenceforward confined to the outside of the house. I baited a large wire rat trap with some corn meal, and placed it outside of the back door, and in the morning, found it literally half filled with these crabs, from the largest sized shell that could enter the trap, down to such as were not larger than a hickory nut. Here was a fine collection made at once, affording a very considerable variety in the size and age of the specimens, and the different shells into which they had introduced themselves.

The soldier, or hermit crab, when withdrawn from his adopted shell, presents about the head and claws a considerable family resemblance to the lobster. The claws, however, are very short and broad, and the body covered with hard shell only in that part which is liable to be exposed or protruded. The posterior or abdominal part of the body, is covered only by a tough skin, and tapers towards a small extremity, furnished with a sort of hook-like apparatus, enabling it to hold on to its factitious dwelling. Along the surface of its abdomen, as well as on the back, there are small projections, apparently intended for the same purpose. When once fairly in possession of a shell, it would be quite a difficult matter to pull the crab out, though a very little heat applied to the shell will quickly induce him to leave it. The shells they select are taken solely with reference to their suitableness, and hence you may catch a considerable number of the same species, each of which is in a different species or genus of shell. The shells commonly used by them, when of larger size, are those of the whilk, which are much used as an article of food by the islanders, or the smaller conch [strombus] shells. The very young hermit crabs are seen in almost every variety of small shell found on the shores of the Antilles. I have frequently been amused by seeing ladies, eagerly engaged in making a collection of these beautiful little shells, and not dreaming of their being tenanted by a living animal, suddenly startled, on displaying their acquisitions, at observing them to be actively endeavouring to escape; or on introducing the hand into the reticule to produce a particular fine specimen, to receive a smart pinch from the claws of the little hermit. The instant the shell is closely approached or touched, they withdraw as deeply into the shell as possible, and the small ones readily escape observation, but they soon become impatient of captivity, and try to make off. The species of this genus (pagurus) are very numerous, and during the first part of their lives are all aquatic. That is, they are hatched in the little pools about the margin of the sea, and remain there until those that are destined to live on land are stout enough to commence their travels. The hermit crabs which are altogether aquatic are by no means so careful to choose the lightest and thinnest shells, as the land troops. The aquatic soldiers may be seen towing along shells of most disproportionate size; but their relatives, who travel over the hills by moonlight, know that all unnecessary incumbrance of weight should be avoided. They are as pugnacious and spiteful as any of the crustaceous class; and when taken, or when they fall and jar themselves, considerably, utter a chirping noise, which is evidently an angry expression. They are ever ready to bite with their claws, and the pinch of the larger individuals is quite painful. It is said, that when they are changing their shells, for the sake of obtaining more commodious coverings, they frequently fight for possession, which may be true where two that have forsaken their old shells meet, or happen to make choice of the same vacant one. It is also said, that one crab is sometimes forced to give up the shell he is in, should a stronger chance to desire it. This, as I never saw it, I must continue to doubt; for I cannot imagine how the stronger could possibly accomplish his purpose, seeing that the occupant has nothing to do but keep close quarters. The invader would have no chance of seizing him to pull him out, nor could he do him any injury by biting upon the surface of his hard claws, the only part that would be exposed. If it be true that one can dispossess the other, it must be by some contrivance of which we are still ignorant. These soldier crabs feed on a great variety of substances, scarcely refusing any thing that is edible; like the family they belong to, they have a decided partiality for putrid meats, and the planters accuse them also of too great a fondness for the sugar cane. Their excursions are altogether nocturnal, in the day time they lie concealed very effectually in small holes, among stones, or any kind of rubbish, and are rarely taken notice of, even where hundreds are within a short distance of each other. The larger soldier crabs are sometimes eaten by the blacks, but they are not much sought after even by them, as they are generally regarded with aversion and prejudice. There is no reason, that we are aware of, why they should not be as good as many other crabs, but they certainly are not equally esteemed.

NO. IX.

Those who have only lived in forest countries, where vast tracts are shaded by a dense growth of oak, ash, chestnut, hickory and other trees of deciduous foliage, which present the most pleasing varieties of verdure and freshness, can have but little idea of the effect produced on the feelings by aged forests of pine, composed in great degree of a single species, whose towering summits are crowned with one dark green canopy, which successive seasons find unchanged, and nothing but death causes to vary. Their robust and gigantic trunks rise an hundred or more feet high in purely proportioned columns, before the limbs begin to diverge; and their tops, densely clothed with long bristling foliage, intermingle so closely as to allow of but slight entrance to the sun. Hence the undergrowth of such forests is comparatively slight and thin, since none but shrubs and plants that love the shade, can flourish under this perpetual exclusion of the animating and invigorating rays of the great exciter of the vegetable world. Through such forests, and by the merest foot paths in great part, it was my lot to pass many miles almost every day; and had I not endeavoured to derive some amusement and instruction from the study of the forest itself, my time would have been as fatiguing to me, as it was certainly quiet and solemn. But wherever nature is, and under whatever form she may present herself, enough is always proffered to fix attention and produce pleasure, if we will condescend to observe with carefulness. I soon found that even a pine forest was far from being devoid of interest, and shall endeavour to prove this by stating the result of various observations made during the time I lived in this situation.

The common pitch, or as it is generally called Norway pine, grows from a seed, which is matured in vast abundance in the large cones peculiar to the pines. This seed is of a rather triangular shape, thick and heavy at the part by which it grows from the cone, and terminating in a broad membranous fan or sail, which, when the seeds are shaken out by the wind, enables them to sail obliquely through the air to great distances. Should an old cornfield or other piece of ground be thrown out of cultivation for more than oné season, it is sown with the pine seeds by the winds, and the young pines shoot up as closely and compactly as hemp. They continue to grow in this manner until they become twelve or fifteen feet high, until their roots begin to encroach on each other, or until the stoutest and best rooted begin to overtop so as entirely to shade the smaller. These gradually begin to fail, and finally dry up and perish, and a similar process is continued until the best trees acquire room enough to grow without impediment. Even when the young pines have attained to thirty or forty feet in height, and are as thick as a man's thigh, they stand so closely together, that their lower branches, which are all dry and dead, are intermingled, sufficiently to prevent any one from passing between the trees without first breaking these obstructions away. I have seen such a wood as that just mentioned, covering an old corn-field, whose ridges were still distinctly to be traced, and which an old resident informed me he had seen growing in corn. In a part of this wood which was not far from my dwelling, I had a delightful retreat, that served me as a private study or closet, though enjoying all the advantages of the open air. A road that had once passed through the field, and was of course more compacted than any other part, had denied access to the pine seeds for a certain distance, while on each side of it they grew with their usual density. The ground was covered with the soft layer or carpet of dried pine leaves which gradually and imperceptibly fall throughout the year, making a most pleasant surface to tread on, and rendering the step perfectly noiseless. By beating off with a stick all the dried branches that projected towards the vacant space, I formed a sort of chamber, fifteen or twenty feet long, which above was canopied by the densely mingled branches of the adjacent trees, which altogether excluded or scattered the rays of the sun, and on all sides was so shut in by the trunks of the young trees, as to prevent all observation. Hither during the hot season, I was accustomed to retire, for the purpose of reading or meditation; and within this deeeper solitude, where all was solitary, very many of the subsequent movements of my life were suggested or devised.

From all I could observe, and all the enquiries I could get answered, it appeared that this rapidly growing tree does not attain its full growth until it is eighty or ninety years old, nor does its time of full health and vigour much exceed an hundred. Before this time it is liable to the attacks of insects, but these are of a kind that bore the tender spring shoots to deposit their eggs therein, and their larvæ appear to live principally on the sap which is very abundant, so that the tree is but slightly injured. But after the pine has attained its acme, it is attacked by an insect which deposits its egg in the body of the tree, and the larva devours its way through the solid substance of the timber; so that after a pine has been for one or two seasons subjected to these depredators, it will be fairly riddled, and if cut down is unfit for any other purpose than burning. Indeed, if delayed too long, it is poorly fit for firewood, so thoroughly do these insects destroy its substance. At the same time that one set of insects is engaged in destroying the body, myriads of others are at work under the bark, destroying the sap vessels, and the foliage wears a more and more pale and sickly appearance as the tree declines in vigour. If not cut down, it eventually dies, becomes leafless, stripped of its bark, and as the decay advances, all the smaller branches are broken off; and it stands with its naked trunk and a few ragged limbs, as if bidding defiance to the tempest which howls around its head. Under favourable circumstances, a large trunk will stand in this condition for nearly a century, so extensive and powerful are its roots, so firm and stubborn the original knitting of its giant frame. At length some storm, more furious than all its predecessors, wrenches those ponderous roots from the soil, and hurls the helpless carcass to the earth, crushing all before it in its fall. Without the aid of fire, or some peculiarity of situation favourable to rapid decomposition, full another hundred years will be requisite to reduce it to its elements, and obliterate the traces of its existence. Indeed, long after the lapse of more than that period, we find the heart of the pitch pine still preserving its original form, and from being thoroughly imbued with turpentine, become utterly indestructible except by fire.

If the proprietor attend to the warnings afforded by the woodpecker, he may always cut his pines in time to prevent them from being injured by insects. The woodpeckers run up and around the trunks, tapping from time to time with their powerful bills. The bird knows at once by the sound whether there be insects below or not. If the tree is sound, the woodpecker soon forsakes it for another; should he begin to break into the bark, it is to catch the worm, and such trees are at once to be marked for the axe. In felling such pines, I found the woodmen alway anxious to avoid letting them strike against neighbouring sound trees, as they said that the insects more readily attacked an injured tree than one whose bark was unbroken. The observation is most probably correct, at least the experience of country folks in such matters is rarely wrong, though they sometimes give very odd reasons for the processes they adopt.

A full grown pine forest is at all times a grand and majestic object to one accustomed to moving through it. Those vast and towering columns, sustaining a waving crown of deepest verdure; those robust and rugged limbs standing forth at a vast height overhead, loaded with the cones of various seasons; and the diminutiveness of all surrounding objects compared with these gigantic children of nature, cannot but inspire ideas of seriousness and even of melancholy. But how awful and even tremendous does such a situation become, when we hear the first wailings of the gathering storm, as it stoops upon the lofty summits of the pine, and soon increases to a deep hoarse roaring, as the boughs begin to wave in the blast, and the whole tree is forced to sway before its power.

In a short time the fury of the wind is at its height, the loftiest trees bend suddenly before it, and scarce regain their upright position ere they are again obliged to cower beneath its violence. Then the tempest literally howls, and amid the tremendous reverberations of thunder, and the blazing glare of the lightning, the unfortunate wanderer hears around him the crash of numerous trees hurled down by the storm, and knows not but the next may be precipitated upon him. More than once have I witnessed all the grandeur, dread, and desolation of such a scene, and have always found safety either by seeking as quickly as possible a spot where there were none but young trees, or if on the main road choosing the most open and exposed situation out of the reach of the large trees. There, seated on my horse, who seemed to understand the propriety of such patience, I would quietly remain, however thoroughly drenched, until the fury of the wind was completely over. To say nothing of the danger from falling trees, the peril of being struck by the lightning, which so frequently shivers the loftiest of them, is so great as to render any attempt to advance at such time highly imprudent.

Like the ox among animals, the pine tree may be looked upon as one of the most universally useful of the sons of the forest. For all sorts of building, for firewood, tar, turpentine, rosin, lampblack, and a vast variety of other useful products, this tree is invaluable to man. Nor is it a pleasing contemplation, to one who knows its usefulness, to observe to how vast an amount it is annually destroyed in this country, beyond the proportion that nature can possibly supply. However, we are not disposed to believe that this evil will ever be productive of very great injury, especially as coal fuel is becoming annually more extensively used. Nevertheless, were I the owner of a pine forest, I should exercise a considerable degree of care in the selection of the wood for the axe.

NO. X.

Among the enemies with which the farmers of a poor or light soil have to contend, I know of none so truly formidable and injurious as the crows, whose numbers, cunning, and audacity, can scarcely be appreciated, except by those who have had long continued and numerous opportunities of observation. Possessed of the most acute senses, and endowed by nature with a considerable share of reasoning power, these birds bid defiance to almost all the contrivances resorted to for their destruction; and when their numbers have accumulated to vast multitudes, which annually occurs, it is scarcely possible to estimate the destruction they are capable of effecting. Placed in a situation where every object was subjected to close observation, as a source of amusement, it is not surprising that my attention should be drawn to so conspicuous an object as the crow; and having once commenced remarking the peculiarities of this bird, I continued to bestow attention upon it during many years, in whatever situation it was met with. The thickly wooded and well watered parts of the state of Maryland, as affording them a great abundance of food, and almost entire security during their breeding season, are especially infested by these troublesome creatures, so that at some times of the year they are collected in numbers which would appear incredible to any one unaccustomed to witness their accumulations.

Individually, the common crow (*corvus corona*) may be compared in character with the brown or Norway rat, being, like that quadruped, addicted to all sorts of mischief, destroying the lives of any small creatures that may fall in its way, plundering with audacity wherever any thing is exposed to its rapaciousness, and triumphing by its cunning over the usual artifices employed for the destruction of ordinary noxious animals. Where food is at any time scarce, or the opportunity for such marauding inviting, there is scarcely a young animal about the farm yards safe from the attacks of the crow. Young chickens, ducks, goslings, and even little pigs, when quite young and feeble, are carried off by them. They are not less eager to discover the nests of domestic fowls, and will sit very quietly in sight, at a convenient distance, until the hen leaves the nest, and then fly down and suck her eggs at leisure. But none of their tricks excited in me a greater interest, than the observation of their attempts to rob a hen of her chicks. The crow, alighting at a little distance from the hen, would advance in an apparently careless way towards the brood, when the vigilant parent would bristle up her feathers, and rush at the black rogue to drive him off. After several such approaches, the hen would become very angry, and would chase the crow to a greater distance from the brood. This is the very object the robber has in view, for as long as the parent keeps near her young, the crow has very slight chance of success; but as soon as he can induce her to follow him to a little distance from the brood, he takes advantage of his wings, and before she can regain her place, has flown over her, and seized one of her chickens. When the cock is present, there is still less danger from such an attack, for chanticleer shows all his vigilance and gallantry in protecting his tender offspring, though it frequently happens that the number of hens with broods renders it impossible for him to extend his care to all. When the crow tries to carry off a gosling from the mother, it requires more daring and skill, and is far less frequently successful than in the former instance. If the gander be in company, which he almost uniformly is, the crow has his labour in vain. Notwithstanding the advantages of flight and superior cunning, the honest vigilance and determined bravery of the former are too much for him. His attempts to approach, however cautiously conducted, are promptly met, and all his tricks rendered unavailing, by the fierce movements of the gander, whose powerful blows the crow seems to be well aware might effectually disable him. The first time I witnessed such a scene, I was at the side of the creek, and saw on the opposite shore a goose with her goslings beset by a crow; from the apparent alarm of the mother and brood, it seemed to me they must be in great danger, and I called to the owner of the place, who happened to be in sight, to inform him of their situation. Instead of going to their relief, he shouted back to me, to ask if the gander was not there too; and as soon as he was answered in the affirmative, he bid me be under no uneasiness, as the crow would find his match. Nothing could exceed the cool impudence and pertinacity of the crow, who, perfectly regardless of my shouting, continued to worry the poor gander for an hour, by his efforts to obtain a nice gosling for his next meal. At length convinced of the fruitlessness of his efforts, he flew off to seek some more easily procurable food. Several crows sometimes unite to plunder the goose of her young, and are then generally successful, because they are able to distract the attention of the parents, and lure them farther from their young.

In the summer the crows disperse in pairs for the purpose of raising their young, and then they select lofty trees in the remotest parts of the forest, upon which with dry sticks and twigs they build a large strong nest, and line it with softer materials. They lay four or five eggs, and when they are hatched, feed, attend, and watch over their young with the most zealous devotion. Should any one by chance pass near the nest while the eggs are still unhatched, or the brood are very young, the parents keep close, and neither by the slightest movement nor noise betray their presence. But if the young are fledged, and beginning to take their first lessons in flying, the approach of a man, especially if armed with a gun, calls forth all their cunning and solicitude. The young are immediately placed in the securest place at hand, where the foliage is thickest, and remain perfectly motionless and quiet. Not so the alarmed parents, both of which fly nearer and nearer to the hunter, uttering the most discordant screams, with an occasional peculiar note, which seems intended to direct or warn their young. So close do they approach, and so clamorous are they as the hunter endeavours to get a good view of them on the tree, that he is almost uniformly persuaded the young crows are also concealed there; but he does not perceive, as he is cautiously trying to get within gun shot, that they are moving from tree to tree, and at each remove are farther and farther from the place where the young are hid. After continuing this trick, until it is impossible that the hunter can retain any idea of the situation of the young ones, the parents cease their distressing outcries, fly quietly to the most convenient lofty tree, and calmly watch the movements of their disturber. Now and then they utter a loud quick cry, which seems intended to bid their offspring lie close and keep quiet, and it is very generally the case that they escape all danger by their obedience. An experienced crow-killer watches eagerly for the tree where the crows first start from; and if this can be observed, he pays no attention to their clamours, nor pretence of throwing themselves in his way, as he is satisfied they are too vigilant to let him get a shot at them; and if he can see the young, he is tolerably sure of them all, because of their inability to fly or change place readily.

The time of the year in which the farmers suffer most from them is in the spring, before their enormous congregations disperse, and when they are rendered voracious by the scantiness of their winter fare. Woe betide the corn field which is not closely watched, when the young grain begins to shoot above the soil! If not well guarded, a host of these marauders will settle upon it at the first light of the dawn, and before the sun has risen far above the horizon, will have plundered every shoot of the germinating seed, by first drawing it skilfully from the moist earth by the young stalk, and then swallowing the grain. The negligent or careless planter, who does not visit his field before breakfast, finds, on his arrival, that he must either replant his corn, or relinquish hopes of a crop; and without the exertion of due vigilance, he may be obliged to repeat this process twice or thrice the same season. Where the crows go to rob a field in this way, they place one or more sentinels, according to circumstances, in convenient places, and these are exceedingly vigilant, uttering a single warning call, which puts the whole to flight the instant there is the least appearance of danger or interruption. Having fixed their sentinels, they begin regularly at one part of the field, and pursuing the rows along, pulling up each shoot in succession, and biting off the corn at the root. The green shoots thus left along the rows, as if they had been arranged with care, offer a melancholy memorial of the work which has been effected by these cunning and destructive plunderers.

Numerous experiments have been made, where the crows are thus injurious, to avert their ravages; and the method I shall now relate I have seen tried with the most gratifying success. In a large tub a portion of tar and grease were mixed, so as to render the tar sufficiently thin and soft, and to this was added a portion of slacked lime in powder, and the whole stirred until thoroughly incorporated. The seed corn was then thrown in, and stirred with the mixture until each grain received a uniform coating. The corn was then dropped in the hills, and covered as usual. This treatment was found to retard the germination about three days, as the mixture greatly excludes moisture from the grain. But the crows did no injury to the field; they pulled up a small quantity in different parts of the planting, to satisfy themselves it was all alike; upon becoming convinced of which, they quietly left it for some less carefully managed grounds, where pains had not been taken to make all the corn so nauseous and bitter.

NO. XI.

It rarely happens that any of the works of nature are wholly productive of evil, and even the crows, troublesome as they are, contribute in a small degree to the good of the district they frequent. Thus, though they destroy eggs and young poultry, plunder the cornfields, and carry off whatever may serve for food, they also rid the surface of the earth of a considerable quantity of carrion, and a vast multitude of insects and their destructive larvæ. The crows are very usefully employed when they alight upon newly ploughed fields, and pick up great numbers of those large and long-lived worms, which are so destructive to the roots of all growing vegetables; and they are scarcely less so, when they follow the seine haulers along the shores, and pick up the small fishes, which would otherwise be left to putrify and load the air with unpleasant vapours. Nevertheless, they become far more numerous in some parts of the country than is at all necessary to the good of the inhabitants, and whoever would devise a method of lessening their numbers suddenly, would certainly be doing a service to the community.

About a quarter of a mile above the house I lived in on Curtis's creek, the shore was a sand bank or bluff, twenty or thirty feet high, crowned with a dense young pine forest to its very edge. Almost directly opposite, the shore was flat, and formed a point extending in the form of a broad sand bar, for a considerable distance into the water, and when the tide was low, this flat afforded a fine level space, to which nothing could approach in either direction, without being easily seen. At a short distance from the water, a young swamp wood of maple, gum, oaks, &c. extended back, towards some higher ground. As the sun descended, and threw his last rays in one broad sheet of golden effulgence over the crystal mirror of the waters, innumerable companies of crows arrived daily, and settled on this point, for the purpose of drinking, picking up gravel, and uniting in one body prior to retiring for the night to their accustomed dormitory. The trees adjacent and all the shore would be literally blackened by those plumed marauders, while their increasing outcries, chattering and screams, were almost deafening. It certainly seems that they derive great pleasure from their social habits, and I often amused myself by thinking the uninterrupted clatter which was kept up, as the different gangs united with the main body, was produced by the recital of the adventures they had encountered during their last marauding excursions. As the sun became entirely sunk below the horizon, the grand flock crossed to the sand bluff on the opposite side, where they generally spent a few moments in picking up a further supply of gravel, and then, rising in dense and ample column, they sought their habitual roost in the deep entanglements of the distant pines. This daily visit to the point, so near to my dwelling, and so accessible by means of the skiff, led me to hope that I should have considerable success in destroying them. Full of such anticipations, I loaded two guns, and proceeded in my boat to the expected place of action, previous to the arrival of the crows. My view was to have my boat somewhere about half way between the two two shores, and as they never manifested much fear of boats, to take my chance of firing upon the main body as they were flying over my head to the opposite side of the river. Shortly after I had gained my station, the companies began to arrive, and every thing went on as usual. But whether they suspected some mischief from seeing a boat so long stationary in their vicinity, or could see and distinguish the guns in the boat, I am unable to say; the fact was, however, that when they set out to fly over, they passed at an elevation which secured them from my artillery effectually, although, on ordinary occasions they were in the habit of flying over me at a height of not more than twenty or thirty feet. I returned home without having had a shot, but resolved to try if I could not succeed better the next day. The same result followed the experiment, and when I fired at one gang, which it appeared possible to attain, the instant the gun was discharged, the crows made a sort of halt, descended considerably, flying in circles, and screaming most vociferously, as if in contempt or derision. Had I been prepared for this, a few of them might have suffered for their bravado. But my second gun was in the bow of the boat, and before I could get it, the black gentry had risen to their former security. While we were sitting at tea that evening, a black came to inform me that a considerable flock of crows, which had arrived too late to join the great flock, had pitched in the young pines not a great way from the house, and at a short distance from the road-side. We quickly had the guns in readiness, and I scarcely could restrain my impatience until it should be late enough and dark enough to give us a chance of success. Without thinking of any thing but the great number of the crows, and their inability to fly to advantage in the night, my notions of the numbers we should bring home were extravagant enough, and I only regretted that we might be obliged to leave some behind. At length, led by the black boy, we sallied forth, and soon arrived in the vicinity of this temporary and unusual roost; and now the true character of the enterprise began to appear. We were to leave the road, and penetrate several hundred yards among the pines, whose proximity to each other, and the difficulty of moving between which, on account of the dead branches, has been heretofore stated. Next, we had to be careful not to alarm the crows before we were ready to act, and at the same time were to advance with cocked guns in our hands. The only way of moving forwards at all, I found to be that of turning my shoulders as much as possible to the dead branches, and breaking my way as gently as I could. At last we reached the trees upon which the crows were roosting; but as the foliage of the young pines was extremely dense, and the birds were full forty feet above the ground, it was out of the question to distinguish where the greatest number were situated. Selecting the trees which appeared by the greater darkness of their summits to be most heavily laden with our game, my companion and I pulled our triggers at the same moment. The report was followed by considerable outcries from the crows, by a heavy shower of pine twigs and leaves upon which the shot had taken effect, and a deafening roar caused by the sudden rising on the wing of the alarmed sleepers. *One* crow at length fell near me, which was wounded too badly to fly or retain his perch, and as the flock had gone entirely off, with this one crow did I return, rather crestfallen from my grand nocturnal expedition. This crow, however, afforded me instructive employment and amusement during the next day, in the dissection of its nerves and organs of sense, and I know not that I ever derived more pleasure from any anatomical examination, than I did from the dissection of its internal ear. The extent and convolutions of its semicircular canals, show how highly the sense of hearing is perfected in these creatures, and those who wish to be convinced of the truth of what we have stated in relation to them, may still see this identical crow skull, in the Baltimore Museum, to which I presented it after finishing the dissection. At least, I saw it there a year or two since, though I little thought, when employed in examining, or even when I last saw it, that it would ever be the subject of such a reference " in a printed book."

Not easily disheartened by preceding failures, I next resolved to try to outwit the crows, and for this purpose prepared a long line, to which a very considerable number of lateral lines were tied, having each a very small fishing hook at the end. Each of these hooks was baited with a single grain of corn, so cunningly put on, that it seemed impossible that the grain could be taken up without the hook being swallowed with it. About four o'clock, in order to be in full time, I rowed up to the sandy point, made fast my main line to a bush, and extending it toward the water, pegged it down at the other end securely in the sand. I next arranged all my baited lines, and then covering them all nicely with sand, left nothing exposed but the bait. This done, I scattered a quantity of corn all around, to render the baits as little liable to suspicion as possible. After taking a final view of the arrangement, which seemed a very hopeful one, I pulled my boat gently homeward, to wait the event of my solicitude for the capture of the crows. As usual, they arrived in thousands, blackened the sand beach, chattered, screamed, and fluttered about in great glee, and finally sailed over the creek and away to their roost, without having left a solitary unfortunate to pay for having meddled with my baited hooks. I jumped into the skiff, and soon paid a visit to my unsuccessful snare. The corn was all gone; the very hooks were all bare, and it was evident that some other expedient must be adopted before I could hope to succeed. Had I caught but one or two *alive*, it was my intention to have employed them to procure the destruction of others, in a manner I shall hereafter describe.

NO. XII.

Had I succeeded in obtaining some living crows, they were to be employed in the following manner. After having made a sort of concealment of brushwood within good gunshot distance, the crows were to be fastened by their wings on their backs, between two pegs, yet not so closely as to prevent them from fluttering or struggling. The other crows, who are always very inquisitive where their species is in any trouble, were expected to settle down near the captives, and the latter would certainly seize the first that came near enough with their claws, and hold on pertinaciously. This would have produced fighting and screaming in abundance, and the whole flock might gradually be so drawn into the fray, as to allow many opportunities of discharging the guns upon them with full effect. This I have often observed, that when a quarrel or fight took place in a large flock or gang of crows, a circumstance by no means infrequent, it seemed soon to extend to the whole, and, during the continuance of their anger, all the usual caution of their nature appeared to be forgotten, allowing themselves at such times to be approached closely and regardless of men, fire-arms, or the fall of their companions, continuing their wrangling with rancorous obstinacy. A similar disposition may be produced among them by catching a large owl, and tying it with a cord of moderate length to the limb of a naked tree in a neighbourhood frequented by the crows. The owl is one of the few enemies which the crow has much reason to dread, as it robs the nests of their young, whenever they are left for the shortest time. Hence, whenever crows discover an owl in the day time, like many other birds, they commence an attack upon it, screaming most vociferously, and bringing together all of their species within hearing. Once this clamour has fairly begun, and their passions are fully aroused, there is little danger of their being scared away, and the chance of destroying them by shooting is continued as long as the owl remains uninjured. But one such opportunity presented during my residence where crows were abundant, and this was unfortunately spoiled by the eagerness of one of the gunners, who, in his anxiety to demolish one of the crows, fixed upon some that were most busy with the owl, and killed it instead of its disturbers, which at once ended the sport. When the crows leave the roost, at early dawn, they generally fly to a naked or leafless tree in the nearest field, and there plume themselves and chatter until the daylight is sufficiently clear to show all objects with distinctness. Of this circumstance I have taken advantage several times to get good shots at them in this way. During the day time, having selected a spot within proper distance of the tree frequented by them in the morning, I have built with brushwood and pine bushes a thick, close screen, behind which one or two persons might move securely without being observed. Proper openings, through which to level the guns, were also made, as the slightest stir or noise could not be made at the time of action, without a risk of rendering all the preparations fruitless. The guns were all in order and loaded before going to bed, and at an hour or two before daylight, we repaired quietly to the field and stationed ourselves behind the screen, where, having mounted our guns at the loop-holes to be in perfect readiness, we waited patiently for the daybreak. Soon after the gray twilight of the dawn began to displace the darkness, the voice of one of our expected visitants would be heard from the distant forest, and shortly after a single crow would slowly sail towards the solitary tree and settle on its very summit. Presently a few more would arrive singly, and in a little while small flocks followed. Conversation among them is at first rather limited to occasional salutations, but as the flock begins to grow numerous, it becomes general and very animated, and by this time all that may be expected on this occasion have arrived. This may be known also, by observing one or more of them descend to the ground, and if the gunners do not now make the best of the occasion, it will soon be lost, as the whole gang will presently sail off, scattering as they go. However, we rarely waited till there was a danger of their departure, but as soon as the flock had fairly arrived and were still crowded upon the upper parts of the tree, we pulled triggers together, aiming at the thickest of the throng. In this way, by killing and wounding them, with two or three guns, a dozen or more would be destroyed. It was of course needless to expect to find a similar opportunity in the same place for a long time afterwards, as those which escaped had too good memories to return to so disastrous a spot. By ascertaining other situations at considerable distances, we could every now and then obtain similar advantages over them.

About the years 1800, 1, 2, 3, 4, the crows were so vastly accumulated and destructive in the state of Maryland, that the government, to hasten their diminution, received their heads in payment of taxes, at the price of three cents each. The store-keepers bought them of the boys and shooters, who had no taxes to pay, at a rather lower rate, or exchanged powder and shot for them. This measure caused a great havoc to be kept up among them, and in a few years so much diminished the grievance, that the price was withdrawn. Two modes of shooting them in considerable numbers were followed and with great success; the one, that of killing them while on the wing towards the roost, and the other attacking them in the night when they had been for some hours asleep. I have already mentioned the regularity with which vast flocks move from various quarters of the country to their roosting places every afternoon, and the uniformity of the route they pursue. In cold weather, when all the small bodies of water are frozen, and they are obliged to protract their flight towards the bays or sea, their return is a work of considerable labour, especially should a strong wind blow against them; at this season also, being rather poorly fed, they are of necessity less vigorous. Should the wind be adverse, they fly as near the earth as possible, and of this the shooters at the time I allude to took advantage. A large number would collect on such an afternoon, and station themselves close along the foot-way of a high bank, over which the crows were in the habit of flying; and as they were in a great degree screened from sight as the flock flew over, keeping as low as possible because of the wind, their shots were generally very effectual. The stronger was the wind, the greater was their success. The crows that were not injured found it very difficult to rise; and those that diverged laterally, only came nearer to gunners stationed in expectation of such movements. The flocks were several hours in passing over, and as there was generally a considerable interval between each company of considerable size, the last arrived, unsuspicious of what had been going on, and the shooters had time to recharge their arms. But the grand harvest of crow heads was derived from the invasion of their dormitories, which are well worthy a particular description, and should be visited by every one who wishes to form a proper idea of the number of these birds, that may be accumulated in a single district. The roost is most commonly the densest pine thicket that can be found, generally at no great distance from some river, bay, or other sheet of water, which is the last to freeze, or rarely is altogether frozen. To such a roost, the crows, which are, during the day-time, scattered over perhaps more than a hundred miles of circumference, wing their way every afternoon, and arrive shortly after sunset. Endless columns pour in from various quarters, and as they arrive pitch upon their accustomed perches, crowding closely together for the benefit of the warmth and the shelter afforded by the thick foliage of the pine. The trees are literally bent by their weight, and the ground is covered for many feet in depth by their dung, which by its gradual fermentation, must also tend to increase the warmth of the roost. Such roosts are known to be thus occupied for years, beyond the memory of individuals; and I know of one or two, which the oldest residents in the quarter state to have been known to their grandfathers, and probably had been resorted to by the crows during several ages previous. There is one of great age and magnificent extent, in the vicinity of Rock Creek, an arm of the Patapsco. They are sufficiently numerous on the rivers opening into the Chesapeake, and are every where similar in their general aspect. Wilson has signalised such a roost at no great distance from Bristol, Pa. and I know by observation, that not less than a million of crows sleep there nightly during the winter season.

To gather crow heads from the roost, a very large party was made up, proportioned to the extent of surface occupied by the dormitory. Armed with double barrelled and duck guns, which threw a large charge of shot, the company was divided into small parties, and these took stations, selected during the day time, so as to surround the roost as nearly as possible. A dark night was always preferred, as the crows could not when alarmed fly far, and the attack was delayed until full midnight. All being at their posts, the firing was commenced by those who were most advantageously posted, and followed up successively by the others, as the affrighted crows sought refuge in their vicinity. On every side the carnage then raged fiercely, and there can scarcely be conceived a more forcible idea of the horrors of a battle, than such a scene afforded. The crows screaming with fright and the pain of wounds, the loud deep roar produced by the raising of their whole number in the air, the incessant flashing and thundering of the guns, and the shouts of their eager destroyers, all produced an effect which can never be forgotten by any one who has witnessed it, nor can it well be adequately comprehended by those who have not. Blinded by the blaze of the powder, and bewildered by the thicker darkness that ensues, the crows rise and settle again at a short distance, without being able to withdraw from the field of danger; and the sanguinary work is continued until the shooters are fatigued, or the approach of daylight gives the survivors a chance of escape. Then the work of collecting the heads from the dead and wounded began, and this was a task of considerable difficulty, as the wounded used their utmost efforts to conceal and defend themselves. The bill and half the front of the skull were cut off together, and strung in sums for the tax-gatherer, and the product of the night divided according to the nature of the party formed. Sometimes the great mass of shooters were hired for the night, and received no shares of scalps, having their ammunition provided by the employers; other parties were formed of friends and neighbours, who clubbed for the ammunition, and shared equally in the result.

During hard winters the crows suffer severely, and perish in considerable numbers from hunger, though they endure a wonderful degree of abstinence without much injury. When starved severely, the poor wretches will swallow bits of leather, rope, rags, in short any thing that appears to promise the slightest relief. Multitudes belonging to the Bristol roost, perished during the winter of 1828–9 from this cause. All the water courses were solidly frozen, and it was distressing to observe these starvelings every morning winging their weary way towards the shores of the sea in hopes of food, and again to see them toiling homewards in the afternoon, apparently scarce able to fly.

In speaking of destroying crows, we have never adverted to the use of poison, which in their case is wholly inadmissible on this account. Where crows are common hogs generally run at large, and to poison the crows would equally poison them; the crows would die, and fall to the ground, where they would certainly be eaten by the hogs.

Crows, when caught young, learn to talk plainly, if pains be taken to repeat certain phrases to them, and they become exceedingly impudent and troublesome. Like all of their tribe, they will steal and hide silver or other bright objects, of which they can make no possible use.

NATURAL SCIENCES IN AMERICA

An Arno Press Collection

Allen, J[oel] A[saph]. **The American Bisons,** Living and Extinct. 1876

Allen, Joel Asaph. **History of the North American Pinnipeds:** A Monograph of the Walruses, Sea-Lions, Sea-Bears and Seals of North America. 1880

American Natural History Studies: The Bairdian Period. 1974

American Ornithological Bibliography. 1974

Anker, Jean. **Bird Books and Bird Art.** 1938

Audubon, John James and John Bachman. **The Quadrupeds of North America.** Three vols. 1854

Baird, Spencer F[ullerton]. **Mammals of North America.** 1859

Baird, S[pencer] F[ullerton], T[homas] M. Brewer and R[obert] Ridgway. **A History of North American Birds:** Land Birds. Three vols., 1874

Baird, Spencer F[ullerton], John Cassin and George N. Lawrence. **The Birds of North America.** 1860. Two vols. in one.

Baird, S[pencer] F[ullerton], T[homas] M. Brewer, and R[obert] Ridgway. **The Water Birds of North America.** 1884. Two vols. in one.

Barton, Benjamin Smith. **Notes on the Animals of North America.** Edited, with an Introduction by Keir B. Sterling. 1792

Bendire, Charles [Emil]. **Life Histories of North American Birds** With Special Reference to Their Breeding Habits and Eggs. 1892/1895. Two vols. in one.

Bonaparte, Charles Lucian [Jules Laurent]. **American Ornithology:** Or The Natural History of Birds Inhabiting the United States, Not Given by Wilson. 1825/1828/1833. Four vols. in one.

Cameron, Jenks. **The Bureau of Biological Survey:** Its History, Activities, and Organization. 1929

Caton, John Dean. **The Antelope and Deer of America:** A Comprehensive Scientific Treatise Upon the Natural History, Including the Characteristics, Habits, Affinities, and Capacity for Domestication of the Antilocapra and Cervidae of North America. 1877

Contributions to American Systematics. 1974
Contributions to the Bibliographical Literature of American Mammals. 1974
Contributions to the History of American Natural History. 1974
Contributions to the History of American Ornithology. 1974
Cooper, J[ames] G[raham]. **Ornithology. Volume I, Land Birds.** 1870
Cope, E[dward] D[rinker]. **The Origin of the Fittest:** Essays on Evolution and **The Primary Factors of Organic Evolution.** 1887/1896. Two vols. in one.
Coues, Elliott. **Birds of the Colorado Valley.** 1878
Coues, Elliott. **Birds of the Northwest.** 1874
Coues, Elliott. **Key To North American Birds.** Two vols. 1903
Early Nineteenth-Century Studies and Surveys. 1974
Emmons, Ebenezer. **American Geology:** Containing a Statement of the Principles of the Science. 1855. Two vols. in one.
Fauna Americana. 1825-1826
Fisher, A[lbert] K[enrick]. **The Hawks and Owls of the United States in Their Relation to Agriculture.** 1893
Godman, John D. **American Natural History:** Part I — Mastology and **Rambles of a Naturalist.** 1826-28/1833. Three vols. in one.
Gregory, William King. **Evolution Emerging:** A Survey of Changing Patterns from Primeval Life to Man. Two vols. 1951
Hay, Oliver Perry. **Bibliography and Catalogue of the Fossil Vertebrata of North America.** 1902
Heilprin, Angelo. **The Geographical and Geological Distribution of Animals.** 1887
Hitchcock, Edward. **A Report on the Sandstone of the Connecticut Valley,** Especially Its Fossil Footmarks. 1858
Hubbs, Carl L., editor. **Zoogeography.** 1958
[Kessel, Edward L., editor]. **A Century of Progress in the Natural Sciences: 1853-1953.** 1955
Leidy, Joseph. **The Extinct Mammalian Fauna of Dakota and Nebraska,** Including an Account of Some Allied Forms from Other Localities, Together with a Synopsis of the Mammalian Remains of North America. 1869
Lyon, Marcus Ward, Jr. **Mammals of Indiana.** 1936
Matthew, W[illiam] D[iller]. **Climate and Evolution.** 1915
Mayr, Ernst, editor. **The Species Problem.** 1957
Mearns, Edgar Alexander. **Mammals of the Mexican Boundary of the United States.** Part I: Families Didelphiidae to Muridae. 1907

Merriam, Clinton Hart. **The Mammals of the Adirondack Region,** Northeastern New York. 1884

Nuttall, Thomas. **A Manual of the Ornithology of the United States and of Canada.** Two vols. 1832-1834

Nuttall Ornithological Club. **Bulletin of the Nuttall Ornithological Club:** A Quarterly Journal of Ornithology. 1876-1883. Eight vols. in three.

[Pennant, Thomas]. **Arctic Zoology.** 1784-1787. Two vols. in one.

Richardson, John. **Fauna Boreali-Americana;** Or the Zoology of the Northern Parts of British America, Containing Descriptions of the Objects of Natural History Collected on the Late Northern Land Expeditions Under Command of Captain Sir John Franklin, R. N. Part I: Quadrupeds. 1829

Richardson, John and William Swainson. **Fauna Boreali-Americana:** Or the Zoology of the Northern Parts of British America, Containing Descriptions of the Objects of Natural History Collected by the Late Northern Land Expeditions Under Command of Captain Sir John Franklin, R. N. Part II: The Birds. 1831

Ridgway, Robert. **Ornithology.** 1877

Selected Works By Eighteenth-Century Naturalists and Travellers. 1974

Selected Works in Nineteenth-Century North American Paleontology. 1974

Selected Works of Clinton Hart Merriam. 1974

Selected Works of Joel Asaph Allen. 1974

Selections From the Literature of American Biogeography. 1974

Seton, Ernest Thompson. **Life-Histories of Northern Animals: An Account of the Mammals of Manitoba.** Two vols. 1909

Sterling, Keir Brooks. **Last of the Naturalists:** The Career of C. Hart Merriam. 1974

Vieillot, L. P. **Histoire Naturelle Des Oiseaux de L'Amerique Septentrionale,** Contenant Un Grand Nombre D'Especes Decrites ou Figurees Pour La Premiere Fois. 1807. Two vols. in one.

Wilson, Scott B., assisted by A. H. Evans. **Aves Hawaiienses:** The Birds of the Sandwich Islands. 1890-99

Wood, Casey A., editor. **An Introduction to the Literature of Vertebrate Zoology.** 1931

Zimmer, John Todd. **Catalogue of the Edward E. Ayer Ornithological Library.** 1926